Coastal Research Library

Volume 38

Series Editor
Charles W. Finkl
Department of Geosciences
Florida Atlantic University
Boca Raton, FL, USA

The aim of this book series is to disseminate information to the coastal research community. The Series covers all aspects of coastal research including but not limited to relevant aspects of geological sciences, biology (incl. ecology and coastal marine ecosystems), geomorphology (physical geography), climate, littoral oceanography, coastal hydraulics, environmental (resource) management, engineering, and remote sensing. Policy, coastal law, and relevant issues such as conflict resolution and risk management would also be covered by the Series. The scope of the Series is broad and with a unique cross-disciplinary nature. The Series would tend to focus on topics that are of current interest and which carry some import as opposed to traditional titles that are esoteric and non-controversial.

Monographs as well as contributed volumes are welcomed.

More information about this series at http://www.springer.com/series/8795

Sughosh Madhav • Sadaf Nazneen
Pardeep Singh
Editors

Coastal Ecosystems

Environmental importance, current challenges and conservation measures

Springer

Editors
Sughosh Madhav
Shaheed Bhagat Singh College
University of Delhi
Delhi, India

Sadaf Nazneen
Department of Civil Engineering
Jamia Millia Islamia
New Delhi, Delhi, India

Pardeep Singh
PGDAV College
University of Delhi
New Delhi, Delhi, India

ISSN 2211-0577 ISSN 2211-0585 (electronic)
Coastal Research Library
ISBN 978-3-030-84254-3 ISBN 978-3-030-84255-0 (eBook)
https://doi.org/10.1007/978-3-030-84255-0

This Springer imprint is published by the registered company Springer Nature Switzerland AG
The registered company address is: Gewerbestrasse 11, 6330 Cham, Switzerland

Contents

Chapter 1
Coastal Ecosystems of India and Their Conservation and Management Policies: A Review

Sadaf Nazneen, Sughosh Madhav, Anusha Priya, and Pradeep Singh

Abstract The present article talks about the coastal ecosystem of India spread over nine states and four union territories which also include two islands under Indian territory. India has diverse coastal features along its vast coastline of 7515.6 km, consisting of mangroves, coral reefs, seagrass meadows, and salt marshes. India also has the largest lagoon in Asia, and there are few more lagoons on both east and west coast. A sizeable population lives near the Indian coast, and there are many prominent cities on the coast. Of the four metropolitan cities, three of them Mumbai, Kolkata and Chennai are coastal cities. However, with the onset of climate change and increasing anthropogenic pressures, Indian coasts are vulnerable and need robust policies for their protection, conservation and management. This study outlines the various prominent coastal features and various laws and policies outlined to protect Indian coasts and their coastal ecosystems and the biodiversity they represent.

Keywords Coastal ecosystems · Seagrasses · Coral reefs · Mangroves · Coastal policies · CRZ

S. Nazneen (✉)
Department of Civil Engineering, Jamia Millia Islamia, New Delhi, India

S. Madhav
School of Environmental Sciences, Jawaharlal Nehru University, New Delhi, India

A. Priya
Department of Management, Jharkhand Rai University, Ranchi, India

P. Singh
Department of Environmental Sciences, PGDAV College, New Delhi, India

S. Madhav et al. (eds.), *Coastal Ecosystems*, Coastal Research Library 38,
https://doi.org/10.1007/978-3-030-84255-0_1

1 Introduction

India has a vast coastline that stretches over nine states and four union territories, including two group of islands (Krishnan et al. 2018). The coastlines of India are on both its eastern and western shores. Indian coastline consists of 5422 km of mainland shore with 2344 km of islands in Indian jurisdiction (Andaman and Nicobar Islands and Lakshadweep islands). This double archipelago total length is 2094 km. Western Indian coast includes Gujarat, Maharashtra, Karnataka, Goa and Kerala, whereas eastern India includes Odisha, West Bengal, Tamil Nadu and Andhra Pradesh. Daman and Diu on the west coast and Puducherry on the east coast are union territories. Andaman and Nicobar Islands (ANI) are located in the Bay of Bengal and Lakshdweep islands in Arabian Sea. Eighteenth longest overall coastal length in the globe. In coastal regions, valuable resources, fertile habitats and high biodiversity contribute greatly to the nation's economy. As a result of rising population, urbanisation, industry and climate change, coastal areas must be looked at and examined in greater depth (Panigrahy and Mohanty 2012; Bhomia et al. 2016).

This chapter examines the coastal states and islands of Indian territory, particularly the coastal ecosystems and wetlands. We also summarise numerous laws and regulations implemented to conserve these ecosystems.

2 The Coastal States, Union Territories and Islands of India

2.1 Gujarat

The western state of Gujarat has the longest coastline of 1600 km. There are 41 ports, both big and little, located along Gujarat's coastline. Gujarat's well-known beaches include Diu, Dwarka and Porbandar. The Gulf of Khambhat and the Gulf of Kachchh are located in Gujarat. Coastal zones include mangroves, coral reefs, seagrasses and salt marshes. Second, to West Bengal, Gujarat has the second greatest mangrove coverage. The Gulf of Kachchh, Kachchh Bay and Saurashtra have a greater density of mangroves (including the Gulf of Khambhat-Dumas Ubharat areas). A maximum of 71.5% mangrove cover exists in the Gulf of Kachchh. The Gulf of Kachchh, Saurashtra and South Gujarat include some mangrove habitats. The Rann of Kachchh is a very rare environment with salt marsh as the primary plant. The GRK base looks like a table with numerous little peaks at ground level, or "islands", some of which are dubbed "bets". In geological time, it's considered an area of instability due to transgression. This body of water has shallow water depth ranging from 0.5 to 1.5 m. During October and November, the area goes through a dry period and is soon overrun by an ocean of salt crystals (Stansley 2004). A quiet stretch of coast in Gujarat is home to mangroves, salt marshes and coral reefs. Coral reefs are abundant throughout the Gujarat coast; however, they are only found in the abundance Gulf of Kachchh, and their diversity is very low in other areas. Gujarat, where reef-forming corals (scleractinian) are few, has coral reefs that have a low variety compared to other main regions of India.

2.2 *Maharashtra*

Maharashtra is the third largest state in India area-wise. It is flanked by the Western Ghats in the east and the Arabian Sea in the west. The Konkan coast has a length of 720 km. Maharashtra features several waterways, tidal mudflats, salt marshes and mangroves. The 720 km coastline along the Konkan area of Maharashtra houses the Great Mumbai Region (GMR). Because of its urban setting, the GMR was chosen as a unique instance when drafting the revised Coastal Regulation Zone (CRZ) notification of 2011. Mangroves in the Mumbai region require monitoring and protection, especially to control coastal pollution and solid waste management (Krishnamurthy et al. 2014). "The Sahyadri" mountain range in Maharashtra has an average elevation of 1000–1200 m above mean sea level (MSL). The Sahyadri hills have numerous offshoots spreading eastwards (Satmala, Ajanta, Harishchandra, Balaghat and Mahadeo). The Konkan coastal strip is a tiny stretch of coastal land that is only 50 km long. Branching rivers from the Sahyadri hills, which join the coastline, bisect the coastline. Konkan comprises four main creeks: Terekhol, Rajapuri, Vijaydurg and Raigad. Mumbai is in the top 20 cities with exposure to extreme sea level and tropical cyclones.

2.3 *Goa*

Goa is the smallest state in India. It is characterised by long sandy beaches well-known for tourism. Goa is known worldwide for its spectacular beaches. Goa is surrounded by Maharashtra on north, Karnataka on the east and the Arabian Sea on the west. Mangrove patches and lush western ghat forest line the state's shoreline.

2.4 *Karnataka*

Karnataka's coastline extends for 320 km, into three districts: Dakshin Kannada (62 km), Udupi (98 km) and Uttara Kannada (160 km). Konkan Peninsula adjoins the Arabian Sea to the west, while the Western Ghats join it to the east, and Kerala plateaus are in the north. Several ridges and branches of the Western Ghats link this region to the ocean. Coastal Karnataka is peppered with rivers, waterfalls, peaks and hill ranges. The coastal plain is 50–80 km in breadth.

2.5 Kerala

Kerala's coastline measures 580 km, being the fifth-longest in India. The Malabar coast in Kerala is well renowned. The Arabian Sea and the Western Ghat mountain range separate Kerala from the east. Malabar's backwaters, beaches and tea and coffee plantations are all well-known. The Kerala coast spans from Manjeswaram in the north to Pozhiyur in the south. The beach infrastructure along the Cochin-Alleppey coast is developed.

2.6 Andhra Pradesh

Andhra Pradesh's coastline runs the second-longest in India for 974 km on the eastern coast. Andhra Pradesh's coastline lies between the Eastern Ghats and the Bay of Bengal. The coastline is home to two large rivers, the Godavari and Krishna, and smaller river deltas with agricultural land. India's second-largest lagoon and India's largest mangrove ecosystem are other key coastal ecosystems. The mangroves of the Godavari estuary (Krishna Delta and Machilipatnam) and Pulicat lagoon are critical coastal habitats in Andhra Pradesh. The largest lagoon in India is connected to the Bay of Bengal and houses both permanent and migratory birds. Seagrass beds and mangrove patches are seen in the lagoon as well.

2.7 Tamil Nadu

Tamil Nadu has a coastline of 1076 km, which is the third longest in the country. The Coromandel Coast is well-known. Every ocean on the Indian subcontinent touches the Tamil Nadu coastline. It runs from the Bay of Bengal to the Arabian Sea. Kaveri Delta is in the east, while the Western Ghats occupy the south, and Utkal plain lies in the north. The state's coastal districts include Ramanathapuram, with 237 km of shoreline, and Chennai, which has 19 km of coastline. The Gulf of Mannar extends for 365 km, Palk Bay extends for 294 km, and the west coast of Tamil Nadu stretches between Kanyakumari and Neerody (60 km). Major seaports such as Chennai and Tuticorin, together with various marine ports, marinas and harbours, are located along the coast of Tamil Nadu. Gulf of Mannar has beautiful seagrass meadows, mudflats and salt marshes. The pichavaram mangroves, Muthupet mangroves, Pulicat lake and Kaliveli backwaters are ecologically vulnerable locations on the Tamil Nadu coast (GIZ Report 2013)

2.8 *Odisha*

Coastal Odisha is also known as Utkal plains and measures 485 km. The shores of the Mahanadi and Brahmani-Baitarni rivers are almost entirely depositional (Ajai et al. 2012a). The lower Ganges plain spans north to south, whereas the Tamilnad lowlands are south of the Bay of Bengal. This area has six main estuaries (including Bhitarkanika and Mahanadi mangroves), a big brackish water lagoon (Chilika Lake), enormous mudflats and sandy beaches. Gopalpur and Chandipur beaches are widely recognised. Nesting grounds for Olive Ridley turtles are located just south of the Bhitarkanika mangroves.

2.9 *West Bengal*

West Bengal's coastline is roughly 157 km in length. The West Bengal shoreline is found in Purba Medinipur and the South 24 Parganas districts. The Sundarbans is on the Ganga-Brahmaputra and Meghna riverbanks, which makes about 40% of the total mangrove coverage in the globe. Sundarbans offers a great diversity of flora, having several species. Sundarbans is famous for the Royal Bengal tiger feeding on fish. The coastline's principal features are mudflats, creeks and tidal flats. West Bengal experiences significant storm activity.

2.10 *Union Territories of India: Coastal Regions*

2.10.1 Puducherry

Union Territory of India, situated on the coromandel coast in the east, was originally known as Puducherry. It is confined by land on three sides, with the Bay of Bengal on the eastern side. The eastern coastal plain of Puducherry is parallel to the Bay of Bengal. The plain is 400–600 m broad, with sand dunes along the coastline.

2.10.2 Daman and Diu

Daman and Diu are a Union Territory (UT) of India located on the west coast of the Arabian Sea. Daman and Diu is India's smallest federal division, which covers just 112 km^2. The territories of Daman and Diu are divided by the Gulf of Khambhat, not one continuous province. The UT is bounded by the state of Gujarat and the Arabian Sea. Daman and Diu feature extensive salt marshes and plentiful fisheries. Daman and Diu were amalgamated with Dadra and Nagar Haveli, an Indian UT, on January 26, 2020.

2.10.3 Andaman and Nicobar Islands (ANI)

Andaman and Nicobar Islands (UT of India) consists of 572 islands, of which 38 are inhabited. East of the archipelago is the Andaman Sea, while west of it is the Bay of Bengal. The capital of the territory is Port Blair. The entire landmass of the islands is 8249 km^2. Ten Degree Channel separates the Andaman Islands from the Nicobar Islands (Bandopadhyay and Carter 2017). ANI is home to the most pristine and vulnerable island ecosystems in the world. This environment has evolved in isolation from the mainland over time, resulting in various flora and fauna. Isolated islands mean many endemic species of flora and wildlife. The majority of the Sundaland biodiversity hotspot includes sections of ANI (Sridhar 2018). Mangroves, coral reefs and seagrasses characterise ANI's rich coastal ecosystems. The whole coastline of the ANI islands is 1962 km, which comprises the huge Exclusive Economic Zone of India. With 96 approved wildlife sanctuaries, ANI manages 9 national parks and biosphere reserve in Great Nicobar. The isolation and remoteness of ANI Islands favour the evolution of endemism of both vegetation and animals.

2.10.4 Lakshadweep Islands

Lakshadweep archipelago is located between 400 and 600 km off the coast of Kerala. It comprises 12 atolls, 3 atoll reefs and 5 coral banks (Nobi et al. 2011; Dalia et al. 2014). Lakshadweep has various ecosystems consisting of mangroves, corals, seagrasses, dunes and seaweeds. The ecosystems of the Lakshadweep islands are rich in biodiversity and productivity. The islands' terrestrial ecosystems are protected from the powerful ocean waves (Robin et al. 2012). The Lakshadweep coral atoll system features enormous shallow lagoons and coral reefs. The island lagoon covers 4200 km^2, and their Exclusive Economic Zone (EEZ) is 4,000,000 km^2 (www.lakshadweep.gov.in). The Lakshadweep Sea is full of fishing resources, especially tuna (Gopi et al. 2021).

3 India's Coastal Ecosystems

3.1 Mangroves

Indian coastal states and Indian territorial islands are rich, diversified mangroves. Indian mangroves occupy only 3.3% of the total mangrove cover of the world. However, when it comes to mangrove species richness, 56% of the global mangrove species are found in India (Ragavan et al. 2019). Salt-tolerant, hardy, and long-lived trees can grow in environments with high salinity and regenerate and proliferate. Through their robustness, these trees give protection against storm surges, stabilise coastline and safeguard coasts. Mangrove forests play a vital role in the carbon, nitrogen, phosphorus and sulphur nutrient cycle. Ecosystems trap organic carbon, debris and sediments that bring along them organic matter from the coastal and river catchments, producing plentiful nourishment for the system. Mangroves provide abundant estuarine and coastal fisheries by harming the nearby oligotrophic systems (Singh et al. 2005; Ranjan et al. 2010). Mangroves provide marine ecosystems with much-needed nutrients because of the deposition of organic-rich, very fine sediments. Mangroves serve as nurseries for the young of marine fish, crabs and shrimp. They offer a wide variety of forest products, helping the local community. Locally used by locals as fuel and fodder, the robust tree wood has several medicinal uses.

Mangrove forest deforestation is one of the main causes of mangrove loss (Ajai et al. 2012b). The deep green mangrove forest with a sufficient food supply also provides habitat for numerous tiny and large creatures. These woodlands are bird species-rich. Some help crocodiles and deer (Bhitarkanika mangroves). Sundarbans mangroves are known for fish-fed tigers. These dense mangrove forest with evergreen trees and tiny rivers has tremendous recreational possibilities. People visit for boating, sightseeing, bird viewing and fishing in mangroves wetland. It has recently been largely valued as a carbon storage and carbon sink. Mangroves in India became important following the 2004 Tsunami because many fringes near mangroves protected from the waters and storm surges because of their immobility.

Around 43,000 km^2 of coastal wetlands include estuaries, lagoons, mangroves and mudflats. These wetlands help slow down runoff, buffer against storm surge, the barrier against storm surge, and shield against tsunami (Bassi et al. 2014). Bees found in mangroves can produce up to 100 pounds of honey per hive in a year. Mangroves on the west coast are considerably less established due to higher coastline and few main rivers moving west (Selvam 2003). East coast Mangroves occupy almost 60% of the mangrove cover, 23% is found on west coast and 17% in ANI . Because of the different geographic locations and rivers that flow east, east coast mangroves are significantly more floristically rich and biodiverse (Nayak et al. 2016; Kathiresan 2018).

This mangrove forest of Sundarbans lies in India and Bangladesh. India's Sundarbans is part of the Ganga-Brahmaputra Delta in the state of West Bengal. The second largest mangrove patch is in Gujarat, followed by Tamil Nadu and Andhra Pradesh (Selvam 2003). ANI islands contain the most variety and rich mangroves, second in number only to Gujarat's.

3.1.1 Sundarbans Mangroves (West Bengal)

Sundarbans mangrove forests are the world's largest lying in India and Bangladesh. The Sundarbans is the result of the merging of the Ganga, Brahmaputra and Meghna rivers before meeting Bay of Bengal. The tidal flooding of the Brahmaputra and Ganga rivers has provided a stable and ideal habitat for mangroves to flourish. Approximately 4100 km^2 of the Sundarbans' mangrove swamp is forested, of which approximately 2125 km^2 are mangrove swamp and 1781 km^2 are water. Over the last three decades, the Sundarbans has been safeguarded by numerous methods.

Included are biosphere reserve, national park and wildlife sanctuary (Gopal and Chauhan 2006). Sundarbans mangroves are a UNESCO World Heritage site. West Bengal has India's greatest mangrove cover, followed by Gujarat and ANI (Ajai et al. 2012b). The Sundarbans supplies a vital livelihood to the locals, yet settlements place unwelcome stresses on this ecology. Several risks have grown alongside the local population: increased tourism, aggressive fishing, loss of mangrove trees for cultivation and overfishing (Saha et al. 2006; Kumar and Ramanathan 2015).

3.1.2 Maharashtra, Karnataka, Kerala and Goa Mangroves

In Maharashtra, 27,092.14 ha of mangrove vegetation exists. Thane, Mumbai, Raigarh, Ratnagiri and Sindhugarh have mangroves (Mugade and Sapkale 2014). *Avicennia*, *Sonneratia* and *Rhizophora* are Maharashtra's main mangrove genus. Mangroves occupy 663 ha in Kerala. Much of Kerala's mangrove region is Valapattanam, Kunhimangalam, Kasargod-Nileshwar, Kavvayi and Puthuvypin. Mangrove forest covers 3463.36 ha in Goa. Mapusa and Zuari rivers all flow through somewhat dense regions of mangroves. Chorao Sanctuary with good mangrove stretches along the Mandovi River. Major mangrove species Avicennia, Sonneratia and Rhizophora are found along the rivers Zuari, Mapusa and Mandovi and cover major mangroves area.

3.1.3 Mangroves Pichavaram and Muthupet (Tamil Nadu)

Mangroves are important in ecologically vulnerable regions along the Tamil Nadu coast. The state mangrove cover area is 5565 ha. The most prominent mangroves in Cauvery Delta are Pichavaram and Muthupet. The Pichavaram mangroves, which have 1100 ha, are constrained by the Vellar and Coleroon rivers. Forty percent of the total area is taken up by waterways and the rest by mangrove vegetation placed in the middle (Ranjan et al. 2010; Sappal et al. 2014). A 75% loss of mangrove forest coverage has been seen in the Pichavaram mangroves (Kathiresan 2002; Ranjan et al. 2010). The 2004 Tsunami damaged the Pichavaram mangroves (Ranjan et al. 2008). Muthupet, meaning "Pearl Land", is part of the Point Calimere Wildlife Sanctuary, Tamil Nadu's only Ramsar site. Both Pichavaram and Muthupet mangroves receive freshwater from October to November, especially during the north-eastern monsoon season. This leads to mangroves experiencing a long dry season and very high water salinity. Besides these two large mangrove patches, most estuaries and backwater system along the Tamil Nadu coast have mangrove forests in patches. The dominant mangrove species present throughout the Tamil Nadu coast is the *Avicennia marina*. Other significant species detected include *Rhizophora*, *Excoecaria* and *Acanthus ilicifolius* (Kathiresan 2000).

3.1.4 Andaman and Nicobar Mangrove (ANI)

ANI contains about 572 islands, islets, creeks, beaches and rocky outcrops with abundant mangrove forest. These islands harbour 34 real mangrove species of 15 genera, 10 orders and 12 families, 50% of the global mangrove species (Ragavan et al. 2015). According to the 2013 India Forest Survey, total mangrove land of 604 km^2 occurs in ANI, which is third in extent after West Bengal and Gujarat. Favourable climate circumstances, including short dry season, heavy rainfall and high tidal amplitude, cause lush green mangrove trees in ANI (Sridhar 2018).

3.1.5 Mahanadi and Bhitarkanika (Odisha)

Mangrove forests on India's east coast are more diversified, healthier, denser and floristically rich than the west coast (Dasgupta and Shaw 2013, 2016). Mangrove patches of Odisha's east coast are located along the Mahanadi, Brahmani, Baitarni, Dhamra and Devi delta. The primary districts with mangroves are Kendrapara, Jagatsinghpur, Bhadrak and Balasore. Bhitarkanika is India's second-biggest mangrove forest. It has rich in plants, birds and animal species. Mangrove comprises about 300 plant species, 174 bird species, 31 animal species and 29 reptile species (Badola and Hussain 2003; Bhomia et al. 2016; Kadaverugu et al. 2021). Bhitarkanika also houses the endangered saltwater crocodile (*Crocodylus porosus*). Bhitarkanika mangroves underwent considerable deforestation pressure during 1951–1961 due to population growth around the forest resulting in human habitation, aquaculture and agriculture mangrove destruction. In 1975, to conserve the surviving mangroves, the Odisha government declared an area of 672 km^2 bounded by Dhamra and Brahmani rivers as Bhitarkanika Wildlife Sanctuary under the 1972 Wildlife (Protection) Act. In 1998, a core area of 145 km^2 was declared as Bhitarkanika National Park within the sanctuary.

3.2 *Salt Marshes*

Salt marshes are halophytic plants growing in marshy environments near the seas and oceans. These plants grow in high saline environment and form an important ecosystem of coastal areas supporting species diversity and providing important ecosystem services. Salt marshes grow in the upper tidal zone and are subject to tidal inundations regularly. The salt marshes vegetation may be herbs, shrubs or grasses which thrive when supplied with saltwater. The salt marshes area in Odisha and West Bengal is yet to be explored. Salt marsh plants are useful in several forms. They are often used as bedding and thatching material and animal fodder by coastal communities. They also help improve the quality and stability of coastal habitats of many aquatic species by stabilising the surrounding environment from various natural forces like storm surge. They help in pollution abetment and water clarity by reducing the quantity of suspended solids in the water column through filtration services (Gopi et al. 2019; Unsworth et al. 2019). The salt marshes like other coastal wetlands are under natural and anthropogenic pressures. In the recent decades, several pressures, increasing salinity and temperature and introduction of exotic species have rendered the natural salt marshes as non-resilient halophytic plants. The salt marshes are vulnerable to sea-level rise as a consequence of climate change which threatens their significant contribution to coastal protection, fisheries support, biodiversity conservation and carbon sequestration. Until few years back, salt marshes have been almost unexplored in India in terms of net area occupied, species diversity, ecological dynamics and ecosystem services they provide (Jagtap and Rodrigues 2004; Patro et al. 2017). According to Viswanathan et al. (2020), the overall area of salt marshes in India is about 1611 km^2 with Gujarat having maximum area of 1443 km^2. Salt marshes have been categorised as one of the Ecologically Sensitive Areas (ESA) under the Coastal Regulation Zone (CRZ) Notification 2019. In India 14 salt marsh species have been reported; while all the 14 are found on the western coastline, 13 species have been reported on the east coast (Patro et al. 2017; Viswanathan et al. 2020).

3.3 *Seagrasses*

Seagrasses are angiosperms growing in shallow coast in estuaries, lagoons and bays. They are considered to be the keystone species as they provide wide range of ecosystem services from food, shelter to carbon storage (Duarte et al. 2013; Stankovic et al. 2021). The seagrass beds are one of the most ecologically important producers in the marine environment as they are the primary producers and provide shelter and food to various organisms, also providing nutrients to coral reefs and storing carbon (Mishra and Apte 2021) Mangrove plants and the seagrass meadows supply a large part of the diets of many large and small marine organisms, including dugongs, sea turtles, fishes and small invertebrates (Thangaradjou et al. 2007; Gopi et al. 2020; Mishra and Apte 2020). Seagrass meadows are the only angiosperms growing under the saline water and exhibit high primary production rates which is closely linked to high fish production of the associated fisheries (Nobi et al. 2011). They are found in shallow isolated coastal ecosystems like lagoons, estuaries and bays. India has reported 16 seagrass species in the coastal areas along the Indian coast with rich seagrass meadows in Palk Bay and Gulf of Mannar coastal areas in Tamil Nadu having 14 species, ANI has 12 species and Lakshadweep Islands have 10 species, and Odisha and Gujarat have 8 species (Bharathi et al. 2014; Thangaradjou and Bhatt 2018; Gopi et al. 2020). Two of the islands in Lakshadweep Agathi and Kadamath have six and five species of seagrasses, respectively, and support large number of green turtles for feeding and nesting (Nobi et al. 2011; Nordlund et al. 2018) Though being one of the most important primary producers in the coastal ecosystems, seagrasses are highly threatened ecosystems owing to several natural causes like cyclones, tsunami, intensive grazing, dieback disease and anthropogenic reasons like illegal fishing, nutrient enrichment, pollution, siltation and dredging techniques. Natural causes for seagrass decline are pollution, nutrient enrichment, intense high waves, cyclones and tsunami, intensive grazing and infestation of fungi and epiphytes (Jagtap and Rodrigues 2004; Nobi and Kumar 2013; Patro et al. 2017). Seagrasses have also been used as heavy metal biomonitors in coastal ecosystems, and there are studies from several locations in India (Thangaradjou et al. 2010; Nobi et al. 2011; Selvaraj et al. 2020). Seagrass meadows in ANI and Lakshadweep islands have been studied well and also in Palk Bay and Gulf of Mannar region in Tamil Nadu coasts which are well documented; however, there are any major studies from states like Gujarat, Andhra Pradesh and Odisha which also have dominant and minor seagrass areas (Thangaradjou and Bhatt 2018).

3.4 Coral Reefs

Coral reefs are an important ecosystem in coastal regions as they provide many services and play a significant role in many tropical and subtropical countries (De Kalyan et al. 2017). They exhibit plentiful diversity, present beautiful picturesque sight and hence a great tourist attraction, form food and habitat for several marine species and provide protection against storms and tsunamis. All the three major reef types, i.e. atoll, fringing and barrier reefs, occur in India, and the region includes some of the most extensive, diverse and least disturbed reef areas of the Indian Ocean. Though being rich and diverse, these reefs have not been studied in detail (Venkatraman 2011; Patterson et al. 2020). Not counting the island territories, India has two well-defined mainland coastal areas containing reefs in the states of Gujarat and Tamil Nadu: the Gulf of Kachchh on the northwestern coast and Palk Bay and Gulf of Mannar on the southeastern coast. Apart from these two main regions, there are patches of reef growth on the West Coast, at Malvan and Redi in Maharashtra coast (Venkatraman 2011; De et al. 2017). The ANI have fringing reefs around many islands and a long barrier reef (329 km) on its western coast. There is very limited study on the ANI reefs, which may be the most diverse and pristine reef building corals in India. Extensive reefs have also been observed in Lakshadweep islands, but again these have been poorly explored.

3.5 Lagoons

Coastal lagoons are shallow water bodies that run along a shoreline but remain separated from the ocean by sand bars/spits, coral reefs and barrier islands and have one or several restricted connections with the sea (Mahapatro et al. 2013; Amir et al. 2019). India has several lagoons. India has the largest lagoon, Chilika lake, in Asia lying on the east coast of India in the state of Odisha. The second largest lagoon in India, Pulicat lake, also lies on the east coast shared between the states of Andhra

Pradesh and Tamil Nadu connected to the Bay of Bengal (Sahu et al. 2014; Nazneen and Raju 2017). Other prominent lagoons on west coast of India are Vembanad Lake and Ashtamudi Lake. Lagoons are fragile ecosystem owing to their shallow depth and restricted connection with the sea. The high productivity encountered in a lagoon is due to the varied salinity regime existing within the ecosystem. However, the mouth of the lagoons tends to close over time, and the inflow of only freshwater can turn them into freshwater ecosystems solely. In the case of Chilika lagoon, the same condition was observed when this lagoon was put under Montreux record of threatened wetlands under the Ramsar Convention. Later with help of Chilika Development Authority (CDA), this lagoon could be revived (Nazneen et al. 2019a, b). Vembanad Lake (Lat. 9° 30′–10° 10′ N and Long. 76° 10′–76° 25′ E) is the largest brackish water system in the state of Kerala on the southwest coast of India. The 96 km-long lake is a major water fowl habitat supporting unique assemblage of marine, brackish and freshwater species and also has mangroves in patches (Selvam et al. 2012).

4 India's Coastal Policies

India is one of the world's mega-biodiversity nations, meaning India is rich in terrestrial biodiversity and rich in marine biodiversity. The three metropolitan cities of Mumbai, Kolkata and Chennai are coastal cities, and the coasts live a substantial population. This section emphasises India's coastal policies for protecting and managing its large coastline with numerous unique, diversified coastal ecosystems. India has a wide coastline characterised by numerous complex ecosystems, with large dependent populations in India's coastal zones. Other Indian beaches suffer natural and manmade issues. Urbanisation, industrialisation and commercial activity and natural disasters directly affecting the coastal ecosystem and marine environment pose a big problem for India (Panigrahy and Mohanty 2012). With increasing human population in India's coastal regions and constant pressure from storms and cyclones, the frequency of which is worsened by climate change, sound policies are needed to manage its wide coast. Marine and coastal deterioration increases pressure on terrestrial and marine natural resources as terrestrial pollution finally finds its way into the oceans and seas.

On the other hand, rising population near the oceans, establishing industries and increasing tourists in coastal areas are primary causes of poor coastal health (Rani et al. 2015). Bay of Bengal and Arab Sea are both dynamic seas with regular cyclones and streams. In recent years, East India has had annual cyclones Phailin, Hudhud, Titli, Phani, Amphan, Tauktae and Yaas (Rani et al. 2015; Barik et al. 2017; Nazneen et al. 2019b; Kumar et al. 2021). Recent yearly cyclonic storms and rising temperatures plainly show that coastal towns like Mumbai and rural communities along the Indian coastline are highly vulnerable to floods, tropical cyclones and tsunamis, causing considerable loss of human life and property devastation (Sindhu and Unnikrishnan 2012). India's coastal wetlands have been protected under various international and national laws and treaties: Convention on Biological Diversity (CBD), Environmental Protection Act 1986, Convention on International Trade in Endangered Species of Wild Fauna and Flora (CITES), Ramsar Convention 1972 and United Nations Convention on the Laws of the Seas (UNCLOS) to name a few (Kumar and Saluja 2019). Some of these statues have been mentioned below.

4.1 Global Conventions and Coastline Protection Treaties

Several global conventions and treaties described below preserve marine and coastal environments. The Meeting of the United Nations on Environment and Development (UNCED) held in Rio de Janeiro in Brazil (also known as the Earth Summit or Rio Summit) in 1992 was an international conference attended by 172 countries, and some of the historic environmental agreements were made here. Three Earth Summit Framework Conventions are relevant to coastal and marine biodiversity. They are the Convention on Biological Diversity (CBD), the UN Framework Convention on Climate Change (UNFCCC) and the UN Convention on Desertification (UNCCD). These agreements cover marine and coastal habitats for protection, management and conservation (Saravanan et al. 2013).

4.1.1 Convention on Biological Diversity (CBD)

India's marine and coastal environment contains dynamic geomorphological features that preserve and represent unique, diverse biodiversity. Country's coastal area sustains a considerable amount of the country's population, imposing tremendous pressure on its coastal and marine resources. Maritime Protected Area Network (MPAN) is a tool to manage natural marine resources for biodiversity conservation and resource-dependent people's well-being. The other laws and provisions include Wildlife (Protection) Act 1972, Environment (Protection) Act 1986, Coastal Regulation Zone Notification 1991, National Biodiversity Act 2002 for the protection and conservation of coastal and marine environment (Saravanan et al. 2013). MPAs cover nearly 7% of the Indian coastline, but most of them are in ANI and fewer on the Indian coast. MPAs in mainland India are 24 in number with around 8214 km^2 area. This represents 4.92% of the total area covered throughout the protected area network. It includes mangroves, estuaries, lagoons, coral reefs, marshes, mudflats, coastal dunes, near-shore ecosystems, gulf waters, creeks, seagrass beds and salt marshes. Most of these MPAs are wildlife sanctuaries, while four are marine parks. Gulf of Mannar National Park, Sundarbans National Park, Bhitarkanika Wildlife Sanctuary and Coringa Wildlife Sanctuary are some of the peninsular India's major MPAs. India is one of the 17 mega-biodiversity countries, and it is well-known that, apart from its land, coastal and marine analogues are rich in biodiversity. There are only a limited number of marine and coastal regions recognised as MPAs. Many marine and coastal ecosystems with great biodiversity along the shoreline are yet unrepresented and identified for conservation actions. A separate, long-term research endeavour is necessary to build a mechanism for identifying sites and prioritising significant conservation areas.

4.1.2 The Convention on the Conservation of Migratory Species of Wild Animals (CMS)

This convention strives to preserve terrestrial, aquatic and migratory bird species throughout their natural range, from many countries to even continents. Under the auspices of the United Nations Environmental Program (UNEP), an international pact was established on the global conservation of migratory wildlife and their habitats (Kumar and Saluja 2019). The convention secretariat is in Bonn, and the decision-making body is the COP.

4.1.3 The Convention on International Trade in Endangered Wildlife (CITES)

CITES is an international agreement between states to ensure wildlife trading does not harm their survival. CITES exercises some degree of supervision in animal and plant trading across countries and borders. CITES secretariat is based in Geneva, Switzerland, and UNEP manages the administration. CITES regulates international trade in numerous marine species to help preserve their wildlife existence, including dolphins, marine tortoises, corals, queen conch, clams, sea horses and whales (Saravanan et al. 2013; Kumar and Saluja 2019). Thus, CITES protects from unregulated trade and use of endangered and rare marine species that would otherwise be extinct.

4.1.4 Ramsar Convention on International Important Wetlands

Ramsar Convention is a treaty between governments that provides the framework for government action and international cooperation on the conservation and wise use of wetlands and their resources. The treaty was born in 1971 in Ramsar, Iranian city. In a broader sense, the definition of wetlands includes lakes and rivers, estuaries, deltas and tidal flats, peatlands, marshes, mangroves, coral reefs and human-made ecosystems such as agricultural fields, ponds, reservoirs and salt pans (Kumar and Saluja 2019; Ragavan et al. 2020). India is a signatory to the Ramsar Convention and currently has 42 Ramsar-declared wetlands. Ramsar Convention in India ensures wetland management to conserve biodiversity and wise use. The scope of the Ramsar Convention extends to a wide variety of habitats, including rivers and lakes, mangroves, coastal lagoons, peatlands, coral reefs and numerous human-made wetlands such as fish and shrimp ponds, farm ponds, irrigated agricultural land, salt pan reservoirs, gravel pits, sewage farms and canals. Ramsar Wetlands in

India include Chilika Lake, Sundarbans Mangroves, Ashtamudi Wetland, Vembanad Kol Lake, Bhitarkanika Mangroves, Kolleru Lake, Point Calimere Wildlife and Bird Sanctuary.

4.1.5 Biosphere Reserves

Biosphere reserve is part of the natural environment representing enormous areas of terrestrial, marine or coastal ecosystems, sometimes comprising both terrestrial and marine ecosystems. In 1971, UNESCO introduced the concept of Biosphere Reserve under its Man and Biosphere Programme. Biosphere reserves are country-specific locations identified under UNESCO's Man and the Biosphere (MAB) Programme without any special legal force. The goal behind classifying a natural area as Biosphere Reserve is to stimulate the engagement of local populations based on sustainable development, coupled with scientific efforts. The objective of forming the biosphere reserve is to preserve all types of life in situ and its natural environment and support system to serve as a small model system for monitoring and analysing changes in natural ecosystems. In the broader resource management and development planning approach, it is regarded from the biodiversity conservation perspective. India was split into ten biogeographic zones, including coasts. The Indian government has now recognised 18 Biosphere Reserves. India's maritime reserves include Tamil Nadu's Gulf of Mannar Biosphere Reserve and West Bengal's Sundarbans Biosphere Reserve and ANI's Great Nicobar Biosphere Reserve. A biosphere reserve often has national parks and sanctuaries within its boundaries. Mannar Marine Park is part of GOM Biosphere Reserve. In 1986, a Tamil Nadu government declaration under the Wildlife Protection Act, 1972, declared the Gulf of Mannar region a marine national park. Mannar National Park Gulf (GOMNP) is 560 km^2. It has 21 islands surrounding Tuticorin and Ramanathapuram, Tamil Nadu. The GOM harbours around 3600 flora and wildlife, making it Asia's biologically richest coastal regions (Arisekar et al. 2021). Although it was declared 20 years ago, there was no fishing ban until 2002. Certain limits have been put in place and maintained by the forest department to avoid overfishing and resource depletion. In addition to other developmental initiatives that pose concerns to the area's biodiversity, such as the impending Sethusamudram canal project and other industrial developments on the Tuticorin coast, extensive fishing is regarded one of the greatest threats to GOM marine resources. These development buildings are not immediately present in the park region; nevertheless they threaten the park's coral reefs and seagrass environment.

4.1.6 Biodiversity Act, 2002

The Biodiversity Act was created in 2002 to preserve and use biological resources sustainably. The statute was developed primarily to fulfil India's CBD responsibilities. The legislation provides provisions enabling a fair and equitable sharing of benefits from the utilisation of biological resources and knowledge. Section 37 of this Act recognises and declares Biodiversity Heritage Sites (BHS). Protecting coastal habitats such as mangroves, coral reefs and seagrasses is provision under this act (Ramesh et al. 2018).

4.1.7 Indian Coastal Zone Regulations

Coastal zone is the area of terrestrial and marine interaction. The phrase coastal zone means coastal seawater, diverse coastal wetlands and marine-influenced coastal lines. The coastal zone comprises the area from high to low tide, up to 10 nautical miles from high tide to sea and up to 20 km from high tide to land. In Indian Coastal Regulation Zone (CRZ), the Ministry of Environment and Forests (MoEF) introduced the Environmental Protection Act of 1986 in 1991. CRZ guidelines prohibit activities, including human exploration and near-coastal industrial activities, to protect vulnerable and fragile ecosystems. They restrict activities that can cause hazard to delicate coastal environments (Panigrahy and Mohanty 2012).

Activities such as establishing new industries, mining, large buildings, storing or disposing of hazardous materials, reclamation and bundling—within a particular distance from the coastline are prohibited. After implementing the Environmental Protection Act in 1986, CRZ rules were first framed in 1991. Before this law came into existence, India had dispersed several. It extended laws and ordinances to oversee near-coast activities, but CRZ 1991 was the first comprehensive policy guideline to manage India's extensive and dynamic coastline harbouring multiple unique and endangered ecosystems. A High Tide Line (HTL) 500 m physical barrier to land was demarcated. This was further separated into four zones specifying either authorised or prohibited activities in these four zones. The entire coastal zone of the country has been classified into four areas: CRZ-I (ecologically sensitive near-shore area where future development activities are not permitted), CRZ-II (an urban coastal area already developed), CRZ-III (a significantly underdeveloped rural or urban coastal area where certain activities are allowed) and CRZ-IV (a special category includes island coasts including the entire Andaman and Nicobar and Lakshadweep islands). With the Ministry of Environment and Forests consent, some vital activities may be permitted (Krishnamurthy et al. 2014). Coastal Regulation Zone (CRZ-I) covers environmentally sensitive, near-shore areas significant from a biodiversity and ecological perspective. These areas include mangroves, coral reefs, marine parks and sanctuaries, reserve forests, wildlife habitats, areas close to fish breeding and spawning grounds, areas of natural beauty, historically important and heritage areas and areas with diverse flora and fauna with abundant genetic diversity areas that are likely to be inundated by rising sea levels as a result of global warming.

Coastal Regulation Zone II (CRZ-II) already covers near-shore metropolitan areas. The region is inside municipal boundaries and was created responsibly with enough roadways, drainage and water supply infrastructure. Coastal Regulation Zone III (CRZ-III) encompasses places that may or may not develop. These areas are also within municipal limits of a not established city or town or other legally declared urban regions. Coastal Regulation Zone IV (CRZ IV) includes Indian island territory. Under CRZ-IV, coastal parts of the Andaman and Nicobar, Lakshadweep and other tiny islands were bounded for their protection and development. Due to its top-down approach, CRZ 1991 notification was modified in 2011 and CRZ notification in 2011 as compared to CRZ 1991 had other faults and met local resistance. The change came into effect, taking into account local communities and stakeholders; however, this resulted in some erosion of their powers. After disasters like the very severe cyclone hitting the Odisha coast in 1999 and the Indian Ocean tsunami in 2004 causing massive property damage and loss of several thousand lives, CRZ had a big influence, and the regulations were seriously followed (Chinnasamy and Parikh 2020). Indian shoreline systems were battered on both east and west coastlines, leading to massive loss of life and infrastructure resulting in harming economy. Post these tragedies, coastal zone protection needs were recognised through stricter restrictions. Integrated Coastal Zone Management (ICZM) concept was taken into account and adjustments made in CRZ 2011 were notified. This included adding policies utilising bottom-up as a strong mechanism of governance. Following CRZ 2011, India increased its potential for disaster management, coastal management and several community-based field projects to enhance stakeholder involvement (Panigrahy and Mohanty 2012; Krishnamurthy et al. 2014). The CRZ also emphasises ecosystem protection including mangroves, coral reefs and seagrasses. In India, seagrass is an "ecologically sensitive area" (ESA) under the Coastal Regulation Zone 2011. States are expected to develop a Coastal Zone Management Plan to defend these regions (Griffiths et al. 2020). Also, ESA includes salt marshes. Coastal Regulation Zone was further altered, and CRZ 2018 was created, resulting in further adjustments. In CRZ 1991, more attention was paid to the safety and livelihood of local fishing communities, but this is lost in CRZ 2018 (Ishan 2019; Chinnasamy and Parikh 2020).

5 Conclusion

This chapter highlights India's large coastline zone including nine states and four Union territories, and two island groups. India's coastal zone has enormous geomorphological features such sandy beaches, mudflats, sand dunes, mangroves, seagrass meadows, lagoons and coral reefs. Mangroves and coral reef ecosystems are well investigated in India; however, seagrasses and salt marshes were not explored. India has increased mangrove cover on the eastern coast due to huge rivers flowing eastwards and building massive deltas. The area beneath the mangrove cover is declining for less awareness of past relevance. Mangrove forests were removed for human habitation, farming, livestock grazing and wood. India also has 14 of South Asia's 16 seagrasses. India's seagrass ecosystems have not been extensively studied, reporting new places with seagrass meadows. The least studied environment is salt marshes. Their multiple critical roles in carbon sequestration, shoreline protection and habitat supply were not examined. India agrees with various international treaties and coastal ecosystem preservation and conservation agreements. Also, India nationally has various laws and policies to safeguard coastal areas by protecting biodiversity laws and other regulations. Coastal Zone Regulation emphasises a comprehensive plan to protect the coastal zone, which divides coastal communities based on their proximity to the sea.

References

Ajai, Nayak S, Bhattacharya S (2012a) Report by Space Applications Centre, ISRO. Sponsored by Ministry of Environment & Forests, Govt. of India

Ajai, Bhaguna A, Chauhan HB, Sen Sarma K, Bhattacharya S, Ashutosh S, Pnadey CN, Thangaradjou T, Gnanppazham L, Selvam V, Nayak SR (2012b) Mangrove inventory of India at community level. Natl Acad Sci Lett. https://doi.org/10.1007/s40009-012-0087-x

Amir M, Paul D, Samal R, N. (2019) Sources of organic matter in Chilika lagoon, India inferred from stable C and N isotopic compositions of particulates and sediments. J Mar Syst 194:81–90

Arisekar U, Shakila RJ, Shalini R, Jeyasekaran G, Sivaraman B, Surya T (2021) Heavy metal concentrations in the macroalgae, seagrasses, mangroves, and crabs collected from the Tuticorin coast (Hare Island), Gulf of Mannar, South India. Mar Pollut Bull 163:111971. https://doi.org/10.1016/j.marpolbul.2021.111971

Badola R, Hussain SA (2003) Valuation of the Bhitarkanika mangrove ecosystem for ecological security and sustainable resource use. Study report. Wildlife Institute of India, Dehradun

Bandopadhyay PC, Carter A (eds) (2017) The Andaman–Nicobar accretionary ridge: geology, tectonics and hazards, vol 47. Geological Society, London, Memoirs, pp 9–18. https://doi.org/10.1144/M47.2

Barik SK, Muduli PR, Mohanty B, Behera AT, Mallick S, Das A, Samal RN, Rastogi G, Pattnaik AK (2017) Spatio-temporal variability and the impact of Phailin on water quality of Chilika lagoon. Cont Shelf Res 136:39–56

Bassi N, Kumar MD, Sharma A, Pardha-Saradhia P (2014) Status of wetlands in India: a review of extent, ecosystem benefits, threats and management strategies. J Hydrol Reg Stud 2:1–19

Bharathi K, Suhasini P, Raja S, Ranjith R, Vanitha K, Thangaradjou T (2014) Spatial variability in distribution of seagrasses along the Tamil Nadu Coast. Int J Curr Res 6(10):8997–9005

Bhomia RK, MacKenzie RA, Murdiyarso D, Sasmito SD, Purbopuspito J (2016) Impacts of land use on Indian mangrove forest carbon stocks: implications for conservation and management. Ecol Appl 26(5):1396–1408

Chinnasamy P, Parikh A (2020) Remote sensing-based assessment of Coastal Regulation Zones in India: a case study of Mumbai, India. Environ Dev Sustain 23:7931–7950. https://doi.org/10.1007/s10668-020-00955-z

Dalia SV, Satheesh KP, Pillai NGK (2014) Biodiversity and seasonal variation of benthic macrofauna in Minicoy Island, Lakshadweep, India. Acta Oceanol Sin 33(10):58–73. https://doi.org/10.1007/s13131-014-0541-3

Dasgupta R, Shaw R (2013) Changing perspectives of mangrove management in India—an analytical overview. Ocean Coast Manag 80:107–118. https://doi.org/10.1016/j.ocecoaman.2013.04.010

Dasgupta R, Shaw R (2016) Perceptive insight into incentive design and sustainability of participatory mangrove management: a case study from the Indian Sundarbans. J For Res 28:815–829. https://doi.org/10.1007/s11676-016-0355-6

De Kalyan, Venkatraman K, Ingole B (2017) Current status and scope of coral reef research in India: A biological perspective. Ind J Geomarine Sci 46(4):642–662

Duarte CM, Losada IK, Hendriks IE, Mazarrasa I, Marbà N (2013) The role of coastal plant communities for climate change mitigation and adaptation. Nat Clim Chang 3:961–968. https://doi.org/10.1038/nclimate1970

GIZ Report (2013) Legal framework for conservation of coastal and marine environment of India: a review. Indo-German Biodiversity Programme Conservation and Sustainable Management of Coastal and Marine Protected Areas

Gopal B, Chauhan M (2006) Biodiversity and its conservation in the Sundarban mangrove ecosystem. Aquat Sci 68:338–354. https://doi.org/10.1007/s00027-006-0868-8

Gopi M, Pravin Kumar M, Joyson Joe Jeevamani J, Raja S, Muruganandam R, Deepak Samuel V, Simon NT, Viswanathan C, Abhilash KR, Krishnan P, Purvaja R, Ramesh R (2019) Distribution and biodiversity of tropical saltmarshes: Tamil Nadu and Puducherry, southeast coast of India. Estuar Coast Shelf Sci. https://doi.org/10.1016/j.ecss.2019.106393

Gopi S, Arulkumar A, Ganeshkumar A, Rajaram R, Miranda JM, Paramasivam S (2020) Heavy metals accumulation in seagrasses collected from Palk Bay, Southeastern India. Mar Pollut Bull 111305

Gopi M, Jeevamani JJJ, Goutham S, Simon NT, Samuel VD, Abhilash KR, Robin RS, Hariharan G, Muruganandam R, Krishnan P, Purvaja R, Ramesh R (2021) Status of health and conservation classification of tropical coral reefs in Lakshadweep archipelago. Wetlands Ecol Manag. https://doi.org/10.1007/s11273-021-09801-z

Griffiths LL, Connollya RM, Brown CJ (2020) Critical gaps in seagrass protection reveal the need to address multiple pressures and cumulative impacts. Ocean Coast Manag 183:104946

Ishan K (2019) Coastal regulation notification: What development are we clearing our coasts for. Down To Earth

Jagtap TG, Rodrigues RS (2004) Seagrasses: the forgotten marine habitat. In: Untawale AG (ed) Know our shore. World Wide Fund for Nature, Goa, pp 57–67

Kadaverugu R, Dhyani S, Dasgupta R, Kumar P, Hashimoto S, Pujari P (2021) Multiple values of Bhitarkanika mangroves for human well-being: synthesis of contemporary scientific knowledge for mainstreaming ecosystem services in policy planning. J Coast Conserv 25:32. https://doi.org/10.1007/s11852-021-00819-2

Kathiresan K (2000) A review of studies on Pichavaram mangrove, southeast India. Hydrobiologia 430:185–205

Kathiresan K (2002) Why are mangroves degrading? Curr Sci 83:1246–1249

Kathiresan K (2018) Mangrove forests of India. Curr Sci 114:976–981

Krishnamurthy RR, Dasgupta R, Chatterjee R, Shaw R (2014) Managing the Indian coast in the face of disasters & climate change: a review and analysis of India's coastal zone management policies. J Coast Conserv. https://doi.org/10.1007/s11852-014-0339-7

Krishnan P, Ananthan PS, Purvaja R, Jeevamani JJJ, John Infantina A, Rao CS, Arur A, Mahendra RS, Kareemulla ISK, Biswas A, Sastry RK, Ramesh R (2018) Framework for mapping the drivers of coastal vulnerability and spatial decision making for climate-change adaptation: a case study from Maharashtra, India. Ambio. https://doi.org/10.1007/s13280-018-1061-8

Kumar A, Ramanathan AL (2015) Speciation of selected trace metals (Fe, Mn, Cu and Zn) with depth in the sediments of Sundarban mangroves: India and Bangladesh. J Soils Sediments. https://doi.org/10.1007/s11368-015-1257-5

Kumar R, Saluja R (2019) Coastal wetlands in India and climate change: status, trends and policy response options. In: Ramesh R, Bhatt JR (eds) Climate change and the vulnerable Indian coast. Ministry of Environment, Forest and Climate Change, Government of India, India, pp 239–258

Kumar R, Rani S, Maharana P (2021) Assessing the impacts of Amphan cyclone over West Bengal, India: a multi-sensor approach. Environ Monit Assess 193:283. https://doi.org/10.1007/s10661-021-09071-5

Mahapatro D, Panigrahy RC, Panda S (2013) Coastal Lagoon: Present Status and Future Challenges. Mar Pollut Bull 3(23):178–186

Mishra AK, Apte D (2020) Ecological connectivity with mangroves influences tropical seagrass population longevity and meadow traits within an island ecosystem. Mar Ecol Prog Ser 644:47–63

Mishra AK, Apte D (2021) The current status of Halophila beccarii: an ecologically significant, yet vulnerable seagrass of India. Ocean Coast Manag 200:105484

Mugade NR, Sapkale JB (2014) A review of mangrove conservation studies in Maharashtra, India. Int J Eng Techn Res 2(11):338–341

Nayak PK, Mohanta RK, Sahu AK, Swain KK (2016) Mangroves of Mahanadi Delta in the state of Odisha and aspects of their conservation. Int J Conserv Sci 7:1095–1104

Nazneen S, Raju NJ (2017) Distribution and sources of carbon, nitrogen, phosphorus and biogenic silica in the sediments of Chilika lagoon. J Earth Syst Sci 126:13. https://doi.org/10.1007/s12040-016-0785-8

Nazneen S, Singh S, Raju NJ (2019a) Heavy metal fractionation in core sediments and potential biological risk assessment from Chilika lagoon, Odisha state, India. Quatern Int 507:370–388. https://doi.org/10.1016/j.quaint.2018.05.011

Nazneen S, Raju NJ, Madhav S, Ahamad A (2019b) Spatial and temporal dynamics of dissolved nutrients and factors affecting water quality of Chilika lagoon. Arab J Geosci 12:243. https://doi.org/10.1007/s12517-019-4417-x

Nobi EP, Kumar PKD (2013) Environmental characteristics of tropical coral reef-seagrass dominated lagoons (Lakshadweep, India) and implications to resilience to climate change. Environ Earth Sci. https://doi.org/10.1007/s12665-013-3020-9

Nobi EP, Dilipan E, Sivakumar K, Thangaradjou T (2011) Distribution and biology of seagrass resources of Lakshadweep group of islands, India. Ind J Geomarine Sci 40(5):624–634

Nordlund LM, Unsworth RKF, Gullström M, Cullen-Unsworth LC (2018) Global significance of seagrass fishery activity. Fish Fish 19:399–412

Panigrahy JK, Mohanty PK (2012) Effectiveness of the Indian coastal regulation zones provisions for coastal zone management and its evaluation using SWOT analysis. Ocean Coast Manag 65:34–50

Patro S, Krishnan P, Deepak Samuel V, Purvaja R, Ramesh R (2017) Seagrass and salt 673 marsh ecosystems in South Asia: an overview of diversity, distribution, threats and 674 conservation status. In: Prusty BAK, Chandra R, Azeez PA (eds) Wetland 675 science: perspectives from South Asia. Springer, New Delhi, pp 87–103

Patterson J, Jeyasanta KI, Sathish N, Edward JKP (2020) Microplastic and heavy metal distributions in an Indian coral reef ecosystem. Sci Total Environ. https://doi.org/10.1016/j.scitotenv.2020.140706

Ragavan P, Alok Saxena A, Mohan PM, Ravichandran K, Jayaraj RSC, Saravanan S (2015) Diversity, distribution and vegetative structure of mangroves of the Andaman and Nicobar Islands, India. J Coast Conserv 19:417–443. https://doi.org/10.1007/s11852-015-0398-4

Ragavan P, Dubey SK, Dagar JC, Mohan PM, Ravichandran K, Jayaraj RSC, Rana TS (2019) Current understanding of the mangrove forests of India. In: Dagar JC et al (eds) Research developments in saline agriculture. Springer, Singapore

Ragavan P, Kathiresan K, Mohan PM, Ravichandran K, Jayaraj RSC, Rana TS (2020) Ensuring the adaptive potential of coastal wetlands of India—the need of the hour for sustainable management. Wetlands Ecol Manag. https://doi.org/10.1007/s11273-020-09742-z

Ramesh R, Banerjee K, Selvam AP, Lakshmi A, Krishnan P, Purvaja R (2018) Legislation and policy options for conservation and management of seagrass ecosystems in India. Ocean Coast Manag 159:46–50. https://doi.org/10.1016/j.ocecoaman.2017.12.025

Rani NNVS, Satyanarayana ANV, Bhaskaran PK (2015) Coastal vulnerability assessment studies over India: a review. Nat Hazards. https://doi.org/10.1007/s11069-015-1597-x

Ranjan RK, Ramanathan AL, Singh G (2008) Evaluation of geochemical impact of tsunami on Pichavaram mangrove ecosystem, southeast coast of India. Environ Geol 55:687–697. https://doi.org/10.1007/s00254-007-1019-9

Ranjan RK, Routh J, Ramanathan AL (2010) Bulk organic matter characteristics in the Pichavaram mangrove—estuarine complex, south-eastern India. Appl Geochem 25:1176–1186

Robin RS, Muduli PR, Kannuri VV, Nagarjuna A, Nallathambi T, Mishra RK, Balasubramanian T (2012) Planktonic communities and trophic interactions in the Kavaratti waters, Lakshadweep Archipelago, India. Int J Ecosyst 2(2):5–18

Saha M, Sarkar SK, Bhattacharya B (2006) Interspecific variation in heavy metal body concentrations in biota of Sundarban mangrove wetland, Northeast India. Environ Int 32:203–207

Sahu BK, Pati P, Panigrahy RC (2014) Environmental conditions of Chilika Lake during pre and post hydrological intervention: an overview. J Coast Conserv 18:285–297

Sappal SM, Ramanathan A, Ranjan RK, Singh G, Kumar A (2014) Rare earth elements as biogeochemical indicators in mangrove ecosystems (Pichavaram, Tamilnadu, India). J Sediment Res 84(9):781–791. https://doi.org/10.2110/jsr.2014.63

Saravanan KR, Sivakumar K, Chodhury BC (2013) Important coastal and marine biodiversity area of India. In: Sivakumar K (ed) Coastal and marine protected area in India: challenges and way forward, ENVIS bulletin: wildlife & protected area, vol 15. Wildlife Institute of India, Dehradun, pp 134–188

Selvam V (2003) Environmental classification of mangrove wetlands of India. Curr Sci 84(6):757–765

Selvam AP, Priya SL, Banerjee K, Hariharan G, Purvaja R, Ramesh R (2012) Heavy metal assessment using geochemical and statistical tools in the surface sediments of Vembanad Lake, Southwest Coast of India. Environ Monit Assess 184:5899–5915. https://doi.org/10.1007/s10661-011-2389-8

Selvaraj G, Abimannan A, Arumugam G, Rajendran R, Jose MM, Sadayan P (2020) Heavy metals accumulation in seagrasses collected from Palk Bay, Southeastern India. Mar Pollut Bull 157:111305

Sindhu B, Unnikrishnan AS (2012) Return period estimates of extreme sea level along the east coast of India from numerical simulations. Nat Hazards 61(3):1007–1028

Singh G, Ramanathan AL, Prasad MBK (2005) Nutrient cycling in mangrove ecosystem: a brief overview. J Ecol Environ Sci 30(3):231–244

Sridhar R (2018) Ecologically sensitive areas and coastal geomorphological features of Andaman & Nicobar Islands, India. Mod Approaches Oceanogr Petrochem Sci. https://doi.org/10.32474/MAOPS.2018.01.000119

Stankovic M, Ambo-Rappe R, Carly F, Dangan-Galon F et al (2021) Quantification of blue carbon in seagrass ecosystems of Southeast Asia and their potential for climate change mitigation. Sci Total Environ. https://doi.org/10.1016/j.scitotenv.2021.146858

Stansley OD (2004) Wetland ecosystems and coastal habitat diversity in Gujarat, India. J Coast Dev 7(2):49–64

Thangaradjou T, Bhatt JR (2018) Status of seagrass ecosystems in India. Ocean Coast Manag 159:7–15

Thangaradjou T, Sridhar R, Senthilkumar S, Kannan S (2007) Seagrass resource assessment in the Mandapam coast of the Gulf of Mannar Biosphere Reserve, India. Appl Ecol Environ Res 6(1):139–146

Thangaradjou T, Nobi EP, Dilipan E, Sivakumar S, Sushila K (2010) Heavy metal enrichment in seagrasses of Andaman Islands and its implication to the health of the coastal ecosystem. Ind J Mar Sci 39(1):85–91

Unsworth RKF, Nordlund LM, Cullen-Unsworth LC (2019) Seagrass meadows support global fisheries production. Conserv Lett 12:e12566. https://doi.org/10.1111/conl.12566

Venkatraman K (2011) Coral reefs of India. In: Hopley D (ed) Encyclopedia of modern coral reefs. Springer, Dordrecht

Viswanathan C, Purvaja R, Jeevamani JJJ, Samuel VD, Sankar R, Abhilash KR, Geevarghese GA, Muruganandam R, Gopi M, Raja S, Das RR, Patro S, Krishnan P, Ramesh R (2020) Salt marsh vegetation in India: species composition, distribution, zonation pattern and conservation implications. Estuar Coast Shelf Sci 242:106792

Chapter 2
Sources and Distribution of Fecal Coliforms in the Coastal Environment: A Case Study from Chilika Lagoon, Odisha, India

Madhusmita Mohapatra, Stiti Prangya Dash, Pratiksha Behera, Sudhakar Panda, and Gurdeep Rastogi

Abstract Worldwide, the contamination of coastal waters by fecal coliforms (FC) is an ongoing public health problem, and the Chilika Lagoon is no exception to it. Chilika, a brackish water coastal lagoon located in the Odisha state of India, is a biodiversity hotspot supporting commercial fisheries, water birds, and wildlife. Fisherman villages densely surround the lagoon, and dumping of solid waste and domestic sewage into the lagoon has become a common practice. We examined the long-term spatiotemporal distribution of FC in a 3-year period from 2017 to 2019, in the Chilika Lagoon and its drainage rivers. FC loads were represented as the most probable number (MPN) which varied seasonally and sectorally ranging from 0 to 2400 MPN/100 ml. The highest average FC load (17 MPN/100 ml) was recorded during monsoon and the lowest (7 MPN/100 ml) during summer. When FC loads of the lagoon were compared with Central Pollution Control Board (CPCB) guidelines for Class SW-II waters, >100 MPN/100 ml values were obtained from 5 (2017), 8 (2018), and 14 (2019) water samples. Kantabania (142 MPN/100 ml) and Kusumi (189 MPN/100 ml) rivers recorded much higher FC loads. Samples collected from Odialpur, a shoreline village, showed an average FC load of 279 MPN/100 ml, indicating a point source of fecal pollution. The runoff from rivers, sewage disposal from villages, birds, and livestock could be the possible sources of FC loads into the lagoon. Overall, FC loads in the lagoon were mostly within safe limits as prescribed for water used for bathing, contact water sports, and commercial fishing. The low FC load in the lagoon could be due to the quick inactivation and rapid mortality of fecal bacteria by the high salinity of the water. Salinity showed a statistically significant negative relationship ($r = -0.06$, p-value < 0.05) with MPN counts. Phylogenetic analysis of 16S rRNA gene sequences amplified from FC isolates revealed that they belonged to *Shigella flexneri* (seven isolates), *Klebsiella pneumoniae* (three

M. Mohapatra · S. P. Dash · P. Behera · S. Panda · G. Rastogi (✉)
Wetland Research and Training Centre, Chilika Development Authority, Balugaon, Odisha, India

S. Madhav et al. (eds.), *Coastal Ecosystems*, Coastal Research Library 38,
https://doi.org/10.1007/978-3-030-84255-0_2

isolates), and *Escherichia fergusonii* (one isolate). Antibiotic resistance profiles showed that all isolates were resistant to one or more antibiotics. The large data set on FC would be useful for wetland management authorities and decision-makers towards pollution control monitoring schemes in Chilika Lagoon.

Keywords Fecal coliforms · Salinity · Antibiotic susceptibility · MAR index · Lagoon · Bird guano

1 Introduction

Coastal lagoons are among the most productive and transitional ecosystems between land, freshwater, and marine waters (Pérez-Ruzafa et al. 2019). These coastal ecosystems are threatened by increasing anthropogenic activities such as land reclamation for agriculture, urbanization, tourism, and aquaculture (Pérez-Ruzafa et al. 2019). These activities have led to the dumping and discharge of solid waste and sewage, leading to microbial pollution of the coastal environments (Kataržytė et al. 2018). The microbial pollutants lead to the deterioration of the sanitary quality of water, creating severe health hazards for wildlife and the public. Therefore, continuous monitoring of microbial pollutants using indicator bacteria can elucidate the sanitary status of water which can ultimately serve as a risk assessment tool for recreational and other human activities (Sugumar et al. 2008).

1.1 Microbial Indicators of Bacteriological Quality of Water

The ideal microbial indicator should be (1) non-pathogenic, (2) present in densities that correlate with pathogens, (3) absent in non-contaminated samples, and (4) abundant and easy to detect (Cabral 2010; Motlagh and Yang 2019). Coliform bacteria fulfill most of these criteria and are considered as indicator species for the bacteriological quality of water. The most routinely used microbial indicators of water quality are total coliforms (TC). The FC, also known as thermotolerant bacteria, are a subset of TC that ferment lactose at 44 °C (Cabral 2010). FC are specifically present in the intestines of humans and other warm-blooded animals (Boyd 2015; Motlagh and Yang 2019). They have a relatively short lifespan in comparison to other coliform bacteria and are used as indicator bacteria for monitoring the sewage contamination in natural water bodies (Motlagh and Yang 2019). The presence of large FC loads in a water body suggests a high probability that other pathogenic bacteria, viruses, and protozoa may be present.

Coliforms are rod-shaped, gram-negative, non-spore-forming, β-galactoside permease-positive, β-galactosidase-positive, and aerobic or facultative anaerobic bacteria (Clesceri et al. 1998; Campbell et al. 2011; Sengupta and Saha 2013). Coliforms of the *Enterobacteriaceae* family belong to the genera *Citrobacter*, *Enterobacter*, *Escherichia*, and *Klebsiella* and can ferment lactose with gas production at 35–37 °C (Cabral 2010). Some coliform bacteria, e.g., *Escherichia coli*, are also common occupants in the bird and mammalian intestinal tracts; others such as *Enterobacter* and *Klebsiella* present on the plant surfaces and in soils are not directly involved with fecal contamination. Therefore, the coliform group comprises both intestinal bacteria and other free-living bacteria that are non-fecal in origin. *E. coli* is constantly found in the feces of human, pets, and farm animals with a much higher abundance (i.e., approximately 10^9 bacteria/g feces) than other coliforms (Sengupta and Saha 2013). Thus, *E. coli* was considered as the only coliform that was directly associated with a fecal source (WHO 2012). A strong positive correlation has been shown between FC and *E. coli* abundances (Town 2001).

The increasing levels of multidrug-resistant bacteria in aquatic environments have been recognized as an emerging global issue. Multiple antibiotic resistances (MAR) have been used to distinguish the fecal pollution sources through antibiotic resistance profiling (Cimenti et al. 2007). The MAR characteristics of FC are useful to understand whether isolates are derived from the high- or low-risk sources of contamination where antibiotics are used frequently or rarely (Krumperman 1983). MAR index ≥0.20 (threshold value) denotes a high-risk source of contamination (Riaz et al. 2011).

1.2 Monitoring and Assessment of FC

FC loads have been used as indicators of fecal contamination and pathogen in natural freshwater sources. For instance, Davis et al. (2005) investigated the spatial and temporal distribution of FC from a drinking water reservoir in California, Canyon Lake. The study found a seasonal variation in the concentration of FC. The study also revealed that the correct interpretation of the fecal contamination was largely influenced by the choice of indicator bacteria and the sampling depth. Mitch et al. (2010) examined the accumulation of FC within the Quinnipiac River during winter when there was no disinfection treatment of wastewater effluents practiced. The study suggested a year-round disinfection process and also control of FC from the non-point sources such as river discharge and upstream or downstream of a wastewater outfall.

Various marine environments have also been assessed for the presence of FC bacteria. For example, Chigbu et al. (2004) examined Mississippi Sound, a coastal water body, for inter-annual variations in FC levels and their relationship with various water quality parameters. The study found a negative correlation between FC loads, salinity, and water temperature, whereas a positive correlation with rainfall suggested that freshwater input was a source of FC bacteria. Wiegner et al. (2017) determined the spatiotemporal variation of fecal indicator bacteria from Hilo Bay,

Hawaii, and demonstrated that fecal indicator bacteria increased alarmingly during the high flow period.

Numerous studies have been conducted on the distribution of FC bacteria from coastal lagoons and estuarine ecosystems. For instance, Konan et al. (2009) studied the spatial and temporal variations of FC from a eutrophic coastal lagoon: Grand-Lahou (south coast of West Africa). FC loadings were higher during the monsoon and lower during the dry season. The FC density was greater in the continental influence zone with anthropogenic inputs than in the oceanic influence zone of the lagoon. Yetis and Selek (2014) analyzed FC levels and their relationship with physicochemical parameters in Akyatan Lagoon (Mediterranean coast of Turkey) and revealed higher FC loads in the drainage channels than inside the lagoon. Cooksey et al. (2019) investigated fecal indicator bacteria, coliphages, and human adenovirus from estuarine recreation sites of a brackish Lake Pontchartrain located in southeast Louisiana, USA. The study found no correlation between fecal indicator bacteria/coliphage and human viral pathogens and suggested direct detection of pathogens using alternative microbial pollution monitoring tool. The spatiotemporal distribution of FC bacteria was assessed from Sontecomapan coastal lagoon (Gulf of Mexico) which revealed that FC exceeded USA-EPA maximum permissible values for services involving direct human contact such as harvesting or extracting shellfish (Soto-Castor and Esquivel-Herrera 2020). Furthermore, the study showed that human settlements and anthropogenic activities (cattle and poultry husbandry) as well as droppings from wildlife such as waterfowl and mammals were the sources for fecal contamination in the lagoon. Blackwater Estuary, UK, has also been examined for FC loads in the overlying water and shellfish (oysters) (Florini et al. 2020). The study found low FC levels in high saline water compared to the freshwater zone which was attributed to the increased bacterial cell inactivation under elevated salt concentrations.

The studies on FC from Indian coastal waters have been conducted mostly at small spatial and temporal scales. Mohandass and Bharathi (2003) studied the FC levels from the coastal water and sediments of Nagore situated on the east coast of India and recorded higher coliform counts in the sediments than the water column. In another study from Mumbai, the west coast of India, samples were collected from coastal areas, creeks, and effluent of wastewater treatment facilities, drains, and ocean outfalls (Vijay et al. 2010). FC levels exceeded the prescribed limits for SW-II class of water. Jayakumar et al. (2013) monitored two estuaries and two coastal lagoons along the southeast coast of India (Chennai) and found a considerable FC count, which evidenced the impact of anthropogenic activities. Latha and Mohan (2013) surveyed the FC levels from Kengeri Lake, Bangalore, India, and concluded that the lake's water was unfit for domestic and agricultural uses. In a study from coastal aquifer of Chennai, groundwater contamination was linked to an on-site sanitation system (Jangam and Pujari 2019).

Chilika, Asia's largest brackish water lagoon, is located in the Odisha state of India. The lagoon's shoreline, especially in the periphery of the central, northern, and southern sectors, is densely populated. The lagoon covers three districts of Odisha which are Puri, Khurdha, and Ganjam. There are 424 villages located within a 2 km range of the lagoon (Kumar and Pattnaik 2012). These villages lack adequate sanitary facilities, and disposal of domestic sewage and open defecation along the shoreline are common. The industries are not well developed in the vicinity of Chilika Lagoon, and fecal pollution is primarily due to the lack of treatment for disposal of domestic sewage and solid waste. This has led to the disposal of considerable quantities of untreated wastes into the lagoon from peripheral villages. Daya River, a major freshwater source to the Chilika Lagoon, transports and disposes approximately 550 million l/day of untreated domestic sewage from the nearby city, Bhubaneswar, India (Ghosh et al. 2006; Joshi and Mishra 2017). Furthermore, wildlife such as birds and buffalo that use Nalabana Bird Sanctuary as a foraging ground may also contribute to FC in Chilika Lagoon.

Mukherjee (2016) used the Modelo Hidrodinâmico (MOHID), a three-dimensional water modeling system, to analyze the intrusion of FC from the Daya River into the Chilika Lagoon. The model predicted that the freshwater influx was mostly responsible for distributing FC in the lagoon, but their propagation was limited due to rapid inactivation. The FC are influenced by several physicochemical factors such as salinity, temperature, sediment texture, and organic matter in estuarine systems (Hassard et al. 2017; Karbasdehi et al. 2017). Parida et al. (2012) have isolated pathogenic bacteria such as *Shigella dysenteriae*, *Bacillus cereus*, *Klebsiella pneumonia*, and *Streptococcus lactis* from the Chilika Lagoon. So far, systematic monitoring of the spatiotemporal distribution of FC has not been carried out; therefore, no baseline data is available from Chilika Lagoon. Considering this knowledge caveat, a long-term monitoring was conducted for the (1) determination of the spatiotemporal distribution of FC in the lagoon, (2) determination of FC loads in major rivers that drain their freshwater into the lagoon, (3) molecular identification and phylogenetic analysis of FC isolates using 16S rRNA gene sequences, and (4) determination of antibiotic resistance profiles of FC isolates.

2 Materials and Methods

2.1 Study Area

Chilika (19° 28′–19° 54′ N and 85° 06′–85° 35′ E) is a coastal brackish water lagoon situated on the east coast of India (Fig. 2.1). The wetland was designated as the first Indian Ramsar site (no. 229) in 1981 due to its rich biodiversity (Srichandan et al. 2015; Behera et al. 2018a). The lagoon is highly productive due to shallow depth and supports an enormous diversity of flora and fauna (Pattnaik et al. 2019, 2020). The lagoon is a well-known wintering ground for thousands of migratory birds and is also home to the Irrawaddy dolphins. The average catchment area of the lagoon is approximately 4146 km^2. The freshwater inflow is mostly brought by Daya, Bhargavi, Luna, and Makara (Srichandan et al. 2015). The lagoon has been divided into four sectors: central, northern, and southern sector, and outer channel

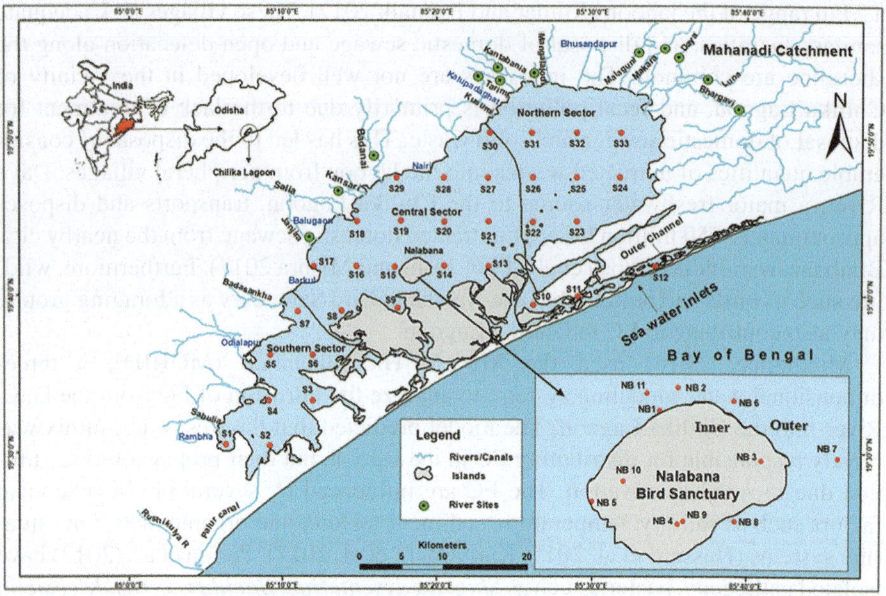

Fig. 2.1 Geographical location of the Chilika Lagoon. Water samples from the lagoon were collected from the 33 GPS fixed stations (shown with red closed circles) located in the four sectors. The sampling sites from the 12 major rivers are shown with green circled dot. The panel shows the detailed location of inner (NB3, NB4, NB5, NB9, and NB10) and outer (NB1, NB2, NB6, NB7, NB8, and NB11) stations in the Nalabana Bird Sanctuary. Shorelines of Barkul, Chandraput, and Odialpur villages were targeted for the FC survey

(Srichandan et al. 2015; Behera et al. 2017). The northern sector is the freshwater zone (salinity 0.5–5) with most of the river influx from Mahanadi River distributaries (Muduli and Pattnaik 2020). The southern sector experiences higher salinity than the central sector due to its connection with the Bay of Bengal (BoB) through the Palur Canal. The central sector is a brackish zone due to the mixing of freshwater and seawater. Both the southern and central sectors experience salinity ranging from 5 to 18. The outer channel is a marine zone with salinity ranging between 18 and 30 due to the direct connectivity to the BoB (Muduli and Pattnaik 2020). The Nalabana Bird Sanctuary covers an area of about 16 km^2 and is situated in the central sector of the lagoon. The sanctuary hosts congregation of millions of migratory and resident birds during winter and act as a nursery and breeding ground.

2.2 *Water Sampling*

Surface water samples (n = 1188) were collected from the 33 GPS (Global Positioning System) fixed stations. Samples were collected monthly for 3 consecutive years, from January 2017 to December 2019 (Fig. 2.1). A total of 84 water samples were collected from 12 major rivers during peak monsoon during the 3 years (Fig. 2.1). Shorelines of Barkul, Chandraput, and Odialpur villages were also targeted for the FC survey with water samples ($n = 30$) collected in June 2017. Water samples ($n = 264$) from 11 GPS fixed positions were collected monthly from outside and inside the Nalabana Bird Sanctuary from January 2018 to December 2019 (Fig. 2.1). Water samples were transported on ice and processed on the same day for MPN assessment. Salinity was measured in situ using a Thermo Scientific™ Orion™ Star A212 Conductivity Benchtop Meter.

2.3 Detection of FC Bacteria

Analysis of FC bacteria was conducted using the multiple tube fermentation method and recorded as the MPN of organisms present in 100 ml of the water sample (WHO 1985). Tubes containing double- and single-strength lactose broth with inverted Durham tubes were sterilized. Ten milliliters of the sample was inoculated in three tubes of double-strength media. 3-3 tubes each for single-strength media were inoculated with 1 and 0.1 ml of sample. For each sample, nine tubes were inoculated and incubated at 44 °C for 48 h. The gas-producing tubes and color change of media from purple to yellow were considered positive, and the MPN index was determined by comparing the presumptive test results with the standard table prescribed by Dubey and Maheshwari (2012) (Table 2.1).

Water samples with the positive presumptive results in the multiple tube fermentation tests were further selected for confirmatory analysis on eosin-methylene-blue (EMB) agar plates. EMB is a differential media that can differentiate between lactose-fermenting and non-lactose-fermenting bacteria. In general, the lactose-fermenting bacteria can be differentiated with purple colonies with dark centers. Further, *E. coli* colonies can be differentiated from other lactose-fermenting colonies due to the distinct metallic green sheen (Fig. 2.2a).

The CPCB, New Delhi, has implemented various regulatory guidelines on water quality for different applications to obtain water quality standards (CPCB 1993). According to the guidelines for primary water quality criteria for Class SW-II (waters for bathing, contact water sports, and commercial fishing), the permissible level of FC load is 100 MPN/100 ml of the sample (CPCB 1993). All samples were compared with the CPCB guidelines for FC load risk assessment.

2.4 Molecular Identification and Phylogenetic Analysis

Lactose-fermenting isolates ($n = 11$) cultured from water samples collected during January to March 2019 were selected from the EMB agar plates, and pure colonies were obtained after streaking on Luria-Bertani (LB) agar plates (Table 2.2). The genomic DNA was extracted from pure cultures using FastDNA SPIN Kit, MP

Table 2.1 MPN index for various combinations of positive presumptive test results of FC when three tubes are inoculated with 10 ml, 1 ml, and 0.1 ml of water samples (Modified from Dubey and Maheshwari, 2012)

Sl no.	Combination of positives	MPN index/100 ml
1	0–0–1	3
2	0–1–0	3
3	1–0–0	4
4	1–0–1	7
5	1–1–0	7
6	1–1–1	1
7	1–2–0	1
8	2–0–0	9
9	2–0–1	14
10	2–1–0	15
11	2–1–1	20
12	2–2–0	21
13	2–2–1	28
14	3–0–0	23
15	3–0–1	39
16	3–0–2	64
17	3–1–0	43
18	3–1–1	75
19	3–1–2	120
20	3–2–0	93
21	3–2–1	150
22	3–2–2	210
23	3–3–0	240
24	3–3–1	460
25	3–3–2	1100
26	3–3–3	2400

Biomedicals (Behera et al. 2018b). The concentration of DNA and its purity were assessed spectrophotometrically using an Epoch™ Microplate Spectrophotometer (BioTek, Mumbai, India) and visualized by gel electrophoresis. PCR amplification of 16S rRNA genes was carried out using S1000 Thermal Cycler (Bio-Rad) with primer sets 8F (5′-AGAGTTTGATCCTGGCTCAG-3′) and 1492R (5′-GGTTACCTTGTTACGACTT-3′) as described earlier by Behera et al. (2018b). PCR conditions include initial denaturation at 95 °C for 1 min, 35 cycles of denaturation at 95 °C for 30 s, annealing at 55 °C for 30 s, extension at 72 °C for 1 min 30 s, and final extension at 72 °C for 10 min. PCR products were confirmed and visualized using 1% agarose gel electrophoresis with ethidium bromide staining. Purification of PCR products was carried out with a HiPurA™ PCR product purification kit (HiMedia) and sequenced using ABI PRISM® 3700 DNA Analyzer (Applied Biosystems).

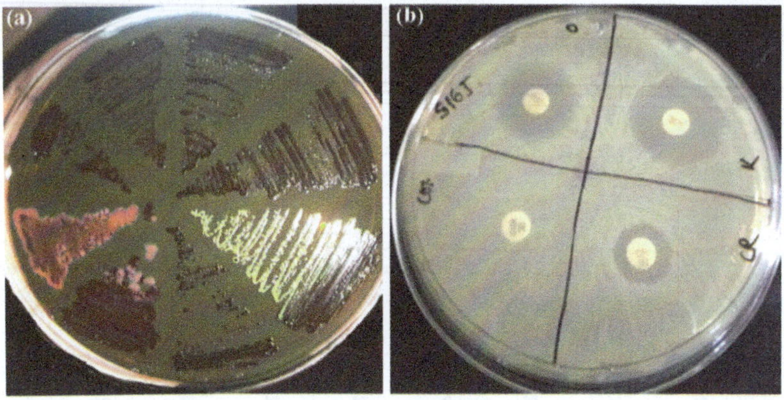

Fig. 2.2 FC isolates including high metallic green sheen producing colonies of *E. coli* on EMB agar plate (**a**) and antibiotic susceptibility testing of isolates showing resistance to cloxacillin and susceptibility to oxytetracycline, kanamycin, and cefaloridine (**b**)

The gene sequences were compared with GenBank sequences using BLASTN tool (http://www.ncbi.nlm.nih.gov/BLAST). Sequence alignments were executed using the ClustalW program (http://www.ebi.ac.uk/clustalw). The editing of aligned sequences and phylogenetic analysis were performed using software MEGA v 6.0 (Tamura et al. 2013). The neighbor-joining method was used for phylogenetic tree construction. Bootstrap test was performed based on 1000 replicates using a Kimura-2 nucleotide evolution model.

2.5 Antibiotic Susceptibility Profiling and MAR Index

The selected FC isolates were analyzed for antibiotic susceptibility profiling against 24 different antibiotics (HiMedia Laboratories, India) using the following discs (μg/disc): ofloxacin (5), trimethoprim (5), sulfadiazine (300), tobramycin (30), cefalexin (30), erythromycin (10), norfloxacin (10), oxytetracycline (30), nalidixic acid (10), nitrofurantoin (300), sulfamethizole (300), bacitracin (10), amoxicillin (30), kanamycin (30), furazolidone (50), amikacin (10), cefadroxil (30), cloxacillin (30), chlortetracycline (30), cefaloridine (30), novobiocin (30), carbenicillin (100), ciprofloxacin (30), and co-trimoxazole (25). The standard disc diffusion method was used to assess the antibiotic resistance pattern on LB agar plates (Bauer 1966). For this, bacterial suspension was inoculated using spread plate technique on solidified LB agar plate, and antibiotic discs were placed individually on the surface. After 24 h of incubation at 30 °C, the isolates were scored either susceptible or resistant for a particular antibiotic based on the appearance of zone around a disc. The experiments were conducted in triplicate, and mean values were considered for antibiotic resistance or susceptibility profiles. Data interpretation was carried out as per the performance standards for antimicrobial disc sensitivity testing recommended by

Table 2.2 Molecular identification, antibiotic susceptibility patterns, and multiple antibiotic resistance (MAR) index of FC isolates. The numbers in the parentheses represent BLAST percent identity scores of 16S rRNA gene sequences

	Closest	Antibiotic susceptibility profile and MAR index																								MAR
Isolates	neighbour	OF	TR	SZ	TOB	CN	E	NX	O	NA	NIT	SM	B	AMX	K	FR	AK	CFR	COX	CT	CR	NV	CB	CIP	COT	index
S5J1	*S. flexneri* (NR026331) (99.24%)	S	S	S	S	S	S	S	**R**	S	S	**R**	**R**	S	S	S	S	S	**R**	S	S	S	S	S	S	0.17
S6J2	*E. fergusonii* (NR074902) (99.70%)	S	S	S	S	S	S	S	S	S	S	S	**R**	S	S	S	S	S	**R**	S	S	S	S	S	S	0.08
S8J3	*S. flexneri* (NR026331) (99.85%)	S	S	S	S	S	S	S	S	S	S	**R**	**R**	**R**	S	S	S	S	**R**	S	S	S	S	S	S	0.17
S13J4	*S. flexneri* (NR026331) (99.77%)	S	S	S	S	S	S	S	S	S	S	S	**R**	**R**	S	S	S	S	**R**	S	S	S	S	S	S	0.13
S16J5	*S. flexneri* (NR026331) (99.29%)	S	S	S	S	S	S	S	S	S	S	S	**R**	S	S	S	S	S	**R**	S	S	S	S	S	S	0.08
S25J6	*K. pneumoniae* (NR114715) (99.58%)	S	S	S	S	S	S	S	S	S	S	S	**R**	**R**	S	S	S	S	**R**	S	S	S	S	S	S	0.13
S26J7	*S. flexneri* (NR026331) (99.02%)	S	S	S	S	S	S	S	**R**	S	S	**R**	**R**	**R**	S	S	S	S	**R**	S	S	S	S	S	S	0.21
S28J8	*K. pneumoniae* (NR114715) (99.48%)	S	S	S	S	S	S	S	S	S	S	S	**R**	**R**	S	S	S	S	**R**	S	S	S	S	S	S	0.13

S33J9	*K. pneumoniae* (NR114715) (99.47%)	S	S	S	S	S	S	S	S	S	S	S	**R**	**R**	S	S	S	S	**R**	S	S	S	S	S	S	0.13
S7F1	*S. flexneri* (NR026331) (99.85%)	S	S	S	S	S	S	S	S	S	S	S	**R**	S	S	S	S	S	**R**	S	S	S	S	S	S	0.08
S16M1	*S. flexneri* (NR026331) (99.63%)	S	S	S	S	S	S	S	S	S	S	**R**	**R**	S	S	S	S	S	**R**	S	S	S	S	S	S	0.13

S: susceptible, *R:* resistant, *S. flexneri: Shigella flexneri*, *E. fergusonii: Escherichia fergusonii*, *K. pneumoniae: Klebsiella pneumoniae*, *OF:* ofloxacin, *TR:* trimethoprim, *SZ:* sulfadiazine, *TOB:* tobramycin, *CN:* cefalexin, *E:* erythromycin, *NX:* norfloxacin, *O:* oxytetracycline, *NA:* nalidixic acid, *NIT:* nitrofurantoin, *SM:* sulfamethizole, *B:* bacitracin, *AMX:* amoxicillin, *K:* kanamycin, *FR:* furazolidone, *AK:* amikacin, *CFR:* cefadroxil, *COX:* cloxacillin, *CT:* chlortetracycline, *CR:* cefaloridine, *NV:* novobiocin, *CB:* carbenicillin, *CIP:* ciprofloxacin, *COT:* co-trimoxazole

CLSI (Clinical and Laboratory Standards Institute) (CLSI 2006). The MAR index for a FC isolate was calculated and interpreted using formula: *a*/*b*, where "*a*" refers to the number of antibiotics for which the isolate was resistant and "*b*" refers to the total number of antibiotics used in antimicrobial sensitivity testing (Krumperman 1983).

2.6 Statistical Analysis

The water samples were grouped based on their locations, seasons, and years and were compared for FC load using one-way Welch's ANOVA, followed by the non-parametric post-hoc Games-Howell test at p-value < 0.05. Pearson's correlation coefficient (r) was computed to determine the relationship between FC and salinity.

2.7 Nucleotide Sequence Accession Numbers

The 16S rRNA gene sequences of FC isolates have been submitted to GenBank database at NCBI under the accession numbers MW527410–MW527420.

3 Results and Discussion

3.1 Distribution of FC Bacteria

The FC distribution showed that the Chilika Lagoon was polluted with a varying degree of fecal contamination; however, FC loads were mostly within the safe limits established by CPCB for class SW-II water. In total 5, 8, and 14 water samples exceeded the threshold value of 100 MPN/100 ml of sample in 2017, 2018, and 2019, respectively (Fig. 2.3). The high FC loads were recorded from station S2 (Palur Canal), S3 (Malud-Talatala), S4 (Honeymoon Island), S5 (Gopakuda), S6 (Budhibaranasi), S9 (Veteswara), S14 (Nuapada), S15, S18 (Kalijugeswar), S22 (Tuagambhari), S23 (Tatebandha), S26 (Teeni Muhani Nali), S28 (Baulabandha), S30 (Sorana), and S31 (Kalupadaghat) which could be due to their proximity to shoreline villages or island with human settlements (Fig. 2.1). The mean FC load was 33 and 13 MPN/100 ml in water samples collected from inside and outside of the Nalabana Bird Sanctuary, respectively.

The average FC load from 33 stations of Chilika Lagoon over the study period was 14 MPN/100 ml. The highest FC load, i.e., 2400 MPN/100 ml, was found in S28 (~4 km from Baulabandha village; November 2019), followed by 1100 MPN/100 ml from both S15 (~1 km from Nalabana Bird Sanctuary; October 2019)

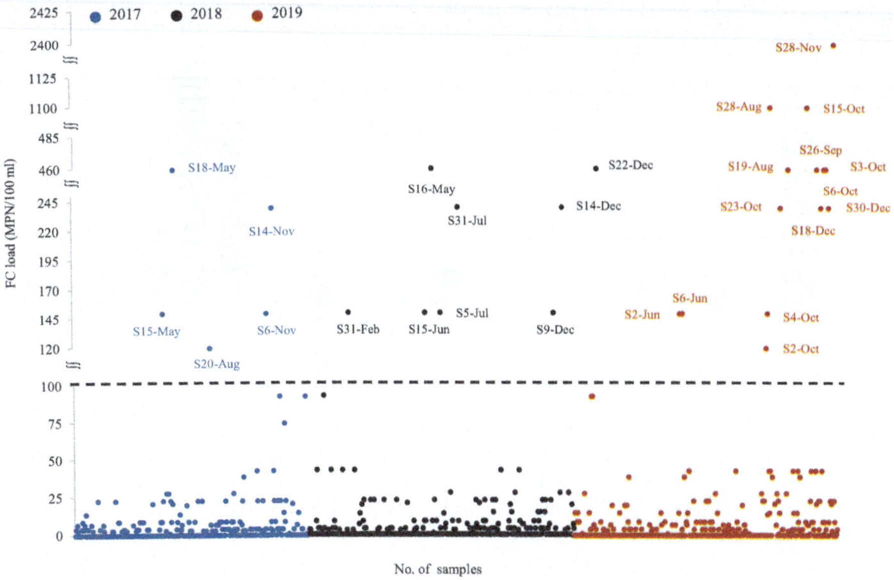

Fig. 2.3 Variation in the FC load in water samples collected during 3-year sampling period from Chilika Lagoon. Each dot represents a sample. Black dashed line denotes the threshold value (100 MPN/100 ml) prescribed as safe limit for FC load according to CPCB guidelines for Class SW-II waters

and S28 (August 2019) (Fig. 2.3). The occurrence of high FC load in S28 could be due to sewage discharge from the Baulabandha village which has a population of ~6660 individuals as per the 2011 census (Fig. 2.1). The monitoring from the peripheral villages showed the highest FC load in samples from Odialpur (279 MPN/100 ml), followed by Chandraput (37 MPN/100 ml) and Barkul (35 MPN/100 ml) (Fig. 2.4a).

The FC load from the rivers was analyzed during the monsoon season when freshwater discharge into the lagoon was the highest. River Kusumi showed the highest FC load (189 MPN/100 ml) followed by Kantabania (142 MPN/100 ml) and Badanai (71 MPN/100 ml) rivers, suggesting that drainage also contributed as a non-point source of FC load into the lagoon (Fig. 2.4b).

3.2 *Spatiotemporal Distribution of FC Bacteria*

Spatially, the mean FC loads varied between 4 (outer channel) and 19 (central sector) MPN/100 ml in the lagoon (Fig. 2.5a). A spatial pattern in FC distribution among monitoring stations reflected that salinity gradient and spatial factors could be an important factor in controlling the distribution. Consistently, the FC loads showed a statistically significant negative correlation ($r = -0.06$, p-value < 0.05)

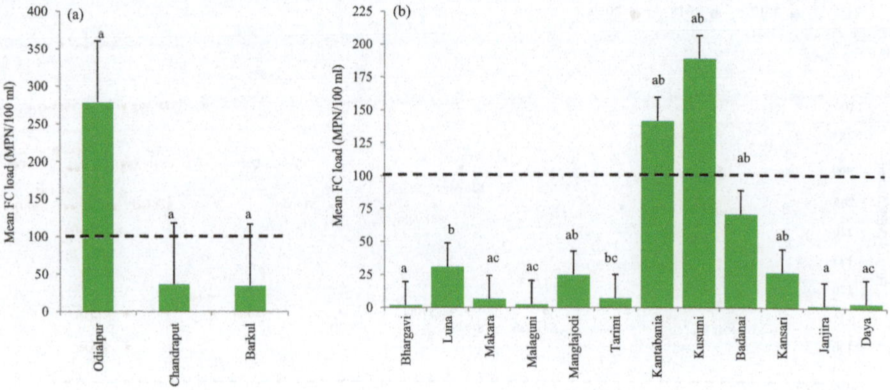

Fig. 2.4 Variation in mean FC load in water samples collected from 3 villages (**a**) and 12 major rivers (**b**) surrounding the Chilika Lagoon. Error bars represent Standard Error. The mean differences were tested by one-way ANOVA. Means with same alphabets are not significantly different (p-value ≥ 0.05). Black dashed line indicates the threshold value (100 MPN/100 ml) prescribed by CPCB guidelines as safe limit for FC load

with the salinity. The higher salinity in outer channel stations and periodic tidal flushing from BoB would lower the FC loads in this sector. The decrease in FC levels with increasing distances from the discharge points in northern sectors could be due to dilution, sedimentation, predation, and inactivation by high salinity and pH (Burkhardt et al. 2000). An earlier study from estuary found that enteric bacteria were typically abundant in head stations which were connected to the river and pollution sources than seawater inlets stations where salinity was high (Mallin et al. 2000). This could further explain the higher FC loads in the central sector where highly populated villages such as Barkul, Baulabandha, and Balugaon are located on the shoreline (Fig. 2.1). The decrease in FC load with increasing salinity was in agreement with studies from the coastal waters of Mississippi Sound (Chigbu et al. 2004), lagoon waters of Grand-Lahou (Konan et al. 2009), Persian Gulf in Bushehr coastal areas (Karbasdehi et al. 2017), and Hilo Bay (Wiegner et al. 2017). Earlier studies also concluded that with an increase in salinity, there is a decrease in the survival rate of FC (Evison 1988; Šolić and Krstulović 1992). This could be due to the increased bacterial cell inactivation because of high salt concentration in water (Hughes 2003; Florini et al. 2020).

Temporal patterns in FC could be influenced by climatic and seasonal factors that affect land runoff, river discharge, and water temperature. The mean FC level ranged between 17 (monsoon and winter) and 7 (summer) MPN/100 ml (Fig. 2.5a). The higher FC load in the monsoon could be attributed to increased land runoff during this season, which may have brought fecal inputs into the lagoon from various sources (Konan et al. 2009; Soto-Castor and Esquivel-Herrera 2020). During winter, the high FC loads in Chilika Lagoon could be due to the resting migratory birds. The lowest FC loads during the summer could be due to an increase in salinity as there is no freshwater flow from rivers into the lagoon. Furthermore, the higher solar

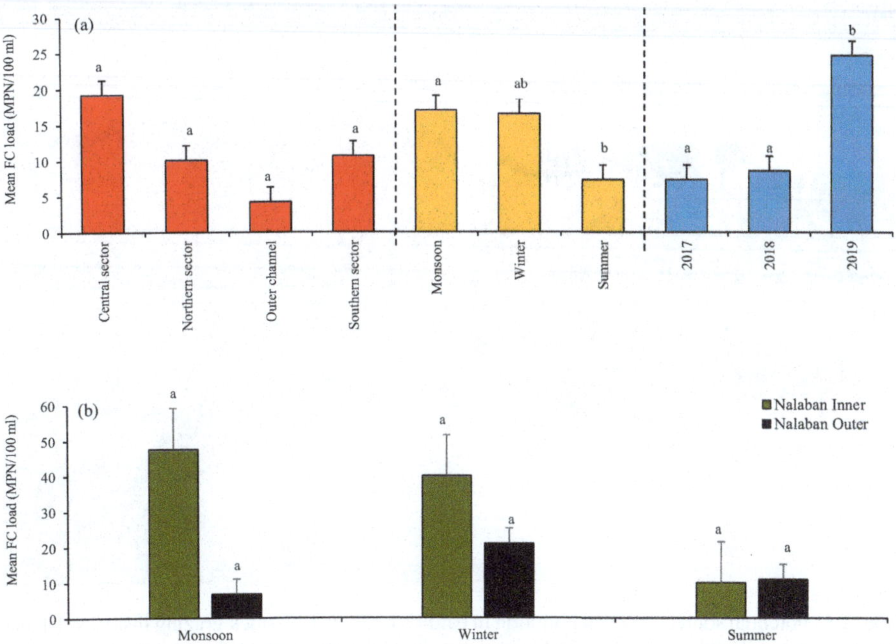

Fig. 2.5 Spatiotemporal variation in the FC load from Chilika Lagoon (**a**) and temporal variation in the FC load from Nalabana Bird Sanctuary (**b**). The value represents mean and error bars represent Standard Error. The mean differences were tested by one-way ANOVA. Means with same alphabets are not significantly different (p-value ≥ 0.05). Nalabana outer includes $n = 132$ samples collected from NB1, NB2, NB6, NB7, NB8, and NB11 stations. Nalabana inner included $n = 120$ samples collected from NB3, NB4, NB5, NB9, and NB10 stations

radiation during summer could also decrease FC viability (Hughes 2003). Seasonal variations in FC levels recorded from Chilika Lagoon were in agreement with earlier studies from marine and estuarine environments (Šolić and Krstulović 1992; Hughes 2003; Florini et al. 2020).

The average FC loads in Nalabana Bird Sanctuary also varied temporally (Fig. 2.5b). The FC levels in samples collected from inner stations of the sanctuary were higher during monsoon (48 MPN/100 ml) followed by winter (40 MPN/100 ml) and summer seasons (10 MPN/100 ml) (Fig. 2.5b). The average FC load varied spatially when compared between the samples collected from outside and inside of the sanctuary. FC loads in samples collected from outside stations (13 MPN/100 ml) of the sanctuary were much lower than inner stations (33 MPN/100 ml). A recent annual bird census carried out during January 2019 estimated a total of 1,047,968 birds from the entire Chilika Lagoon, of which 391,764 were sighted from the sanctuary. The higher FC load during winter could be due to thousands of birds that congregate in the sanctuary during winters for feeding and reproduction. Over 600 families in the vicinity of Chilika rely on the domestic and livestock animals (e.g., Chilika buffalo, cattle) for their livelihood. The total Chilika buffalo population has

Fig. 2.6 Potential sources of fecal pollution in Chilika Lagoon. Livestock grazing inside Nalabana Bird Sanctuary during summer (**a**) and on the shoreline of Barkul village (**b**), bird flocks resting in the sanctuary (**c**), and their fecal guanos on the mudflats (**d**)

been estimated to be approximately 30,000 (Singh et al. 2017). These buffaloes stand out for their distinct habitat such as consumption of saline water and vegetations and their ability to cope well with high temperature (38–40 °C) during summer. These buffaloes are abundantly present in Bhusandpur, Satapada, Krushnaprasad, Rambha, Parikud, Malud, and Palur area and enter into the lagoon during dry season (Singh et al. 2017). Studies have shown the existence of FC in a wide variety of warm-blooded animals that congregate in coastal wetlands (Chigbu et al. 2004; Siewicki et al. 2007; Yetis and Selek 2014; Soto-Castor and Esquivel-Herrera 2020). Thus, guano and dung inputs from wildlife such as birds and buffaloes could also be one of the potential sources of FC (Fig. 2.6).

3.3 *Inter-annual Variation in FC Bacteria*

Since FC were monitored over 3 consecutive years from Chilika Lagoon, inter-annual variation in their abundances was also examined. The mean FC load (MPN/100 ml) of the lagoon was the highest (25) in 2019 and the lowest (7) in 2017 (Fig. 2.5a). Inter-annual variability was also observed in FC load from the water samples collected from inside stations of Nalabana Bird Sanctuary. The mean FC load was higher in 2019 (45 MPN/100 ml) compared to 2018 (20 MPN/100 ml). River Kusumi recorded the highest FC loads (534 MPN/100 ml), followed by River

Kantabania (377 MPN/100 ml) and River Badanai (153 MPN/100 ml) during 2019. The higher FC load during 2019 could be attributed to an extremely severe cyclonic storm *Fani* (a Category 4 cyclone) that made landfall on May 3, 2019. The cyclone *Fani* was accompanied by a high precipitation and runoff which could have brought sewage from a variety of sources that resulted in an increase in the FC load of the lagoon. The impact of cyclone *Fani* was consistent with other studies that demonstrated higher FC levels after cyclonic events (Mohandass and Bharathi 2003; Mosley et al. 2004; Wiegner et al. 2017). The mean annual salinity of the Chilika Lagoon was the lowest during 2019 (8.70), whereas the highest annual salinity was recorded during 2017 (13.05). This could further account for the recorded maximum FC load during 2019 when the salinity was the lowest.

3.4 Molecular Identification and Phylogenetic Analysis

Nearly complete 16S rRNA gene sequences (~1300 bp) were obtained from 11 FC isolates and used to generate a phylogenetic tree (Fig. 2.7). All FC isolates were affiliated to family *Enterobacteriaceae* that comprises enteric bacteria and are frequently isolated from water bodies with high fecal contamination (Singh et al. 2020). Isolates S25J6, S28J8, and S33J9 exhibited > 99% sequence similarity with *K. pneumoniae* (NR114715) (Table 2.2 and Fig. 2.7). Isolates S16M1, S7F1, S5J1, S26J7, S16J5, S13J4, and S8J3 displayed > 99% sequence similarity with *S. flexneri* (NR026331). S6J2 isolate showed maximum homology (99.70%) with *E. fergusonii* (NR074902) (Table 2.2, Fig. 2.7). *S. flexneri* can cause shigellosis that leads to death and morbidity in infants and immunosuppressed adults (Ranganathan et al. 2019). *K. pneumoniae* are the most common nosocomial pathogens that can cause various diseases such as pneumonia and urinary tract infections (Cabral 2010). *E. fergusonii* are seldom emerging pathogens related to intestinal and extra-intestinal infections in humans and animals (Wragg et al. 2009).

3.5 Antibiotic Susceptibility Profile and MAR Index

All FC isolates were susceptible to ofloxacin, trimethoprim, sulfadiazine, tobramycin, cefalexin, erythromycin, norfloxacin, nalidixic acid, nitrofurantoin, kanamycin, furazolidone, amikacin, cefadroxil, chlortetracycline, cefaloridine, novobiocin, carbenicillin, ciprofloxacin, and co-trimoxazole (Table 2.2). All FC isolates were resistant to bacitracin and cloxacillin. Six isolates were resistant to amoxicillin, four to sulfamethizole, and two to oxytetracycline. Multidrug resistance is defined as resistance to at least three classes of antibiotics and was recorded in S5J1, S8J3, S26J7, and S16M1 isolates (Table 2.2). Majority of the isolates showed MAR scores < 0.20 indicating that they were derived from low-risk contamination sources where antibiotics are rarely used (Krumperman 1983). However, isolate S26J7 had MAR

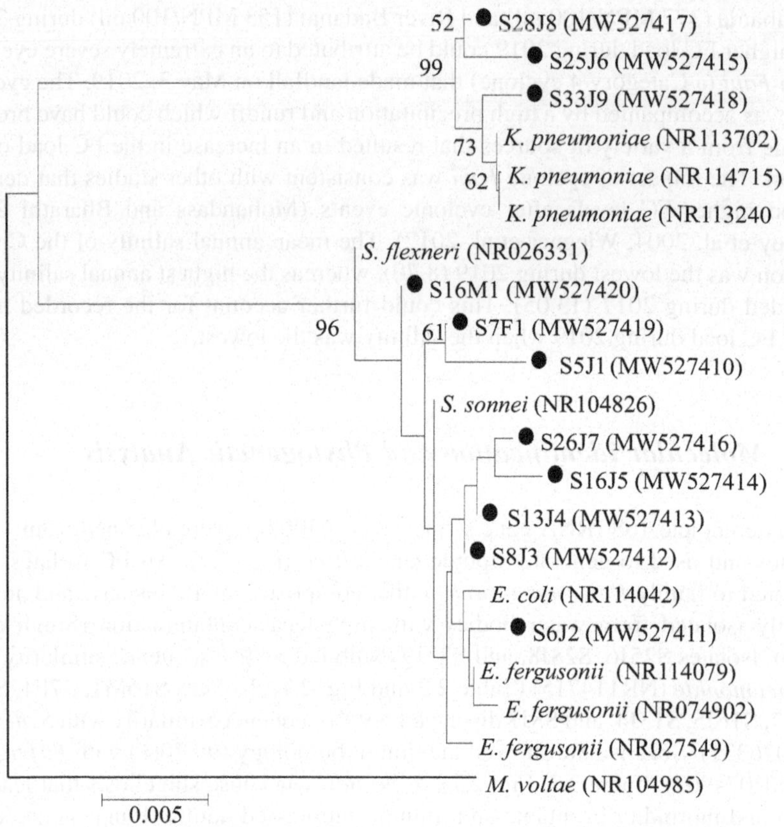

Fig. 2.7 Neighbor-joining phylogenetic tree derived from 16S rRNA gene sequences showing the positions of FC isolates and related organisms. The tree is based on a 1300 bp alignment of 16S rRNA gene sequences. *M. voltae* was used as an out-group to root the tree. GenBank accession numbers are given in parentheses. Numbers at the node points are bootstrap values (%) based on 1000 resamplings. Bootstrap values < 50% are not shown. Bar, 0.005 substitutions per nucleotide position. Black closed circles indicate fecal coliforms isolated in this study. *K. pneumoniae*: *Klebsiella pneumoniae*; *S. flexneri*: *Shigella flexneri*; *S. sonnei*: *Shigella sonnei*; *E. coli*: *Escherichia coli*; *E. fergusonii*: *Escherichia fergusonii*; *M. voltae*: *Methanococcus voltae*

index value of 0.21 indicating that it could have originated from a high-risk contamination sources (e.g., human wastes, commercial poultry farms, aquaculture farms, and dairy cattle and swine farms) where antibiotics are often used. Antibiotic-resistant FC bacteria can enter into the lagoon through sewage, land runoff, river discharge, and open defecation. Studies have also shown that migratory birds and livestock can also be a source of antibiotic resistance genes (ARGs) and bacteria (ARBs) in aquatic habitats (Huang et al. 2019; Cao et al. 2020). The ARGs may be transmitted to other microorganisms through horizontal gene transfer (Mishra et al. 2018). Furthermore, the ARB can enter into humans through contact with water during fishing and recreational activities causing a potential threat to human health. The high prevalence of antibiotic resistance in the FC bacteria raises concerns about their continued use as safe indicator species.

4 Conclusion

The present study, for the first time, investigated the distribution of fecal bacteria in the Chilika Lagoon, major drainage rivers, and shoreline villages. The results indicated that the FC responded to salinity regimes (marine versus freshwater), locations (sectors), and seasons. Direct discharge from rivers (specifically Kusumi, Kantabania, and Badanai), wildlife, and untreated sewage and open defecation from shoreline villages were among the primary sources of fecal contamination in the Chilika Lagoon. A total of 27 water samples from the lagoon exceeded the MPN index when compared with the CPCB, New Delhi guidelines for Class SW-II waters. A total of 13 water samples from Nalabana Bird Sanctuary and 4-4 water samples each from the river and village sites exceeded the CPCB threshold value. Overall, the number of samples from the lagoon that exceeded the prescribed CPCB guidelines was not high, but continuous monitoring of FC should be practiced. Furthermore, the study revealed an inverse correlation between the FC loads and salinity which could be one of the reasons for the low abundances of FC in the lagoon. The high salinity of the lagoon combined with dynamic changes in physicochemical factors and climatic factors (e.g., solar radiation) could be the reason for the rapid inactivation of FC in the lagoon. The antibiotic-resistant profiles should be monitored on a continuous basis to foresee the emergence and widespread of MAR. Effective measures must be taken to prevent the development and spread of new resistance from various point and non-point sources of pollution. Further research should include FC assessment from sediments which could also act as reservoirs of FC.

Chilika Lagoon is a major resource for the state's commercial fisheries, recreation, tourism, biodiversity, and aesthetics; therefore, disease interception scheme must be implemented in order to protect the public. Thus, achieving and ensuring good water quality is a prime concern for wetland managers and policymakers. The microbial pollution from fecal bacteria is an emerging issue considering the lagoon's large size and lack of sanitary facilities in the shoreline villages. The present study provided a baseline data on spatiotemporal distributions, molecular identification, and MAR indexing of FC, which would be useful in formulating conservation plans and monitoring schemes in Chilika Lagoon.

Acknowledgments The study was supported by the funding of the World Bank (Credit No. 4765-IN) awarded to Chilika Development Authority (Government of Odisha) under the Integrated Coastal Zone Management Project (ICZMP). The financial support received from National Plan for Conservation of Aquatic Ecosystems (NPCA) to Chilika Development Authority (CDA) is also acknowledged. Bibhuti Bhusan Dora, Research Fellow (GIS cell), is acknowledged for the preparation of the sampling map.

References

Bauer AT (1966) Antibiotic susceptibility testing by a standardized single disc method. Am J Clin Pathol 45:149–158

Behera P, Mahapatra S, Mohapatra M, Kim JY, Adyta TK, Raina V, Suar M, Pattnaik AK, Rastogi G (2017) Salinity and macrophyte drive the biogeography of the sedimentary bacterial communities in a brackish water tropical coastal lagoon. Sci Total Environ 595:472–485

Behera P, Mohapatra M, Adhya TK, Suar M, Pattnaik AK, Rastogi G (2018a) Structural and metabolic diversity of rhizosphere microbial communities of *Phragmites karka* in a tropical coastal lagoon. Appl Soil Ecol 125:202–212

Behera P, Mohapatra M, Seuylemezian A, Vaishampayan P, Ramana VV, Joseph N, Joshi A, Shouche Y, Suar M, Pattnaik AK, Rastogi G (2018b) Taxonomic description and draft genome of *Pseudomonas sediminis* sp. nov., isolated from the rhizospheric sediment of Phragmites karka. J Microbiol 56(7):458–466

Boyd CE (2015) Microorganisms and water quality. In: Water quality. Springer, Cham, pp 189–222

Burkhardt W III, Calci KR, Watkins WD, Rippey SR, Chirtel SJ (2000) Inactivation of indicator microorganisms in estuarine waters. Water Res 34(8):2207–2214.5

Cabral JPS (2010) Water microbiology. Bacterial pathogens and water. Int J Environ Res Public Health 7(10):3657–3703

Campbell JW, Watson A, Watson C, Ball H, Pirkle R (2011) *Escherichia coli*, other coliform, and environmental chemoheterotrophic bacteria in isolated water pools from six caves in northern Alabama and northwestern Georgia. J Cave Karst Stud 73(2):75–82

Cao J, Hu Y, Liu F, Wang Y, Bi Y, Lv N, Li J, Zhu B, Gao GF (2020) Metagenomic analysis reveals the microbiome and resistome in migratory birds. Microbiome 8(1):1–18

Chigbu P, Gordon S, Strange T (2004) Influence of inter-annual variations in climatic factors on fecal coliform levels in Mississippi sound. Water Res 38(20):4341–4352

Cimenti M, Hubberstey A, Bewtra JK, Biswas N (2007) Alternative methods in tracking sources of microbial contamination in waters. Water SA 33(2). http://sci-hub.tw/10.4314/wsa.v33i2.49059

Clesceri LS, Greenberg AE, Eaton AD (eds) (1998) Standard methods for the examination of water and wastewater, 20th edn. American Public Health Association (APHA), Washington, DC, p 1325

CLSI (2006) Methods for antimicrobial dilution and disk susceptibility testing of infrequently isolated or fastidious bacteria: approved guideline (M45-A). Clinical and Laboratory Standards Institute, Wayne, PA

Cooksey EM, Singh G, Scott LC, Aw TG (2019) Detection of coliphages and human adenoviruses in a subtropical estuarine lake. Sci Total Environ 649:1514–1521

CPCB (1993) Criteria for classification and zoning of coastal waters (sea waters SW-II)—a coastal pollution series: COPOCS/6/1993. CPCB, New Delhi

Davis K, Anderson MA, Yates MV (2005) Distribution of indicator bacteria in Canyon Lake, California. Water Res 39(7):1277–1288

Dubey RC, Maheshwari DK (2012) Practical microbiology. S. Chand Pvt. Limited, New Delhi

Evison LM (1988) Comparative studies on the survival of indicator organisms and pathogens in fresh and sea water. Water Sci Technol 20(11–12):309–315

Florini S, Shahsavari E, Ngo T, Aburto-Medina A, Smith DJ, Ball AS (2020) Factors influencing the concentration of fecal coliforms in oysters in the River Blackwater Estuary, UK. Water 12(4):1086

Ghosh AK, Pattnaik AK, Ballatore TJ (2006) Chilika Lagoon: restoring ecological balance and livelihoods through re-salinization. Lakes Reserv Res Manag 11(4):239–255

Hassard F, Andrews A, Jones DL, Parsons L, Jones V, Cox BA, Daldorph P, Brett H, McDonald JE, Malham SK (2017) Physicochemical factors influence the abundance and culturability of human enteric pathogens and fecal indicator organisms in estuarine water and sediment. Front Microbiol 8:1996

Huang Z, Zhao W, Xu T, Zheng B, Yin D (2019) Occurrence and distribution of antibiotic resistance genes in the water and sediments of Qingcaosha Reservoir, Shanghai, China. Environ Sci Eur 31(1):1–9

Hughes KA (2003) Influence of seasonal environmental variables on the distribution of presumptive fecal coliforms around an Antarctic Research Station. Appl Environ Microbiol 69(8):4884–4891

Jangam C, Pujari P (2019) Impact of on-site sanitation systems on groundwater sources in a coastal aquifer in Chennai, India. Environ Sci Pollut Res 26(3):2079–2088

Jayakumar R, Steger K, Chandra TS, Seshadri S (2013) An assessment of temporal variations in physicochemical and microbiological properties of barmouths and lagoons in Chennai (Southeast coast of India). Mar Pollut Bull 70(1–2):44–53

Joshi A, Mishra SP (2017) Anthropocene effects on the river Daya and the lagoon Chilika by the effluents of Bhubaneswar city India: a physico-chemical study. Int J Adv Res 5(10):1370–1384

Karbasdehi VN, Dobaradaran S, Nabipour I, Ostovar A, Arfaeinia H, Vazirizadeh A, Khalifei F (2017) Indicator bacteria community in seawater and coastal sediment: the Persian Gulf as a case. J Environ Health Sci Eng 15(1):6

Kataržytė M, Mėžinė J, Vaičiūtė D, Liaugaudaitė S, Mukauskaitė K, Umgiesser G, Schernewski G (2018) Fecal contamination in shallow temperate estuarine lagoon: source of the pollution and environmental factors. Mar Pollut Bull 133:762–772

Konan KS, Kouassi AM, Adingra AA, Gnakri D (2009) Spatial and temporal variation of fecal contamination indicators in Grand-Lahou lagoon, Côte d'Ivoire. J Appl Biosci 23:1422–1435

Krumperman PH (1983) Multiple antibiotic resistance indexing of Escherichia coli to identify high-risk sources of fecal contamination of foods. Appl Environ Microbiol 46(1):165–170

Kumar R, Pattnaik AK (2012) Chilika—an integrated management planning framework for conservation and wise use. Wetlands International—South Asia, Chilika Development Authority, New Delhi, Bhubaneswar

Latha N, Mohan MR (2013) Microbial pollution- total coliform and fecal coliform of Kengeri lake, Bangalore region Karnataka, India. Int J Sci Res Publ 3(11):1–3

Mallin MA, Williams KE, Esham EC, Lowe RP (2000) Effect of human development on bacteriological water quality in coastal watersheds. Ecol Appl 10(4):1047–1056

Mishra M, Arukha AP, Patel AK, Behera N, Mohanta TK, Yadav D (2018) Multi-drug resistant coliform: water sanitary standards and health hazards. Front Pharmacol 9:311

Mitch AA, Gasner KC, Mitch WA (2010) Fecal coliform accumulation within a river subject to seasonally-disinfected wastewater discharges. Water Res 44(16):4776–4782

Mohandass C, Bharathi PA (2003) Representation, dispersion, and variation of bacterial indicators in the coastal waters of Nagore (east coast of India). Water Environ Res 75(1):66–72

Mosley LM, Sharp DS, Singh S (2004) Effects of a tropical cyclone on the drinking-water quality of a Remote Pacific Island. Disasters 28(4):405–417

Motlagh AM, Yang Z (2019) Detection and occurrence of indicator organisms and pathogens. Water Environ Res 91(10):1402–1408

Muduli PR, Pattnaik AK (2020) Spatio-temporal variation in physicochemical parameters of water in the Chilika Lagoon. In: Finlayson C, Rastogi G, Mishra D, Pattnaik A (eds) Ecology, conservation, and restoration of Chilika Lagoon, India, Wetlands: ecology, conservation and management, vol 6. Springer, Cham, pp 203–229

Mukherjee S (2016) Influence of discharge of river Daya in Chilika Lagoon: a brief study using Mohid simulation model. Suresh Gyan Vihar Univ J Eng Technol (Int Bi-Annual J) 2(2):10–21

Parida S, Jena RC, Samal KC, Chand PK (2012) Isolation and identification of pathogenic bacteria from brackish waters of Chilika Lagoon, Odisha, India for pharmaceutical use. Malaysian J Microbiol 8:197–202

Pattnaik R, Tarafdar L, Mohapatra M, Behera P, Mohanty P, Rastogi G (2019) Biodiversity of Chilika through the lens of foldscope. J Emerg Technol Innov Res 6:865–874

Pattnaik AK, Panda PC, Rastogi G (2020) Survey, characterization, ecology, and management of macrophytes in Chilika Lagoon. In: Finlayson C, Rastogi G, Mishra D, Pattnaik A (eds)

Ecology, conservation, and restoration of Chilika Lagoon, India, Wetlands: ecology, conservation and management, vol 6. Springer, Cham, pp 415–438

Pérez-Ruzafa A, Pérez-Ruzafa IM, Newton A, Marcos C (2019) Coastal lagoons: environmental variability, ecosystem complexity, and goods and services uniformity. In: Coasts and estuaries. Elsevier, Amsterdam, pp 253–276

Ranganathan S, Doucet M, Grassel CL, Delaine-Elias B, Zachos NC, Barry EM (2019) Evaluating Shigella flexneri pathogenesis in the human enteroid model. Infect Immun 87(4):e00740-18

Riaz S, Faisal M, Hasnain S (2011) Antibiotic susceptibility pattern and multiple antibiotic resistances (MAR) calculation of extended spectrum β-lactamase (ESBL) producing Escherichia coli and Klebsiella species in Pakistan. Afr J Biotechnol 10(33):6325–6331

Sengupta C, Saha R (2013) Understanding coliforms—a short review. Int J Adv Res 1:16–25

Siewicki TC, Pullaro T, Pan W, McDaniel S, Glenn R, Stewart J (2007) Models of total and presumed wildlife sources of fecal coliform bacteria in coastal ponds. J Environ Manag 82(1):120–132

Singh R, Mishra SK, Rajesh C, Dash SK, Niranjan SK, Kataria RS (2017) Chilika—a distinct registered buffalo breed of India. Int J Livest Res 7(9):259–266

Singh AK, Das S, Kumar S, Gajamer VR, Najar IN, Lepcha YD, Tiwari HK, Singh S (2020) Distribution of antibiotic-resistant enterobacteriaceae pathogens in potable spring water of eastern Indian Himalayas: emphasis on virulence gene and Antibiotic resistance genes in Escherichia coli. Front Microbiol 11:581072

Šolić M, Krstulović N (1992) Separate and combined effects of solar radiation, temperature, salinity, and pH on the survival of faecal coliforms in seawater. Mar Pollut Bull 24(8):411–416

Soto-Castor R, Esquivel-Herrera A (2020) Evaluation anthropic fecal contamination through enteropathogenic genera identification in Sontecomapan coastal lagoon, Veracruz. In: Ecology of the Sontecomapan Lagoon, Veracruz. IRD Éditions, Marseille, p 237

Srichandan S, Kim JY, Bhadury P, Barik SK, Muduli PR, Samal RN, Pattnaik AK, Rastogi G (2015) Spatiotemporal distribution and composition of phytoplankton assemblages in a coastal tropical lagoon: Chilika, India. Environ Monit Assess 187(2):47

Sugumar G, Chrisolite B, Velayutham P, Selvan A, Ramesh U (2008) Occurrence and seasonal variation of bacterial indicators of faecal pollution along Thoothukudi coast, Tamil Nadu. J Environ Biol 29(3):387–391

Tamura K, Stecher G, Peterson D, Filipski A, Kumar S (2013) MEGA6: molecular evolutionary genetics analysis version 6.0. Mol Biol Evol 30(12):2725–2729

Town DA (2001) Historical trends and concentrations of fecal coliform bacteria in the Brandywine Creek Basin, Chester County, Pennsylvania, vol 1, No. 4026. US Department of the Interior, US Geological Survey

Vijay R, Khobragade PJ, Sohony RA (2010) Water quality simulation of sewage impacts on the west coast of Mumbai, India. Water Sci Technol 62(2):279–287

WHO (1985) Guidelines for drinking-water quality, Drinking water quality control in small community supplies, vol 3. World Health Organization, Geneva, p 136

WHO (2012) In: Dufour A, Bartram J, Bos R, Gannon V (eds) Animal waste, water quality and human health. World Health Organization by IWA Publishing, London

Wiegner TN, Edens CJ, Abaya LM, Carlson KM, Lyon-Colbert A, Molloy SL (2017) Spatial and temporal microbial pollution patterns in a tropical estuary during high and low river flow conditions. Mar Pollut Bull 114(2):952–961

Wragg P, La Ragione RM, Best A, Reichel R, Anjum MF, Mafura M, Woodward MJ (2009) Characterisation of Escherichia fergusonii isolates from farm animals using an *Escherichia coli* virulence gene array and tissue culture adherence assays. Res Vet Sci 86(1):27–35

Yetis AD, Selek Z (2014) Determination of total and fecal coliforms of Akyatan lagoon in terms of microbiological pollution. Arab J Geosci 8(2):1125–1132

Chapter 3
Seagrass Ecosystems of India as Bioindicators of Trace Elements

Amrit Kumar Mishra, Rajalaxmi Sahoo, Saumya S. Samantaray, and Deepak Apte

Abstract Seagrasses are considered efficient bioindicators of coastal trace element contamination. This chapter provides an overview of the trace element accumulation, tolerance, and biomonitoring capacity of the various seagrass species along the coast of India. A total of 10 (Cd, Co, Cr, Cu, Fe, Mg, Mn, Ni, Pb, and Zn) trace elements are reported in seagrasses, 11 in sediment (As, Cd, Co, Cr, Cu, Fe, Mg, Mn, Ni, Pb, and Zn) and 9 (Cd, Cr, Cu, Fe, Mg, Mn, Ni, Pb, and Zn) in the water column from India. From the eleven seagrass species studied, 60% of research has focused on *Syringodium isoetifolium*, *Cymodocea serrulata*, *Cymodocea rotundata*, and *Halophila ovalis*. Seventy-eight percent of seagrass trace element research in India is from the Palk Bay and the Gulf of Mannar (GOM) of Tamil Nadu and 16% from Lakshadweep Islands. Of the ten trace elements, Cd, Cu, Pb, and Zn are the most studied in seagrass; Fe, Mn, Ni, and Pb in sediment; and Cu, Fe, Mg, Ni, and Zn in the water column. The accumulation capacity of various trace elements in seagrass was species-specific. *Syringodium isoetifolium* has the highest concentration of Cd and Mg at Palk Bay and Lakshadweep Islands, respectively. The concentration of Cu was higher in *C. serrulata* at GOM. *Halodule uninervis* and *Halophila decipiens* have the highest concentration of Co and Cr, Ni, Pb, and Zn from Lakshadweep Islands. The concentration of Fe and Mn was highest in *Halophila beccarii* and *H. ovalis* from the coast of Goa and Palk Bay, respectively. Threshold levels (>10 mg L^{-1}) of Cd, Cu, Pb, and Zn were observed for *C. serrulata, H. ovalis, H. uninervis*, and *T. hemprichii*, which can affect the Photo System II of these seagrasses and exert cellular stress leading to seagrass loss and die-off. The high

A. K. Mishra (✉) · D. Apte
Department of Marine Conservation, Bombay Natural History Society, Mumbai, Maharashtra, India
e-mail: a.mishra@bnhs.org

R. Sahoo
Department of Botany, Bangalore University, Bengaluru, Karnataka, India

S. S. Samantaray
National Institute of Science Education and Research, Homi Bhaba National Institute, Jatni, Odisha, India

S. Madhav et al. (eds.), *Coastal Ecosystems*, Coastal Research Library 38,
https://doi.org/10.1007/978-3-030-84255-0_3

concentration of these elements can exert negative impacts on seagrass-associated trophic assemblages and ecosystem functioning. Seagrasses of India can be utilized as bioindicators of coastal trace element contamination, but the associated toxicity and human health risks need further investigation.

Keywords Seagrass · Trace metals · Coastal ecosystems · Anthropogenic pollution · Bioindicators · India

1 Introduction

Seagrass ecosystems are distributed worldwide, covering the five important bioregions of the world oceans except for Antarctica (Hemminga and Duarte 2002; McKenzie 2020). Seagrasses form complex interlinkage between salt marsh and mangrove ecosystems that are important in maintaining a wide range of ecological functions (Medina-Gomez 2016; Mishra and Apte 2020). This interlinkage forms complex food webs that support herbivore grazing and detrital food chains and provide habitat and nurseries for various species (Unsworth and Cullen-Unsworth 2018). Seagrasses provide 24 different types of ecosystem services (Nordlund et al. 2016), including habitat and nurseries for commercially important fish populations and endangered animals. They also help in carbon sequestration and storage (Duarte et al. 2013), shoreline protection from storm surges and prevention of coastal erosion (Ondiviela et al. 2014; Potouroglou et al. 2017), and regulation of nutrient cycles (Costanza et al. 2017) which are critical in maintaining the coastal biodiversity and ecosystem functioning. These ecosystem services support millions of coastal communities by supplying livelihood and food security (Nordlund et al. 2018; Unsworth et al. 2017). Like coral reefs, seagrass ecosystems are also declining worldwide (Waycott et al. 2009) due to various anthropogenic factors. However, the most relevant factors include habitat modification, dredging, wastewater discharge, nutrient enrichment, fishing, coastal developmental activities, and boat anchoring (Lewis and Richard 2009). These various anthropogenic activities act as sources of various harmful chemicals and trace elements that enter the marine ecosystem (Machowski et al. 2019; Serrano et al. 2011).

Trace elements occur in very low concentrations in the marine environment. At these low levels, trace elements are not toxic and play a critical role in marine ecosystem functioning (Avelar et al. 2013; Mishra et al. 2019). However, some are nonessential and toxic to organisms (As, Cd, Cr, Hg, and Pb), whereas others act as essential micronutrients (Cu, Mn, and Zn), provided that their concentrations do not exceed the threshold levels (Millero et al. 2009; Stockdale et al. 2016). These trace elements pose a serious risk to seagrass ecophysiology because of their persistent nature in the marine sediment (Stockdale et al. 2016). Once accumulated in the seagrass roots, their bioavailability increases (Bonanno and Borg 2018; Govers

2014), and trace elements get absorbed into the root plasmalemma at the root/soil interface. These elements are then translocated to the leaves via rhizomes. Once threshold levels of trace elements are reached, it affects both root cellular structure and plant photosynthesis (Ambo-Rappe et al. 2011; Prange and Dennison 2000). Consequently, once concentrated in the seagrass tissues, through bioaccumulation, these trace elements can move up the food chain through seagrass-associated organisms and get biomagnified at higher trophic levels and pose a serious risk for humans through marine food intake (Roberts et al. 2008; Vizzini et al. 2013).

This chapter aims to provide state-of-the-art information about trace element concentration in the seagrass ecosystems of India and their bioindicator potential. In the early 1990s, studies on trace element accumulation patterns in seagrasses of India started, when Jagtap (1983) first reported about the trace element levels in the seagrass *Halophila beccarii*. Thereafter, in the last few decades, a considerable amount of data has been generated on trace elements in various seagrass species of India (Govindasamy and Arulpriya 2011; Nobi et al. 2010; Sachithanandam et al. 2020; Sudharsan et al. 2012; Thangaradjou et al. 2013). However, these studies have focused on few locations of India, mostly in the Palk Bay and Gulf of Mannar (GOM) region of Tamil Nadu and the islands of Lakshadweep and Andaman and Nicobar, even though seagrasses have a pan-India distribution. These studies have mostly recorded the trace element levels in water-seagrass, sediment-seagrass or seagrass without focusing on accumulation capacity or seagrass bioindicator potential.

1.1 Distribution and Ecology of Indian Seagrasses

Seagrass ecosystems are distributed around India's west and east coast, including the islands of Andaman and Nicobar and Lakshadweep (Fig. 3.1). These seagrass ecosystems of India are also part of South Asian countries, such as Pakistan, Sri Lanka, Bangladesh, and the Maldives, and the Southeast Asian countries due to the exclusive economic zone of Andaman and Nicobar Islands (ANI) in the Indian Ocean region (Fortes 2018; Patro et al. 2017). Seventeen species of seagrasses belonging to three families, i.e., Hydrocharitaceae, Cymodoceaceae, and Ruppiaceae, have been recorded from India. These 17 seagrass species are part of the 19 seagrass species found in Southeast Asia (Short et al. 2011). These 17 seagrass species of India cover an area of 516.59 km^2 up to a depth limit of 21 m (Bayyana et al. 2020; Geevarghese et al. 2018).

These various seagrass species of India occupy sandy, muddy, or mixed habitats, in the intertidal region to increased depth (Parthasarathy et al. 1991). For example, small seagrass species like *Halophila beccarii* and *Halophila ovalis* is found in the muddy or sandy-muddy habitat of the intertidal region (Mishra and Apte 2021; Parthasarathy et al. 1991), whereas other seagrass species like *Thalassia hemprichii* and *H. beccarii* are found associated with mangroves (Jagtap et al. 2003; Mishra and Apte 2020; Mishra and Mohanraju 2018). Consequently, bigger seagrass plants

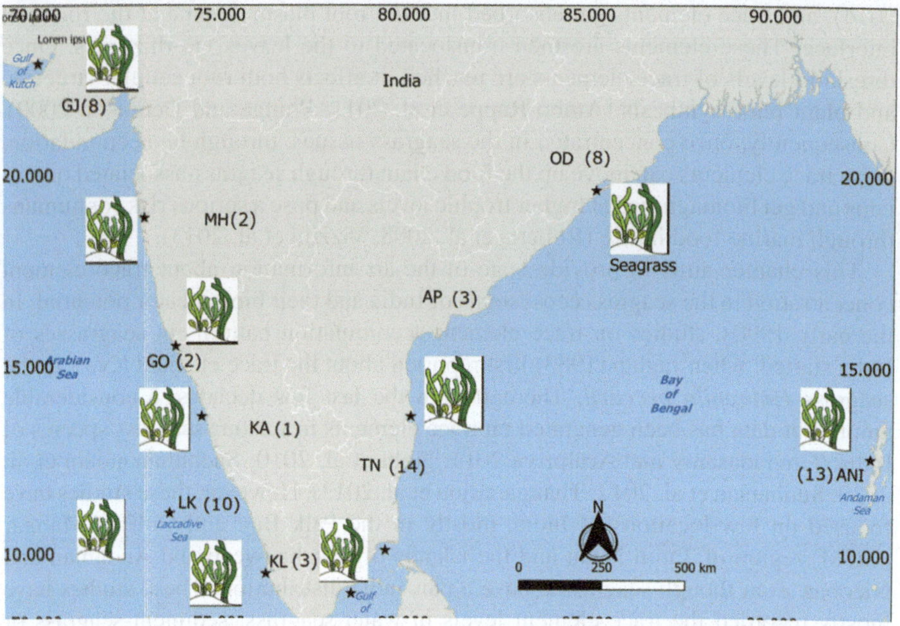

Fig. 3.1 Map showing the states and islands that have seagrass ecosystems around the east and west coast of India. Numbers in bracket for each state indicate the number of seagrass species found in that state. Odisha (OD), Andhra Pradesh (AP), Andaman and Nicobar Islands (ANI), Tamil Nadu (TN), Kerala (KL), Lakshadweep Island (LK)*, Karnataka (KA), Goa (GO), Maharashtra (MH), Gujarat (GJ)

like *Enhalus acoroides* are found at increased depths (Patankar et al. 2018). However, this distribution of seagrass plants is dependent upon various limiting factors such as turbidity, light penetration, and nutrient availability (Arumugam et al. 2013) that affect seagrass photo-physiology and reproductive processes (Patankar et al. 2018; Mishra and Apte 2020). Secondly, this distribution of seagrass species is also influenced by the presence of other seagrasses or mangroves or coral reefs that determine distribution patterns and ecological connectivity with surrounding ecosystems (Apte et al. 2016; Mishra and Apte 2020).

The presence of various seagrass species at the land and sea interface makes them suitable bioindicators of coastal metal contamination (Bonanno and Borg 2018; Mishra et al. 2019). This suitability of seagrass as bioindicators has been extensively used by the European Water Framework Directive using the endemic seagrass *Posidonia oceanica* and *Cymodocea nodosa* of the Mediterranean Sea (Bonanno and di Martino 2016; Bonanno and Orlando-Bonaca 2018; Bonanno and Raccuia 2018). However, in India, few studies have explored the potential of seagrass as a bioindicator of coastal pollution (Gopi et al. 2020; Govindasamy and Azariah 1999; Sudharsan et al. 2012). This chapter will provide valuable information about the various metal studies carried out using different seagrass species, the efficiency of seagrass in accumulating trace elements, and the toxic effects of these trace elements on seagrass physiology above threshold levels, and the bioindicator potential of seagrass to these trace elements.

2 Trace Element in Coastal Water, Sediment, and Seagrasses

2.1 Trace Element in the Water Column above Seagrass Meadows

In general, it is thought that the trace element concentration is higher in the water column and readily available for the leaves of the seagrass for uptake. However, it has been observed that this is not the case always (Bonanno and di Martino 2017). The accumulation of trace elements from the water column is species-specific among seagrasses and depends on the plant physiology and the nature of the trace element, i.e., toxic or essential (Millero et al. 2009; Mishra et al. 2019). In India, the trace element studies of the water column above seagrass ecosystems are very less compared to that of the sediment and seagrasses. Only 4 studies have reported the 9 out of 11 elements (As, Cd, Co, Cr, Cu, Fe, Mg, Mn, Ni, Pb, Zn) in sediment and 10 in seagrass (Fig. 3.2). However, on the west coast, trace elements like Cd, Cr, Pb, and Zn are not reported (Table 3.1). These studies on trace elements are restricted to five locations, i.e., Palk Bay and Gulf of Mannar (GOM) of Tamil Nadu, Goa, Maharashtra, and Lakshadweep Islands, consisting of only eight seagrass species.

Out of the nine trace elements, Mg was the most studied trace element in the water column of seagrass ecosystems in India and Mn the least (Fig. 3.2a). The concentration of Mg in the water column was highest in the seagrass meadows of *Cymodocea rotundata*, *Syringodium isoetifolium*, *Halodule uninervis*, *Thalassia hemprichii,* and *Halophila ovalis* of Lakshadweep Islands. The concentration of Fe and Mn was highest in *Halophila beccarii* from the west coast in Goa. The water column above *T. hemprichii* meadows has a higher concentration of Cd and Pb in

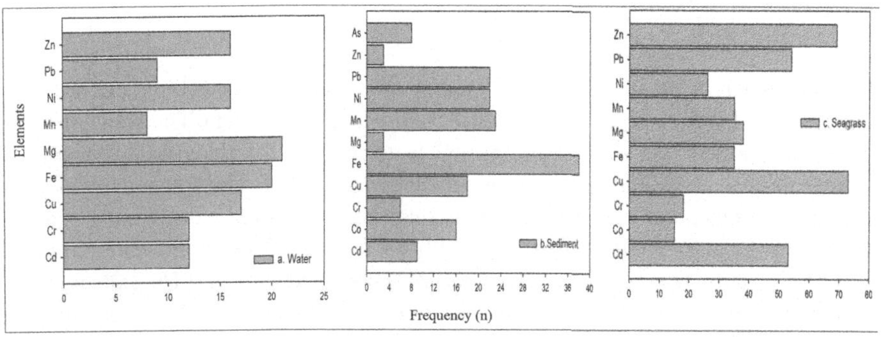

Fig. 3.2 Frequency distribution of number of studies on various trace elements in the (**a**) water, (**b**) sediment, and (**c**) seagrasses from the coast of India

Table 3.1 Mean or range of trace element concentration (ppb) in marine water associated with various seagrass species of India. Gulf of Mannar (GOM), Tamil Nadu (TN), Maharashtra (MH), Lakshadweep Island (LK)

Seagrass	Location	Trace elements									Ref.
		Cd	Cr	Cu	Fe	Mg	Mn	Ni	Pb	Zn	
C. rotundata	Palk Bay, TN	–	–	–	0.12–0.31	–	–	–	–	–	a
C. serrulata	Palk Bay, TN	0.09–0.11	0.31–0.67	0.50–1.02	0.12–1.02	0.28–0.33	0.57–0.89	0.19–0.37	0.01–0.12	0.11–6.38	a, b
S. isoetifolium	Palk Bay, TN	0.09–0.12	0.31–0.67	0.32–1.12	0.19–0.37	0.28–0.35	–	0.20–0.39	0.01–0.13	2.06–6.38	b
T. hemprichii	Palk Bay, TN	0.10–0.15	0.31–0.67	0.32–1.02	0.19–0.37	0.28–0.33	–	0.20–0.39	0.01–0.13	2.06–6.38	b
C. serrulata	GOM, TN	0.02–0.06	0.26–2.03	0.117	4.60–5.30	0.16–2.06	–	0.22–0.56	0.007	4.74–11.6	b
E. acoroides	GOM, TN	0.02–0.06	0.26–2.03	0.117	4.60–5.30	0.16–2.06	–	0.22–0.56	0.007	4.74–11.6	b
S. isoetifolium	GOM, TN	0.02–0.06	0.26–2.03	0.117	4.60–5.30	0.16–2.06	–	0.22–0.56	0.007	4.74–11.6	b
H. beccarii	Goa	–	–	0.42	7.4	614	0.76	0.22	–	–	c
H. beccarii	Malvan, MH	–	–	0.32	1.05	990	0.18	0.46	–	–	c
H. beccarii	Vijayagiri, MH	–	–	0.27	2.75	774	0.17	0.30	–	–	c
H. beccarii	Ratnagiri, MH	–	–	0.19	0.55	1100	0.09	0.54	–	–	c
C. rotundata	LK	–	–	–	–	18,318	–	–	–	–	d
S. isoetifolium	LK	–	–	–	–	18,317	–	–	–	–	d
H. uninervis	LK	–	–	–	–	18,318	–	–	–	–	d
T. hemprichii	LK	–	–	–	–	18,318	–	–	–	–	d
H. ovalis	LK	–	–	–	–	18,318	–	–	–	–	d

[a]Govindasamy et al. (2011)
[b]Baby et al. (2017)
[c]Jagtap (1983)
[d]Jagtap and Untawale (1984)

Palk Bay, whereas the water above *S. isoetifolium* meadows has a higher concentration of Cu and similar levels of Pb with *T. hemprichii* (Table 3.1). The concentration of Cr, Ni, and Zn was similar among the water column of *Cymodocea serrulata*, *S. isoetifolium*, and *Enhalus acoroides* meadows at GOM. The trace element concentration in the water column of various seagrasses followed a decreasing pattern, Mg > Zn > Fe > Cr > Cu> > Mn > Ni > Pb > Cd (Table 3.1). In the water column, Mg concentration was very low in the east coast within a range of 0.16–2.06 mg kg^{-1}, while that in the west coast was 550-fold higher (Table 3.1). The Mg concentration (18,318 mg kg^{-1}) in the Lakshadweep Islands water column was highest in India's coastal waters. The Zn levels were in the range of 0.11–11.6 mg kg^{-1}, with higher levels in the water column of GOM, Tamil Nadu. The Cd concentration was in the range of 0.02–0.15 mg kg^{-1}, while that of Cr is 0.31–2.03 mg kg^{-1} on the east coast. Copper levels in the water column were twofold higher on the east coast than that of the west coast. Iron concentrations were higher in the coastal waters of Goa. Manganese and Ni levels were similar among both coasts. The concentration of Pb was 0.007–0.13 mg kg^{-1} and that of Zn was 0.11–11.6 mg kg^{-1} (Table 3.1). Trace elements such as Co and Hg are not reported in the water column, even though they are reported from the sediment and seagrass tissues (Fig. 3.2b, c).

The source of trace elements in the coastal waters of India is mostly through riverine input, which varies according to the monsoon-dependent seasonal runoff and subsequent erosion from river catchment area (Tripathy et al. 2014). Consequently, local anthropogenic discharge from industrial and domestic wastewater also leads to the input of these trace elements into the coastal waters (Baby et al. 2017; Nobi et al. 2010; Thangaradjou et al. 2009; Thangaradjou and Bhatt 2018). Other than these inputs, the release of trace elements from the sediment to the water column within the seagrass ecosystems also plays an important role in varying concentrations of trace elements in the water column (Govindasamy and Azariah 1999; Baby et al. 2017). The low concentration of most of the elements like Cd, Cr, Cu, Ni, and Pb is a result of settling of the organic matter content that inflows with the land runoff. Seagrass ecosystems are considered as efficient ecosystem engineers, and they help in settling a small fraction of this organic matter content on their leaf surface or into the sediment, thus reducing water turbidity and enhancing their photosynthetic activity (Gillis et al. 2017; Guannel et al. 2016). This seasonal and local variation of input of trace elements is reflected in seagrass ecosystems (Govindasamy and Azariah 1999). This seasonal influence of high concentration of Co, Cd, Cu, Fe, Ni, Mn, and Zn in the water column has been observed at GOM of Tamil Nadu (Govindasamy and Azariah 1999) and the east coast of India (Vinithkumar et al. 1999).

2.2 *Trace Metals in the Sediment of Seagrass Meadows*

A total of 11 trace elements has been reported in the sediment of seagrass meadows of India. This includes As and Co that have not been reported in the water column of seagrass meadows (Fig. 3.2b). In sediment, the trace element concentration of Cd, Cr, Cu, Fe, Mn, Ni, Pb, and Zn was multifold higher than their water column values, except Mg, which was higher in the water column (Untawale and Jagtap 1984). The concentration of trace elements in the sediment of seagrass meadows followed the decreasing pattern of Fe > Mg > Mn > Cr > Ni > Cu > Zn > Co > As > Pb > Cd (Table 3.2).

Five locations of India on both the east and west coast within seagrass meadows have been used for trace element studies, i.e., Palk Bay and GOM, Tamil Nadu, Goa, Maharashtra, Lakshadweep, and Andaman and Nicobar Islands (ANI). The sediment of seven seagrass species, such as *S. isoetifolium*, *T. hemprichii*,

Table 3.2 Mean or range of trace element concentration (mg kg^{-1}) in sediment associated with various seagrass species of India. Gulf of Mannar (GOM), Tamil Nadu (TN), Maharashtra (MH), Lakshadweep Island (LK), Andaman and Nicobar Islands (ANI)

Seagrass	Loc.	Trace elements											Ref.
		As	Cd	Co	Cr	Cu	Fe	Mg	Mn	Ni	Pb	Zn	
C. serrulata	Palk Bay, TN	3.28	0.23–0.27	–	4.81–4.92	1.55–45.5	16.5–1040	42	29.4–64.8	1.22–1.41	0.61–0.81	2.01–49.5	a, b
S. isoetifolium	Palk Bay, TN	3.28	0.23–0.27	–	4.82–4.92	1.55–45.5	16.5–1040	42	29.4–64.8	1.22–1.41	0.61–0.81	2.01–49.5	a, b
T. hemprichii	Palk Bay, TN	3.21	0.24–0.28	–	4.81–4.92	1.55–1.68	965–1040	42	–	1.22–1.41	0.61–0.81	2.01–2.24	c
C. serrulata	GOM, TN	7.66–8.62	0.23–0.31	–	4.86–6.44	1.94–2.16	1095–1195	52.95–64.2	–	1.56–2.02	0.54–0.67	1.34–2.72	b
E. acoroides	GOM, TN	7.66–8.62	0.23–0.31	–	4.86–6.44	1.94–2.16	1095–1195	52.95–64.2	–	1.56–2.02	0.54–0.67	1.34–2.72	b
S. isoetifolium	GOM, TN	7.66–8.62	0.23–0.31	–	4.86–6.44	1.94–2.16	1095–1195	52.95–64.2	–	1.56–2.02	0.54–0.67	1.34–2.72	b
Seagrass	GOM, TN	–	–	–	–	11.3–18.2	1756–5756	–	52.1–128	–	–	20.5–30.8	d
H. beccarii	Goa	–	–	–	–	26	37,750	3200	767	25			e
H. beccarii	Malvan, MH	–	–	–	–	2	15,750	1500	260	4	–	–	e
H. beccarii	Vijayagiri, MH	–	–	–	–	83	73,900	2700	402	35	–	–	e
H. beccarii	Ratnagiri, MH	–	–	–	–	121	75,500	2750	862	37	–	–	e
C. rotundata	LK	–	–	–	–	–	–	988.3	–	–	–	–	c
H. ovalis	LK	–	–	–		–	–	986.2	–	–	–	–	c
H. uninervis	LK	–	–	–	–	–	–	988.3	–	–	–	–	c

S. isoetifolium	LK	–	–	–	–	–	–	988.3	–	–	–	–	c
T. hemprichii	LK	–	–	–	–	–	–	988.3	–	–	–	–	c
Seagrass	LK	–	0.52–5.72	0.04–0.16	2.32–12	2.76–21.64	45.76–316	–	4–11.32	0.64–3.08	4.4–10.36	10.3–127	f
Seagrass	ANI	7–21	0.69–1.96	0.82–100	5.76–887	6.64–130	508–32,370	2018–6204	23–940	2–607	10–29	10.4–127	g, h

[a]Govindasamy et al. (2011)
[b]Baby et al. (2017)
[c]Jagtap and Untawale (1984)
[d]Kumaresan et al. (1998)
[e]Jagtap (1983)
[f]Thangaradjou et al. (2014)
[g]Nobi et al. (2010)
[h]Sachithanandam et al. (2020)

C. serrulata, *C. rotundata*, *H. beccarii*, *H. ovalis*, and *H. uninervis*, has been used for trace element studies (Jagtap 1983; Untawale and Jagtap 1984; Govindasamy and Azariah 1999; Vinithkumar et al. 1999; Thangaradjou et al. 2013; Baby et al. 2017). On the east coast, the sediment within the seagrass meadows of GOM had higher levels of As than the seagrass sediment of the Palk Bay region of Tamil Nadu (Table 3.2). However, the highest concentration of As in the sediment of seagrass meadows was recorded from ANI. This higher concentration of As can be due to the volcanic origin of this island, where As enters the coastal ecosystem through seasonal land runoff, as these islands are far from industrially polluted (Nobi et al. 2010; Sachithanandam et al. 2020). Other than arsenic, Co, Cr, Cu, Mg, Mn, Ni, and Pb concentration in the sediment of seagrass meadows are the highest in ANI (Table 3.2). The Fe concentration in the sediment was higher on the west coast at Vijayagiri and Ratnagiri of Maharashtra within the *H. beccarii* meadows (Table 3.2). However, these Fe values in the sediment of *H. beccarii* are more than three decades old, and this high concentration of Fe in *H. beccarii* sediment compared to other seagrass ecosystems of India can be due to its presence within proximity of mangrove sediments, which act as a sink of trace elements (Apte et al. 2016; Mishra and Kumar 2020). Though most of the trace element levels in the sediment of *S. isoetifolium*, *T. hemprichii*, and *C. serrulata* meadows of Palk Bay are similar (Govindasamy et al. 2013; Thangaradjou et al. 2013; Baby et al. 2017), the concentration of Cu and Zn are 20-fold lower in the sediment of *T. hemprichii* meadows (Jagtap and Untawale 1984). On the west coast, the sediment of *H. beccarii* meadows was found with high levels of Cu at Vijayagiri and Ratnagiri, Maharashtra (Jagtap 1983). There is clear evidence that the sediment of seagrass meadows acts as a sink of various trace elements. Consequently, the continuous persistence of these trace elements (particularly trace elements like As, Cu, Pb) can result in potential toxicity to seagrass rhizosphere and the seagrass-associated biota (Ambo-Rappe et al. 2011; Richir 2016; Richir and Gobert 2014). However, for the trace elements to be toxic, it has to be bioavailable to the seagrass root systems and reach above threshold levels. This bioavailability depends on trace elements' mobility in the sediment, their chemical speciation (Usero et al. 2005), and sediment characteristics such as pH, organic matter content, and redox potential (Yang and Ye 2009).

2.3 Role of Sediment Characteristics in Making Trace Elements Bioavailable

The sediment within seagrass meadows acts as a storehouse of trace elements, where the influx of land runoff and anthropogenic chemicals recharge this storehouse. Other than this input, trace element recycling happens within the seagrass meadows (Sanz-Lázaro et al. 2012), releasing trace elements bound to the fine-grain sediment fraction of seagrass meadows. The pH of the sediment and the overlying water plays a major role in the release of this sediment-bound trace metals, as low pH can alter the metal speciation and favor the release of metals from sediment pore waters (Atkinson et al. 2007; Simpson et al. 2005) that are generally not bioavailable. The trace metals in the water column above the sediment are absorbed onto sediments where redox stratification of metal-bound particles with depth occurs (Basallote et al. 2014, 2020; Eggleton and Thomas 2004), until resuspension of these particles happens due to physical processes and bioturbation. Resuspension with oxygenated overlying waters results in metal speciation in the dissolved phase (Simpson et al. 2005), making the metals bioavailable in pore waters (Batley et al. 2004). Once released from pore waters into the water column, these metals are bioavailable to seagrass and associated organisms till the precipitation of these metals is initiated by the fine fraction (<63 μm) of the sediments suspended in the water column (Zoumis et al. 2001).

2.4 Trace Element Accumulation in Seagrasses

A total of ten trace elements are reported in seagrass tissues of India, excluding As (Fig. 3.2c). Six seagrass species from Palk Bay, seven from GOM, and eight from Lakshadweep and *H. beccarii* from Goa and Maharashtra are studied for trace metal research in India (Table 3.3). For trace elements in seagrass, the Palk Bay region is the most studied region followed by Lakshadweep Islands, whereas GOM and ANI have similar levels of studies (Fig. 3.3a). In general, *S. isoetifolium* is the most studied seagrass for various trace element levels followed by *C. serrulata* and *C. rotundata* (Fig. 3.3b). There are only four studies in India, which have studied trace elements in water, sediment, and seagrass (Jagtap 1983; Jagtap and Untawale 1984; Govindasamy et al. 2011; Baby et al. 2017), and there are six studies including the above four, which have reported about trace element levels in sediment and seagrasses (Nobi et al. 2010; Vinithkumar et al. 1999), and the rest of studies have reported only about the trace elements in seagrass ecosystems, excluding the trace elements in water or sediment.

The accumulation capacity of trace elements in the various seagrass species is different, which is reflected by the highest concentration of each trace element observed in a different seagrass species. For example, Cd concentrations were highest in the tissues of *S. isoetifolium* in Palk Bay, with similar levels of Cd in *H. pinifolia* and *H. uninervis* of Lakshadweep Islands. The concentration of Mg was also highest in *S. isoetifolium* of Lakshadweep Islands (Table 3.3). The concentration of Cu in seagrasses of India was highest in *C. serrulata*, at GOM, even though the highest levels were reported from mixed seagrass species of ANI (Nobi et al. 2010; Thangaradjou et al. 2014). The *H. uninervis* of Lakshadweep Islands had the highest concentration of Co, whereas Cr concentration was highest in *H. decipiens* of Lakshadweep Islands. However, the highest levels of Cr were reported from ANI, but the authors have not specified any seagrass species (Nobi et al. 2010; Arumugam et al. 2013; Thangaradjou et al. 2013). Other than Cr, the concentration of Ni, Pb, and Zn levels in *H. decipiens* were also the highest in India. The concentration of Fe

Table 3.3 Mean or range of trace element concentration (mg kg^{-1}) in tissues of various seagrass species of India. Gulf of Mannar (GOM), Tamil Nadu (TN), Maharashtra (MH), Lakshadweep Island (LK), Andaman and Nicobar Islands (ANI), whole plant (WP), leaves (L)

Seagrass	Loc.	Tissue	Trace elements										Ref.
			Cd	Co	Cr	Cu	Fe	Mg	Mn	Ni	Pb	Zn	
C. rotundata	Palk Bay, TN	WP	0.34–1.53	–	–	1.86–17.35	–	–	–	–	0.59–6.14	4.19–52.66	a
C. serrulata	Palk Bay, TN	WP	0.12–2.84	–	–	0.13–18.59	0.33–2.85	0.24–14.38	–	0.18	0.26–2.88	0.33–42.03	a, b, c, d
E. acoroides	Palk Bay, TN	WP	0.22–0.31			3.23–36.5	909	–	304	–	1.42–1.74	13.66–35	a, e
H. ovalis	Palk Bay, TN	WP	0.31–0.36	–	–	3.79–21.67	795	–	2235		0.56–1.96	9.45–45	a, e
H. pinifolia	Palk Bay, TN	WP	–	–	–	37–60.83	1085–1886	–	220–491	–	–	35–69.17	e
S. isoetifolium	Palk Bay, TN	WP	0.3–2.83	–	–	0.53–56.67	0.22–708	–	0.3–553	0.18	0.26–3.12	0.15–53.3	a, b, d, f
T. hemprichii	GOM, TN	L	0.21	0.37	3.98	2.94	67.58	591	9.28	1.44	1.52	4.56	g
S. isoetifolium	GOM, TN	L	0.35	0.43	3.47	3.18	82.01	912.73	14.93	1.44	2.04	6.54	g
S. isoetifolium	GOM, TN	WP	0.32	–	0.28	0.75	1074	918	–	9.18	7.5	38.36	h
H. pinifolia	GOM, TN	L	0.14	0.32	1.21	6.65	156	863	23.42	0.67	1.28	17.59	g
H. uninervis	GOM, TN	WP	0.3	–	0.29	4.36	945	920	–	6.35	3.09	22.21	h
E. acoroides	GOM, TN	L	0.12	0.41	3.87	3.59	90.45	732.88	7.17	1.51	1.44	5.24	g
C. serrulata	GOM, TN	L	0.23	0.283	0.98	7.49–69	–	905.57	12.83	0.41	0.81	13.57	h
C. serrulata	GOM, TN	WP	0.30	–	0.28	7.11	540	1015	–	12.3	1.22	29.74	g
C. rotundata	GOM, TN	L	0.26	0.31	1.47	7.8	71.22	91.45	13.46	0.51	0.82	13.95	g
H. beccarii	GOA	WP	–	–	–	11	32,562	9800	575	10	–	4	i

H. beccarii	MH	WP	–	–	–	3–6	8650–27,000	7225–9400	500–560	4–6	–	–	i
C. serrulata	LK	WP	1.58	0.21	2.95	8	124.38	6000	54	2.11	7	52	j
C. rotundata	LK		2	0.2	5	4.87	178.9	1412–5123	42	1.91	5.18	55	j, k
T. hemprichii	LK		0.1–1.85	0.47	3	4.4–13	172–221	1147–6200	26–63	2.35	12	0.6–30	j, l, m
S. isoetifolium	LK	WP	2	0.16	2.95	5	224.3	1279–80,050	78	2.25	10	27.36	j, l
H. pinifolia	LK	WP	2.01	8.12	18	12	5992	9700	143.2	13	12	45	j
H. uninervis	LK	WP	2.25	11.21	18.4	18	5980	1330–6300	160	8	23.12	31	j, l
H. decipiens	LK	WP	2	10	19.8	21.57	5990	10,368	1024	19.49	15	60.96	j
H. ovalis	LK	WP	–	–	–	–	–	1732.6	–	–	–		l
Seagrass	ANI	WP	1.04–3.88	0.44–2.43	34.8–138.2	44.36–86.76	604–7308	762–7468	349–1180	1.8–5	3.08–6.64	21.64–48.72	f, n

[a]Gopi et al. (2020)
[b]Govindasamy et al. (2011)
[c]Baby et al. (2017)
[d]Sudharsan et al. (2012)
[e]Kannan et al. (1992)
[f]Nobi et al. (2010)
[g]Kannan et al. (2011)
[h]Immaculate et al. (2018)
[i]Jagtap (1983)
[j]Thangaradjou et al. (2013)
[k]Kumaresan et al. (1998)
[l]Jagtap and Untawale (1984)
[m]Gopinath et al. (2011)
[n]Thangaradjou et al. (2010)

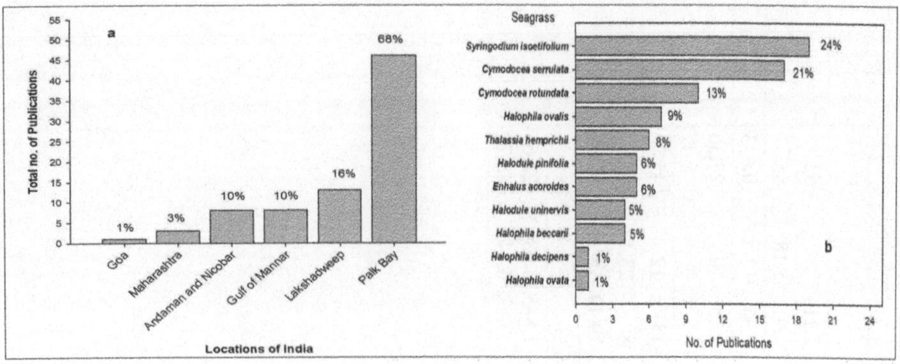

Fig. 3.3 (**a**) The various location of India from where seagrass studies on trace elements have been reported and (**b**) the various seagrass species used for trace element studies. Percentage values represent the contribution of each location or seagrass to the total studies on trace elements

was highest in the tissues of *H. beccarii*, at Goa, whereas the highest concentration of Mn was recorded from *H. ovalis* at Palk Bay (Table 3.3). This accumulation capacity of different seagrass species of India indicates trace element accumulation in seagrass is a species-specific phenomenon, and the various seagrass species of India are potential indicators of different trace element concentrations of their respective local environment. However, regarding the kind of investigated organs/ tissues of seagrass, most of the trace element studies in India have used the whole seagrass plants, except Kannan et al. (2011) and Immaculate et al. (2018), who have reported trace element levels in the leaves of *C. rotundata*, *C. serrulata*, *T. hemprichii*, *S. isoetifolium*, *H. pinifolia*, and *E. acoroides* from GOM of Tamil Nadu.

3 Effects of Trace Elements on Seagrass Physiology

In seagrasses, the concentration of elements varies within the tissues, leaves, rhizomes, and roots. Where roots accumulate the maximum concentrations and the leaves accumulate less (Bonanno and Orlando-Bonaca 2018; Mishra et al. 2019) as a higher metal concentration in leaves can lead to trace metal toxicity and damage photosynthetic apparatus of seagrass (Govers 2014; Prange and Dennison 2000). However, in India, trace metals and their toxicity on seagrass physiology or growth have not been reported. Globally, there is some toxicity assessment of trace elements on seagrass physiology of *C. serrulata* (Aljahdali and Alhassan 2020; Prange and Dennison 2000), *H. ovalis* (Prange and Dennison 2000; Ambo-Rappe et al. 2011), *H. uninervis* (Prange and Dennison 2000), and *T. hemprichii* (Lei et al. 2012) which can be compared to the seagrass species of India. Trace elements such as Cd (10 mg L^{-1}), Cu (1–10 mg L^{-1}), Pb (10 mg L^{-1}), and Zn (10 mg L^{-1}) are toxic to *C. serrulata*, *H. ovalis*, *T. hemprichii*, and *H. uninervis* photosynthetic apparatus: Photo System II (PS-II). Other than damaging PS-II, Cu concentrations reduced leaf growth and width of *H. ovalis* and amino acid levels in *C. serrulata* and *H. uninervis* (Prange and Dennison 2000; Ambo-Rappe et al. 2011). Zinc toxicity reduced photosynthetic pigments of *T. hemprichii* (Lei et al. 2012). The antioxidant activity of *H. ovalis* and *C. serrulata* was decreased by Cd and Pb toxicity (Ambo-Rappe et al. 2011; Aljahdali and Alhassan 2020).

For the above mentioned four seagrass species in India, Cu concentration is 6-fold and 1.7-fold higher than toxic levels in the tissues of *C. serrulata* at GOM and Palk Bay of Tamil Nadu (Govindasamy et al. 2013; Baby et al. 2017; Immaculate et al. 2018). In Lakshadweep Islands, *H. uninervis* and *T. hemprichii* have 1.8-fold and 1.3-fold higher Cu levels than toxic concentrations (Untawale and Jagtap 1984; Gopinath et al. 2011; Thangaradjou et al. 2013), whereas *H. ovalis* has twofold higher Cu levels than toxic levels at Palk Bay region (Kannan et al. 2011; Gopi et al. 2020). Lead levels are 1.2-fold and 2.3-fold higher in *T. hemprichii* and *H. uninervis* at Lakshadweep Islands (Jagtap and Untawale 1984; Gopinath et al. 2011; Thangaradjou et al. 2013). *T. hemprichii* and *H. uninervis* have threefold higher Zn levels than toxicity levels (Jagtap and Untawale 1984; Gopinath et al. 2011; Thangaradjou et al. 2013), whereas *C. serrulata* has three- to fivefold higher levels of Zn concentration which can exert toxicity on its PS-II at Palk Bay and GOM, Tamil Nadu, and at Lakshadweep Islands (Govindasamy et al. 2011; Sudharsan et al. 2012; Thangaradjou et al. 2013; Gopi et al. 2020). Trace element levels above toxic concentration for these seagrasses suggest that these four seagrass species are under stress from metal toxicity, which needs further research and attention from the scientific community of India.

The high concentration of trace elements in these seagrass species will result in trophic transfer of these elements and exert toxicity to the associated trophic assemblages (Prange and Dennison 2000; de los Santos et al. 2019), such as gastropods, mollusks, fish, and invertebrates that depend on seagrass for direct and indirect food sources (Manikandan et al. 2011). Consequently, metal toxicity can lead to seagrass population loss and die-offs, which will have negative consequences on the coastal ecosystem functioning.

4 Future Scenarios and Metal Toxicity on Seagrass

Global changes, such as ocean acidification due to increased CO_2 concentrations, and low pH will affect the trace metal chemistry, speciation, and bio-availability (Millero et al. 2009; Zeng et al. 2015) and can have possible negative impacts on the seagrass ecosystem. Low pH can increase the bioavailability of trace elements bound to seagrass sediment and even increase their concentrations as trace metal speciation in seawater is strongly dependent on seawater chemistry, with several metals known to be sensitive to speciation changes within the pH range projected for the near future (Byrne et al. 1988; Richards et al. 2011). Changes in ocean carbon chemistry may also alter the behavior of metals bound to sediments, influencing metal fluxes from contaminated sediments (Millero et al. 2009; Zeng et al. 2015).

Low pH is predicted to increase the toxic-free ion concentration of metals in coastal waters by as much as 115% in the next 100 years (Millero et al. 2009; Lewis et al. 2016). Saying that most of the studies on metal concentrations have been focused on marine animals (e.g., marine invertebrates, mussels, planktons, fish larva) with very few studies on seagrass ecosystems, which needs to be addressed in India.

5 Conclusions

Globally, seagrasses are used as bioindicators of coastal contamination (Lewis and Richard 2009; Bonanno and Orlando-Bonaca 2017, 2018). In India, though seagrass is found to be an efficient indicator of the environmental concentration of trace elements in their tissues, it has not been used as a bioindicator of coastal pollution. However, the National Action Plan for seagrass ecosystems that have been launched in 2018 (Koshy et al. 2018) plans to address these issues and provides guidelines that will use this bioindicator potential of vast seagrass ecosystems of India to facilitate their conservation and management issues.

References

Aljahdali MO, Alhassan AB (2020) Heavy metal accumulation and anti-oxidative feedback as a biomarker in Seagrass Cymodocea Serrulata. Sustainability (Switzerland) 12. https://doi.org/10.3390/su12072841

Ambo-Rappe R, Lajus DL, Schreider MJ (2011) Heavy metal impact on growth and leaf asymmetry of seagrass, Halophila ovalis. Environ Chem 3:149–159

Apte DA, Bhave V, Have A, Negate P, Pitale R (2016) Ecologically important marine and coastal biodiversity areas of Maharashtra, India. Bombay Natural History Society, Bombay, p 36

Arumugam R, Kannan RRR, Saravanan KR, Thangaradjou T, Anantharaman P (2013) Hydrographic and sediment characteristics of seagrass meadows of the Gulf of Mannar Marine Biosphere Reserve, South India. Environ Monit Assess 185:8411–8427. https://doi.org/10.1007/s10661-013-3183-6

Atkinson CA, Jolley DF, Simpson SL (2007) Effect of overlying water pH, dissolved oxygen, salinity and sediment disturbances on metal release and sequestration from metal contaminated marine sediments. Chemosphere 69:1428–1437. https://doi.org/10.1016/j.chemosphere.2007.04.068

Avelar M, Bonilla-Heredia B, Merino-Ibarra M, Herrera-Silveira JA, Ramirez J, Rosas H, Valdespino J, Carricart-Ganivet JP, Martínez A (2013) Iron, cadmium, and chromium in seagrass (Thalassia testudinum) from a coastal nature reserve in karstic Yucatán. Environ Monit Assess 185:7591–7603. https://doi.org/10.1007/s10661-013-3121-7

Baby L, Gireesh Kumar TR, Remyakumari KR, Varkey J, Sankar TV, Chandramohanakumar N (2017) Comparison of hydrographic and sediment characteristics of seagrass meadows of Gulf of Mannar and Palk Bay, South West Coast of India. Int J Fish Aquat Stud 5:80–84

Basallote MD, de Orte MR, DelValls TÁ, Riba I (2014) Studying the effect of CO_2-induced acidification on sediment toxicity using acute amphipod toxicity test. Environ Sci Technol 48:8864–8872. https://doi.org/10.1021/es5015373

Basallote MD, Borrero-Santiago AR, Cánovas CR, Hammer KM, Olsen AJ, Ardelan MV (2020) Trace metal mobility in sub-seabed sediments by CO2 seepage under high-pressure conditions. Sci Total Environ 700:134761. https://doi.org/10.1016/j.scitotenv.2019.134761

Batley GE, Apte SC, Stauber JL (2004) Speciation and bioavailability of trace metals in water: progress since 1982. Aust J Chem 57:903–919. https://doi.org/10.1071/CH04095

Bayyana S, Pawar S, Gole S, Dudhat S, Pande A, Mitra D, Johnson JA, Sivakumar K (2020) Detection and mapping of seagrass meadows at Ritchie's archipelago using Sentinel 2A satellite imagery. Curr Sci 118:1275–1282. https://doi.org/10.18520/cs/v118/i8/1275-1282

Bonanno G, Borg JA (2018) Comparative analysis of trace element accumulation in seagrasses Posidonia oceanica and Cymodocea nodosa: biomonitoring applications and legislative issues. Mar Pollut Bull 128:24–31. https://doi.org/10.1016/j.marpolbul.2018.01.013

Bonanno G, di Martino V (2016) Seagrass Cymodocea nodosa as a trace element biomonitor: bioaccumulation patterns and biomonitoring uses. J Geochem Explor 169:43–49. https://doi.org/10.1016/j.gexplo.2016.07.010

Bonanno G, di Martino V (2017) Trace element compartmentation in the seagrass Posidonia oceanica and biomonitoring applications. Mar Pollut Bull 116:196–203. https://doi.org/10.1016/j.marpolbul.2016.12.081

Bonanno G, Orlando-Bonaca M (2017) Trace elements in Mediterranean seagrasses: accumulation, tolerance and biomonitoring. A review. Mar Pollut Bull 125:8–18. https://doi.org/10.1016/j.marpolbul.2017.10.078

Bonanno G, Orlando-Bonaca M (2018) Trace elements in Mediterranean seagrasses and macroalgae. A review. Sci Total Environ 618:1152–1159. https://doi.org/10.1016/j.scitotenv.2017.09.192

Bonanno G, Raccuia SA (2018) Comparative assessment of trace element accumulation and bioindication in seagrasses Posidonia oceanica, Cymodocea nodosa and Halophila stipulacea. Mar Pollut Bull 131:260–266. https://doi.org/10.1016/j.marpolbul.2018.04.039

Byrne RH, Kump LR, Cantrell KJ (1988) The influence of temperature and pH on trace metal speciation in seawater. Mar Chem 25:163–181. https://doi.org/10.1016/0304-4203(88)90062-X

Costanza R, de Groot R, Braat L, Kubiszewski I, Fioramonti L, Sutton P, Farber S, Grasso M (2017) Twenty years of ecosystem services: how far have we come and how far do we still need to go? Ecosyst Serv 28:1–16. https://doi.org/10.1016/j.ecoser.2017.09.008

de los Santos CB, Arenas F, Neuparth T, Santos MM (2019) Interaction of short-term copper pollution and ocean acidification in seagrass ecosystems: toxicity, bioconcentration and dietary transfer. Mar Pollut Bull 142:155–163. https://doi.org/10.1016/j.marpolbul.2019.03.034

Duarte CM, Kennedy H, Marbà N, Hendriks I (2013) Assessing the capacity of seagrass meadows for carbon burial: current limitations and future strategies. Ocean Coast Manag 83:32–38. https://doi.org/10.1016/j.ocecoaman.2011.09.001

Eggleton J, Thomas K v (2004) A review of factors affecting the release and bioavailability of contaminants during sediment disturbance events. Environ Int 30:973–980. https://doi.org/10.1016/j.envint.2004.03.001

Fortes MD (2018) Seagrass ecosystem conservation in Southeast Asia needs to link science to policy and practice. Ocean Coast Manag 159:51–56. https://doi.org/10.1016/j.ocecoaman.2018.01.028

Geevarghese GA, Akhil B, Magesh G, Krishnan P, Purvaja R, Ramesh R (2018) A comprehensive geospatial assessment of seagrass distribution in India. Ocean Coast Manag 159:16–25. https://doi.org/10.1016/j.ocecoaman.2017.10.032

Gillis LG, Jones CG, Ziegler AD, van der Wal D, Breckwoldt A, Bouma TJ (2017) Opportunities for protecting and restoring tropical coastal ecosystems by utilizing a physical connectivity approach. Front Mar Sci 4:374. https://doi.org/10.3389/fmars.2017.00374

Gopi S, Arulkumar A, Ganeshkumar A, Rajaram R, Miranda JM, Paramasivam S (2020) Heavy metals accumulation in seagrasses collected from Palk Bay, South-eastern India. Mar Pollut Bull 157:111305. https://doi.org/10.1016/j.marpolbul.2020.111305

Gopinath A, Muraleedharan NS, Chandramohanakumar N, Jayalakshmi K v (2011) Statistical significance of biomonitoring of marine algae for trace metal levels in a coral environment. Environ Forensic 12:98–105. https://doi.org/10.1080/15275922.2011.547440

Govers LL (2014) The effects of biogeochemical stressors on seagrass ecosystems
Govindasamy C, Arulpriya M (2011) Seasonal variation in seagrass biomass in Northern Palk Bay, India. Biodiversity 12:223–231. https://doi.org/10.1080/14888386.2011.653542
Govindasamy C, Azariah J (1999) Seasonal variation of heavy metals in coastal water of the Coromandel coast, Bay of Bengal, India. Indian J Mar Sci 28:249–256
Govindasamy C, Arulpriya M, Ruban P, Francisca JL, Ilayaraja A (2011) Concentration of heavy metals in Seagrasses tissue of the Palk Strait, Bay of Bengal. Int J Environ Sci 2:145–153. https://doi.org/10.6088/ijes.00202010016
Govindasamy C, Anantharaj K, Srinivasan R (2013) Seasonal variations in seagrass biomass and productivity in Palk Bay, Bay of Bengal, India. Int J Biodivers Conserv 5:408–417. https://doi.org/10.5897/IJBC12.132
Guannel G, Arkema K, Ruggiero P, Verutes G (2016) The power of three: coral reefs, seagrasses and mangroves protect coastal regions and increase their resilience. PLoS One 11:e0158094. https://doi.org/10.1371/journal.pone.0158094
Hemminga MA, Duarte CM (2002) Seagrass ecology, limnology and oceanography. Cambridge University Press, Cambridge
Immaculate JK, Lilly TT, Patterson J (2018) Macro and micro nutrients of seagrass species from Gulf of Mannar, India. MOJ Food Process Technol 6:391–398. https://doi.org/10.15406/mojfpt.2018.06.00193
Jagtap TG (1983) Metal distribution in Halophila beccarii (Aschers) and surrounding environment along the central west coast of India. Oceanography 16:429–434
Jagtap TG, Untawale A (1984) Chemical composition of marine macrophytes and their surrounding water and sediments, from Minicoy, Lakshadweep. Indian J Mar Sci 13:123–125
Jagtap TG, Komarpant DS, Rodrigues RS (2003) Status of a seagrass ecosystem: an ecologically sensitive wetland habitat from India. Wetlands 23:161–170. https://doi.org/10.1672/0277-5212(2003)023[0161:SOASEA]2.0.CO;2
Kannan R, Ganesan M, Govindasamy C, Rajendran K, Sampathkumar P, Kannan L (1992) Tissue concentration of heavy metals in seagrasses of Palk Bay, Bay of Bengal. Int Ecol Environ Sci 18:29–34
Kannan R, Ganesan M, Govindasamy C, Rajendran K, Sampathkumar P, Kannan L (2011) Tissue concentration of heavy metals in seagrasses of the Palk Bay, Bay of Bengal. Int J Environ Sci 2:145–153
Koshy NE, Bhatt JR, Vakily JM (2018) Synthesis of the conference on management and conservation of seagrass ecosystems in India. Ocean Coast Manag 159:3–6. https://doi.org/10.1016/j.ocecoaman.2017.11.001
Kumaresan S, Vinith Kumar NV, Balasubramanian T, Subramanian AN (1998) Trace metals in sediments from the gulf of mannar region, southeast coast of India. Indian J Geo Mar Sci 256–258
Lei LI, Xiaoping H, Borthakur D, Hui NI (2012) Photosynthetic activity and antioxidative response of seagrass Thalassia hemprichii to trace metal stress. Acta Oceanol Sin 31:98–108. https://doi.org/10.1007/s13131-012-0210-3
Lewis MA, Richard D (2009) Nonnutrient anthropogenic chemicals in seagrass ecosystems: fate and effects. Environ Toxicol Chem 28:644–661. https://doi.org/10.1897/08-201.1
Lewis C, Ellis RP, Verno E, Elliot K, Newbatt S, Wilson RW (2016) Ocean acidification increases copper toxicity differentially in two key marine invertebrates with distinct acid-base responses. Sci Rep 6:1–10. https://doi.org/10.1038/srep21554
Machowski R, Rzetala MA, Rzetala M, Solarski M (2019) Anthropogenic enrichment of the chemical composition of bottom sediments of water bodies in the neighborhood of a non-ferrous metal smelter (Silesian Upland, Southern Poland). Sci Rep 9:1–13. https://doi.org/10.1038/s41598-019-51027-w
Manikandan S, Ganesapandian S, Singh M, Kumaraguru AK (2011) Distribution and Spatial variation of seagrass in the northern part of Gulf of Mannar, Southeastern India. Asian J Plant Sci 10:80–85. https://doi.org/10.3923/ajps.2011.80.85
McKenzie N (2020) The global distribution of seagrass meadows. Environ Res Lett 15:074041

Medina-Gomez I (2016) Response of Thalassia testudinum morphometry and distribution to environmental drivers in a pristine tropical lagoon. PLoS One 2007:1–24. https://doi.org/10.1371/journal.pone.0164014
Millero F, Woosley R, DiTrolio B, Waters J (2009) Effect of ocean acidification on the speciation of metals in seawater. Oceanography. https://doi.org/10.5670/oceanog.2009.98
Mishra A, Apte D (2020) Ecological connectivity with mangroves influences tropical seagrass population longevity and meadow traits within an island ecosystem. Mar Ecol Prog Ser 644:47–63. https://doi.org/10.3354/meps13349
Mishra AK, Apte D (2021) The current status of Halophila beccarii: an ecologically significant, yet Vulnerable seagrass of India. Ocean Coast Manag 200:1–37. https://doi.org/10.1016/j.ocecoaman.2020.105484
Mishra AK, Kumar M (2020) Andaman mangrove sediments: source of nutrients and sink of heavy metals. Indian J Geomarine Sci 49:156–166. https://doi.org/10.1101/431262
Mishra AK, Mohanraju R (2018) Epiphytic bacterial communities in seagrass meadows of oligotrophic waters of Andaman Sea. OALib 5:1–12. https://doi.org/10.4236/oalib.1104388
Mishra AK, Santos R, Hall-Spencer JM (2019) Elevated trace elements in sediments and seagrasses at CO2 seeps. Mar Environ Res:104810. https://doi.org/10.1016/j.marenvres.2019.104810
Nobi EP, Dilipan E, Thangaradjou T, Sivakumar K, Kannan L (2010) Geochemical and geostatistical assessment of heavy metal concentration in the sediments of different coastal ecosystems of Andaman Islands, India. Estuar Coast Shelf Sci 87:253–264. https://doi.org/10.1016/j.ecss.2009.12.019
Nordlund ML, Koch EW, Barbier EB, Creed JC (2016) Seagrass ecosystem services and their variability across genera and geographical regions. PLoS ONE 11(10):e0163091. https://doi.org/10.1371/journal.pone.0163091
Nordlund LM, Unsworth RKF, Gullström M, Cullen-Unsworth LC (2018) Global significance of seagrass fishery activity. Fish Fish 19:399–412. https://doi.org/10.1111/faf.12259
Ondiviela B, Losada IJ, Lara JL, Maza M, Galván C, Bouma TJ, van Belzen J (2014) The role of seagrasses in coastal protection in a changing climate. Coast Eng 87:158–168. https://doi.org/10.1016/j.coastaleng.2013.11.005
Parthasarathy N, Ravikumar K, Ganesan R, Ramamurthy K (1991) Distribution of seagrasses along the coast of Tamil Nadu, southern India. Aquat Bot 40:145–153. https://doi.org/10.1016/0304-3770(91)90092-J
Patankar V, Wagh T, Tyabji Z (2018) Observations on the female flowers and fruiting of tape grass Enhalus acoroides from South Andaman Islands, India. J Threat Taxa 10:11391–11398. https://doi.org/10.11609/jott.3138.10.1.11126-11146
Patro S, Krishnan P, Deepak Samuel V, Purvaja R, Ramesh R (2017) Seagrass and saltmarsh ecosystems in South Asia: an overview of diversity, distribution, threats and conservation. In: Wetland science: perspectives from South Asia. Springer, New Delhi, pp 1–587. https://doi.org/10.1007/978-81-322-3715-0
Potouroglou M, Bull JC, Krauss KW, Kennedy HA, Fusi M, Daffonchio D, Mangora MM, Githaiga MN, Diele K, Huxham M (2017) Measuring the role of seagrasses in regulating sediment surface elevation. Sci Rep 7:1–11. https://doi.org/10.1038/s41598-017-12354-y
Prange JA, Dennison WC (2000) Physiological responses of five seagrass species to trace metals. Mar Pollut Bull 41:327–336. https://doi.org/10.1016/S0025-326X(00)00126-0
Richards R, Chaloupka M, Sanò M, Tomlinson R (2011) Modelling the effects of "coastal" acidification on copper speciation. Ecol Model 222:3559–3567. https://doi.org/10.1016/j.ecolmodel.2011.08.017
Richir J (2016) Trace elements in marine environments: occurrence, threats and monitoring with special focus on the coastal Mediterranean. J Environ Anal Toxicol 6:1–19. https://doi.org/10.4172/2161-0525.1000349
Richir J, Gobert S (2014) A reassessment of the use of Posidonia oceanica and Mytilus galloprovincialis to biomonitor the coastal pollution of trace elements: new tools and tips. Mar Pollut Bull 89:390–406. https://doi.org/10.1016/j.marpolbul.2014.08.030

Roberts DA, Johnston EL, Poore AGB (2008) Contamination of marine biogenic habitats and effects upon associated epifauna. Mar Pollut Bull 56:1057–1065. https://doi.org/10.1016/j.marpolbul.2008.03.003

Sachithanandam V, Parthasarathy P, Sai Elangovan S, Kasilingam K, Dhivya P, Mageswaran T, Mohan PM (2020) A baseline study on trace metals concentration and its ecological risk assessment from the coast of South Andaman Island, India. Reg Stud Mar Sci 36:101242. https://doi.org/10.1016/j.rsma.2020.101242

Sanz-Lázaro C, Malea P, Apostolaki ET, Kalantzi I, Marín A, Karakassis I (2012) The role of the seagrass Posidonia oceanica in the cycling of trace elements. Biogeosciences 9:2497–2507. https://doi.org/10.5194/bg-9-2497-2012

Serrano O, Mateo MA, Dueñas-Bohórquez A, Renom P, López-Sáez JA, Martínez Cortizas A (2011) The Posidonia oceanica marine sedimentary record: a holocene archive of heavy metal pollution. Sci Total Environ 409:4831–4840. https://doi.org/10.1016/j.scitotenv.2011.08.001

Short F, Prathep A, Carpenter KE, Erftemeijer PLA, Calumpong HP, Orth RJ, Waycott M, Coles RG, Dennison WC, van Tussenbroek B, Bujang JS, la Nafie YA, Short FT, Nasution IM, Fortes MD, Polidoro B, Livingstone SR, Judson Kenworthy W, Freeman AS, Jagtap TG, Carruthers TJB, Bandeira S, Zieman JC, Kamal AHM, Kendrick GA, Vergara SG, Sanciangco JC (2011) Extinction risk assessment of the world's seagrass species. Biol Conserv 144:1961–1971. https://doi.org/10.1016/j.biocon.2011.04.010

Simpson BSL, Batley GE, Chariton AA, Stauber JL, King CK, Chapman JC, Hyne RV, Gale SA, Roach AC, Maher WA, Simpson SL (2005). Handbook for sediment quality assessment quality assessment. Design, p 126.

Stockdale A, Tipping E, Lofts S, Mortimer RJG (2016) Effect of ocean acidification on organic and inorganic speciation of trace metals. Environ Sci Technol 50:1906–1913. https://doi.org/10.1021/acs.est.5b05624

Sudharsan S, Seedevi P, Ramasamy P, Subhapradha N, Vairamani S, Shanmugam A (2012) Heavy metal accumulation in seaweeds and sea grasses along southeast coast of India. J Chem Pharm Res 4:4240–4244

Thangaradjou T, Bhatt JR (2018) Status of seagrass ecosystems in India. Ocean Coast Manag 159:7–15. https://doi.org/10.1016/j.ocecoaman.2017.11.025

Thangaradjou T, Nobi EP, Change C, Sivakumar K (2009) Seagrass-Watch the official magazine of the Seagrass-Watch global assessment and monitoring program Coastal Canaries. Seagrass Watch.

Thangaradjou T, Nobi EP, Dilipan E, Sivakumar K, Susila S (2010) Heavy metal enrichment in seagrasses of Andaman Island and its implication to coastal ecosystem. Indian J Geo Mar Sci 85–91

Thangaradjou T, Raja S, Subhashini P, Nobi EP, Dilipan E (2013) Heavy metal enrichment in the seagrasses of Lakshadweep group of islands—a multivariate statistical analysis. Environ Monit Assess 185:673–685. https://doi.org/10.1007/s10661-012-2583-3

Thangaradjou T, Subhashini P, Raja S, Dilipan E, Nobi EP (2014) Evidences for heavy metal contamination in surface sediments of seagrass ecosystem of Lakshadweep archipelago, India. Environ Earth Sci 71:1135–1146. https://doi.org/10.1007/s12665-013-2517-6

Tripathy GR, Singh SK, Ramaswamy V (2014) Major and trace element geochemistry of Bay of Bengal sediments: implications to provenances and their controlling factors. Palaeogeogr Palaeoclimatol Palaeoecol 397:20–30. https://doi.org/10.1016/j.palaeo.2013.04.012

Unsworth RKF, Cullen-Unsworth LC (2018) A call for seagrass protection. Science. https://doi.org/10.1126/science.aat7318

Unsworth RKF, Gullström M, Cullen-Unsworth LC (2017) Global significance of seagrass fishery activity. Fish Fish. https://doi.org/10.1111/faf.12259

Untawale AG, Jagtap TG (1984) Marine macrophytes of minicoy (Lakshadweep) coral atoll of the Arabian Sea. Aquat Bot 19:97–103. https://doi.org/10.1016/0304-3770(84)90010-X

Usero J, Morillo J, Gracia I (2005) Heavy metal concentrations in molluscs from the Atlantic coast of southern Spain. Chemosphere 59:1175–1181. https://doi.org/10.1016/j.chemosphere.2004.11.089

Vinithkumar N v, Kumaresan S, Manjusha M, Balasubramanian T (1999) Organic matter, nutrients and major ions in the sediments of coral reefs and seagrass beds of Gulf of Mannar biosphere reserve, southeast coast of India. Indian J Mar Sci 28:383–393

Vizzini S, Costa V, Tramati C, Gianguzza P, Mazzola A (2013) Trophic transfer of trace elements in an isotopically constructed food chain from a semi-enclosed marine coastal area (Stagnone di Marsala, Sicily, Mediterranean). Arch Environ Contam Toxicol 65:642–653. https://doi.org/10.1007/s00244-013-9933-1

Waycott M, Duarte CM, Carruthers TJB, Orth RJ, Dennison WC, Olyarnik S, Calladine A, Fourqurean JW, Heck KL, Hughes AR, Kendrick GA, Kenworthy WJ, Short FT, Williams SL (2009) Accelerating loss of seagrasses across the globe threatens coastal ecosystems. Proc Natl Acad Sci 106:12377–12381. https://doi.org/10.1073/pnas.0905620106

Yang J, Ye Z (2009) Metal accumulation and tolerance in wetland plants. Front Biol China 4:282–288. https://doi.org/10.1007/s11515-009-0024-7

Zeng X, Chen X, Zhuang J (2015) The positive relationship between ocean acidification and pollution. Mar Pollut Bull 91:14–21. https://doi.org/10.1016/j.marpolbul.2014.12.001

Zoumis T, Schmidt A, Grigorova L, Calmano W (2001) Contaminants in sediments: remobilisation and demobilisation. Sci Total Environ 266:195–202. https://doi.org/10.1016/S0048-9697(00)00740-3

Chapter 4
Phosphorus Availability and Speciation in the Intertidal Sediments of Sundarbans Mangrove Ecosystem of India and Bangladesh

Alok Kumar, Swati Mohan Sappal, and AL. Ramanathan

Abstract The intertidal sediments of Sundarbans mangrove ecosystem from the Indian and Bangladesh side were studied for the sedimentary phosphorus cycling and phosphorus bioavailability. A total of seven sediment cores were collected from the Sundarbans mangroves which were subjected to sequential extraction procedure to study the different sedimentary pools of phosphorus, namely, loosely sorbed P, Fe-bound P, authigenic P, detrital P and organic P. The total sedimentary P in the cores varied between 8.36 and 11.20 μmol/g for the Indian Sundarbans and from 8.80 to 11.40 μmol/g for the Bangladesh Sundarbans. The average percentage of respective P fractions in the three cores collected from Indian Sundarbans followed the order: authigenic P (39.96%) > detrital P (30.58%) > organic P (22.44%) > sorbed P (4.86%) > Fe-bound P (1.96%), and in the four cores from Bangladesh, it followed the sequence: detrital P (33.43%) > authigenic P ~ organic P (31.49%) > Fe-bound P (12.43%) > sorbed P (3.58%). Diagenetic redistribution of P was attributed to be the dominant factor responsible for the conversion of organic and Fe-bound P to authigenic P. A considerable fraction of the total sedimentary P was constituted by the bioavailable P which accounted for <33.70% and <41.07% in Indian and Bangladesh Sundarbans, respectively. The results are suggestive of the internal loading of P being an important but a limiting factor which governs the biological productivity in the Sundarbans mangrove ecosystem.

A. Kumar (✉)
Department of Environmental Studies, University of Delhi, New Delhi, India
e-mail: akumar1@es.du.ac.in

S. M. Sappal
Coastal Environment Impact Assessment Division, National Centre for Sustainable Coastal Management, Anna University Campus, Chennai, India

AL. Ramanathan
School of Environmental Sciences, Jawaharlal Nehru University, New Delhi, India
e-mail: alr0400@mail.jnu.ac.in

S. Madhav et al. (eds.), *Coastal Ecosystems*, Coastal Research Library 38,
https://doi.org/10.1007/978-3-030-84255-0_4

Keywords Phosphorus · Sequential extraction · Mangroves · Bioavailability · Productivity · Sundarbans

1 Introduction

Coastal wetlands including estuaries and mangrove settings are important sites for nutrient exchange from the marine as well as continental environment (Moutin et al. 1993; Cardoso et al. 2002; Coelho et al. 2004; Twilley et al. 2019). Mangroves are highly productive ecosystems which cover approximately 3/4th of the global coastline in the tropics and subtropics (Alongi 1996; Marchand et al. 2006). Phosphorus is an important limiting nutrient which plays a critical role in governing the coastal and marine productivity (Sanudo-Wilhelmy et al. 2001; Hou et al. 2009; Abdallah 2011). It is well established that the coastal bodies like estuaries, bays and seas receive and capture large amounts of P, thus acting as potential sinks (Slomp 2011; Yang et al. 2016). However, the fate of these sinks is principally governed by how reactive are the different fractions in which P occurs (Ruttenberg 1992; Andrieux-Loyer and Aminot 1997; Yang et al. 2016). There is a marked difference in the pH of the sediments from the coastal ecosystems which is higher when compared to the terrestrial soils as a result of the processes like release from the root exudates, decomposition of leaf litter, proton extrusion, etc. which makes the soil naturally acidic (Oxmann and Schwendenmann 2015). This distinct variation in pH, along with the prevailing redox conditions in the varying salinity regimes of the coastal ecosystems, is responsible for the processes like desorption and precipitation of phosphate which ultimately control the P availability (Van Beusekom and De Jonge 1997).

Phosphorus occurs in multiple compound sedimentary phase (Emsley 1980) and is bound to the sediment by means of adsorption onto Fe and Al compounds, by precipitation with Ca compounds, by means of microbial immobilization processes or by association with organic matter (Föllmi 1996; Benitez-Nelson 2000; Vepraskas and Faulkner 2001; Song 2010). This attachment of P onto the sediments via different mechanisms plays a key role in determining its fate in the coastal ecosystems (Zhuang et al. 2014). The release of the sediment-bound P contributes to the presence of phosphates in the overlying water column which significantly impacts coastal eutrophication (Duhamel et al. 2017). The bioavailability and cycling of P are determined not just by the total P in the sedimentary pool but also by the different forms in which it exists which impact its biological uptake (Psenner and Puckso 1988; Ruttenberg 1992; Andrieux-Loyer and Aminot 2001; Coelho et al. 2004; Yang et al. 2016). Chemical sequential extraction techniques have been widely used to examine the different forms in which P exists which helps in a clear understanding of the phosphorus dynamics and other geochemical processes in the coastal sediments (Zabel et al. 1998; Eijsink et al. 2000; Zhuang et al. 2014). It is found that a host of prevailing conditions, viz., pH, changes in salinity, redox conditions and granulometry of the sediments, governs the relative abundance of different P forms

(Lebo 1991; Paludan and Morris 1999; Coelho et al. 2004). Fractionation of the sedimentary P into different forms has proven to be an important and a useful tool to understand phosphorus cycling at different spatial scales (Ruttenberg 1993; Filippelli 2001; Slomp 2011; Yang et al. 2016). Furthermore, estuaries and coastal wetlands act as an important sink for phosphorus (Nixon et al. 1996; Hou et al. 2009), and during the process of burial, significant diagenetic reorganization of phosphorus may take place (Schenau and De Lange 2001; Fang et al. 2007; Hou et al. 2009). Therefore, these different sedimentary forms in which P occurs in the coastal settings are an outcome of the complex biogeochemical reworking of the sediments. In the Indian subcontinent, few studies have been undertaken with respect to the P speciation in the estuaries, coastal wetlands and mangroves. Previously, only Pichavaram mangroves have been investigated along the east coast of India (Prasad and Ramanathan 2010; Ranjan et al. 2011), while Cochin estuary has been studied on the west (Renjith and Chandramohanakumar 2007; Joseph et al. 2010; Renjith et al. 2011; Gireeshkumar et al. 2013).

The present study was undertaken in the Sundarbans mangrove sediments of India and Bangladesh which have come under intense anthropogenic pressure in the recent decades. The main objectives of this study were to (a) investigate the geochemical fractionation of P in sediment cores of Sundarbans mangroves, (b) delineate the factors that control the geochemical behaviour of different sedimentary P fractions and (c) determine the possible bioavailability of sedimentary P.

2 Study Area

Sundarbans is the single largest block of littoral mangrove ecosystem and the largest delta formed by the Ganges, Brahmaputra and Meghna rivers. It covers an area of 10,200 km^2 (Fig. 4.1) extending geographically between latitudes 21°31′ N and 22°30′ N and longitudes 89°01′ E and 90°18′ E (Katebi 2001; Islam 2003). The Ganges-Brahmaputra delta plain comprises a complex network of tidally influenced estuaries and islands. The delta is underlain by quaternary sediments brought down to the hinterland via erosion and deposited by Brahmaputra, Ganges and Meghna rivers along with their distributaries. In addition to this, fine silt and clay from the Bay of Bengal nourish the delta (Umitsu 1997; Islam 2001). Out of the 1.76 million km^2, 62% of the GBM drainage basin falls in India and 7.5% in Bangladesh (Elahi et al. 1998).

Sundarbans falls under Ganges-Brahmaputra delta, and delta formation is very rapid in this region (Morgan and McIntire 1959). The tectonic activity controls the geomorphology of the region which in turn influences the biota (Junk et al. 2006). The terrestrial load of sediments via rivers and creeks results in the formation of mangrove swamps, salt flats and mud flats. These landforms help in the propagation of mangrove propagules which form a barrier between land and sea to provide coastal protection as they grow (Selvam et al. 2003). The hydrodynamics of the region is ruled by the tidal flow as Sundarbans experiences semi-diurnal tides, i.e. occurrence of two

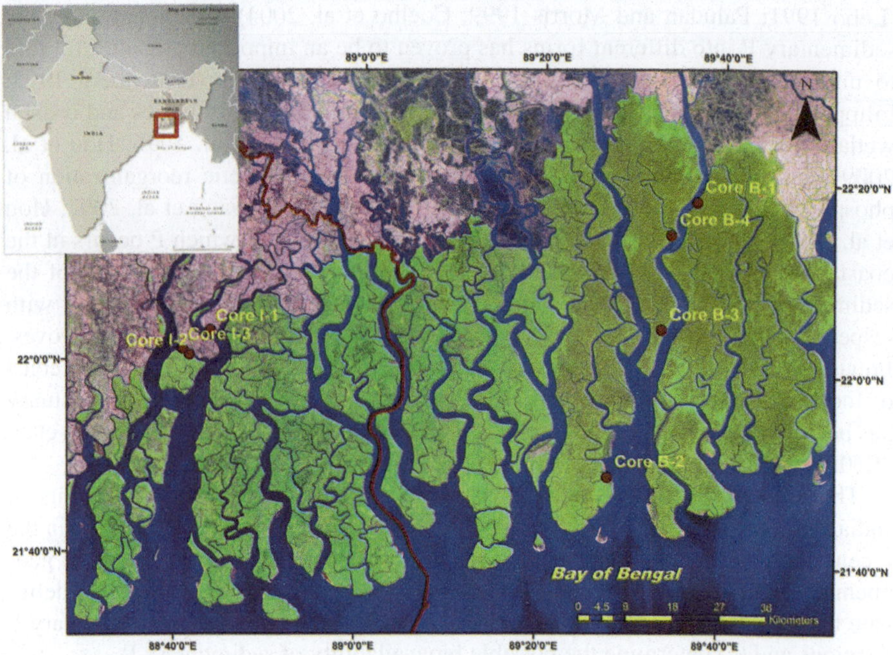

Fig. 4.1 Map showing study area and sampling locations in the Indian and Bangladesh Sundarbans

flood and two ebb tides within 24 h. These tides are macrotidal in nature, and their amplitude ranges from 3 to 5 m which rises further up to 8 m during spring tide (Mandal and Ghosh 1989; Banerjee 1998) and the extent of tidal flow reaches up to 100 km upstream (Rogers et al. 2013). The western region of the Sundarbans experiences higher tidal amplitude than the eastern (Untawale 1987). The Indian part of the Sundarbans is fed by estuarine rivers which have mostly lost their upstream connection as a result of neo-tectonic movement in the east (Morgan and McIntire 1959) and to massive siltation in the west (Gopal and Chauhan 2006; Mitra et al. 2009). The major tributaries in the Indian part are Muriganga, Saptamukhi, Thakuran, Bidya, Matla, Bhangaduni, Gosaba, Haribhanga, Raimangal, etc., while Bangladesh has Raimangal, Malancha, Kunga, Passur, Sibsa, Bangra, Baleshwar, etc. The tropical southwest monsoon controls the freshwater discharge in the region. Majority of the sediment load is deposited between May and September in a year (Coleman 1969; Goodbred et al. 2003; Rogers et al. 2013). Annual range of temperature varies from 20 °C (December–January) to 34 °C (June–July) with high humidity, and the average annual rainfall varies between 1800 and 1920 mm (Kumar and Ramanathan 2015).

With respect to vegetation, *Heritiera fomes* (Sundari) is the dominant species, and others which include *Excoecaria agallocha* (Gewa), *Bruguiera* sp. (Kankara), *Aegialitis rotundifolia*, *Avicennia* sp., *Sonneratia apetala* (Keora), *Ceriops decandra* (Goran), *Rhizophora apicultura*, *Xylocarpus* sp. and *Rhizophoraceae.* are of minor importance. Other species include *Nypa fruticans* (Golpata palm), *Phoenix paludosa*

(Hental) and *Carapa obovata* (Karim 1995). *Avicennia* is common in Indian part and *Excoecaria* in Bangladesh, while *Ceriops* is equally dominant in both regions (Chaudhuri and Choudhury 1994). Salinity controls the succession of mangrove species and other physiological growth and distribution of mangroves. These mangrove species have an immense role to play in the heavy metal sequestration (Sarkar et al. 2008; Rahman et al. 2009). Sundarbans is extensively rich in faunal species with more than 1500 recorded species which include vertebrates, chordates, invertebrates, protozoans, etc. Sundarbans is home to Royal Bengal Tiger (*Panthera tigris tigris*) (Banerjee 1998) whose presence on one hand makes sample collection life-threatening and on the other hand provides protection to the mangroves which are under threat from human interferences (Sen and Naskar 2003). Other notable species include Blyth's kingfisher (*Alcedo hercules*), yellow's monitor (*Varanus flavescens*), Irrawaddy dolphin (*Orcaella brevirostris*) and estuarine crocodile (*Crocodylus porosus*) (Siddiqi 2001). The Indian Sundarbans, despite being very rich in species, has a lower complexity and structure when compared with the Bangladesh Sundarbans, which may be due to variability of salinity in the two regions (FAO 2007). There is large freshwater inflow in the north-eastern part and elevated level of the ground surface. Sundarbans serves as a barrier from storm surges caused by tropical cyclones that recur on sub-decadal time scales in the Bay of Bengal, thus protecting both India and Bangladesh (Giri et al. 2007). Increased anthropogenic activities in the last few decades have resulted in rapid degradation of Sundarbans (Hussain and Acharya 1994). The biodiversity and biogeochemical processes in the Sundarbans have been seriously affected due to rapid changes in land use pattern, effluent discharges, reduced inflow of river water and run-off from agricultural land (Erwin 2009; Rahman et al. 2009).

3 Material and Methods

3.1 Sample Collection

Seven vertical profiles of sediments (cores) were collected from Sundarbans mangrove ecosystem. Three of these vertical sediment profiles were collected along the river Bidya in Indian Sundarbans (IS) and four along the river Passur from Bangladesh Sundarbans (BS) with the help of a plexiglass corer having a length of 1.5 m and an inner diameter of 7 cm. The choice of locations was based on prevailing vegetation cover, land-use pattern and presence of stress (anthropogenic). The collected core samples were sectioned at 5 cm uniform interval and immediately stored in an ice chest before being transported to the field laboratory. Due to the difference in the nature of substratum, the maximum length of the vertical profile of sediments varied between sampling locations (up to 45 cm). The vertical profiles of the sediment were uniformly sub-sectioned, packed in polythene bags and stored at 4 °C until analysis. Before extraction and analyses, the samples were lyophilized and powdered to a homogenized consistency. The three sediment cores collected from Indian Sundarbans (I-1, I-2 and I-3) were mostly dominated by *Avicennia* species, and core I-1 was at the backdrop of village (200 m) Moitrakhand and had sparse mangrove vegetation owing to anthropogenic activities. Core I-2 was collected from Treepnagar mangroves and core I-3 from Jharkhali mangroves; both sites were dominated by *Avicennia marina*. The salinity of estuarine river varied from 20 to 28 psu in the Indian part of the sampling transect. In Bangladesh, four sediment cores were collected along the Passur river transect; core B-1 and B-4 were collected from dense mangrove forests. *Heritiera fomes* and associated species (e.g. *Bruguiera* sp., *Nypa* sp.) dominated core B-1 site at Harbaria and *Excoecaria* species dominated the core B-4 site at Bhadra. Core B-2 was taken from the intertidal plain of Nilkomol near the jetty (anthropogenically influenced), and core B-3 was taken from the macrotidal plain of a tidal creek at Pashakhali. Core B-1 and B-4 fall in the oligohaline zone, while B-3 in the mesohaline and B-2 in the polyhaline zone of salinity (Siddiqi 2001).

3.2 *Chemicals and Solutions*

All the laboratory working solutions were prepared using Milli-Q water (Milli Q Plus system; Millipore, Bedford, MA) having a resistivity of 18.2 MΩ/cm. The dilutions of working standard solutions were prepared on daily basis before analyses. All the reagents were of analytical reagent grade (Merck), and NIST SRM 1646a (Estuarine sediment) was used as sediment reference material to validate our results. The precision and bias for reagent blanks and replicate samples were <5% of the mean analytical concentrations. The recovery rates in the standard reference material (NIST SRM 1646a) were around 95–105% for all elements.

3.3 *Total Sedimentary Phosphorus (TSP)*

Sediment samples were treated with 1 M $Mg(NO_3)_2$ solution and then combusted at 550 °C for 2 h to determine total sedimentary phosphorus in each sample. After combustion at 550 °C, the samples were brought to the room temperature and then agitated and extracted in 1 M HCl for 16 h at 25 °C (Zhang et al. 2004). Using spectrophotometric phospho-molybdenum blue method, the orthophosphate concentration in the extracts was determined (Zhang et al. 1999) with the help of HACH DR 2800 spectrophotometer.

3.4 *Sequential Extraction Procedure*

The sequential extraction procedure (SEDEX) used in this study is derived from the scheme developed by Ruttenberg (1992) for marine sediments and further modified by Zhang et al. (2004) to separate and quantify different forms of phosphorus in solid samples. Total sedimentary phosphorus was fractioned into five different geochemical P pools: (1) Exch-P which is adsorbed and exchangeable inorganic phosphorus; (2) Fe-P which is Fe-bound inorganic phosphorus; (3) Auth-P which is authigenic carbonate fluorapatite, biogenic apatite and calcium carbonate-bound inorganic phosphorus; (4) Det-P which is detrital apatite of igneous and metamorphic origin; and (5) Ref-Org-P which is refractory organic phosphorus. The target phase and the reaction mechanism of each extractant are shown in Table 4.1. 0.5 gram of dry sediment (<125 μm size) was taken in a 75 ml Oak Ridge centrifuge tubes, and 50 ml of $MgCl_2$ solution (pH adjusted to 8 with dilute NaOH solution) was added in step 1. The solution was agitated for 2 h at 25 °C followed by centrifugation and filtration using Millipore GF/F filters. The filtrate collected was analysed for dissolved phosphate, and the filter paper containing residue of sediments was used for subsequent extraction step. In order to improve the stability of phosphomolybdenum blue complex in $MgCl_2$ solution, the molybdate reagent was prepared as suggested by Zhang et al. (2004). In step 2, the residue from the previous step 1 was extracted with 50 ml BD solution (bicarbonate dithionite mixed solution, containing 0.11 M $Na_2S_2O_4$ and 0.11 M $NaHCO_3$ with final pH adjusted to 7) for 4 h. Amorphous $Fe(OH)_3$ and $Mn(OH)_3$ were reduced to soluble ferrous and Mn^{2+} with the help of dithionite which is a strong reducing agent leading to the release of Fe-bound phosphorus. Fe-bound P gets separated more effectively from $CaCO_3$-bound P with the help of BD solution than dithionate-citrate reagent as citrate interferes with spectrophotometric determination and acts as a strong complexing agent with a pK of 4.7 for both Fe and Ca (Hieltjes and Lijklema 1980). It is for this reason, citrate was not used as an extraction reagent for Fe-bound P (Jensen et al. 1998). The filtrate from step 2 was divided into two aliquots: (1) used for total Fe analyses and (2) used for P analysis using ICP-AES method. Analyses for total Fe were performed immediately after collection, whereas samples for P analysis were performed with ICP-AES analysis on decomposition of excess dithionite to SO_2 in 72 h. In step 2 for phosphate analysis, the ammonium molybdate solution was modified as done by Zhang et al. (2004). In step 3, to extract the authigenic carbonate fluorapatite, biogenic apatite and calcium carbonate-bound inorganic P, an acetate buffer solution of pH 4 was used which further prevents dissolution of detrital

Table 4.1 Protocol for the sequential extraction of sedimentary phosphorus into its various forms

Fraction	Extraction reagent	Extraction method
Exchangeable P (Sorb-P)	50 ml of 1 M $MgCl_2$ (pH 8)	Shaking at 25 ± 2 °C for 2 h
DB-extractable P (Fe-P)	Dithionite bicarbonate solution (final pH adjusted to 7) Followed by 0.5 M NaCl rinse	Shaking at 25 ± 2 °C for 4 h Shaking at 25 ± 2 °C for 2 h
Authigenic CFAP (Auth-P)	Acetate buffer with its pH adjusted to 4 Followed by 1 M $MgCl_2$	Shaking at 25 ± 2 °C for 6 h Shaking at 25 ± 2 °C for 2 h
Detrital P (Det-P)	1 M HCl	Shaking at 25 ± 2 °C for 16 h
Ref-Organic P (Org-P)	Sediment mixed with 1 M $Mg(NO_3)_2$ 1 M HCl	Ashed at 550 °C for 2 h Shaking at 25 ± 2 °C for 16 h

apatite. The sediment residues from step 2 were placed in 50 ml acetate buffer solution and agitated for 6 h at 25 °C. In step 3 during phosphate analysis of the extracted solution, the acetate in the solution was titrated by addition of HCl to individual samples in 1:1 ratio for optimal colour development. In order to recover any P re-adsorbed on the sediment surface during extraction, sample residues from step 2 and 3 were washed with 1 M $MgCl_2$ solution. Further for the presence of dissolved phosphate, $MgCl_2$ wash solutions were analysed using molybdate reagent as suggested by Zhang et al. (2004). In step 4, the sediment residue from step 3 was extracted using 1 M HCl solution to dissolve detrital apatite from sediments. For this, the sediment residues were agitated for 16 h at 25 °C in 50 ml 1 M HCl solution and filtered, and the filtrates were neutralised using NaOH and further analysed using molybdate reagent which was used as rinse solutions. In step 5, refractory organic P was obtained from the sediment residues of step 4 after wetting with 1 M $Mg(NO_3)_2$ and ashed for 2 h at 550 °C (Solorzano and Sharp 1980). Once the samples cooled to room temperature, they were extracted for 16 h at 25 °C in 50 ml of 1 M HCl solution. Samples were filtered, neutralised with dilute NaOH and analysed for dissolved phosphate using molybdate reagent used for rinse solutions.

3.5 *Reactive Fe Analyses*

Samples from step 2 were further analysed for reactive Fe analysis immediately after sample collection using atomic absorption spectroscope (Thermo Fisher, M series).

3.6 *Porewater Solute Analyses*

Porewater pH was analysed using HACH pH sension electrode. Porewater Cl^- and SO_4^{2-} were analysed using Metrohm Ion chromatograph, whereas porewater PO_4^{3-} was analysed using spectrophotometric phospho-molybdenum blue complex method (Murphy and Riley 1962). Salinity of porewater was calculated with the help of chloride content in it.

4 Results

4.1 *Spatial Variability of Phosphorus Fractions*

Total sedimentary phosphorus in the vertical profile for different sediment cores varied between 9.83 and 10.76 μmol/g (10.32 ± 0.28) for I-1; 10.12 and 11.20 μmol/g (10.66 ± 0.32) for I-2; 8.36 and 9.40 μmol/g (8.80 ± 0.37) for I-3;

8.80 and 11.30 μmol/g (10 ± 1) for B-1; 11 and 11.40 μmol/g (11.09 ± 0.15) for B-2; 9.40 and 11.20 μmol/g (10.21 ± 0.54) for B-3; and 10.40 and 11.20 μmol/g (10.76 ± 0.35) for B-4 (Fig. 4.2). Furthermore, depth-wise trend for total sedimentary phosphorus showed little change and followed irregular trends. It was observed that the total sedimentary phosphorus concentrations decreased down the length of the cores, except core B-1 and B-4, where it increased in an otherwise inconsistent manner. Mean concentration suggests dominance of Auth-P in I-1, I-2 and B-1 sediment cores followed by Det-P, whereas in sediment cores of I-3, B-2, B-3 and B-4, Det-P was the most dominant fraction followed by Auth-P suggesting any change in TSP is driven by these fractions. Org-P is third in abundance in the vertical profiles of all sediment cores, and there were no apparent depth trends in any of these sediment cores. The mean concentration for sorb-P and Fe-P in different sediment cores varied as I-1, 6.17%; I-2, 6.85%; I-3, 7.43%; B-1, 15.30%; B-2, 15.34%; B-3, 17.47%; and B-4, 15.92% with no apparent vertical trend (Fig. 4.3a, b) and minor variation in concentrations with depth in all sediment cores. A higher percentage of Fe-P was observed in the Bangladesh Sundarbans (BS) than the Indian Sundarbans (IS). The depth trends for various phosphorus fractions were relatively constant for all the sediment cores. Thus, in order to simplify the comparisons, average phosphorus concentration for the entire vertical sediment profiles was used. At locations where the

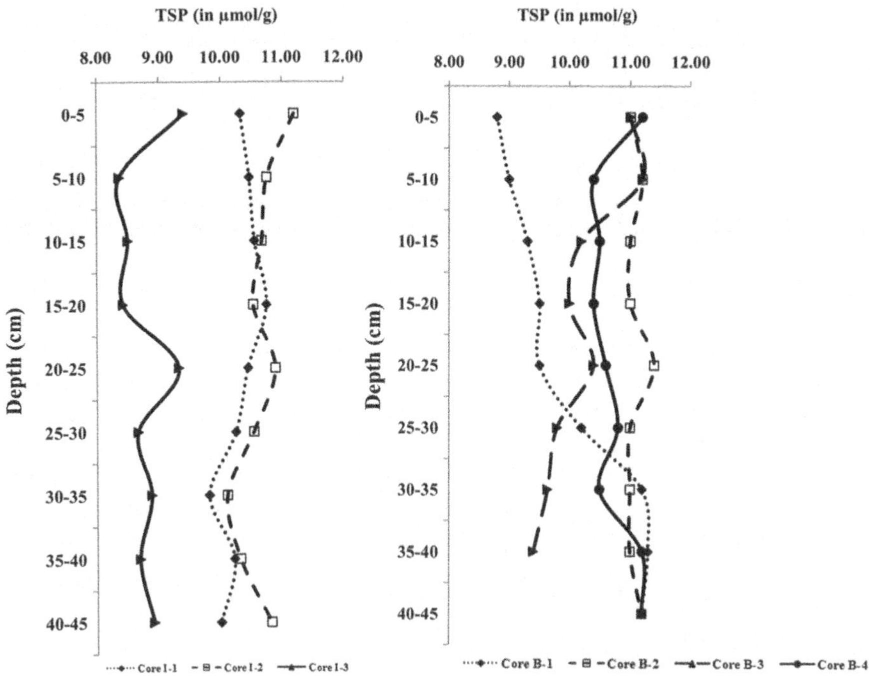

Fig. 4.2 Vertical profile of the total sedimentary phosphorus in sediment cores collected from the Indian and Bangladesh Sundarbans

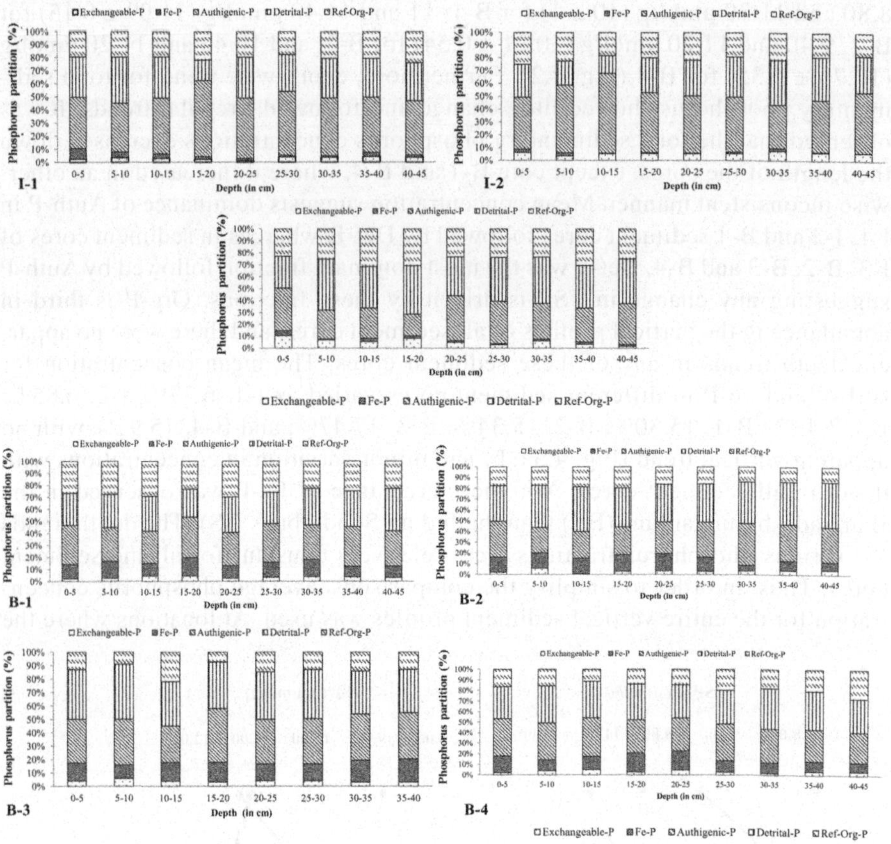

Fig. 4.3 (**a**) Depth-wise % occurrence of phosphorus species in the sediment profile of Indian Sundarbans. (**b**) Depth-wise % occurrence of phosphorus species in sediment profile of Bangladesh Sundarbans

salinity was high, there were higher mean TSP values observed but no consistent trend with varying salinity. Though there was a general declining trend in salinity with depth, porewater salinity showed increasing trend with depth in core I-3, and sediment cores from I-3 showed lowest TSP amongst all the studied cores. The mean bioavailable fraction which is the sum of Sorb-P, Fe-P and Org-P together accounted <33.7% in IS and <41.07% in BS of the TSP in the Sundarbans mangroves suggesting higher availability in BS. The mean concentrations for DB-Fe observed in the sediment cores were, I-1, 33.22 μmol/g; I-2, 32.70 μmol/g; I-3, 33.54 μmol/g; B-1, 30.27 μmol/g; B-2, 27.84 μmol/g; B-3, 43.64 μmol/g; and B-4, 31.36 μmol/g (Fig. 4.4a, b), and no apparent trend was observed.

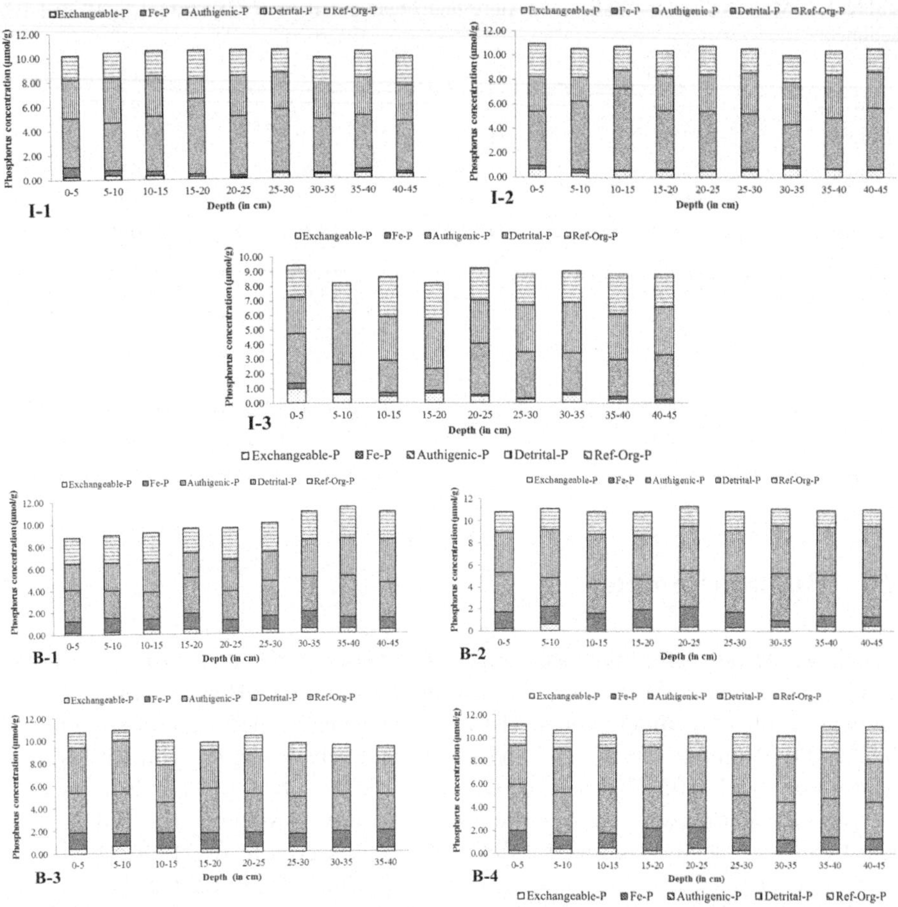

Fig. 4.4 (**a**) Depth-wise concentration of phosphorus (μmol/g) species in sediment profile of Indian Sundarbans. (**b**) Depth-wise concentration of phosphorus (μmol/g) species in sediment profile of Bangladesh Sundarbans

4.2 *Sediment Characteristics*

The sediment texture exhibits dominance of fine-grained fractions of silt and clay with sand being least in proportion (Table 4.2). The % sand content was higher in the Bangladesh Sundarbans (BS) when compared with Indian Sundarbans (IS), and the values ranged from 3.93% to 11.51% and 2.46% to 7.62%, respectively. Similar abundance of silt fraction was observed in BS when compared with IS, and the values were 67.20% to 80.63% and 64.14% to 73.92%, respectively. The clay content was higher for the IS (ranging between 19.60% and 31.22%) when compared with BS (ranging between 12.11% and 27.74%). Sediment carbonate

Table 4.2 Mean values for sediment texture and organic carbon (OC) content in IS and BS sediments

ID	Sand%	Silt%	Clay%	%OC
Core I-1	6.49	66.69	26.82	0.811
Core I-2	4.56	71.08	24.36	0.65
Core I-3	5.33	67.84	26.82	0.618
Core B-1	5.64	76.57	17.79	0.527
Core B-2	7.37	74.51	18.11	0.953
Core B-3	5.89	70.32	23.78	0.859
Core B-4	7.70	78.32	13.99	0.669

concentration ranged from 2.16% to 2.43% in IS and 1.58% to 3.70% in BS. The variability OC content in the vertical profile of sediment cores is shown in Table 4.2.

4.3 *Porewater Solutes*

The mean pH values for porewater ranged from 6.14 to 6.63 and 6.31 to 6.66 for IS and BS, respectively (Fig. 4.5a, b). No apparent variation in pH was observed in the vertical profile of sediment cores. The mean phosphate concentrations were 0.01, 0.02 and 0.01 mmol/L for I-1, I-2 and I-3, respectively, in the IS, while in BS, the values were 0.02, 0.02, 0.01 and 0.003 mmol/L for B-1, B-2, B-3 and B-4, respectively. Porewater phosphate concentration and salinity showed no clear and consistent trend with depth. The mean salinity values observed were 24.89, 31.93 and 25.86 ppt for I-1, I-2 and I-3, respectively, in IS. Similarly, in BS they were 2.28, 21.66, 14.90 and 2.43 ppt for B-1, B-2, B-3 and B-4, respectively (Fig. 4.5a, b). As per the salinity profile, core B-1 and B-4 belong to low salinity zones, while core B-3 falls under mesohaline zone, and all three sediment cores in IS and B-2 in BS fall under polyhaline category. There was no apparent trend in total sedimentary phosphorus distribution observed with respect to salinity. The mean sulphate concentration in porewater ranged between 14.75 and 21.01 mmol/L for IS and 5.12 and 18.12 mmol/L for the BS (Fig. 4.5a, b). An increase in porewater sulphate concentration was observed along the salinity gradient, but there was no consistent trend observed along the depth profile. Reactive Fe (Fe extracted in DB fraction) content in the sediments ranged from 24.10 to 37.43 and 16.59 to 51.15 μmol/g in IS and BS sediments, respectively (Fig. 4.6).

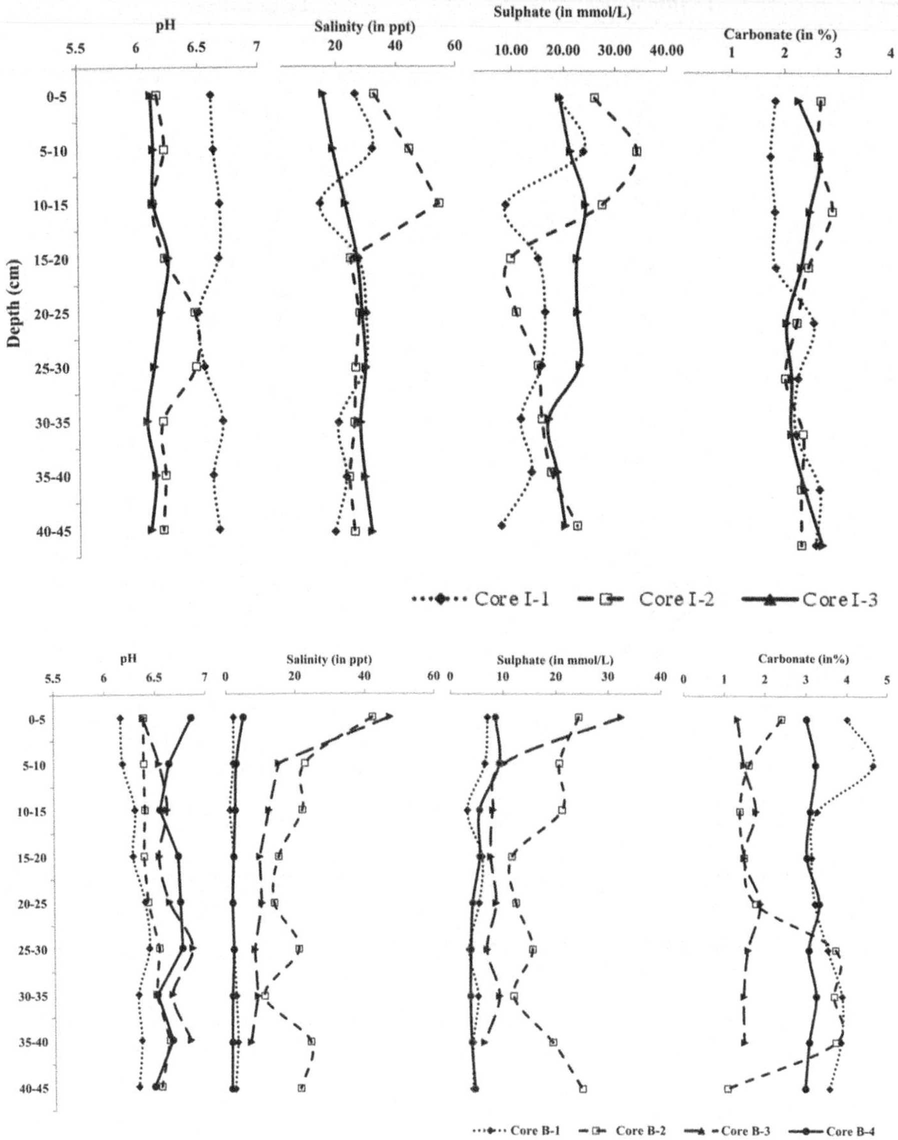

Fig. 4.5 (**a**) Vertical profiles of porewater pH, salinity, sulphate and sedimentary carbonate content in the cores collected from Indian Sundarbans. (**b**) Vertical profiles of porewater pH, salinity, sulphate and sedimentary carbonate content in the cores collected from Bangladesh Sundarbans

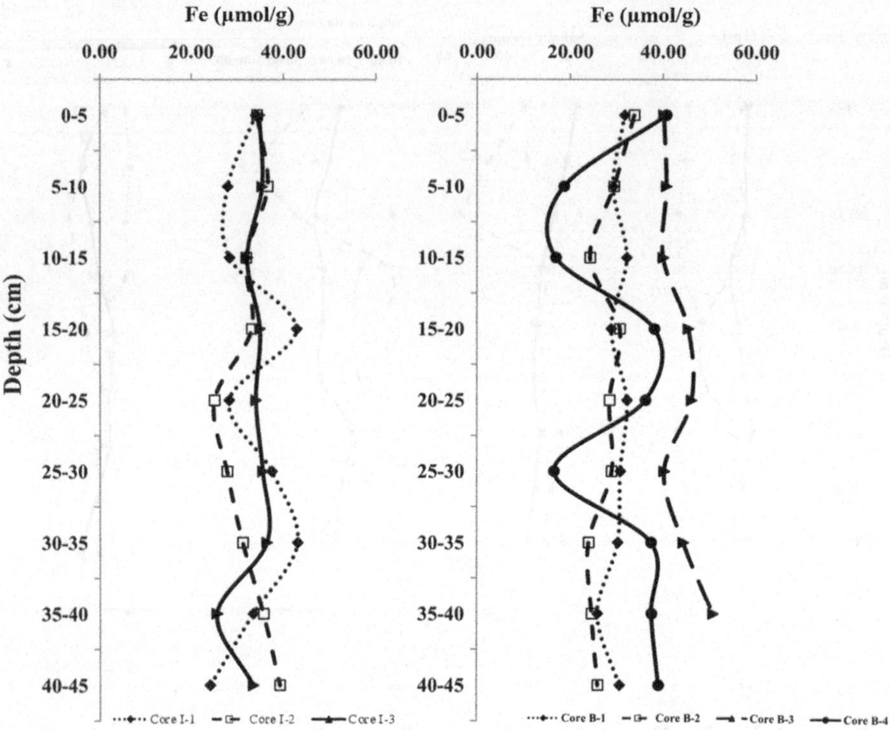

Fig. 4.6 Depth-wise variation of reactive iron concentration (µmol/g) in Indian and Bangladesh Sundarbans sediments

5 Discussion

5.1 Sedimentary Phosphorus Species

Speciation of the sedimentary P depends on factors like the sediment composition and overlying water column chemistry. Total sedimentary phosphorus in the sediments collected from the Sundarbans correlates well with the range reported from the global mangrove ecosystems (3.22 to 51.61 µmol/g; Alongi et al. 1992). A few studies have been conducted in the Indian estuarine and mangrove settings on P speciation in sediments using different extraction protocols. For example, TSP ranged from 14.80 to 17.09 µmol/g for Pichavaram mangroves (Prasad and Ramanathan 2010); 10.29 to 94.80 µmol/g in the Cochin estuary (Renjith et al. 2011); and 11.51 to 28.54 µmol/g in the Hooghly estuary (Vaithiyanathan et al. 1993). The average TSP content in the Sundarbans mangrove sediments was lower than reported TSP contents of Pichavaram mangroves, Cochin estuary and Hooghly estuary (Table 4.3). However, the speciation of the sedimentary P in these mangrove and estuarine sediments was

Table 4.3 The range of total sedimentary phosphorus (TSP) (μmol/g) reported in various Indian estuaries and mangroves

Location	TSP (μmol/g)	References
Indian Sundarbans	8.36–11.2	This study
Bangladesh Sundarbans	8.80–11.40	This study
Pichavaram, Tamil Nadu	14.80–17.09	Prasad and Ramanathan (2010)
Ganges estuary, West Bengal	11.51–28.54	Vaithiyanathan et al. (1993)
Cochin estuary, Kerala	10.29–94.80	Renjith et al. (2011)
Global range	3.22–51.61	Slomp (2011)

characteristically different. The speciation pattern for Sundarbans sediments is discussed further.

5.1.1 Detrital P (Det-P) and Exchangeable P (Sorb-P)

The mean % concentration for detrital fraction was abundant in core I-3, B-2, B-3 and B-4 (<37.5%). High levels of Det-P indicate fluvial contribution in the sediments and not available for release under normal conditions (Hou et al. 2009). Sorp-P accounts for the easily available/exchangeable P in the sediments may be regulated by the Auth-P as a result of low pH content which helps in the release of P from the sediments.

5.1.2 Authigenic CFA

Authigenic P represents one of the largest fractions of TSP at both IS and BS (Fig. 4.3a, b). Authigenic P constitutes authigenic CFA + biogenic apatite + $CaCO_3$ reservoir and is considered authigenic as it is formed by the exclusion of reactive phosphorus from the overlying waters or the sedimentary porewater. Authigenic P reservoir increased with depth which could be attributed to the formation of authigenic CFA, and higher Auth-P in the surface sediments may have resulted from non-CFA background phase as finely divided biogenic apatite, and P associated with smectite (Ruttenberg and Berner 1993) which has been reported in the Sundarbans sediments (Datta and Subramanian 1997; Rajkumar et al. 2012). Several studies have reported the similar formation of authigenic CFA in coastal and marine sediments (Ruttenberg and Berner 1993; Slomp et al. 1996; Louchouarn et al. 1997; Hartzell et al. 2010). Our study also suggests "sink switching" or diagenetic redistribution from organically bound or Fe-bound P to authigenic P in the Sundarbans sediments. The depth trends for Auth-P suggest sink switching, but organic P alone may not be responsible for the authigenic CFA formation as the depth trends do not perfectly match each other. This indicates that there are processes other than mineralization of the organically bound P such as Fe-bound P (e.g. vivianite) operating at different rates for authigenic CFA formation (Ruttenberg and Berner 1993; Anderson et al. 2001; Filippelli 2001). Vivianite formation in sulphate-rich coastal sediments has been found to take place by the conversion of available sulphate to Fe sulphides (Slomp 2011). There is a characteristic decline in Fe-P concentration with depth except for core B-1 and B-3, whereas Org-P concentrations showed no apparent declining trend but still allow for diagenetic sink switching. Porewater data for phosphate showed no apparent trend owing to mixing and disturbance in the region. The porewater studies were used in the past to infer processes occurring in the solid phase (Goldhaber et al. 1977; Froelich et al. 1982; Jahnke and Christiansen 1989; Ruttenberg and Berner 1993; Mort et al. 2010). Auth. CFA formation is known to be triggered by the precipitation of a precursor Ca-P phase and its subsequent conversion to apatite (Slomp 2011). Porewater diagnostic studies for Auth. CFA formation may not be used in the present study as steady-state assumption cannot be made in the deltaic environment (Ruttenberg and Berner 1993). Previous study (Vaithiyanathan et al. 1993) regarding P distribution in the Ganges estuary also reported dominance of Ca-associated P in the estuarine sediments. However, the fate of the Ca-P form in view of its reactive or stable nature needs to be assessed (Golterman 1988; Moutin et al. 1993; Gomez et al. 1999). It has been found that for marine systems, apatites represent an insoluble and hence stable pool of phosphorus (Andrieux-Loyer and Aminot 2001). Since seawater is considered to be largely undersaturated or in close saturation with CFAP, it can be concluded that formation and dissolution of CFAP in seawater are both possible (Atlas and Pytkowicz 1977; Faul et al. 2005; Lyons et al. 2011; Oxmann and Schwendenmann 2015).

5.1.3 Organic P

Coastal sediments are generally rich in organic matter which acts as a major carrier of reactive P (Slomp 2011). Except in core B-2, Org-P showed no apparent trend with depth and was quantitatively third most abundant sink for P in the Sundarbans sediments. Previous studies in the past have reported both decline in Org-P with depth (Ruttenberg and Berner 1993; Cha et al. 2005) and no apparent decline with depth (Slomp et al. 1996; Van der Zee et al. 2002; Hartzell et al. 2010) in the coastal sediments. Organic P content in coastal and marine sediments has been poorly characterized and may be found associated with high molecular weight organic matter, phosphonates and phospholipids in sediments which determine its biological availability (Reitzel et al. 2007; Slomp 2011). Besides, bacterial decomposition also helps in sequestering P as polyphosphates in oxidised sediments and releases P as DIP in anoxic sediments (Gächter et al. 1988). Anoxic conditions ensure release of polyphosphates which results in elevated concentrations of DIP in porewaters and authigenic apatite formation via sulphur-oxidizing bacteria (Schulz and Schulz 2005). Further studies targeting Org-P forms need to be conducted in order to better understand the burial of P in coastal sediments.

5.1.4 DB-Extractable P (Fe-P)

Fe-bound P is considered as an important sedimentary sink for P and shows consistent decline with depth at all sites and salinities except in core B-3. Both crystalline and amorphous forms of Fe oxides represent DB-extractable P in the sediments. The sulphide content in the sediments has been reported to influence the ability of sediments to take up P as the sulphate may either compete with phosphate for anion sorption sites or sulphides from sulphate reduction may bind with Fe in anoxic sediments and prevent ferrous phosphate compound formation within the anoxic sediments (Caraco et al. 1989). The low values of Fe-bound P in the Sundarbans sediments can be due to any of the above two mechanisms for Fe binding and its non-availability for Fe-P compound formation. In order for ferrous phosphate mineral to precipitate in iron-rich mangrove sediments, there needs to be a limited supply of sulphides and greater reduction of Fe oxides to Fe (Burns 1997; Hartzell et al. 2010). Fe/P ratio varied from 135.13 to 171.85 for IS, whereas for BS it varied from 95.75 to 133.37 indicating that Fe-rich matrix exhibits lesser adsorption capacity for P (Andrieux-Loyer and Aminot 2001). Also, majority of the iron is present in immobile form (i.e. bound to silicates) which decreases its association with P (Raiswell and Canfield 1998). The higher DB-Fe/P ratio (>2.51) in the Sundarbans sediments indicates that the trapping efficiency of the reactive Fe (DB-Fe) is impeded (Moutin et al. 1993; Andrieux-Loyer and Aminot 2001). Further studies are needed to ascertain if pyrite formation which has been reported in the region (Roychoudhury et al. 2003) is iron limited in the Sundarbans sediments which will help us in understanding availability of Fe to form ferrous phosphates.

5.2 *Bioavailable Phosphorus*

P speciation helps in establishing the possible bioavailability of P in an ecosystem (Penn et al. 1995; Andrieux-Loyer and Aminot 1997; Prasad and Ramanathan 2010). Mean sum of Auth-P and Det-P accounts for 66.04–73.67% in IS and 58.52–67.74% in BS in the sedimentary pool of Sundarbans, and they are not available under normal physico-chemical conditions in the sedimentary settings. Exchangeable P, Fe-bound P and Ref-Organic P may form part of the bioavailable fraction (Hou et al. 2009) as with variation in redox conditions, Fe-bound P can be reduced and released from the sediments, whereas Ref-Org-P could become bioavailable by microbial remineralization (Jensen and Thamdrup 1993; Andrieux-Loyer and Aminot 1997; Rozan et al. 2002; Coelho et al. 2004; Álvarez-Rogel et al. 2007). In the sedimentary pool of Sundarbans, 26.74–33.70% and 30.43–41.07% phosphorus was found to be bioavailable for IS and BS, respectively. Thus, BS sediments were found to be an important autochthonous source of P as compared to IS which poses a higher risk of release of P from the sediments to the overlying water column. However, further research is needed to understand the role of different fractions in internal loading of P in mangrove ecosystem.

6 Conclusions

Sequential extraction technique along with porewater analyses was used to study the dominance and distribution of specific phosphorus-containing phases in the sedimentary environment of Sundarbans mangroves. Total sedimentary phosphorus was homogenously distributed in the Sundarbans sediments and varied between 8.36 and 11.20 μmol/g in the Indian Sundarbans and between 8.80 and 11.40 μmol/g in the Bangladesh Sundarbans, and solid-phase P forms varied slightly amongst sites. Det-P was the most dominant form of TSP closely followed by Auth-P highlighting terrigenous input and showing trend for diagenetic sink switching of P form. Ref-Org-P was the third most abundant pool for phosphorus in the sedimentary form, and P_{bio} accounted for less than 41.07% of TSP. Vertical profile of DB-P (reactive P species) along with dissolved porewater phosphate indicated redistribution of P into Auth-P as a result of diagenetic sink switching. It is suggested that adsorption, dissolution and precipitation processes play a part in distributing the sedimentary P into different forms/fractions. Besides these processes, role of particle size and mixing due to processes such as bioturbation may control the elevated dissolved phosphate concentration from the Sundarbans sediments which needs to be further elaborated in the future studies to understand the benthic and pelagic P processes.

Acknowledgement The authors would like to acknowledge the Asia-Pacific Network for Global Change Research, Japan, for funding this study (Project ref. no. ARCP2012-07CMY-Ramanathan; previously ARCP2011-17NMY-Mathukumalli).

References

Abdallah MAM (2011) Potential for internal loading by phosphorus based on sequential extraction of surficial sediment in a shallow Egyptian Lake. Environ Monit Assess 178(1–4):203–212

Alongi DM (1996) The dynamics of benthic nutrient pools and fluxes in tropical mangrove forests. J Mar Res 54(1):123–148

Alongi DM, Boto KG, Robertson AI (1992) Nitrogen and phosphorus cycles. In: Robertson AI, Alongi DM (eds) Tropical mangrove ecosystems. Coastal and estuarine studies. American Geophysical Union, Washington, DC, pp 251–292

Álvarez-Rogel J, Jiménez-Cárceles FJ, Egea-Nicolás C (2007) Phosphorus retention in a coastal salt marsh in SE Spain. Sci Total Environ 378(1–2):71–74

Anderson LD, Delaney ML, Faul KL (2001) Carbon to phosphorus ratios in sediments: implications for nutrient cycling. Glob Biogeochem Cycles 15(1):65–79

Andrieux-Loyer F, Aminot A (1997) A two-year survey of phosphorus speciation in the sediments of the Bay of Seine (France). Cont Shelf Res 17(10):1229–1245

Andrieux-Loyer F, Aminot A (2001) Phosphorus forms related to sediment grain size and geochemical characteristics in French coastal areas. Estuar Coast Shelf Sci 52(5):617–629
Atlas EL, Pytkowicz RM (1977) Solubility behavior of apatites in seawater. Limnol Oceanogr 22:290–300
Banerjee A (1998) Environment, population and human settlements of Sunderban Delta, 1st edn. Concept Publishing Company, New Delhi
Benitez-Nelson CR (2000) The biogeochemical cycling of phosphorus in marine systems. Earth Sci Rev 51(1–4):109–135
Burns SJ (1997) Early diagenesis in Amazon fan sediments. In: Flood RD, Piper DJW, Klaus A, Peterson LC (eds) Proceedings of the ocean drilling program scientific results. Ocean Drilling Program, College Station, pp 497–504
Caraco NF, Cole JJ, Likens GE (1989) Evidence for sulphate-controlled phosphorus release from sediments of aquatic systems. Nature 341(6240):316–318
Cardoso PG, Lillebø AI, Pardal MA, Ferreira SM, Marques JC (2002) The effect of different primary producers on *Hydrobia ulvae* population dynamics: a case study in a temperate intertidal estuary. J Exp Mar Biol Ecol 277(2):173–195
Cha HJ, Lee CB, Kim BS, Choi MS, Ruttenberg KC (2005) Early diagenetic redistribution and burial of phosphorus in the sediments of the southwestern East Sea (Japan Sea). Mar Geol 216(3):127–143
Chaudhuri AB, Choudhury A (1994) Mangroves of the Sundarban, India, vol 1. The IUCN Wetlands Programme, IUCN, Bangkok
Coelho JP, Flindt MR, Jensen HS, Lillebø AI, Pardal MA (2004) Phosphorus speciation and availability in intertidal sediments of a temperate estuary: relation to eutrophication and annual P-fluxes. Estuar Coast Shelf Sci 61(4):583–590
Coleman JM (1969) Brahmaputra River: channel processes and sedimentation. Sediment Geol 3(2–3):129–239
Datta DK, Subramanian V (1997) Texture and mineralogy of sediments from Ganga-Brahmaputra-Meghna river system in the Bengal Basin, Bangladesh and their environmental implications. Environ Geol 30(3–4):181–188
Duhamel S, Nogaro G, Steinman AD (2017) Effects of water level fluctuation and sediment–water nutrient exchange on phosphorus biogeochemistry in two coastal wetlands. Aquat Sci 79:57–72
Eijsink LM, Krom MD, Herut B (2000) Speciation and burial flux of phosphorus in the surface sediments of the eastern Mediterranean. Am J Sci 300(6):483–503
Elahi KM, Das SC, Sultana S (1998) Geography of coastal environment: a study of selected issues. In: Bayes A, Mahammad A (eds) Bangladesh at 25, An analytical discourse on development. The University Press Limited, Dhaka, pp 336–368
Emsley J (1980) The phosphorus cycle, The handbook of environmental chemistry, vol 1(part A). Springer-Verlag, Berlin, pp 147–167
Erwin KL (2009) Wetlands and global climate change: the role of wetland restoration in a changing world. Wetl Ecol Manag 17(1):71–84
Fang TH, Chen JL, Huh CA (2007) Sedimentary phosphorus species and sedimentation flux in the East China Sea. Cont Shelf Res 27(10–11):1465–1476
FAO (Food and Agricultural Organization) (2007) The world's Mangroves 1980–2005, vol 12. FAO, Rome
Faul KL, Paytan A, Delaney ML (2005) Phosphorus distribution in sinking oceanic particulate matter. Mar Chem 97:307–333
Filippelli GM (2001) Carbon and phosphorus cycling in anoxic sediments of the Saanich Inlet, British Columbia. Mar Geol 174(1–4):307–321
Föllmi KB (1996) The phosphorus cycle, phosphogenesis and marine phosphate-rich deposits. Earth Sci Rev 40(1–2):55–124
Froelich PN, Bender ML, Luedtke NA, Heath GR, DeVries T (1982) The marine phosphorus cycle. Am J Sci 282(4):474–511

Gächter R, Meyer JS, Mares A (1988) Contribution of bacteria to release and fixation of phosphorus in lake-sediments. Limnol Oceanogr 33(6, part 2):1542–1558

Gireeshkumar TR, Deepulal PM, Chandramohanakumar N (2013) Phosphorous speciation in surface sediments of the Cochin estuary. Environ Monit Assess 185(3):2535–2552

Giri C, Pengra B, Zhiliang Z, Singh A, Tieszen LL (2007) Monitoring mangrove forest dynamics of the Sundarbans in Bangladesh and India using multitemporal satellite data from 1973 to 2000. Estuar Coast Shelf Sci 73(1–2):91–100

Gomez E, Durillon C, Rofes G, Picot B (1999) Phosphate adsorption and release from sediments of brackish lagoons: pH, O2 and loading influence. Water Res 33(10): 2437–2447

Goldhaber MB, Aller RC, Cochran JK, Rosenfeld JK, Martens CS, Berner RA (1977) Sulfate reduction, diffusion, and bioturbation in Long Island Sound sediments; report of the FOAM Group. Am J Sci 277(3):193–237

Golterman HL (1988) The calcium- and iron bound phosphate phase diagram. Hydrobiologia 159(2):149–151

Goodbred SL Jr, Kuehl SA, Steckler MS, Sarker MH (2003) Controls on facies distribution and stratigraphic preservation in the Ganges– Brahmaputra delta sequence. Sediment Geol 155(3–4):301–316

Gopal B, Chauhan M (2006) Biodiversity and its conservation in the Sundarban Mangrove Ecosystem. Aquat Sci 68(3):338–354

Hartzell JL, Jordan TE, Cornwell JC (2010) Phosphorus burial in sediments along the salinity gradient of the Patuxent River, a sub-estuary of the Chesapeake Bay (USA). Estuar Coasts 33(1):92–106

Hieltjes AHM, Lijklema L (1980) Fractionation of inorganic phosphates in calcareous sediments. J Environ Qual 9(3):405–407

Hou LJ, Liu M, Yang Y, Ou DN, Lin X, Chen H, Xu SY (2009) Phosphorus speciation and availability in intertidal sediments of the Yangtze Estuary, China. Appl Geochem 24(1):120–128

Hussain Z, Acharya G (eds) (1994) Mangroves of the Sundarbans, Bangladesh, vol 2. IUCN—The World Conservation Union, Bangkok, p 257

Islam MS (2001) Sea-level changes in Bangladesh: the last ten thousand years. Asiatic Society of Bangladesh, Asiatic Civil Military Press, Dhaka, p 185

Islam SN (2003) Sustainable eco-tourism as a practical management policy. AH Development Publishing House, Dhaka, pp 34–47

Jahnke RA, Christiansen MB (1989) A free-vehicle benthic chamber instrument for sea floor studies. Deep Sea Res A Oceanogr Res Papers 36(4):625–637

Jensen HS, Thamdrup B (1993) Iron-bound phosphorus in marine sediments as measured by bicarbonate–dithionite extraction. Hydrobiologia 253(1–3):47–59

Jensen HS, McGlathery KJ, Marino R, Howarth RW (1998) Forms and availability of sediment phosphorus in carbonate sand of Bermuda seagrass beds. Limnology and Oceanography, 43(5):799–810

Joseph MM, Ratheesh Kumar CS, Renjith KR, Gireesh Kumar TR, Chandramohanakumar N (2010) Phosphorus fractions in the surface sediments of three mangrove systems of southwest coast of India. Environ Earth Sci 62(6):1209–1218

Junk WJ, Brown M, Campbell IC, Finlayson M, Gopal B, Ramberg L, Warner BG (2006) The comparative biodiversity of seven globally important wetlands: a synthesis. Aquat Sci 68(3):400–414

Karim A (1995) Mangrove silviculture, Vol. 1. Draft report of FAO/ UNDP project BGD/84/056-integrated resource development of the Sundarbans reserve forest, Khulna

Katebi MNA (2001) Sundarbans and forestry. In: Haider R (ed) Cyclone'91—an environmental and perceptional study. BCAS, Dhaka, pp 79–100

Kumar A, Ramanathan AL (2015) Speciation of selected trace metals (Fe, Mn, Cu and Zn) with depth in the sediments of Sundarban mangroves: India and Bangladesh. J Soils Sediments 15(12):2476–2486

Lebo ME (1991) Particle-bound phosphorus along an urbanized coastal plain estuary. Mar Chem 34(3–4):225–246

Louchouarn P, Lucotte M, Duchemin E, de Vernal A (1997) Early diagenetic processes in recent sediments of the Gulf of St-Lawrence: phosphorus, carbon and iron burial rates. Mar Geol 139(1–4):181–200

Lyons G, Benitez-Nelson CR, Thunell RC (2011) Phosphorus composition of sinking particles in the Guaymas Basin, Gulf of California. Limnol Oceanogr 56(3):1093–1105

Mandal AK, Ghosh RK (1989) Sunderban—a socio bio-ecological study, 1st edn. Bookland Pvt. Ltd., Calcutta

Marchand C, Lallier-Verges E, Baltzer F, Alberic AP, Cossa D, Baillif P (2006) Heavy metals distribution in mangrove sediments along the mobile coastline of French Guiana. Mar Chem 98(1):1–17

Mitra A, Gangopadhyay A, Dube A, Schmidt ACK, Banerjee K (2009) Observed changes in water mass properties in the Indian Sundarbans (northwestern Bay of Bengal) during 1980–2007. Curr Sci 97(10):1445–1452

Morgan JP, McIntire WG (1959) Quaternary geology of the Bengal Basin, East Pakistan and India. Bull Geol Soc Am 70(3):319–342

Mort HP, Slomp CP, Gustafsson BG, Andersen TJ (2010) Phosphorus recycling and burial in Baltic Sea sediments with contrasting redox conditions. Geochim Cosmochim Acta 74(4):1350–1362

Moutin T, Picot B, Ximenes MC, Bontoux J (1993) Seasonal variations of P compounds and their concentrations in two coastal lagoons (Herault, France). Hydrobiologia 252(1):45–59

Murphy J, Riley JP (1962) A modified single solution method for the determination of phosphate in natural waters. Anal Chim Acta 27:31–36

Nixon SW, Ammerman JW, Atkinson LP, Berounsky VM, Billen G, Boicourt WC, Boynton WR, Church TM, Ditoro DM, Elmgren R, Garber JH, Giblin AE, Jahnke RA, Owens NJP, Pilson MEQ, Seitzinger SP (1996) The fate of nitrogen and phosphorus at the land–sea margin of the North Atlantic Ocean. Biogeochemistry 35(1):141–180

Oxmann JF, Schwendenmann L (2015) Authigenic apatite and octacalcium phosphate formation due to adsorption–precipitation switching across estuarine salinity gradients. Biogeosciences 12(3):723–738

Paludan C, Morris JT (1999) Distribution and speciation of phosphorus along a salinity gradient in intertidal marsh sediments. Biogeochemistry 45(2):197–221

Penn MR, Auer T, Van Orman EL, Korienek JJ (1995) Phosphorus diagenesis in lake sediments: investigations using fractionation techniques. Mar Freshw Res 46(1):89–99

Prasad MBK, Ramanathan AL (2010) Characterization of phosphorus fractions in the sediments of a tropical intertidal mangrove ecosystem. Wetl Ecol Manag 18(2):165–175

Psenner R, Puckso R (1988) Phosphorus fractionation: advantages and limits of the method for the study of sediment P origins and interactions. Arch Hydrobiol Adv Limnol 30:43–59

Rahman MM, Chongling Y, Islam KS, Haoliang L (2009) A brief review on pollution and ecotoxicologic effects on Sundarbans mangrove ecosystem in Bangladesh. Int J Environ Eng 1(4):369–383

Raiswell R, Canfield DE (1998) Sources of iron for pyrite formation in marine sediments. Am J Sci 298(3):219–245

Rajkumar K, Ramanathan AL, Behera PN, Chidambaram S (2012) Preliminary studies on the characterization of clay minerals in the Sundarbans mangrove core sediments, West Bengal, India. Arab J Geosci 7(2):537–544

Ranjan RK, Ramanathan AL, Chauhan R, Singh G (2011) Phosphorus fractionation in sediments of the Pichavaram mangrove ecosystem, south-eastern coast of India. Environ Earth Sci 62(8):1779–1787

Reitzel K, Ahlgren J, DeBrabandere H, Waldebäck M, Gogoll A, Tranvik L, Rydin E (2007) Degradation rates of organic phosphorus in lake sediment. Biogeochemistry 82(1):15–28

Renjith KR, Chandramohanakumar N (2007) Geochemical characteristics of surficial sediments in a tropical estuary, south-west India. Chem Ecol 23(4):337–343

Renjith KR, Chandramohanakumar N, Joseph MM (2011) Fractionation and bioavailability of phosphorus in a tropical estuary, Southwest India. Environ Monit Assess 174(1–4):299–312

Rogers KG, Goodbred SL Jr, Mondal DR (2013) Monsoon sedimentation on the 'abandoned' tide-influenced Ganges-Brahmaputra delta plain. Estuar Coast Shelf Sci 131:297–309

Roychoudhury AN, Kostka JE, VanCappellen P (2003) Pyritization: a palaeoenvironmental and redox proxy reevaluated. Estuar Coast Shelf Sci 57(5–6):1183–1193

Rozan TF, Taillefert M, Trouwborst RE, Glazer BT, Ma S, Herszage J, Valdes LM, Price KS, Luther GW III (2002) Iron–sulfur–phosphorus cycling in the sediments of a shallow coastal bay: implications for sediment nutrient release and benthic macroalgal blooms. Limnol Oceanogr 47(5):1346–1354

Ruttenberg KC (1992) Development of a sequential extraction method for different forms of phosphorus in marine sediments. Limnol Oceanogr 37(7):1460–1482

Ruttenberg KC (1993) Reassessment of the oceanic residence time of phosphorus. Chem Geol 107(3–4):405–409

Ruttenberg KC, Berner RA (1993) Authigenic apatite formation and burial in sediments from non-upwelling, continental margin environments. Geochim Cosmochim Acta 57(5):991–1007

Sanudo-Wilhelmy SA, Kustka AB, Gobler CJ, Hutchins DA, Yang M, Lwiza K, Burns J, Capone DG, Raven JA, Carpenter EJ (2001) Phosphorus limitation of nitrogen fixation by *Trichodesmium* in the central Atlantic Ocean. Nature 411(6833):66–69

Sarkar SK, Cabral H, Chatterjee M, Cardoso I, Bhattacharya AK, Satpathy KK, Alam MA (2008) Biomonitoring of heavy metals using the bivalve molluscs in Sundarban mangrove wetland, northeast coast of Bay of Bengal (India): possible risks to human health. Clean Soil Air Water 36(2):187–194

Schenau SJ, De Lange GJ (2001) Phosphorus regeneration vs burial in sediments of the Arabian Sea. Mar Chem 75(3):201–217

Schulz HN, Schulz HD (2005) Large sulfur bacteria and the formation of phosphorite. Science 307(5708):416–418

Selvam V, Ravichandran KK, Gnanappazham L, Navamuniyammal M (2003) Assessment of community-based restoration of Pichavaram mangrove wetland using remote sensing data. Curr Sci 85(6):794–798

Sen N, Naskar KR (2003) Algal flora of Sundarbans Mangals. Daya Publishing House, New Delhi, p 318

Siddiqi NA (2001) Mangrove forest in Bangladesh. Institute of Forestry and Environmental Science, University of Chittagong, Chittagong, p 201

Slomp CP (2011) Phosphorus cycling in the estuarine and coastal zones: sources, sinks, and transformations. In: Wolanski E, McLusky DS (eds) Treatise on estuarine and coastal science, vol 5. Academic Press, Waltham, pp 201–229

Slomp CP, Epping EHG, Helder W, van Raaphorst W (1996) A key role for iron-bound phosphorus in authigenic apatite formation in North Atlantic continental platform sediments. J Mar Res 54(6):1179–1205

Solorzano L, Sharp JH (1980) Determination of total dissolved phosphorus and particulate phosphorus in natural waters. Limnol Oceanogr 25(4):754–758

Song JM (2010) Biogeochemical processes of biogenic elements in China marginal seas. Springer, Berlin, p 676

Twilley RR, Rivera-Monroy VH, Rovai AS, Castañeda-Moya E, Davis S (2019) Mangrove biogeochemistry at local to global scales using ecogeomorphic approaches. In: Coastal wetlands. Elsevier, Amsterdam, pp 717–785

Umitsu M (1997) Landforms and floods in the Ganges delta and coastal lowland of Bangladesh. Mar Geod 20(1):76–87

Untawale AG (1987) Conservation in Indian mangroves a national perspective. In: Rao TSS, Natarajan R, Desai BN, Narayanaswamy G, Bhat SR (eds) A special collection of papers to felicitate Dr. S. Z. Qasim on his sixtieth birthday. National Institute of Oceanography, India, Dona Paula, pp 85–104

Vaithiyanathan P, Jha PK, Subramanian V (1993) Phosphorus distribution in the sediments of the Hooghly (Ganges) estuary, India. Estuar Coast Shelf Sci 37(6):603–614

Van Beusekom JEE, De Jonge VN (1997) Transformation of phosphorus in the Wadden Sea: apatite formation. Deutsche Hydrographische Zeitschrift 49:297–305

Van der Zee C, Slomp CP, Van Raaphorst W (2002) Authigenic P formation and reactive P burial in sediments of the Nazare canyon on the Iberian margin (NE Atlantic). Mar Geol 185(3–4):379–392

Vepraskas MJ, Faulkner SP (2001) Redox chemistry of hidrics soils. In: Richarson JL, Vepraskas MJ (eds) Wetland soils. Lewis Publishers, Boca Raton, FL, pp 85–107

Yang B, Liu SM, Wu Y, Zhang J (2016) Phosphorus speciation and availability in sediments off the eastern coast of Hainan Island, South China Sea. Cont Shelf Res 118:111–127

Zabel M, Dahmke A, Schulz HD (1998) Regional distribution of diffusive phosphate and silicate fluxes through the sediment-water interface: the eastern South Atlantic. Deep-Sea Res I Oceanogr Res Pap 45(2–3):277–300

Zhang JZ, Fischer CJ, Ortner PB (1999) Optimization of performance and minimization of silicate interference in continuous flow phosphate analysis. Talanta 49(2):293–304

Zhang JZ, Fischer CJ, Ortner PB (2004) Potential availability of sedimentary phosphorus to sediment resuspension in Florida Bay. Glob Biogeochem Cycles 18(4):GB4008, 1–14. https://doi.org/10.1029/2004GB002255

Zhang JZ, Fischer CJ, Ortner PB (1999) Optimization of performance and minimization of silicate interference in continuous flow phosphate analysis. Talanta 49(2):293–304

Zhuang W, Gao X, Zhang Y, Xing Q, Tosi L, Qin S (2014) Geochemical characteristics of phosphorus in surface sediments of two major Chinese mariculture areas: the Laizhou Bay and the coastal waters of the Zhangzi Island. Mar Pollut Bull 83(1):343–351

Chapter 5
Phytoplankton Ecology in Indian Coastal Lagoons: A Review

Sambit Singh, Tamoghna Acharyya, and Anu Gopinath

Abstract Phytoplankton composition, diversity, and distribution pattern serve as efficient bioindicators in determining a lagoon's health. In this regard, this review was carried out to understand the phytoplankton ecology vis-à-vis environmental factors (biotic and abiotic) that regulate phytoplankton biomass and diversity in Indian coastal lagoons. Indian subcontinent houses eight coastal lagoons on the eastern seaboard and nine on the western seaboard. Phytoplankton ecology in Indian lagoons is principally determined by nutrient availability and light penetration, that promote phytoplankton biomass gain and factors that contribute to biomass loss, such as tidal flushing and zooplankton grazing. The phytoplankton floral spectra of Indian lagoons are represented by diverse algal divisions such as Bacillariophyta, Dinophyta, Cyanophyta, Chlorophyta, Euglenophyta, Chrysophyta, Cryptophyta, and Xanthophyta. This review revealed that the phytoplankton ecology of Chilika Lagoon is relatively well investigated compared to the other Indian lagoons. A total of 739 phytoplankton species have been reported from Chilika, followed by 141 from Muthukadu, 101 from Muthupet, and 53 from Pulicat, on the eastern seaboard. While on the western seaboard, 181 genera from Vembanad, 53 genera from Veli, and 53 species of phytoplankton from Ashtamudi Lagoon have been documented. Bacillariophyta is the most diverse and abundant phytoplankton group in coastal lagoons of both Indian east and west coast, which may be attributed to their high growth rates and positive correlation with regulating environmental factors. Indian coastal lagoons, which are hubs of fisheries and tourist attractions, are undergoing rapid changes due to natural and anthropogenic forcing such as littoral drift, climate change, agricultural runoff, industrial waste discharge, and

S. Singh
XIM University, Harirajpur, Odisha, India

Kerala University of Fisheries and Ocean Studies (KUFOS), Kochi, Kerala, India

T. Acharyya (✉)
XIM University, Harirajpur, Odisha 752050, India
e-mail: acharyyat@xim.edu.in

A. Gopinath
Kerala University of Fisheries and Ocean Studies (KUFOS), Kochi, Kerala, India

S. Madhav et al. (eds.), *Coastal Ecosystems*, Coastal Research Library 38,
https://doi.org/10.1007/978-3-030-84255-0_5

domestic sewage. Hence, long-term but rapid analysis of phytoplankton communities across the broad spatial scales of lagoons are needed to decipher how these factors potentially influence lagoon phytoplankton ecology. This review recommends adoption of rapid and more robust pigment chemotaxonomy and remote sensing techniques to study phytoplankton ecology in Indian coastal lagoons, in addition to conventional microscopy.

Keywords Phytoplankton · Coastal lagoon · Anthropogenic influence · Pollution

1 Introduction

Coastal lagoons are shallow water bodies that run along a shoreline but remain separated from the ocean by sand bars/spits, coral reefs, or barrier islands (Kjerfve 1994; Duarte et al. 2002). They are widely distributed from the arctic to the tropics (Nichols and Boon 1994) constituting about 13% of the world's coastline (Kjerfve 1994). These shallow, nutrient-rich, turbulent, and light-attenuated ecosystems are particularly common along the east coasts of continents where tidal ranges are moderate to low (<2 m). Three distinct physical features characterize costal lagoons; first, continuous mixing of the shallow water column coupled with bottom friction that favor vertical homogeneity and sediment-water exchange; second, periodic tidal motion that facilitates material transfer with the adjacent continental shelf water; and third, water circulation is regulated primarily by tide in the mouth and wind in the interior part of the lagoon. Coastal lagoons are also subjected to rapid salinity changes year-round due to precipitation, evaporation, and wind action. Geomorphologic evolution of costal lagoons is driven by the balance between the rate of sedimentation and relative sea-level rise. Based on the degree of connectivity to the adjoining ocean, coastal lagoons are classified as choked (narrow inlet prevents tidal mixing with the sea), restricted (multiple channels and wind allow limited tidal exchange), and leaky (featuring the highest tidal mixing due to wider channels and faster water currents) (Kjerfve 1994). Coastal lagoons are among the most productive ecosystems globally because of their shallowness, relative isolation from the sea, and strong physicochemical gradients. Many of them support rich fisheries and act as a wintering ground for migratory birds. They extend a myriad of ecological services such as hydro-chemical maintenance, water treatment, oxygen production, recreation, climate regulation, flood protection, and ecotourism (Newton et al. 2018). Coastal lagoons are undergoing frequent environmental disturbances, fluctuations, habitat loss and modification, physical alteration, pollution, and over-exploitation (Borja et al. 2010).

The lagoon's variable hydrological conditions due to floods fresh water runoff and marine intrusion cause spatiotemporal variations in Phytoplankton community composition and their production in a lagoon vary spatially and temporally due to dynamic hydrological conditions and marine intrusion through tide. Phytoplankton are microscopic algae primarily found in the upper sunlit layer of the aquatic ecosystem occupying about 1% of the global biomass. Phytoplankton are incredibly diverse and are majorly grouped into Bacillariophyta, Dinophyta, Cyanophyta, Chlorophyta, Euglenophyta, Chrysophyta, Cryptophyta, and Xanthophyta. They constitute base of all aquatic food webs, play an essential role in nutrient cycling, and contribute to the climatic processes by regulating the carbon cycle. They are also excellent indicators of health and productivity of aquatic environment (Brutemark et al. 2006; Goebel et al. 2013). Similar to other aquatic ecosystems, phytoplankton distribution in a lagoon is controlled by various physicochemical and biological processes which can be broadly grouped into abiotic (physical, chemical, geological, and climatic) and biotic (zooplankton grazing, photo adaptation, sinking rate) factors. In the lagoon environment, biogeochemical processes are interlinked with each other; for example, the impact of nutrients on phytoplankton communities in lagoon is dependent on other factors like light availability (Bledsoe and Phlips 2000), sedimentation, death and decomposition, hydraulic flushing (Richardson and Jørgensen 1996), and grazing (Frost 1980) of phytoplankton biomass. The challenge of declining the phytoplankton diversity remains a global problem in aquatic ecosystems (Cloern and Dufford 2005). Coastal lagoons in India, dotted across 7500 km-long coastline, are increasingly being subjected to anthropogenic influences that can affect phytoplankton and other biota up in the food chain. Several surveys in the past have been conducted to decipher diversity and environmental factors affecting phytoplankton in Indian costal lagoon, but no comprehensive synthesis of the existing literature has been conducted so far. This chapter summarizes the phytoplankton literature conducted on Indian coastal lagoons and points out future research directions.

2 Coastal Lagoons of India

The lagoons are highly productive and valuable aquatic ecosystem with an abundance of flora and fauna including migratory birds, thus acting as blue economy hub by supporting fisheries and coastal tourism. The Indian lagoons are shallow with an average depth of 2 m (Mahapatro et al. 2013) and remain well mixed by waves and currents. Their primary production is between ~50 and >500 g C/m^2/year and hence grouped into eutrophic (300–500 g C/m^2/year), mesotrophic (100–300 g C/m^2/year), and oligotrophic (<100 g C/m^2/year) lagoons (Nixon 1995). Broadly there are 17 lagoons present along the Indian coast, out of which 8 are on the eastern coast (Chilika Lagoon, Pulicat Lagoon, Pennar Lagoon, Bendi Lagoon, Nizampatnam Lagoon, Muttukadu Lagoon, Muthupet Lagoon, Lagoon of Gulf of Mannar) and 9 are in the western coast (Vembanad Lagoon, Ashtamudi Lagoon, Paravur Lagoon, Ettikulam Lagoon, Murukumpuzha Lagoon, Veli Lagoon, Talapady Lagoon, Lagoons of Mumbai, and Lagoons of Lakshadweep) (Ingole 2005; Mahapatro et al. 2013). The Indian lagoon ecosystems face challenges from two different sources:

natural stressors and anthropogenic stressors that potentially affect phytoplankton assemblage and dynamics. Ecological processes of the Indian coastal lagoons are significantly less understood due to the lack of extensive studies. A combination of the keywords 'Coastal lagoon' AND 'India' and 'Coastal lagoon' AND 'Phytoplankton' and 'Specific name of the coastal lagoon present in east and west coast' AND 'India' was used to extract the Google Scholar database's bibliographic information till February 1, 2021. A total of 103 numbers of research articles were found. Out of India's total 17 coastal lagoons, we were able to get the information related to phytoplankton ecology only in 8 lagoons (four in the east coast and four in west coast). There are still 9 coastal lagoons, in which phytoplankton diversity-related information are yet to be explored. A brief description of the climatic and geomorphological features of Indian coastal lagoons has been provided in Table 5.1. Their location along the India's coastline has been shown in Fig. 5.1.

3 Phytoplankton Diversity and Seasonal Dynamics in Indian Lagoons

Some of the most important phytoplankton groups found in Indian coastal lagoons include Bacillariophyta, Dinophyta, Cyanophyta, Chlorophyta, Euglenophyta, Chrysophyta, and Xanthophyta. A brief description of them can be found in Table 5.2. A brief account of dominant phytoplankton groups, their relative abundance, and major controlling factors from Indian coastal lagoons have been shown in Table 5.3 and Fig. 5.2 and are discussed briefly below.

The seasonal abundance of phytoplankton has been described as per the prevailing monsoon wind system in Indian subcontinent where pre-monsoon (hot summer months comprising of Mar - May), monsoon (rainy season comprising June - September) and post-monsoon (relatively cooler months October to February) offers distinct temperature, humidity and rainfall pattern.

This is also to be noted that this chapter describes phytoplankton dynamics only in the coastal lagoon attached to the mainland. Atolls and lagoons in the islands have been excluded from this literature synthesis.

3.1 *East Coast of India*

There are eight coastal lagoons present on India's east coast, and among them, Chilika is the largest lagoon with brackish water environment supporting a rich diversity of phytoplankton. A total of 739 species have been reported from the groups Bacillariophyta (270) > Chlorophyta (178) >Cyanophyta (103) > Euglenophyta (92) > Dinophyta (88) > Chrysophyta (5), and > Xanthophyta (3)

Table 5.1 Climate and geomorphological features of the lagoons

Lagoon (state)	Lat/Long	Area (km2)	Depth (m)	Climate	Geological setting	Ecological Significance if any	Anthropogenic setting and threat	References
Chilika Lagoon (Odisha)	19°28'–19°54' N to 85°6'–85°35'S	906–1165	0.3–4.2	Temperature range: 14°C–39.9 °C Average annual rainfall 1238.8 mm (driven by SW monsoon from July to September)	• Shallow bar-built estuary • Pear shaped having length 64.3 km and width of 20.1 km • Narrow channels connect the lagoon with the Bay of Bengal	• Largest brackish water lagoon in Asia • Designated Ramsar site (Ramsar Convention of Wetlands in 1981) • Largest wintering ground for migratory birds	• Siltation • Shrinkage of water surface area • Choking of the inlet channel leading to decrease in salinity • Proliferation of freshwater invasive species • Overfishing and illegal prawn farming	Panigrahi et al. (2009)
Pullicat Lagoon (Andhra Pradesh 96%, Tamil Nadu 3%)	13.33°–13.66° N to 80.23°–80.25° E	250–450	1.8	Air temperature (15 °C to 45 °C)	• A barrier island separates the lagoon from the Bay of Bengal • Hypersaline (salinity can reach up to 55 psu)	• Second largest brackish water lagoon in India • Harbors Pulicat Lake Bird Sanctuary	• Siltation and periodic closure of the mouth • Thermal pollution, oil spill from mechanized boats, pollution from municipal sewage, agricultural chemicals and industrial effluents • Unsustainable fishing practice	Basuri et al. (2020), Basha et al. (2012)
Muthukadu Lagoon (Tamil Nadu)	12°49′N, 80°15′E	0.87	1–2	North east monsoon	• Connected to the sea by a bar-built mouth • Remain cut off from the sea during May-September	NA	• Fishing and boating activities • Oil spill from mechanized boat • Aquaculture	Prasath et al. (2019)

(continued)

Table 5.1 (continued)

Lagoon (state)	Lat/Long	Area (km2)	Depth (m)	Climate	Geological setting	Ecological Significance if any	Anthropogenic setting and threat	References
Muthupet Lagoon (Tamil Nadu)	10.33°N, 79.55°E	68	1	North east monsoon	• Semi enclosed coastal lagoon • The bottom of the lagoon is formed of silt clay substratum	• This lagoon is known for oyster population and acts as a nursery ground for marine fishes	• Aquaculture • Agricultural runoff	Gupta et al. (2006)
Vembanad Lagoon (Kerala)	9.5968°N, 76.3958°E	2114	1–9	Rainfall varies from 250 to 400 cm (southwest and northeast monsoon)	• Separated from the Lakshadweep Sea by a narrow barrier island • Connected to the Arabian sea by two permanent inlets	• Designated Ramsar site • Home to more than 20,000 waterfowls • Wintering ground for the migratory birds	• Plastic littering • Oil spill from motorized boat • Discharge from the sea food processing units • Receive heavy load of organic material and sewage • Tourism activity	Asha et al. (2016)
Asthamudi Lagoon (Kerala)	8.98°N, 76.6°E	1700	Max 6.4	Max temp (27.5 °C), Min temp (25.5 °C) Average annual rainfall 2400 mm	• It is a palm-shaped extensive water body with eight prominent arms	• Designated Ramsar site • Hosts many migratory birds	• Solid waste dumping • Oil spills from thousands of fishing boats • Disposal of large quantities of untreated effluents from industries	
Talapady Lagoon (Karnataka)	12.78°N, 74.85°E	0.2	1	Southwest monsoon	• It is a blind lagoon • Seawater seeps into the lagoon across the sand bar or during the high tide	• Support rich fish and bird biodiversity	• Fishing and boating activities • Sewage discharge	Nayar (2006)
Veli Lagoon (Kerala)	8.46°N, 76.95°E	135	2–3		• It has no permanent connection with the sea, remains separated by a narrow strip of sandy beach	NA	Pollution from agricultural runoff, industrial effluents, and domestic sewage	Sajinkumar et al. (2017)

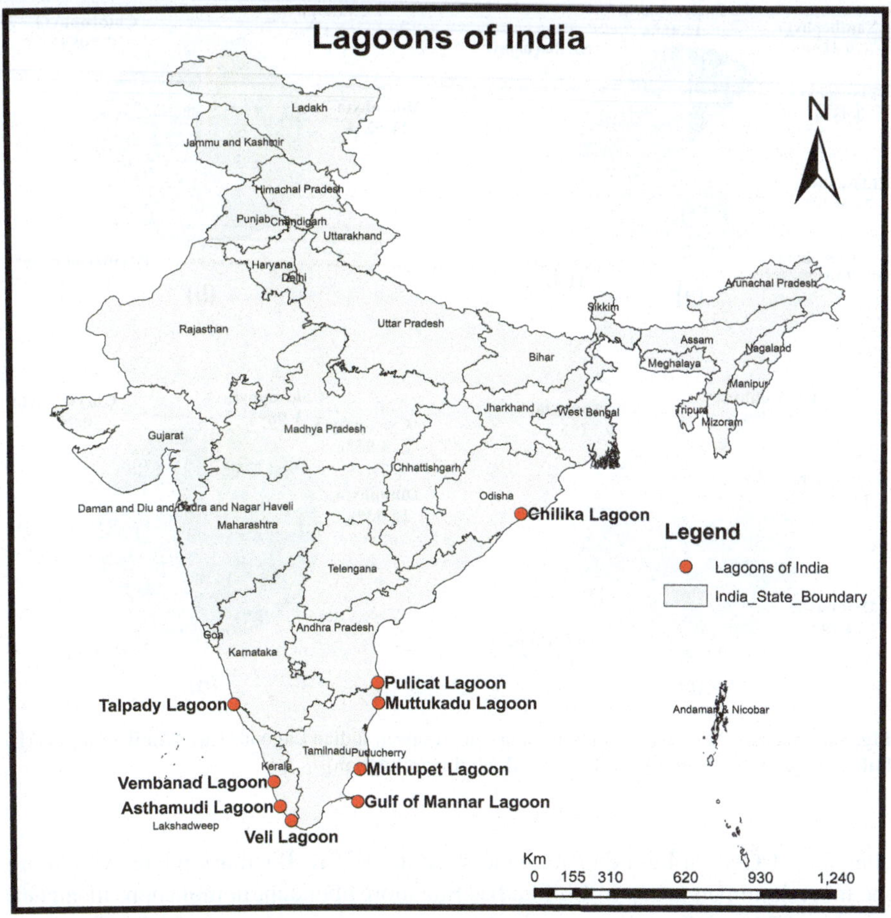

Fig. 5.1 Costal lagoons of India

during 2000–2014 (Srichandan and Rastogi 2020). Among the recorded 270 species of Bacillariophyta, *Pleurosigma* sp., *P. normanii*, *Synedra* sp., *Thalassionema nitzschioides*, *Surirella* sp., *Chaetoceros* sp., *Coscinodiscus* sp., *Lithodesmium undulatum*, *Hemiaulus sinensis*, and *Paralia sulcata* dominate all over the lagoon (Srichandan et al. 2015a; Srichandan and Rastogi 2020). Phytoplankton population density varies between 2000 and 12,000 cells/L registering higher values during the pre-monsoon period (Srichandan and Rastogi 2020). Euglenophyta predominated among other phytoplankton at low salinity sectors, whereas Bacillariophyta predominated in high salinity sectors (i.e., southern sector, central sector, and outer

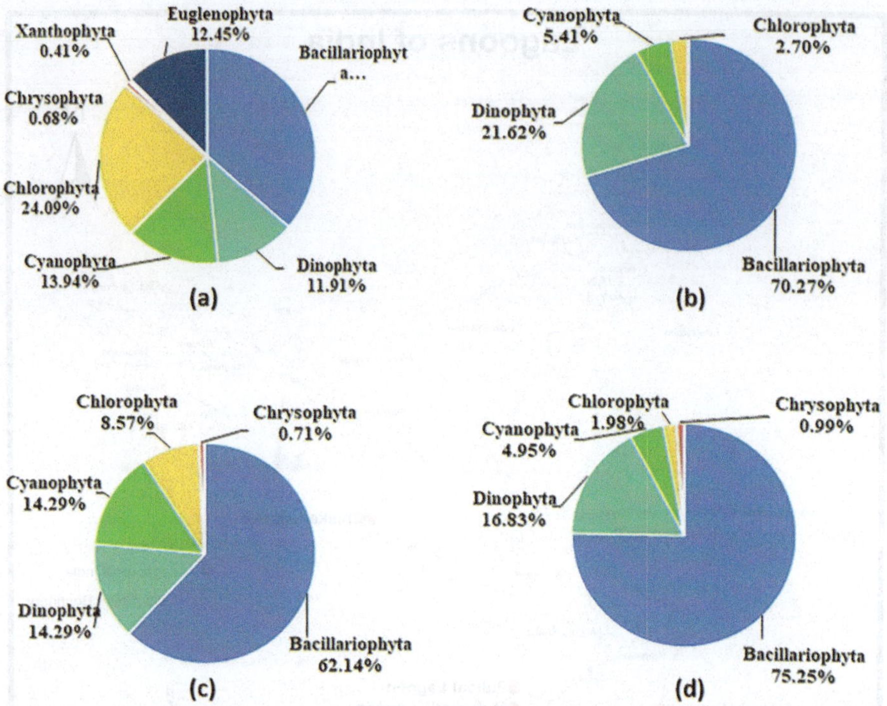

Fig. 5.2 Relative Abundance of phytoplankton groups in Indian Lagoons [(**a**): Chilika Lagoon (**b**) Pulicat Lagoon (**c**) Mutthukadu Lagoon (**d**)Muthupet Lagoon]

channel) of Chilika Lagoon (Srichandan et al. 2015a). Bacillariophytes which are the most dominant group in Chilika become more abundant in monsoon (mean cell density 1879 cells/L), subsequently decreasing in post-monsoon (710 cells/L) and further increasing in pre-monsoon (1134 cells/L) (Srichandan et al. 2015a).

Pulicat is the second largest brackish water lagoon of India. Earlier, a total of 53 phytoplankton species were documented (Basha et al. 2012). Later, 37 species have been reported in 2016 (http://jscmwr.org/research-2/page/2/) constituted by Bacillariophyta (26), Dinophyta (8), Cyanophyta (2), and Chlorophyta (1). Maximum phytoplankton diversity is recorded during monsoon, followed by the post-monsoon period. Phytoplankton population density varies between 1.02×10^5 and 5.94×10^5 cells/L in monsoon and 6.79×10^4 and 6.28×10^5 cells/L in post-monsoon. Group abundance follows the order Bacillariophyta > Dinophyta > Cyanophyta > Chlorophyta. According to Basha et al. (2012), invasive phytoplankton species such as *Spirulina major*, *Oscillatoria* sp., and *Anabaena* sp. have been found in Pulicat. A bloom has been reported in this lagoon consisting of *Biddulphia pulchella*, *Biddulphia biddulphiana*, and *Biddulphia laevis*, which indicates Pulicat is experiencing pollution (Santhanam and Farooqui 2018).

Table 5.2 Identifying characters of major phytoplankton groups in lagoon

Phytoplankton groups	Characters
Bacillariophyta	• Unicellular but often found to form chains/colony; size range 15–400 mm • Spectrum of shape varies from pennate (bilateral symmetry) to centric (radial symmetry) • The cell wall, called frustule, has three parts: base (hypotheca), cap (epitheca), and belt (singulum). Frustule is composed of silicon and is often ornamented • Key pigments found: chlorophyll *a* and *c*, beta carotene, and fucoxanthin. Autotrophs • Specialized vegetative cell division • Ubiquitous; found across all salinity ranges; fresh > brackish > marine. Both benthic and pelagic forms are found
Dinophyta	• Mainly (90%) marine • Size range 15–40 μm • Can swim by means of two flagella • Key pigment: Peridinin; Autotrophs and heterotrophs; some are mixotrophs • Most common sources of bioluminescence and toxic algal bloom
Cyanophyta (blue green algae)	• Cyanophyta are unicellular and filamentous organisms that are ubiquitous in all the aquatic medium • Key pigment: chlorophyll *a*, phycobilin, carotenoids, phycocyanin, and zeaxanthin; Autotrophs • Chlorophyll is scattered throughout the protoplasm but not contained within the chloroplasts • Prokaryotes; cell wall contains pectin, hemicellulose, and cellulose sometimes in the form of mucus
Chlorophyta (green algae)	• Largest among the eight algal divisions • Unicellular or multicellular in nature • Green-colored phytoplankton with chlorophyll and xanthophyll and carotenes as the dominant photosynthetic pigment • Photosynthetic pigments localized in chloroplasts in which usually pyrenoids are present • These are common inhabitants of marine, freshwater, and terrestrial environments
Euglenophyta	• Unicellular flagellates found in freshwater and marine environment • Possess definite nucleus • Chlorophyll localized in chromatophores giving grass green color; few species lack chromatophores and are colorless • One, two, or rarely three flagella being attached to the anterior end of the cell • Reproduction is by cell division along the long axis
Chrysophyta (golden-brown algae)	• Often unicellular pigmented heterokonts and have a flagellum, allowing them to be mobile in the water • Commonly referred to as due to their coloration from specific photosynthetic pigments; Autotrophs • Major component of coastal and estuarine water
Xanthophyta (yellow green algae)	• Non-motile, unicellular, or colonial eukaryotic algae • These are commonly found in freshwater and most of them are free floating • The cell wall is often absent, but when present it contains more pectic compounds • The motile forms usually bear two flagella but rarely one. They are unequal and inserted at the anterior end • The chromatophores are discoid in shape and are numerous in each cell • The photosynthetic pigments are chlorophyll a, P-carotene, diadinoxanthin, violaxanthin, and lutein

Table 5.3 Dominant phytoplankton groups and their controlling factors in Indian lagoons

Location	Name of the lagoon	Phytoplankton relative abundance	Phytoplankton Total # of sp.	Major controlling factors
East Coast	Chilika	Bacillariophyta>Chlorophyta>Cyanophyta>Euglenophyta>Dinophyta>Chrysophyta>Xanthophyta	739	• Salinity • Turbidity • Nutrient stoichiometry • Zooplankton grazing
	Pullicat	Bacillariophyta>Dinophyta>Cyanophyta>Chlorophyta	37	• Nutrient • Water temperature
	Muttukadu	Bacillariophyta>Cyanophyta> Dinophyta>Chlorophyta>Chrysophyta	141	• Salinity • Solar radiation • Zooplankton grazing
	Muthupet	Bacillariophyta>Dinophyta>Cyanophyta>Chlorophyta>Chrysophyta	101	• Salinity • Nutrient stoichiometry
West Coast	Vembanad	Bacillariophyta>Chlorophyta>Cyanophyta>Dinophyta>Euglenophyta>Eusstigmatophyta>Rhodophyta>Haptophyceae>Chrysophyceae>Cryptophyceae>Dictyochophyceae	181(Genera)	• Nutrient • Water temperature • Salinity • Organic matter
	Asthamudi	Cyanophyta> Chlorophyta>Bacillariophyta>Dinophyta	53	• Salinity • Nutrient • Organic matter • Turbidity
	Talapady	Chlorophyta>Bacillariophyta>Cyanophyta>Dinophyta		• Salinity • Nutrient
	Veli	Bacillariophyta>Chlorophyta>Cyanophyta>Dinophyta>Euglenophyta	53 (Genera)	• Turbidity • Salinity

In Muttukadu, 141 phytoplankton species have been recorded, including Bacillariophyta (87), Cyanophyta (21), Dinophyta (20), Chlorophyta (12), and Chrysophyta (1) during 2010–2012 (Prasath et al. 2019). Phytoplankton population density ranges between 1.78×10^4 and 3.26×10^7 cells/L (Prasath et al. 2019), with maximum population density recorded during pre-monsoon and minimum during post-monsoon. The following abundance order Bacillariophyta > Cyanophyta > Dinophyta > Chlorophyta > Cryptophyta has been documented. Muttukadu witnessed bloom of *Microcystis aeruginosa* in January–April 2014 due to the high concentration of nutrient input from the catchment area (Vasudevan et al. 2015).

One hundred one (101) species of phytoplankton have been recorded in the Muthupet lagoon. The species comprised Bacillariophyta (76), Dinophyta (17), Cyanophyta (5), Chlorophyta (2), and Chrysophyta (1) as per the survey conducted in 2009–2010 (Babu et al. 2013). The most abundant species were *Nitzschia seriata*,

Coscinodiscus centralis, *Thalassiothrix frauenfeldii*, and *Ceratium furca* (Babu et al. 2013). Diatoms (*Nitzschia seriata*, *Thalassiothrix frauenfeldii*, *Odontella sinensis*) dominated during the pre-monsoon, and freshwater algae (*Anabaena* sp., *Oscillatoria* sp., *Chlorella* sp., *Nostoc* sp., *Lyngbya* sp., *Spirogyra* sp.) dominated in the monsoon. Phytoplankton population density ranged between 5.91×10^3 and 7.63×10^5 cells/L. Following relative abundance was recorded; Bacillariophyta > Dinophyta > Cyanophyta > Chlorophyta > Chrysophyta. A bloom of *Trichodesmium erythraeum*, a filamentous cyanobacteria capable of fixing atmospheric nitrogen, was reported by Santhanam et al. (2013) in May 2011 from Muthupet.

3.2 West Coast of India

There are nine coastal lagoons present on the western coast of India. Fifty-three genera of phytoplankton have been reported from Veli with the following distribution order: Bacillariophyta (22), Chlorophyta (17), Cyanophyta (9), Dinophyta (2), and Euglenophyta (1). Bacillariophyta showed preference to pre-monsoon months but were barely present during peak monsoon (Mathew and Nair 1981). Chlorophyta such as *Oedogonium*, *Spirogyra*, and Desmidiales dominated in monsoon. Among Cyanophyta, *Oscillatoria* sp. occurred almost throughout the year.

Fifty-three (53) species of phytoplankton belonging to 38 genera have been documented from Ashtamudi during summer month of 2017 (Badusha and Santhosh 2018). Following relative dominance has been reported: Cyanophyta > Chlorophyta > Bacillariophyta> Dinophyta (Badusha and Santhosh 2018). Blue-green algal members such as *Oscillatoria* dominated the lagoon's phytoplankton community (Badusha and Santhosh 2018; Mathew and Nair 1980).

One-hundred and eighty-one genera of phytoplankton belonging to 11 classes have been documented from Vembanad during 2010–2012 (Vidya et al. 2020). Abundance order is Bacillariophyta > Chlorophyta > Cyanophyta > Dinophyta > Euglenophyta > Eustigmatophyta > Rhodophyta > Haptophyta > Chrysophyta > Cryptophyta > Dictyochophyta (Nandan and Sajeevan 2018; Vidya et al. 2020).

Phytoplankton density varied from 0.01 to 65.55×10^5 cells/L in Talapady Lagoon during 1996–1997 (Nayar 2006). Bacillariophyta dominated the assemblage during pre-monsoon, while Cyanophyta dominated during monsoon and post-monsoon season. High temperature, salinity, and organic waste promote cyanophytes' growth in Vembanad (Babu et al. 2013). Seasonal variation showed Bacillariophyta (*Thalassiosira subtilis*, *Nitzschia closterium*, *Navicula henneydii*, *Coscinodiscus marginatus*, *Chaetoceros indicus*, and *Campylodiscus cribrosus*) dominating the assemblage during pre-monsoon while Cyanophyta (*Oscillatoria limosa*, *Gomphosphaeria aponia*, and *Agomenellum quadruplicatum*) dominating during monsoon and post-monsoon season.

4 Factors Controlling Phytoplankton Distribution and Dynamics in Indian Lagoons

The phytoplankton abundance, distribution, and diversity are mostly influenced by the habitat heterogeneity in lagoons (Clegg et al. 2007). The coastal lagoon's heterogeneity is due to freshwater input coming from river discharge and monsoon rainfall and tidal influx that affect the lagoon's water chemistry. We have grouped various controlling factors that regulate phytoplankton distribution in Indian coastal lagoons under (1) physical factors, (2) chemical factors, (3) geological factors, (4) biological factors, (5) anthropogenic factors, and (6) meteorological factors which are discussed below. Range values of various physicochemical parameters have been given in Table 5.4.

Table 5.4 Range of various parameters in Indian Coastal lagoons (from published literature)

East coast					
	Parameters	Chilika lagoon	Pullicat lagoon	Muttukadu lagoon	Muthupet lagoon
1	Water Temperature (°C)	18.9–35.9	25.2–32.8	22–34	22–31
2	Salinity (psu)	0–37	12.6–61.1	0–36	1–34
3	pH	6.1–10.35	7.9–8.8	7.23–8.95	7.6–8.2
4	DO (mg/L)	0.3–14	2.7–7.8	2.1–6.93	2.87–8.64
5	NO_2 (μmol/L)	0.01–2.01	0.0–0.7	0.23–1.57	0.16–3.33
6	NO_3 (μmol/L)	0.12–19.88	0.1–4.9	16.21–42.6	1.42–6.15
7	PO_4 (μmol/L)	0.01–2.85	0.2–2	10.4–24.58	0.16–1.6
8	SiO_2 (μmol/L)	0.1–363		22–188	
West coast					
		Vembanad lagoon	**Asthamudi lagoon**	**Veli lagoon**	**Talapady lagoon**
1	Water Temperature (°C)	28–33	27.40–31.80	27.1–32.3	26.18–34.50
2	Salinity (psu)	0.1–33	7.9–32.5	0.2–4.5	0.2–18.5
3	pH	6.25–10.2	6.70–8.20	7.5–8.15	
4	DO (mg/L)	4–9.6	3.16–8.86	4.29–6.37	5.07–8.06
5	NO_2 (μmol/L)			0.13–1.2	0.1–9.8
6	NO_3 (μmol/L)	74.0–95.80	0.69–3.99	6.0–12.3	2.5–16.1
7	PO_4 (μmol/L)	18.67–25.80	0.09–0.78	0.66–2.41	3.4–6.5
8	SiO_2 (μg-at/L)			46.3–121.5	15.3–132.9

4.1 *Physical Factors*

Coastal lagoons being shallow water bodies are subjected to physical forcing like air and water temperature, water level, photic depth, water currents, turbidity and transparency, precipitation, and evaporation that affects lagoon ecosystem directly or indirectly.

4.1.1 Water Temperature

Water temperature is a crucial component of coastal lagoons, influenced mainly by solar radiation, heat transfer from the atmosphere, stream confluence, thermal pollution, and turbidity. Each phytoplankton species can only survive within a specific temperature range. 25 °C is the optimal temperature range where the species grows best, and it grows less at lower and higher temperatures. The optimum growth temperature for a given species is different at different light regime or PCO_2 (Partial pressure of carbon dioxide in water) conditions. Some phytoplankton grows faster during warmer conditions, but high growth rates are not always a good indicator of cell health. Usually, phytoplankton cells grow very fast only during the stressed condition, which only continues for a short period. Based on the published literature, the water temperature lies between 18 and 36 °C in eastern coastal lagoons and 26 and 35 °C in India's western coastal lagoons. The highest water temperature usually is observed in the pre-monsoon period with an increase in solar energy and resulting stable water column (Saravanakumar et al. 2008). Low water temperature is recorded during monsoon, possibly due to prevailing sea breeze, rainfall, and cloudy sky (Rajkumar et al. 2009). Water temperature influences the phytoplankton's metabolic rates and photosynthetic production (US Environmental Protection Agency 2012). For example, Cyanophyta dominates due to its growth preference in the higher temperature range (Kosten et al. 2012). However, the effect of temperature is not uniform across phytoplankton groups. For example, in Chilika, Bacillariophyta can tolerate an extensive range of water temperature (Sasamal et al. 2005), whereas growth of benthic Cyanobacteria occurs only at favorable temperature (Srichandan and Rastogi 2020). Basha et al. (2012) reported the adverse effect of thermal pollution on the phytoplankton population in Pulicat Lagoon. Muthupet and Muttukadu's phytoplankton abundance was lowest during monsoon months because of decreased water temperature due to overcast sky and cool conditions (Babu et al. 2013). The effect of water temperature on phytoplankton distribution was not very prominent in Veli (Mathew and Nair 1981).

4.1.2 Photic Depth and Turbidity

Ninety percent of the aquatic lives live in the photic zone due to the availability of abundant solar energy. Photic depth is the uppermost layer of the water body, which allows phytoplankton to photosynthesize. In this zone, the photosynthesis rate exceeds the respiration rate. The high intensity of solar radiation may damage the algae's light-harvesting system, so they develop photoprotective compounds. Phytoplankton diversity, abundance, and spatial variation change according to photic depth (Flöder et al. 2002). Photic depth in coastal lagoons is a function of solar radiation intensity, water turbidity, presence of submerged aquatic vegetation, and pollution. Turbidity, which inversely varies with the photic depth, is the amount of cloudiness in water (Gallegos 1992). A high amount of suspended matter occurs due to flooding and intense rainfall, common during monsoon seasons. Turbidity is the key controlling factor which regulates the photic depth of the lagoon and affects the phytoplankton production in Chilika (Srichandan et al. 2015a, b). According to Srichandan et al. (2015b), phytoplankton bloom could not happen in Chilika for high turbidity even though a large amount of dissolved nutrient was brought inside the lagoon during cyclone Phailin. Badusha and Santhosh (2018) observed that in Ashtamudi lagoon, phytoplankton species diversity reached its maximum (41 species) when turbidity was minimum. According to Mathew and Nair (1981), high turbidity during monsoon resulted in fewer phytoplankton counts in Veli. Similar trend was also noticed in both Muttukadu and Muthupet in the east coast, where the phytoplankton abundance was the lowest during monsoon months due to high turbidity caused by river runoff and phytoplankton abundance was highest during summer/pre-monsoon season due to low turbidity (Babu et al. 2013; Prasath et al. 2019).

4.1.3 Water Current

The water current is the rate of movement of the water. It plays very important role for its influence in transportation of sediment, pollutant, phytoplankton species, and nutrient (Mohanty and Panda 2009). Water current inside the coastal lagoon is mainly controlled by freshwater runoff, tidal effect, strong winds, water density difference, and difference in temperature and salinity. Water current has significant importance in driving the phytoplankton from one place to another in an aquatic ecosystem (Phlips et al. 2002). The southwest and northeast monsoons influence the current pattern by playing a key role in determining the phytoplankton species diversity inside the lagoon (Murty and Varma 1964; Rao et al. 2011; Jyothibabu et al. 2013). According to Mohanty and Panda (2009), wind, tide, and freshwater input regulate the circulation and mixing pattern in Chilika.

4.2 *Chemical Factor*

The critical chemical factors responsible for phytoplankton distribution in coastal lagoons are pH, total alkalinity, salinity, dissolved oxygen, biochemical oxygen demand, and nutrients which are discussed below.

4.2.1 Nutrients

Nutrient concentrations and stoichiometry, specifically of silicate, nitrate, nitrite, phosphate, and ammonia, act as one of the most significant determinants of phytoplankton biomass and distribution in a lagoon environment. Any shortage of nutrients causes a decrease in the photosynthetic rate in phytoplankton. The source of nutrients in coastal lagoons can be both autochthonous (decomposition of organic matter, upwelling, wind-driven resuspension) and allochthonous (river discharge, weathering, atmospheric deposition). Concentration of nitrate and phosphate was recorded maximum during the pre-monsoon season due to the higher residence time of water and low-flow period in Chilika Lagoon (Muduli et al. 2013). In Chilika concentration of silicate was higher throughout the monsoon period due to high land runoff which went down in the pre-monsoon period for low freshwater influx and consumption by the Bacillariophyta (Srichandan et al. 2015a, b). In Chilika, nitrate and phosphate greatly influence phytoplankton abundance and diversity, especially of Dinophyta (Srichandan et al. 2015a). Coastal lagoons highly loaded with sewage (organic matter) may manifest algal blooms or sometimes toxic algal blooms, as shown in Pulicat Lagoon (Santhanam and Farooqui 2018). In Muthukadu, a high nutrient input concentration leads to a very high phytoplankton population density dominated by blue-green algae *Microcystis aeruginosa*. Vembanad receives enough nutrients from Kuttanad paddy fields that cause bacillariophytes and chlorophytes' dominance. In Veli, a clear inverse relationship of phosphate concentration with Chlorophyta has been established. Bacillariophyta shows silicate dependence as it has been observed that there was a decrease in silicate concentration with increased Bacillariophyta population (Mathew and Nair 1981).

4.2.2 Salinity

Salinity is an important determining factor for phytoplankton diversity and abundance that decreases with an increase in salinity. Salinity in a coastal lagoon is mainly controlled by freshwater influx through riverine water, local rainfall, tidal amplitude, high solar radiation, and water residence time. In coastal lagoons, phytoplankton are generally euryhaline (able to sustain in an array of salinity) in nature. Salinity fluctuation causes osmotic stress in phytoplankton cells. In Chilika, salinity plays a critical role by controlling the phytoplankton abundance and distribution (Panigrahi et al. 2009; Srichandan et al. 2015a, b). Raman et al. (1990) and

Srichandan et al. (2015a) observed that southern sector (i.e., stable salinity region) is dominated by the phytoplankton groups of Dinophyta and Chrysophyta, whereas northern sector (i.e., low salinity region) is mostly dominated by the group of Euglenophyta. Bacillariophyta were prevalent all over the high salinity regimes (southern sector, central sector, and outer channel) in Chilika. The outer channel (i.e., high salinity zone) is dominated by marine phytoplankton species because of its direct connection with the Bay of Bengal. Due to high freshwater influx from rivers and land runoff, the northern sector is usually dominated by freshwater forms of phytoplankton. Siltation caused the complete closing of mouths of Pulicat leading to salinity fluctuation and water level changes of the lagoon, which had noticeable effect on the phytoplankton abundance (Basha et al. 2012). In Muttukadu, the minimum population density was recorded during low salinity months, whereas the maximum population was recorded during high salinity months (Prasath et al. 2019). Widely changing salinity at the Muthupet is responsible for the predominance of bacillariophytas at the mouth (Mishra et al. 1993; Ramakrishnan et al. 1999; Senthilkumar et al. 2002). According to Nayar et al. (1999), in Talapady, Bacillariophyta counts were higher in the summer season and lower in monsoon and were moderate during the post-monsoon season.

4.2.3 pH

pH is the measurement of the hydrogen ion concentration in water. Coastal lagoons are prone to fluctuating pH concentrations due to regular intermixing of fresh and saline water, phytoplankton's photosynthetic activity, respiration, and decomposition of organic matters (Ganguly et al. 2015; Muduli et al. 2013). pH concentration in the aquatic medium can regulate the growth and diversity of phytoplankton and nutrient availability. It regulates the carbon cycle and has a crucial role in the growth and survival of phytoplankton (Sculthorpe 1967). Srichandan et al. (2015a) reported the pH variability in Chilika is due to the CO_2 assimilation by phytoplankton and macrophytes. Srichandan et al. (2015a) reported a strong positive correlation of pH with abundance of Cyanophyta in Chilika. Similarly, in Ashtamudi, phytoplankton abundance dropped in highly polluted water with a lower pH value (6.70) (Badusha and Santhosh 2018).

4.2.4 Dissolved Oxygen

Dissolved oxygen (DO) represents the total amount of oxygen present in dissolved form in water, and it is an essential determinant of the aquatic ecosystem's health. Oxygen is produced in the process of photosynthesis by the phytoplankton, macrophytes, and submerged vegetation. Wind flow drives the mixing of oxygen in the water from the atmosphere. According to Garnier et al. (1999), oxygenation in aquatic systems results from a variation between photosynthesis, organic matter degradation, re-aeration, and physicochemical characteristics of water. Changes in freshwater flow may have important effects on dissolved oxygen (Mallin et al. 1993). Respiration and denitrification of bacteria in the lagoon act as a sink of oxygen. According to Barik et al. (2017), water temperature and wind maintain a linear relation with dissolved oxygen in Chilika. Dissolved oxygen level 3 mg/L must be maintained to protect aquatic life (CPCB 1986). It was reported that DO concentration varied from 0.3 to 14 mg/L in the Chilika Lagoon, which is well saturated due to the large area of the lagoon, high rate of photosynthesis, and wind churning effect (Barik et al. 2017; Srichandan and Rastogi 2020), Dissolved oxygen varied from 2.7 to 7.8 mg/L in Pulicat Lagoon (Basuri et al. 2020). In lagoon ecosystems, concentration of DO decreases due to the rise in water temperature and organic matter decomposition. If the concentration of dissolved oxygen increases, it tends to increase the phytoplankton number.

4.2.5 Biochemical Oxygen Demand (BOD)

BOD is defined as the amount of dissolved oxygen used by aerobic bacteria to oxidize organic matters. It is an important indicator of the microbial load and amount of organic pollution in aquatic ecosystems (Ndimele 2012). High concentration of BOD is an indicator of high demand of oxygen and low water quality and vice versa. According to Badusha and Santhosh, (2018), maximum phytoplankton species diversity occurred in less polluted areas, while the lowest number of species was observed at more polluted (high BOD) parts of the Ashtamudi Lagoon contributed by oil spills from tourist boats and fecal contamination.

4.3 Geological Factor

The geological factors impact coastal lagoon through littoral drift, groundwater discharge, catchment influx, marine water intrusion, the coastal geomorphological process, inlet configuration and dimension, lagoon size, orientation to prevailing winds, and water depth (Mahapatro et al. 2013). Chilika characterizes a complex geological process involving the deposition of beach ridges and spits. The lagoon spreads from northeast to southwest in parallel to the Bay of Bengal, with a variable width of 20 km in the east coast of India. The lagoon gets its share of saline water through a narrow tidal inlet connected to the Bay of Bengal in the east and freshwater through various rivers and rivulets from the north, leading to various salinity regimes in the lagoon. This has created a diverse niche for inhabiting phytoplankton to exploit resources efficiently. Hence, marine phytoplankton dominates in the lagoon mouth than the northern sector, where mostly freshwater phytoplankton dominates (Srichandan et al. 2015a, b). The central sector is inhabited by freshwater and marine phytoplankton because of the mixing of freshwater and seawater. Maintaining connection between lagoon and sea is challenging as littoral drift, basin sedimentation, depletion in tidal prism, and tidal influence change water chemistry of a lagoon and also affect species migration pattern between sea and lagoon (Kumar and Pattnaik 2012).

The Pulicat has connection with the Bay of Bengal by an inlet channel at the north and outflow channel at its southern end. Muttukadu is a bar-built coastal lagoon where the effect of littoral drift is prominent. The Muttukadu is typically separated from the sea during May to September because of the flood stream's inundation from the upper reaches. Ashtamudi is a palm-shaped extensive water body and has eight prominent arms (Nagaraj 2014). Talapady is a semi-enclosed water body lying parallel with the Arabian Sea (Nayar et al. 1999). The lagoon opens into the Arabian Sea during southwest monsoon through an opening in the sand bar. During the rest of the year, seawater seeps into the lagoon across the sand bar, or the high tide water flows over the sand bar into the lagoon. During the southwest, the river's influx transports silt-laden freshwater into the lagoon. During summer, the river dries up and becomes an extension of the lagoon (Nayar et al. 1999, 2001; Nayar 2006). Due to the above geological features, the resulting water chemistry strongly influences the spatial distribution of phytoplankton.

4.4 Biological Factor

Biological factors such as zooplankton grazing, heterotrophy, shading, community shift, community succession, parasitism, and microbial loop play an important role in spatiotemporal distribution of phytoplankton in the lagoon ecosystem. Grazing, deposition, and washout can effectively remove phytoplankton biomass from the water column (Underwood and Kromkamp 1999). Dube and Jayaraman (2008) and Dube et al. (2010), through modeling study, showed zooplankton grazing is a significant biotic regulating factor of phytoplankton abundance in Chilika. According to Jyothibabu et al. (2006), microzooplankton are responsible for the grazing rate of $43 \pm 1\%$ for the daily phytoplankton standing stock during the high saline condition (27.5). Eutrophication causes phytoplankton community shifts by changing their physiology. Studies related to biological control of phytoplankton on Indian coastal lagoons are generally lacking.

4.5 Meteorological Factor

Some important meteorological factors that control phytoplankton biomass, distribution, and abundance in Indian coastal lagoons are monsoon (discussed before in this chapter), regional climate change, and cyclones. It was reported that regional climate change leads to changes in the precipitation pattern in the Mahanadi River basin region which impacts salinity gradient in the Chilika Lagoon by altering freshwater flow (Kumar and Pattnaik 2012). Monsoon-driven rainfall appears to be an essential cyclic phenomenon, and the hydrographical parameter in the Indian lagoon system showed a distinct pattern of variations leading to predictable phytoplankton community succession. It was reported that in a lagoon ecosystem, abundance of phytoplankton remains high in the dry period (pre-monsoon) and low during the wet period (monsoon) (Perumal et al. 2009). Phytoplankton abundance during pre-monsoon increases due to the increased water temperature and penetration of high intensity solar irradiance and was lowest throughout the monsoon seasons due to heavy rainfall, high turbidity, decrease in temperature, an overcast sky, and cool conditions (Saravanakumar et al. 2008). The high density during the pre-monsoon might be ascribed to more stable hydrological conditions prevailing during the season (Babu et al. 2013). Based on the published literature, it appears that Bacillariophyta dominates in almost all coastal lagoons of India due to its ability to withstand wide ranges of salinity and temperature. Tropical cyclones frequently impact coastal lagoons in India's eastern peninsula, bringing rapid changes in salinity, turbidity, and nutrient stoichiometry (Srichandan et al. 2015b). Increased number of freshwater Cyanophyta, viz., *Cylindrospermum* sp., and toxic dinophyta species, viz., *Alexandrium* sp., *Gonyaulax* sp., *and Prorocentrum cordatum*, have been reported in the central and northern sectors of Chilika in the post-Phailin (an Extremely severe cyclonic storm that made landfall in the vicinity of Chilika lagoon) period (Srichandan et al. 2015b).

4.5.1 Aerosol

Aerosol is the suspension of solid or liquid particles in air or other gases (Hinds 1999). It consists of volcanic ash, biological particles, and mineral dust, black carbon, and water vapor. Aerosols act as an essential source of nutrients and trace metal to the aquatic system and regulate climate system by influencing radiation budget. These particles regulate earth's temperature by attenuating solar radiation through absorption, scattering and cloud droplet formation (Goosse et al. 2010). So, aerosol particles are directly or indirectly responsible for the growth and diversity of phytoplankton in India's coastal lagoons. But there is a lack of research regarding aerosol particles' impact on phytoplankton diversity in Indian coastal lagoon.

4.6 Anthropogenic Factor

Marine pollution has long been a problem in the world's coastal zones, and there is ever-increasing pressure on coastal ecosystems. Lagoons are important areas of economic activity. Population growth and industrial developments alongside the lagoons have greatly influenced pollution loads into the lagoon environment. Pollutants enter into lagoons through various point and non point sources, such as industrial effluents, untreated municipal effluents sediment from catchment, agricultural runoff. Fertilizers and pesticides from agricultural runoff, tourism waste, urban and industrial waste, and aquaculture act as direct sources of contaminants while leaching from antifouling boat paints act as an indirect source of pollutants (Ahner et al. 1997). It has been reported that Chilika Lagoon is receiving heavy silt load because of the changes in land use pattern within the Chilika basin. Fragmentation of floodplains, and aggraded channels lead to loss of water holding capacity of the lagoon thus directly affecting the phytoplankton diversity (Kumar and Pattnaik 2012). Increasing tourist pressure and fishing boats in Chilika can add to the pollution burden leading to stress on the ecosystem. Malpracticed aquaculture activity leads to organic pollution in the lagoons like Chilika, which has an adverse effect on the phytoplankton community. Amir et al. (2019) have reported that the northern sector of Chilika Lagoon is heavily affected by urban and industrial wastewater input. The contaminants pose severe threat to the coastal lagoon ecosystems stimulating eutrophication, algal bloom occurrence, increased mortality of fishery resources and decreased fishery production, and significant economic loss (Gao et al. 2014). Heavy loading of organic matter in tropical coastal lagoons (Balasubramanian et al. 2004) can potentially disrupt the Redfield ratio in the lagoon environment. For example, Pulicat receives industrial effluents from the Arani and Kalangi rivers, which dump high concentrations of nutrients into the lagoon (Basha et al. 2012). Hence toxic phytoplankton blooms are not uncommon in Pulicat (Santhanam and Farooqui 2018). Pulicat also experiences thermal pollution due to the release of hot coolant water and discharge of toxic fly ash in the form of a slurry from the nearby thermal power plant, which causes a rise of water temperature by 5 °C resulting in a decline in phytoplankton species. It has been reported that phytoplankton in Pulicat may have reduced to just half from the original level due to pollution (Basha et al. 2012). Badusha and Santhosh (2018) observed that, in Ashtamudi Lagoon, phytoplankton species diversity reached its minimum, where oil pollution from tourist boats and fecal contamination is more, while maximum diversity reached up to 41 species, where pollution and turbidity were minimum. The Vembanad receives an excess amount of nutrients from adjacent Kuttanad paddy fields leading to accelerated growth of bacillariophytes and chlorophytes. It is important to note that cyanophytes are favored in organic wastes, potentially replacing Dinophyta and bacillariophytes (Nandan and Sajeevan 2018; Vidya et al. 2020). High BOD designates unhealthy and poor water quality of an aquatic system and high microbial load.

5 Conclusion and Future Research Directions

Few trends are apparent from this literature review on phytoplankton in the Indian coastal lagoon. First, Indian coastal lagoons are not as intensively studied as other coastal estuarine systems, although coastal lagoons are crucial for providing various ecosystem services. Most of the studies relied on either one-time sampling or seasonal sampling. No long-term (more than 5 years) studies on phytoplankton biomass and productivity pattern change over the year have been established. Second, investigation and publication frequency from Indian coastal lagoons are disproportionately skewed than others, Chilika in the East Coast, for example has the highest number of scientific publication. Third, when it comes to studying phytoplankton in Indian lagoons, all are focused on microscopic studies. Microscopy has the distinct advantage of identifying phytoplankton at the species level, but it is time-consuming, prone to subjective error and needs trained man power. Fourth, the impact of monsoon-driven rainfall is distinct in shallow coastal aquatic systems such as lagoons. Monsoon brings freshwater, sediments, and nutrients in the system from the catchment, but it does not coincide with the maximum phytoplankton growth and biomass accumulation, which instead happens a few months later after water column gets stabilized. Fifth, anthropogenic impact such as thermal pollution and organic and inorganic matter loading in the Indian coastal lagoons is on the rise, leading to cascading effects on the food web. Sixth, for sustainable management of lagoon ecosystem, it is needed to understand the ecological components (physical form, lagoon soils, physicochemical characteristic of water biota, climate, geomorphology, hydrobiology, energy-nutrient dynamics), ecological processes (the process that maintains animal and plant population, species interaction, physical processes), and ecological services (provisioning services and cultural services) in a long-term basis. Increased frequency of monitoring can ensure sustainable management of these vulnerable ecosystems in Indian coastal waters. Remote sensing observation for case-2 waters, remote sensing algorithms for observation and detection of multiple phytoplankton types from satellite data, ecological algorithm, and modeling (primary productivity, carbon budget, nutrient budget, trophic status indexing), now-casting, and forecasting water quality parameters are possible future research scopes. The importance of deploying high-frequency remote sensors for cost-effective, long-term monitoring of large lagoons is gaining attention. Besides, HPLC to characterize pigment as a chemical marker for the determination of phytoplankton size classes and functional types can also give valuable information to study phytoplankton at the interface of biogeochemistry and management. More such studies need to be undertaken in the future.

Acknowledgments SS (the JRF) and TA (the PI) acknowledge financial support from Science and Engineering Research Board (SERB) funded project "Phyto-Health" (File #ECR/2017/001005).

References

Ahner BA, Morel FMM, Moffett JW (1997) Trace metal control of phytochelatin production in coastal waters. Limnol Oceanogr 42(3):601–608. https://doi.org/10.4319/lo.1997.42.3.0601

Amir M, Paul D, Samal RN (2019) Sources of organic matter in Chilika Lagoon, India inferred from stable C and N isotopic compositions of particulates and sediments. J Mar Syst 194:81–90. https://doi.org/10.1016/j.jmarsys.2019.03.001

Asha CV, Retina IC, Suson PS (2016). Ecosystem analysis of the degrading Vembanad wetland ecosystem, the largest Ramsar site on the South West Coast of India—Measures for its sustainable management. Regional Studies in Marine Science 8:408–421

Babu A, Varadharajan D, Perumal NV, Thilagavathi B, Manikandarajan T, Sampathkumar P, Balasubramanian T (2013) Diversity of phytoplankton in different stations from Muthupettai, South East Coast of India. J Mar Sci Res Dev 3(3). https://doi.org/10.4172/2155-9910.1000128

Badusha M, Santhosh S (2018) Diversity and abundance of phytoplankton with respect to physico-chemical parameters in ashtamudi wetland, Kerala, India. Indian J Environ Sci 22(2):89–94

Balasubramanian CP, Pillai SM, Ravichandran P (2004) Zero-water exchange shrimp farming systems (extensive) in the periphery of Chilka lagoon, Orissa, India. Aquac Int 12(6):555–572. https://doi.org/10.1007/s10499-004-0350-3

Barik SK, Muduli PR, Mohanty B, Behera AT, Mallick S, Das A, Samal RN, Rastogi G, Pattnaik AK (2017) Spatio-temporal variability and the impact of Phailin on water quality of Chilika Lagoon. Cont Shelf Res 136:39–56. https://doi.org/10.1016/j.csr.2017.01.019

Basha S, Laksmi ER, Rao BR, Murthy CVN, Savithramma N (2012) Biodiversity and conservation of Pulicat Lake—Andhra Pradesh. Int J Geol Earth Environ Sci 2(2):129–135

Basuri CK, Pazhaniyappan E, Munnooru K, Chandrasekaran M, Vinjamuri RR, Karri R, Mallavarapu RV (2020) Composition and distribution of planktonic ciliates with indications to water quality in a shallow hypersaline lagoon (Pulicat Lake, India). Environ Sci Pollut Res 27(15):18303–18316. https://doi.org/10.1007/s11356-020-08177-6

Bledsoe EL, Phlips EJ (2000) Relationships between phytoplankton standing crop and physical, chemical, and biological gradients in the Suwannee River and plume region, USA. Estuaries 23(4):458–473. http://www.jstor.org/stable/1353139

Borja Á, Elliott M, Carstensen J, Heiskanen AS, van de Bund W (2010) Marine management—towards an integrated implementation of the European marine strategy framework and the water framework directives. Mar Pollut Bull 60(12):2175–2186. https://doi.org/10.1016/j.marpolbul.2010.09.026

Brutemark A, Rengefors K, Anderson NJ (2006) An experimental investigation of phytoplankton nutrient limitation in two contrasting low arctic lakes. Polar Biol 29(6):487–494. https://doi.org/10.1007/s00300-005-0079-0

Clegg MR, Maberly SC, Jones RI (2007) Behavioral response as a predictor of seasonal depth distribution and vertical niche separation in freshwater phytoplanktonic flagellates. Limnol Oceanogr 52(1):441–455. https://doi.org/10.4319/lo.2007.52.1.0441

Cloern JE, Dufford R (2005) Phytoplankton community ecology: principles applied in San Francisco Bay. Mar Ecol Prog Ser 285(Cembella 2003):11–28. https://doi.org/10.3354/meps285011

CPCB (1986) Environment protection rules. CPCB, New Delhi. https://scholar.google.com/scholar?hl=en&as_sdt=0,5&q=CPCB+New+Delhi+(1986)+Environment+protection+rules.+https://www.cpcb.nic.in/+Water+quality+Criteria.php.+Accessed+5+Jan+2016

Duarte P, Bernardo JM, Costa AM, Macedo F, Calado G, Cancela da Fonseca L (2002) Analysis of coastal lagoon metabolism as a basis for management. Aquat Ecol 36(1):3–19. https://doi.org/10.1023/A:1013394521627

Dube A, Jayaraman G (2008) Mathematical modelling of the seasonal variability of plankton in a shallow lagoon. Nonlinear Anal Theory Meth Appl 69(3):850–865. https://doi.org/10.1016/j.na.2008.02.122

Dube A, Jayaraman G, Rani R (2010) Modelling the effects of variable salinity on the temporal distribution of plankton in shallow coastal lagoons. J Hydro Environ Res 4(3):199–209. https://doi.org/10.1016/j.jher.2010.03.003

Flöder S, Urabe J, Kawabata ZI (2002) The influence of fluctuating light intensities on species composition and diversity of natural phytoplankton communities. Oecologia 133(3):395–401. https://doi.org/10.1007/s00442-002-1048-8

Frost BW (1980) Grazing. In: The physiological ecology of phytoplankton. University of California Press, Berkeley, CA, pp 465–491. https://scholar.google.com/scholar?hl=en&as_sdt=0%2C5&q=Frost%2C+B.+W.+1980+Grazing.+In+The+Physiological+Ecology+of+Phytoplankton+%28Morris%2C+I.%2C+ed.%29.+University+of+California+Press%2C+Berkeley%2C+California%2C+USA%2C+pp.+465–492&btnG=

Gallegos CL (1992) Phytoplankton photosynthesis, productivity, and species composition in a eutrophic estuary: comparison of bloom and non-bloom assemblages. Mar Ecol Prog Ser 81(3):257–267. https://doi.org/10.3354/meps081257

Ganguly D, Patra S, Muduli PR, Vishnu Vardhan K, Abhilash KR, Robin RS, Subramanian BR (2015) Influence of nutrient input on the trophic state of a tropical brackish water lagoon. J Earth Syst Sci 124(5):1005–1017. https://doi.org/10.1007/s12040-015-0582-9

Gao X, Zhou F, Chen CTA (2014) Pollution status of the Bohai Sea: an overview of the environmental quality assessment related trace metals. Environ Int 62:12–30. https://doi.org/10.1016/j.envint.2013.09.019

Garnier J, Billen G, Palfner L (1999) Understanding the oxygen budget and related ecological processes in the river Mosel: the RIVERSTRAHLER approach. Hydrobiologia 410:151–166. https://doi.org/10.1023/A:1003894200796

Goebel NL, Edwards CA, Zehr JP, Follows MJ, Morgan SG (2013) Modeled phytoplankton diversity and productivity in the California current system. Ecol Model 264:37–47. https://doi.org/10.1016/j.ecolmodel.2012.11.008

Goosse H, Barriat P, Loutre M, Zunz V (2010) Chapter 5: Brief history of climate: causes and mechanisms. Introduction to climate dynamics and climate modelling. http://www.climate.be/textbook

Gupta GVM, Natesan U, Murthy MR, Kumar VS, Viswanathan S, Bhat MS, ... Subramanian, BR (2006). Nutrient budgets for Muthupet lagoon, southeastern India. Current Science 967–972

Hinds WC (1999) Aerosol technology properties, behavior, and measurement of airborne particles. Wiley, New York

Ingole BS (2005) Indian Ocean coasts, coastal ecology. In: Encyclopedia of coastal science, Jan, pp 546–554. https://doi.org/10.1007/1-4020-3880-1

Jyothibabu R, Madhu NV, Jayalakshmi KV, Balachandran KK, Shiyas CA, Martin GD, Nair KKC (2006) Impact of freshwater influx on microzooplankton mediated food web in a tropical estuary (Cochin backwaters—India). Estuar Coast Shelf Sci 69(3–4):505–518. https://doi.org/10.1016/j.ecss.2006.05.013

Jyothibabu R, Mohan AP, Jagadeesan L, Anjusha A, Muraleedharan KR, Lallu KR, Kiran K, Ullas N (2013) Ecology and trophic preference of picoplankton and nanoplankton in the Gulf of Mannar and the Palk Bay, southeast coast of India. J Mar Syst 111:29–44

Kjerfve B (1994) Coastal lagoons. Coastal Lagoon Proc 1952:1–8. https://doi.org/10.2307/634562

Kosten S, Huszar VLM, Bécares E, Costa LS, van Donk E, Hansson LA, Jeppesen E, Kruk C, Lacerot G, Mazzeo N, De Meester L, Moss B, Lürling M, Nõges T, Romo S, Scheffer M (2012) Warmer climates boost cyanobacterial dominance in shallow lakes. Glob Chang Biol 18(1):118–126. https://doi.org/10.1111/j.1365-2486.2011.02488.x

Kumar R, Pattnaik AK (2012) Chilika: an integrated management planning framework for conservation and wise use. Wetlands International-South Asia, New Delhi and Chilika Development Authority, Bhubaneswar. https://idl-bnc-idrc.dspacedirect.org/handle/10625/52100

Mahapatro D, Panigrahy RC, Panda S (2013) Coastal lagoon: present status and future challenges coastal lagoon: present status and future challenges. Int J Mar Sci 3:178–186. https://doi.org/10.5376/ijms.2013.03.0023

Mallin MA, Paerl HW, Rudek J, Bates PW (1993) Regulation of estuarine primary production by watershed rainfall and river flow. Mar Ecol Prog Ser 93(1–2):199–203. https://doi.org/10.3354/meps093199

Mathew T, Nair B (1980) Phytoplankton of the Ashtamudi Estuary, Kerala. Indian J Mar Sci 9:253–257

Mathew T, Nair NB (1981) Phytoplankton of the Veli Lake, a lagoon on the south-west coast of India. Mahasagar Bull Natl Inst Oceanogr 14(1):45–54

Mishra S, Panda D, Panigrahy RC (1993) Physico-chemical characteristics of the Bahuda estuary (Orissa), east coast of India. Indian J Mar Sci 22:75–77

Mohanty PK, Panda BUS (2009) Chilika: an integrated management planning framework for conservation and wise use. Indian J Mar Sci 38(2):209–214

Muduli PR, Vardhan Kanuri V, Robin RS, Charan Kumar B, Patra S, Raman AV, Nageswara Rao G, Subramanian BR (2013) Distribution of dissolved inorganic carbon and net ecosystem production in a tropical brackish water lagoon, India. Cont Shelf Res 64:75–87. https://doi.org/10.1016/j.csr.2013.05.014

Murty AVS, Varma PU (1964) The hydrographical features of the waters of Palk Bay during March, 1963. Mar Biol Assoc India 6:207–216

Nagaraj S (2014) Impact of urbanisation on water quality parameters—a case study of Ashtamudi Lake Kollam. Int J Res Eng Technol 3(18):140–147. https://doi.org/10.15623/ijret.2014.0318023

Nandan SB, Sajeevan K (2018) Distribution and abundance of phytoplankton in vembanad estuary, a Ramsar site on south west coast. Int J Eng Technol Manag Res 5(3):75–87. https://doi.org/10.5281/zenodo.1207427

Nayar S (2006). Spatio-temporal fluxes in particulate organic carbon in a tropical coastal lagoon. Environmental monitoring and assessment 112(1):53–68

Nayar S (2006) Spatio-temporal fluxes in particulate organic carbon in a tropical coastal lagoon. Environ Monit Assess 112(1–3):53–68. https://doi.org/10.1007/s10661-006-0310-7

Nayar S, Gowda G, Gupta TRC (1999) Size-fractionated primary production of a tropical coastal lagoon on the south west coast of India. Asian Fish Sci 12:217–222

Nayar S, Gowda G, Lakshmipathi MT (2001) Diel variation of biotic parameters in a tropical coastal lagoon on the south west coast of India. J Inland Fish Soc India 33(1):32–37

Ndimele PE (2012) The effects of water hyacinth (Eichhornia crassipes [Mart.] Solms) infestation on the physico-chemistry, nutrient and heavy metal content of Badagry Creek and Ologe Lagoon, Lagos, Nigeria. J Environ Sci Technol 5(2):128–136

Newton A, Brito AC, Icely JD, Derolez V, Clara I, Angus S, Schernewski G, Inácio M, Lillebø AI, Sousa AI, Béjaoui B, Solidoro C, Tosic M, Cañedo-Argüelles M, Yamamuro M, Reizopoulou S, Tseng HC, Canu D, Roselli L et al (2018) Assessing, quantifying and valuing the ecosystem services of coastal lagoons. J Nat Conserv 44:50–65. https://doi.org/10.1016/j.jnc.2018.02.009

Nichols MM, Boon JD (1994) Sediment transport processes in coastal lagoons. Elsevier Oceanogr Ser 60(C):157–219. https://doi.org/10.1016/S0422-9894(08)70012-6

Nixon SW (1995) Coastal marine eutrophication: a definition, social causes, and future concerns. Ophelia 41(1):199–219. https://doi.org/10.1080/00785236.1995.10422044

Panigrahi S, Wikner J, Panigrahy RC, Satapathy KK, Acharya BC (2009) Variability of nutrients and phytoplankton biomass in a shallow brackish water ecosystem (Chilika Lagoon, India). Limnology 10(2):73–85. https://doi.org/10.1007/s10201-009-0262-z

Perumal NV, Rajkumar M, Perumal P, Rajasekar KT (2009) Seasonal variations of phytoplankton diversity in the Kaduviyar estuary, Nagapattinam, southeast coast of India. J Environ Biol 30:1035–1046

Phlips EJ, Badylak S, Grosskopf T (2002) Factors affecting the abundance of phytoplankton in a restricted subtropical lagoon, the Indian River Lagoon, Florida, USA. Estuar Coast Shelf Sci 55(3):385–402. https://doi.org/10.1006/ecss.2001.0912

Prasath BB, Nandankumar R, Santhanam P (2019) Biodiversity and aquatic research: an international journal seasonal variations in diversity and relative abundance of phytoplankton from the Muttukkadu backwaters, southeast coast of India. Biodiver Aquat Res 1(2):1–10

Rajkumar M, Perumal P, Prabu VA, Perumal NV, Rajasekar KT (2009) Phytoplankton diversity in Pichavaram mangrove waters from south-east coast of India. J Environ Biol 30(4):489–498

Ramakrishnan R, Perumal P, Santhanam P (1999) Spatio-temporal variations of hydrographical features in the Pichavaram mangroves and Mohi aqua farm, southeast coast of India. In: Proceedings in international seminar for application in hydrogeochemistry. Annamalai University, Chidambaram, pp 197–203

Raman AV, Satyanarayana C, Adishasai K, Prakash KP (1990) Phytoplankton characteristics of Chilka Lake, a brackish water lagoon along east coast of India. Indian J Mar Sci 19(4):274–277

Rao RR, Kumar MSG, Ravichandran M (2011) Author version 32(22):1–22

Richardson K, Jørgensen BB (1996) Eutrophication: definition, history and effects. Vol 52, pp 1–19. https://doi.org/10.1029/ce052p0001

Sajinkumar KS, Revathy A, Rani VR (2017). Hydrogeochemistry and spatio-temporal changes of a tropical coastal wetland system: Veli-Akkulam Lake, Thiruvananthapuram, India. Applied Water Science 7(3):1521–1534.

Santhanam H, Farooqui A (2018) Bloom of the diatom, Biddulphia sp. and ecology of Pulicat lagoon, Southeast India in the aftermath of the 2015 north east monsoonal rainfall. Environ Monit Assess 190:636. https://doi.org/10.1007/s10661-018-7020-9

Santhanam P, Kumar N, Selvaraju A (2013) Impact of Trichodesmium erythraeum Ehrenberg bloom on water quality and productivity of Muthupet lagoon, Southeast coast of India. Seaweed Res Util Assoc 35(1 & 2):178–186

Saravanakumar A, Rajkumar M, Thivakaran GA, Serebiah JS (2008) Abundance and seasonal variations of phytoplankton in the creek waters of western mangrove of Kachchh-Gujarat. J Environ Biol 29(2):271–274

Sasamal SK, Panigrahy RC, Misra S (2005) Asterionella blooms in the northwestern Bay of Bengal during 2004. Int J Remote Sens 26(17):3853–3858. https://doi.org/10.1080/01431160500185391

Sculthorpe C (1967) Biology of aquatic vascular plants. https://agris.fao.org/agris-search/search.do?recordID=US201300598176

Senthilkumar S, Santhanam P, Perumal P (2002) Diversity of phytoplankton in Vellar estuary, southeast coast of India. In: The 5th Indian fisheries forum proceedings. AFSIB, Mangalore, pp 245–248

Srichandan S, Rastogi G (2020) Spatiotemporal assessment of phytoplankton communities in the Chilika Lagoon. In: Finlayson CM, Rastogi G, Mishra DR (eds) Ecology, conservation, and restoration of Chilika Lagoon, India, Wetlands: ecology, conservation and management, vol 6, 6th edn. Springer, New York, pp 251–294. https://doi.org/10.1007/978-3-030-33424-6_11

Srichandan S, Kim JY, Bhadury P, Barik SK, Muduli PR, Samal RN, Pattnaik AK, Rastogi G (2015a) Spatiotemporal distribution and composition of phytoplankton assemblages in a coastal tropical lagoon: Chilika, India. Environ Monit Assess 187:47. https://doi.org/10.1007/s10661-014-4212-9

Srichandan S, Kim JY, Kumar A, Mishra DR, Bhadury P, Muduli PR, Pattnaik AK, Rastogi G (2015b) Interannual and cyclone-driven variability in phytoplankton communities of a tropical coastal lagoon. Mar Pollut Bull 101(1):39–52. https://doi.org/10.1016/j.marpolbul.2015.11.030

Underwood GJC, Kromkamp J (1999) Primary production by phytoplankton and Microphytobenthos in estuaries

US Environmental Protection Agency (2012) 2012 guidelines for water reuse, vol 26, Sept. http://www.epa.gov/nrmrl/pubs/625r04108/625r04108.pdf

Vasudevan S, Arulmoorthy MP, Gnanamoorthy P, Ashokprabu V, Srinivasan M (2015) Continuous blooming of harmful microalgae Microcystis Aeruginosa Kutzing, 1846 in Muttukadu estuary, Tamilnadu, southeast coast of India. Int J Sci Invent Today 4(1):15–23

Vidya V, Prasad G, Moses A (2020) Assessment of threats to a Ramsar site from seafood processing operation effluents. Lakes Reserv Res Manag 2019:1–18. https://doi.org/10.1111/lre.12321

Chapter 6
Growing Menace of Microplastics in and Around the Coastal Ecosystem

Moharana Choudhury, Anu Sharma, Asma Pervez, Prachi Upadhyay, and Joystu Dutta

Abstract Microplastics have emerged as a major threat for the aquatic ecosystems in recent decades. They have been a great danger to the coastal ecosystems where they pose serious harmful effects on the water quality, fishes, dolphins, crabs, planktons, and other benthic organisms. Urbanization and industrialization have brought with them a deteriorating impact on the environment. Such harmful impacts are very much visible in various forms like air pollution, water pollution, toxic chemicals, waste generation and management issues, and degradation of land. Among many such cases, plastic pollution in general and microplastics in particular is the one among many. According to sizes, plastics are grouped as macroplastics and microplastics. The difference is because of size difference of fragments. Microplastics are generally less than 5 mm, and above this are the macroplastics. They are widely spread across the marine ecosystems. They have become part of the system. They pose many dangers to marine biodiversity because whatever be the source and site of their generation, they are carried away with rivers, streams, and floodwaters to oceans. Seas and oceans are their final destination, where they remain in the form of debris and contaminate the water quality and affect marine biodiversity as well. Often, the fishes, whales, and other sea animals eat plastics considering them to be their food. Microplastics are chemical formulation. When ingested by

M. Choudhury
Voice of Environment (VoE), Guwahati, Assam, India

A. Sharma (✉)
Govt. Degree College, Bhaderwah, Doda, University of Jammu, Jammu, Jammu and Kashmir, India

A. Pervez
Department of Environmental Sciences, Central University of Jammu, Jammu, Jammu and Kashmir, India

P. Upadhyay
Forest Research Institute Deemed to be University, Dehradun, Uttarakhand, India

J. Dutta
Department of Environmental Science, Sant Gahira Guru University, Ambikapur, Chhattisgarh, India

S. Madhav et al. (eds.), *Coastal Ecosystems*, Coastal Research Library 38,
https://doi.org/10.1007/978-3-030-84255-0_6

sea animals, they suffer from many health issues. Many times, they die out of suffocation. As far as microplastics are concerned, they remain suspended in the water because of their minimal size, which ultimately hinders the marine animals' various life processes—problems related to the coastal regions and their great significance in the overall ecology. This chapter is designed to focus on the spread of microplastics' menace across the globe and the ill effects caused by it on marine life and overall system, the challenges involved in tackling such a situation, and possible recommendations.

Keywords Coastal ecosystems · Harmful impacts · Microplastics · Urbanization · Waste management

1 Introduction

Environmentally persistent plastics are graded according to their size, although the size limits are not yet accepted (Gilgault et al. 2018; Frias and Nash 2019). They are commonly categorized as microplastics (particles >2.5 cm), mesoplastics (2.5 cm–5 mm), and macroplastics. The upper and lower limits (5 mm) and nanoplastics (1–100 nm) of these groups are constantly discussed. Macroplastic products, like bags made up of plastic, bottles, discarded fishing nets, plastic toys, etc., are considered the biggest marine debris component that can be seen in beaches, sea beds, or floats that eventually split into smaller pieces (Litter MTSOM 2013). Microplastics cannot be easily detected in any ecosystem, especially when it gets mixed with sediments. Microplastics have formed from several distinct molecules, which refer to different product forms like polypropylene (PP), LDPE, polyurethane polyethylene terephthalate (PET), HDPE, polystyrene (PS), polyvinyl chloride (PVC), and polyamide terephthalate (PET) which are the most widely generated and consumed polymer forms of microplastics. They contain chemicals that can be ingested and incorporated into the food network, such as phthalates, organic contaminants, and pathogens (Barcelo and Pico 2020).

Microplastic pollution in the coastal ecosystem has become a threat now. This has received ample recognition not only from scientists and policymakers but also from the common public. Microplastics affect the breeding and reproductive behavior of marine animals. Impacts can be observed in different varieties of organisms, including vertebrates and invertebrates, and, of course, in tiny organisms (Galloway et al. 2017; Law 2017). Several studies record marine microplastic ingestion (Lusher 2015). There can be seen many species growing on massive floating plastic waste, and then these are carried to novel places they had not occupied before (Kiessling et al. 2015). Since vertebrate species are somewhat more extensive, they are more recognized, noticed, and identified than small aquatic invertebrates. The seabirds and other marine animals get trapped in massive plastic debris, networks, chains, etc., and reports of such happenings have existed since the early 1970s (Derraik 2002). In the same manner, the effect of microplastics on fishes and seabirds has been well-known. The species affected, including the seabirds, are becoming larger and larger (Kenyon and Kridler 1969; Carpenter et al. 1972; Ryan 1988) and continue to expand (Wilcox et al. 2016). Waste generated from different land sources is potentially caught up in the water of coastal areas and in areas closer to the origin (Hinojosa and Thiel 2009; Rech et al. 2014). Owing to their minimal size, microplastics are easily ingested by zooplankton, thus causing negative impacts on their biological processes. Since zooplanktons are the primary source of food for other organisms, especially secondary consumers, microplastics enter food web and spread across the trophic level from base to top (Botterell et al. 2019). Microplastics show a capability to absorb various organic contaminants, metals, and pathogens into the bodies of organisms from the environment, thereby increasing the toxicity and related effects. Microplastic readily gets translocated into gastrointestinal membranes through endocytosis and ultimately is distributed into different organs (Alimba and Faggio 2019).

The plastic load that is added in the oceans from land is extensive and voluminous, and several reports say that there has been an increase in it, with estimates ranging between 4 and 12 million tonnes annually (Jambeck et al. 2015). Polymers are located all over the seas and oceans. They continue to be negatively buoyant in ocean waters. Furthermore, plastics and chemical residue gather at the ocean bottom (Backhurst and Cole 2000; Angiolillo et al. 2015). Several explorers and researchers have discovered microplastics in almost all oceans across the world. Various studies conducted across water bodies include the Pacific (Moore et al. 2001), Mediterranean (Collignon et al. 2012; Vianello et al. 2013), North Sea (Claessens et al. 2011; Dubaish and Liebezeit 2013) and polar waters in Arctic region. All environmental forms, including surface water (Moore et al. 2001) and columns of water (Lattin et al. 2004; Ng and Obbard 2006), are polluted (Murray and Cowie 2011; Fossi et al. 2012; Lusher et al. 2013; Devriese et al. 2015). About 1400 species of aquatic organisms in the marine water is seriously impacted by pollution. If the elimination rate of microplastic particles is prolonged or not at all, it can, in turn, lead to a reduced intake rate, thus leading to decreased intake of food, therefore leading to hunger and hence disastrous impacts (Cole et al. 2011; Jemec et al. 2016; Windsor et al. 2019). Among many researches by several scientists aiming at the bottom-dwelling marine debris on the Antalya shelf, Olguner et al. (2018) recorded 72% of debris in marine ecosystem as microplastics. Gundogdu et al. (2017) observed that Levantine Coast situated in northeast Turkey is highly microplastic-polluted. The dispersal of microplastic on the Marmara sea surface was analyzed by Tuncer et al. (2018), and it was reported that microplastic concentrations are far higher than neighboring areas. While extensive studies have illustrated the spread and abundance of microplastics along the Turkish coasts, the transport mechanism of microplastic to the Turkish coast is still to be studied.

Microplastics can arise from the primary plastics often used as resin pellets or in customized care items deliberately (Fendall and Sewell 2009). When tiny molecules enter the system, their suspected bioavailability for aquatic species is the primary risk associated with them (Wright et al. 2013; Desforges et al. 2015). Microplastics are taken up by various types of animals, including invertebrates, fish, birds, turtles, and other animals found on the sea (Di Beneditto and Awabdi 2014; Lavers et al. 2014; Lusher et al. 2015; Nadal et al. 2016; Peters and Bratton 2016; Welden and Cowie 2016). The most ordinarily contained in the stomachs or intestines of marine tortoises include medium-sized plastic parts, like plastic containers, rope, nylon monofilament, and fishing fillets (Brito 2001; Guerra-Correa et al. 2007; IMARPE 2011; Jiménez et al. 2017). Several scholars have reported the death of turtles which were stranded in Ecuador and Chile due to plastic use (Brito et al. 2007; Silva et al. 2007; Alemán 2014). Marine plastic debris act as a sink for various toxic chemicals and metals present in water and can also act as a transmitter for long-term chemical transportation (Ogata et al. 2009; Engler 2012; Holmes et al. 2012). Persistent organic pollutants (POPs) like polychlorinated biphenyls, polybrominated diphenyl ether, polycyclic aromatic hydrocarbons (PAHs), and other compounds are listed in many countries as primary pollutants (e.g., mercury, lead, and nonylphenols) and have been identified in chemicals that absorb marine plastic debris from coastal waters (Ogata et al. 2009; Hirai et al. 2011; Holmes et al. 2012). Microplastics are mainly composed of different pieces assembled and thus differ in their shape, size, color, and composition (Hidalgo-Ruz 2012). The environmental risk of primary importance associated with microplastic is its bioavailability for aquatic species. Among marine species, bivalves because of their substantial filter-feeding activity are of prime interest, and this activity introduces them to microplastics (Li et al. 2016).

Microplastics are rapidly identified in Chinese coastal ecosystems and quantified, and legislation to contain such contamination is strongly recommended (Wang et al. 2019). Approximately 39 tonnes of primary microplastics are estimated to be released into China's atmosphere based on available data (Lei et al. 2017). In recent research, China's coast was proposed as a hotspot for microplastic contamination (Zhao et al. 2014; Yu et al. 2016). The local land and ocean sources of microplastic include debris from various anthropogenic activities around the South Pacific (Thiel et al. 2013; Kiessling et al. 2017). Litter contamination is primarily from land, beach walkers, and marine activities such as aquaculture in the eastern Pacific (Astudillo et al. 2009; Hinojosa and Thiel 2009; Kiessling et al. 2017). In river water, huge quantity of plastics both macro and micro are also found (Rech et al. 2014, 2015). Floating microplastics' distribution in southeast Pacific shows the average distribution reported in other ocean basins and is the highest in the subtropical gyre (Eriksen et al. 2014; Law 2017).

India and her islands have a coastline of nearly 7500 km. Indian coasts have different habitats like mangrove forests, coral reefs, wetlands, shores, etc. (Malakar et al. 2019; Veerasingam et al. 2020). But these coasts face threats from increasing urbanization and industrialization, fishing activities, and induction of non-native species (Daniel et al. 2019; Imran et al. 2019; Naik et al. 2019; Sharma et al. 2020; Veerasingam et al. 2020). The plastic industry in India has grown enormously (FICCI 2014; Davis and Raja 2020), and with this the quantity of plastic waste is also very high contributing in pollution. As per the Central Pollution Control Board (CPCB) in year 2015, 82% out of 62 million tonnes waste was collected, and out of that only 28% was treated, whereas rest was dumped (Joshi and Ahmed 2016). It has been found that Indian Ocean flooring has heavy presence of fibers with 4 billion fibers per km^2 (Davis and Raja 2020). India is among the countries which contribute to enormous environmental pollution because of dumping of plastic waste in oceans and landfills each year (Porecha 2015; Davis and Raja 2020). This chapter attempts to assess the effect of microplastic pollution on marine animals and suggests measures to address the issue.

2 Few Sources of Microplastics

- The microplastics mainly originate from sources based on land such as sewage, storm water, as well as ocean-based sources such as fishing items that are lost or discarded (Li 2018).
- Another source of microplastic pollution in marine environment is the cosmetic industry which uses microbeads in their products especially face scrubs.
- Microplastics arise when microfibers are shredded during the washing of synthetic clothing (Thompson 2015; Napper and Thompson 2016).
- Besides, microplastics are washed by rainfall from the wearing of tyres on roads and finally are accumulated and block the drainage systems (Kole et al. 2017).

3 Menace Across the Globe

3.1 Effects on Turtles and Dolphins

Plastics and microplastics are the most popular waste on nearly 2000 km of coastline area of Rio de la Plata (Colombini et al. 2008; Esteves et al. 1997). Green and leatherback (*Dermochelys coriacea*) turtles that live on Rio de la Plata are considered plastic ingested, as per Gonzalez Carman et al. (2011, 2012, 2014) and are the most frequently affected animals. Gonzalez Carman et al. (2014) have shown that the turtles have been exposed and regularly consumed high amounts of plastic. More than 90% of the tortoises which were tested showed presence of plastics in their digestive tract, mostly plastic bags. The Franciscana dolphins (*Pontoporia blainvillei*) were also reported in resident plastics (Mendez et al. 2008; Denuncio et al. 2011) and are vulnerable as per IUCN Red List (IUCN 2013). Denuncio et al. (2011) studied and estimated that out of the dolphins caught nearly 88% were captured through nets which were of plastics, of which cellophane, plastic bags, and bands are the usual packaging debris. The source from where waste is ingested by sea turtles and dolphins can be both from urban and industrial and from shipping. A report from Turkey talks about the abundant availability and dissemination of plastic waste near Turkey's coastal areas. Many SE Pacific marine vertebrates have encountered sea litter, including fish, seabirds, marine turtles, etc. (Miranda-Urbina et al. 2015; Ory et al. 2017). Also, along the coasts, populations of Chondrichthyes are associated with marine litter present. Skates belonging to the genus *Sympterygia* lay extended tendril egg capsules (Oddone and Vooren 2002, 2008; Hernández et al. 2005; Flammang et al. 2007; Concha et al. 2013). Sea turtles are exposed to numerous anthropogenic stressors such as marine plastic contamination through different habitats, migration patterns, and nuanced life stories (Nelms et al. 2016).

3.2 *Effects on Coral Reefs*

The menace of microplastics has also affected the coral reefs. A recent study has revealed that some species of Scleractinia show the ability of ingesting microplastics (Hall et al. 2015). The study was conducted in the coasts of Australia. It was found that these microplastics are quickly accumulated within the bodies of coral reefs, which ultimately cause serious damage to them. It was found that these coral reefs get mistaken and consider microplastics as their prey and thus can consume up to 50 μg of plastic and the rate is close to their plankton consumption rate. However, the impact of ingesting microplastics and causing changes in the development of corals are still unknown (Hall et al. 2015). In a study of deep-sea corals, the results were also apparent. A study on different coral species (Reichert et al. 2018) has shown that each species responded differently to microplastic exposure. The responses by ingestion of microplastic polymers are mucus formation on reef corals, overgrowth, and microplastic attachment to tentacles (37–163 μm polyethylene) or mesenterial filaments. Coral bleaching and tissue necrosis, however, are the most damaging outcomes of microplastic consumption. Also, microplastic exposure leads to the stimulation of the stress response resulting in the repression through signal pathways, namely, Jun N-terminal kinase (JNK) and extracellular signal regulated kinase (ERK) pathway (Tang et al. 2018). Microplastics affect coral polyps through direct or indirect involvement or by disturbing photosynthesis as microplastics cover the coral surface (Syakti et al. 2019). The Indian coastal areas along South India (Chennai coast) have many well-listed and recorded coral species by the Zoological Survey of India. Brush coral (*Acropora hyacinthus*) and Biral coral (*Anacropora reticulata*) on Tamil Nadu's coast are also affected by chemicals from microplastics. The discharge coming from the nematocysts and ingestion of various microplastics (including PE) are discovered in examining hard corals, suggesting that the presence of phago-stimulants is potentially linked to toxic compounds. The study also proved that microplastics are stored in the intestinal cavity as well as inside mesenterial tissue (Allen et al. 2017; Nagarajan et al., 2020). Polystyrene in benthic invertebrates was analyzed using spectroscopic techniques in Kerala (Kochi, India).

3.3 *Effects on Crabs, Fishes*

In a study conducted along the California coast, Pacific mole crab (*Emerita analoga*) has shown to contain more than 15% of microplastics (Horn et al. 2019). It was shown that with increase in number of microplastics, the death rate of such crabs grew. Variability in embryos and reduction in egg-carrying capacity were other possible consequences of eating plastics (Horn et al. 2020). The fish bodies from the United Kingdom were also examined for microplastics presence (Lusher 2015). The gastrointestinal tracts of fishes such as *Basilichthys australis* (pejerrey) and *Aplodactylus punctatus* (jerguilla) of central Chile also showed presence of microplastics (Pozo et al. 2019). Researchers have also seen evidence of the scads that were exposed to synthetic plastic in Chile's Easter Island (Ory et al. 2017). According to US studies, fishes' adverse effects are among the microplastic exposure routes (Athey et al. 2020). The European bass is found with more microplastics in their systems than regular fish (Peda et al. 2016). The same aquatic species showed neurological changes linked to microplastic consumption. A concern for increase in mercury bioconcentration in gills and its bioaccumulation in the liver has also been linked to microplastics (Barboza et al. 2018). Microplastics observed in nine commercial fish market bivalve species and wild mussels caught along China's coastal waters are found to be in higher concentrations (Li et al. 2015, 2016). Higher amount of microplastics have been related to intensive anthropogenic activities. Microplastic contamination in estuarine and freshwater environments has also been found (Zhao et al. 2015; Su et al. 2016). However, data is not sufficiently available on the plastic contamination of coastal or freshwater fish in China. For the first time, in 21 sea fish and 6 freshwater fish in Shanghai, China, microplastic contamination was documented. In the stomach and intestine of researched fish, microplastic emissions were widespread and comparatively large. In most animals, microplastics concentration was high. The plastic abundance of the intestines in some animals was also more significant when compared to the belly (Jabeen et al. 2016). Studies reveal that many fish species ate microplastics during suction feeding. The microplastic is likely to be absorbed in fishes without them being directly exposed. Microplastics also accumulate in gills of a fish. The size, concentration, and toxicity are some of the factors that will dictate the extent to which microplastics will affect animals in question (Lusher 2015).

3.4 *Effects on Seabirds and Oysters*

Seabirds are another species primarily affected by interconnection and debris ingestion (Fossi et al. 2018). Seabirds are another species primarily affected by debris ingestion (Gallo et al. 2018). Microplastic ingestion is needed to be looked after especially for seabirds mainly because almost half seabird species are endangered and plastic affects the birds by altering its feeding behavior, reproductive ability, and mortality (Wilcox et al. 2015; Chatterjee and Sharma 2019). In six seabirds species, namely, *Phalacrocorax bougainvillii*, *Pelecanoides urinatrix*, *Pelecanoides garnotii*, *Spheniscus humboldti*, *Pelecanus thagus*, and *Larus dominicanus*, there were fragments of plastic (microplastics) in their stomach. *Larus dominicanus* showed maximum capacity to ingest as this bird commonly feeds on fishing nets, disposed products, and containers made of plastic (Thiel et al. 2018; Chatterjee and Sharma 2019). The amount of plastic ingested depends on factors like size of the bird, its weight, and its habitat. For example, *Spheniscus* penguins and *Thalassarche* albatross with small body size have lower ingestion rates, whereas species like *Fulmarus* fulmars, *Cyclorhynchus* auklets, *Oceanodroma*, *Pachyptila* prions, and *Pelagodroma* due to large body size ingest plastic debris at a faster rate (Wilcox et al. 2015; Chatterjee and Sharma 2019).

Oysters are also heavily influenced by microplastic waste. Oysters that have consumed microplastic particles collected them in their body and showed noticeable physical effects. Ward and Kach (2009) studied oysters and mussels (*Mytilus edulis*) to assess their capacity to absorb polystyrene beads in their experiment (Ward and Kach 2009).

3.5 Effects on Benthic Organisms

A study investigated microplastic particles' presence within *Sternaspis scutata*, *Magelona cinta*, and *Tellina* sp. (Naidu et al. 2017). It was found that microplastics are washed off by the seas into beaches and sedimentary formations along the shores. Unfortunately, these microplastics accumulate unintentionally or selectively within benthic organisms and deposit feeders (Thompson et al. 2004; Cole et al. 2011; Wright et al. 2013). Even though plastic might move slowly through the ocean, it continues to persist near the bottom of the ocean floor as it accumulates from the water (Chubarenko and Stepanova 2017). Nevertheless, understanding the benthic habitats and how they are influenced by microplastic contamination remains a field of significant research scope (Thompson et al. 2004; Cole et al. 2011; Maximenko et al. 2012; Wagner et al. 2014; Dris et al. 2015). The plastic bag is then crushed and changed into microplastics and mesoplastics (Andrady 2011; OSPAR 2014).

4 Challenges and Recommendations

Various studies indicate that because of ever-increasing population and extreme consumerism, microplastics have become difficult to handle and manage. De Frond et al. (2019) worked on selected chemical additives which are entering in oceans with conventional plastic waste, and the bulk of sorbed chemicals along with microplastics were worked on in a designated area. The weight of additives entering the oceans as components of seven common plastic waste materials bottle containers, caps, extended polystyrene (EPS), cutlery and food bags, food packets, and straws or stirrers was calculated in 2015. Some 190 tonnes of 20 chemical additives of these chemicals were reported from the oceans in 2015. The mass of PCBs correlated with microplastics was also measured in comparative case studies at two locations on the coasts of Hong Kong and Hawaii and at two on the west coasts of Hong Kong and Hawaii (North Pacific and South Atlantic gyres). The mass of chemicals where PCBs are closer to the source has been related to plastics' mass. The estimated number of plastic-related PCBs was approximately 85,000 times the average beach length in the North Pacific gyre in Hong Kong.

Structure and scale of plastic debris have been studied in surface waters on the Mediterranean Sea. In the northwest Mediterranean, coastal waters data from the previously published studies were combined with intensive samples. In regions far from the earth and at the first kilometer along the coast, the highest plastic concentration was found. Plastic concentrations were strongly linked to the proximity of a human coastal community. The nearshore water belt and local areas close to major anthropological habitations showed the massive number of plastic components every square km. The plastic-plankton ratio in coastal waters was exceptionally high. Three plastics, namely, polyethylene, polyamides, and polypropylene, were most prevalent at all locations off the coast. Though there were variety of polymers found in the coastal water strip, polystyrene or polyacrylic fibers were highly prevalent. A steady increase in the total to smaller sizes was shown by the plastic size distributions, suggesting that small plastics were effectively eliminated from the soil. However, the relative abundance of smaller particles less than 2 mm was much more extensive within 1-km strip of water, suggesting fragmentation down the shoreline which is probably related to washing ashore on the beaches. The paper is an effort to investigate the aftermaths of plastic waste in the Mediterranean region. If plastic concentrations are high for small plastic products, significant environmental, health, and economic effects could occur. With a coastal population of 466 million, the region is vast and enclosed (UNEP/MAP 2014), and significant demographic pressures support it. In 1976, as a matter of great importance, the Barcelona Convention for the Protection of the Mediterranean Sea highlighted anthropogenic marine litter's problem. Visual surveys were carried out in the central part of basin in detailed manner in 2013, and they revealed that much of the non-submersible litter consisted of plastics, often dominating the total debris observed. Sample's combined data set with those reported from the Mediterranean Sea previously of inland and offshore regions (de Lucia et al. 2014; Cozar et al. 2015) showed greater plastic abundance. The debris mainly consisting of plastic was typically determined in the 1-km water strip which was adjoining to the shore.

Mediterranean Sea is in critical status because it contains more than 115,000 particles/km^2. Plastic products are among the major kinds of debris in marine water. Microplastics can infiltrate the marine ecosystem and contaminate the whole food chain. But still there is a long way to go to reduce plastic in the Mediterranean Sea as well as protect marine biodiversity. Also, plastic debris include some chemicals that tend to assemble in the ocean (Fossi et al. 2016). If the reality is accurate, the outcome will pose a significant threat to endangered species which are part of the Mediterranean biodiversity. In the most biodiversity-rich habitats of the Mediterranean Sea, it is seen that cetaceans coexist with high human density, and these species are often affected due to microplastics in the Mediterranean Sea.

Microplastics can infiltrate the marine ecosystem and contaminate the food chain (Fossi et al. 2016). There, still, is a need of uniform methodology to detect or quantify microplastics in environment. Present methods have some limitation based on size or color (Bhattacharya and Khare 2019). Lack for effective sample design and processing methods often affects the spatial temporal patterns of these contaminants (Galgani et al. 2013; Veerasingam et al. 2020). So it is needed to work to standardize the methods for accurate microplastic assessment in environmental samples (Bhattacharya and Khare 2019) and come up with integrated sampling and processing methodologies (Galgani et al. 2013; Veerasingam et al. 2020). Majority of the work on microplastic till date is based in Europe, North America, and Australia. A few reports can also be found for countries like Brazil, India, and Japan. But since the issue of microplastic is global, extensive evaluation of microplastic is needed. This will help us not only to identify the extent and seriousness of issue but also to design and develop better technology to clean up microplastic from contaminated environment (Bhattacharya and Khare 2019).

5 Indian Context

"Plastic is very much on the menu," Prince Charles said at a recent Our Ocean summit. Our waterways are dwindling, and plastic waste has clogged floodplains. Thousands of quintals of plastic as waste have clogged India's waterways. They are starting from Assam's Barak Valley to the Ganges' riverbeds, from the Himalayan valleys in north to the rain-fed rivers in the southern India. We are in the middle of a disaster. The first report on pollution due to microplastic in India was the study conducted in Vembanad Lake, and sediments from lake were tested for microplastic, and low-density polyethylene microplastic was found dominantly (Sruthy and Ramasamy 2017; Davis and Raja 2020). Industries contribute a lot in plastic pollution. Plastic litter has been reported from the beaches of Karnataka (Sridhar et al. 2007; Davis and Raja 2020) and Caranzalem beach, Goa (Nigam 1982; Davis and Raja 2020). Reports on presence of resin pellets and debris have come from Chennai and Tennakkara Island (Mugilarasan et al. 2017; Davis and Raja 2020) and Great Nicobar, respectively (Dharani et al. 2003; Davis and Raja 2020). The floods of Chennai which occurred in 2015 have also influenced the plastic and micropellets present in the coasts of Chennai (Veerasingam et al. 2016; Pradhan et al. 2018; Davis and Raja 2020). Microplastics are toxic for the environment, and their pollution directly affects the fresh and marine waters. In India different health problems have arisen among people due to this (Raju et al. 2018; Davis and Raja 2020). India's government has initiated the "*Swachha Bharat Abhiyaan*" and "*Namami Gange*" initiatives, which are devoted for seeking long-term solutions to this pressing problem. At this crucial point, all aspiring environmentalists are responsible to raise massive awareness about plastic contamination in society. Strict rules, legislation, and regulations are in effect, but are not followed because of lack of awareness among the common people and carelessness by the administration. All disposable plastic types are banned in New Delhi and the NCR (National Capital Region). However, such initiatives do not last long again because of lack of awareness among the common people and carelessness by the administration.

According to reports, in India plastic waste of approximately 5.6 million tonnes is produced every year. It is approximately 60% of the total plastic waste being added in the world's oceans. As per an article published in the journal *Environmental Science and Technology* in October 2017, among the world's ten rivers, three rivers that bring 90% of plastic and add in the world's oceans are in India, namely, the Indus, the Ganga, and the Brahmaputra. Plastics have been prohibited in most Indian states since the beginning of 2018, but still, they clog rivers and landfills and impact overall environment.

In all of India's major cities, including Chennai, Delhi, Mumbai, and Kolkata, clogged drainage systems are causing water logging problems. Coastal cities are particularly vulnerable because they witness water upsurges regularly due to heavy flooding during monsoon, depression, tidal cycles, ocean circulations, and wind movements, and the clogging of all water sources is by plastic. The need of the hour is to become more cohesive in sustainable urban development and take the anti-plastics movement to every section of society to raise the collective consciousness. A nationwide campaign to ban plastics is going on by Voice of Environment, a Guwahati-based youth environmental organization (Dutta and Choudhury 2018). To make the premises plastic-free, recent awareness campaigns have been organized in Guwahati which is situated in northeastern state of Assam of India, at the popular *Maa Kamakhya*, *Basistha* Temple, and *Umananda* Island. *Shri Basistha* Temple and other historical and religious monuments also participate in related practices. At all levels of society, there is a need to develop model campaigns against plastic waste. If different bodies, government departments, and institutions refuse to use plastic, then the positive message will pass through all such initiatives. It is also crucial to comprehend plastic pollution at local level. Plastic pollution has a low level of public knowledge. Our heavy dependence on plastic is a significant impediment to combating the problem. Except for a few individuals and organizations who are environmentally conscious, the general public and their concern about the situation are generally marred by apathy and indifference. By 2022, India wants to be plastic-free. Plastic clean-up campaigns in public parks, forests, and biosphere reserves and also beach clean-up events are at work. Besides, the commitment calls for converting 100 monuments throughout the country into plastic- and litter-free zones. The Ministry of Tourism has responded by pledging to eliminate plastic straws and its use in public places. Several non-governmental organizations (NGOs) are working on local projects across the nation. Plastic has aroused as a hurdle in sustaining the indices for sustainable development, despite its reputation as a wonder material. In India and around the world, a complete and step-by-step ban on plastic is required. Various socioeconomic and political problems should resolve for the burning crisis to remain viable over time. However, the situation has recently deteriorated. It is the need of the hour to look for environmentally friendly alternatives like natural fibers (cotton, jute) and other forest-based resources to aid in long-term forest management (Khan et al. 2018). Using environmentally friendly alternatives to plastic will help to reduce their use. There are no such issues when using options such as plates of sal leaves or paper, cloth, and jute bags. Small-scale businesses and MSME will benefit from these environmentally sustainable alternatives. Making sal *thali* or plate is a traditional rural economic activity in India, particularly in Bihar, Jharkhand, Chhattisgarh, Odisha, and Madhya Pradesh. In the industry, it costs between Rs 20 and Rs 30 per hundred. Sal leaf value addition is a significant income source in rural areas of Uttar Pradesh, West Bengal, Arunachal Pradesh, Jharkhand, and Odisha. Microplastic and related studies in India are in progress and provide scope for future research (Veerasingam et al. 2020)

6 Suggestions

It is predicted that by 2025 plastic in marine ecosystem will see a substantial increase (Jambeck et al. 2015; Chatterjee and Sharma 2019). Seeing the seriousness of the issue, it was brought into notice in the "16th Global Meeting of the Regional Seas Conventions and Action Plans." This was organized to educate nation on plastic and how plastic pollution is damaging the marine ecosystem. The large creatures in marine ecosystem include fishes like sharks, reptiles, sea turtles and mammals like whales, and polar bears, and they are also prone to consume microplastics. Monetary value of the damage caused to marine ecosystem is calculated to be of approximately US $13 billion every year (Chatterjee and Sharma 2019). The various ways of managing the menace of microplastics range from less consumption to safe disposal to the level of awareness of people. The public's attitude is very significant in the overall plastic waste management. There are varieties of challenges that should be addressed before we plan for any strategy. Some of them are:

- The clear and pan vision of the policymakers and planners dealing with the policymaking is the need of the hour. We just cannot frame policies and let them be in the files. The guidelines should be suitable as per the local conditions of upstream and downstream.
- Attitude has a remarkable role to manage this problem extensively. The mindset of people of the area decides how the problem will be tackled. The sense of responsibility, accountability, and belongingness is an important part.
- Consumption patterns, in daily life as well as during festive seasons and marriages, are yet another challenge. The conspicuous way of consumption is also responsible for this situation.
- Disposal techniques play a prominent role. Proper and scientific disposal of plastics waste at the source of generation can play an essential part in the overall management of plastic.

- Management strategies should be well taken care of right from the stage of policy framing to its implementation.
- Manufacturing plastic should also be limited. It should be manufactured according to its final disposal. The principle of extended producer responsibility should be a primary part before any license is given to the manufacturer.
- The large population should be an asset and not a burden.
- Unplanned production and also consumption should be avoided.
- A scientific plan for safe and environmentally sound disposal should be there.
- Unsustainable tourist activities must not be allowed. Tourism and coastal regulation should be maintained through different coastal regulation zones.
- Strict implementation of legal norms should be there. The mere formation of laws is not sufficient. Their performance is significant to tackle the issue.
- Making public aware about the issue is also important.
- Control and regulation up-streams are an important part of the whole program.
- Use of biodegradable plastic in place of non-biodegradable ones is an important option to reduce plastic waste generated.

Today plastic waste and related microplastic pollution is increasing at an alarming rate. Its ill effects on environment are now easily identified (Bhattacharya and Khare 2019). Awaring people would make them think about innovative measures to manage plastic and related products which will assist in collection and reuse of the waste. To prevent the harm being caused by microplastic pollution, it is needed that we come up with alternative (Chatterjee and Sharma 2019) and check the situation before it's too late (Bhattacharya and Khare 2019).

7 Conclusion

The problem should be observed and addressed at the global level. Seas and oceans do not belong to one particular country. The activities carried out in one country affect world's different parts. Therefore, it is needed to be taken care that one country's activities are not affecting other parts of the world. The international agreements are currently in place to resolve significant environmental challenges globally. Manufacturing to sale to use-collection and final disposal forms a critical chain that involves all important plastic management steps. Microplastics' composition is yet another concern that should be analyzed, identified, and taken care of while handling and disposing them. Some steps that can become a part of the solution are strict coastal regulation zones, like a declaration of sensitive zones, declaration of buffer zones, and no permission or limited permission in regions close to the seas and oceans. The coastal cities are already overcrowded, and they suffer this threat from the local population as well as from the tourists. With the increasing population and changing lifestyle, a significant influx of tourism in the coastal areas is seen. A large population is dependent on seafood. Increasing tourism and eventually the waste in the coastal regions especially plastics are the most significant issues that need a deep introspection and understanding before formulating any plans and policies for the coastal areas. While developing strategies, we cannot ignore microplastic pollution as it has played havoc with marine biodiversity. When all said, everything is depending on the state of mind. It's the people of the land who have the responsibility to save and protect the oceans. *Homo sapiens* being at the center of the helm of affairs are wholly responsible for everything. They should ensure the survival of other species living on the earth. In the waters, too, since we are accountable for generating plastic generation problems, so we should also solve them.

References

Alemán R (2014) Informe de Actividades realizadas en Conservación de Tortugas Marinas durante el año 2013. Informe Técnico MAE-PN-PNMRA 010: 9

Alimba CG, Faggio C (2019) Microplastics in the marine environment: current trends in environmental pollution and mechanisms of toxicological profile. Environ Toxicol Pharmacol 68:61–74

Allen AS, Seymour AC, Rittschof D (2017) Chemoreception drives plastic consumption in a hard coral. Mar Pollut Bull 124:198–205. https://doi.org/10.1016/j.marpolbul.2017.07.030

Andrady AL (2011) Microplastics in the Marine Environment. Mar Pollut Bull 62: 1596–1605.

Angiolillo M, Lorenzo B, di Farcomeni A, Bo M, Bavestrello G, Santangelo G, Cau A, Mastascusa V, Cau A, Sacco F, Canese S (2015) Distribution and assessment of marine debris in the deep Tyrrhenian Sea (NW Mediterranean Sea, Italy). Mar Pollut Bull 92(1–2): 149–159

Astudillo JC, Bravo M, Dumont CP, Thiel M (2009) Detached aquaculture buoys in the SE Pacific: potential dispersal vehicles for associated organisms. Aquat Biol 5:219–231

Athey SN, Albotra SD, Gordon CA, Monteleone B, Seaton P, Andrady AL, Taylor AR, Brander SM (2020) Trophic transfer of microplastics in an estuarine food chain and the effects of a sorbed legacy pollutant. Limnol Oceanogr Lett 5:154–162. https://doi.org/10.1002/lol2.10130

Backhurst MK, Cole RG (2000) Subtidal benthic marine litter at Kawau Island, north-eastern New Zealand. J Environ Manag 60:227–237

Barboza LGA, Vieira LR, Branco V, Carvalho C, Guilhermino L (2018) Microplastics increase mercury bioconcentration in gills and bioaccumulation in the liver, and cause oxidative stress and damage in *Dicentrarchus labrax* juveniles. Sci Rep 8:1–9. https://doi.org/10.1038/s41598-018-34125-z

Barcelo D, Pico Y (2020) Case studies of macro- and microplastics pollution in coastal waters and rivers: is there a solution with new removal technologies and policy actions? Case Stud Chem Environ Eng. https://doi.org/10.1016/j.cscee.100019

Bhattacharya A, Khare S (2019) Microplastic pollution: an overview of current scenario, challenges, and research gaps. Adv Biotechnol Microbiol 12(3):555836

Botterell ZL, Beaumont N, Dorrington T, Steinke M, Thompson RC, Lindeque PK (2019) Bioavailability and effects of microplastics on marine zooplankton: a review. Environ Pollut 245:98–110

Brito JL (2001) "Informe preliminar de tortugas marinas en Chile: su situación actual," in Taller Nacional de Trabajo Para Definir las Líneas de Acción Prioritarias. de un Programa Para la Conservación de las Tortugas Marinas (Valparaíso), 95

Brito C, Brito JL, Zúñiga M, Campos M, Toro S (2007) "Tortugas marinas en el centro de rescate y rehabilitación de fauna silvestre del museo de San Antonio," in VII Simposio Sobre Medio Ambiente: Estado Actual y Perspectivas de la Investigación y Conservación de las Tortugas Marinas en las Costas del Pacífico Sur Oriental (Antofagasta) 96

Carpenter EJ, Anderson SJ, Harvey GR, Miklas HP, Peck BB (1972) Polystyrene spherules in coastal waters. Science 178(4062):749–750. http://www.ncbi.nlm.nih.gov/pubmed/4628343
Chatterjee S, Sharma S (2019) Microplastics in our oceans and marine health. Field Actions Sci Rep, Special Issue 19:54–61
Chubarenko I, Stepanova N (2017) Microplastics in sea coastal zone: Lessons learned from the Baltic amber, Environmental Pollution. https://dx.doi.org/10.1016/j.envpol.2017.01.085
Claessens M, Meester SD, Landuyt LV, Clerck KD, Janssen CR (2011) Occurrence and distribution of microplastics in marine sediments along the Belgian coast. Mar Pollut Bull 62:2199–2204. https://doi.org/10.1016/j.marpolbul.2011.06.030
Cole M, Lindeque P, Halsband C, Galloway TS (2011) Microplastics as contaminants in the marine environment: a review. Mar Pollut Bull 62:2588–2597
Collignon A, Hecq JH, Glagani F, Voisin P, Collard F, Goffart A (2012) Neustonic microplastic and zooplankton in the North western Mediterranean sea. Mar Pollut Bull 64:861–864. https://doi.org/10.1016/j.marpolbul.2012.01.011
Colombini M, Alderete S, Musmeci JM, Caille G, Harris G, Esteves JL (2008) Censo Nacional de Contaminacion Costera de la Republica Argentina. Proyecto Consolidacion e Implementacion del Plan de Manejo de la Zona Costera Patagonica para la Conservacion de la Biodiversidad. Informe Tecnico N_7. http://www.patagonianatural.org/informes-tecnicos.html. Accessed 18 Jun 14
Concha F, Morales N, Larraguibel J (2013) Egg capsules of the Filetail fanskate Sympterygia lima (Poeppig 1835) (Rajiformes, Arhynchobatidae) from the southeastern Pacific Ocean, with observations on captive egg-laying. Ichthyol Res 60:203–208
Cozar A, Sanz-Martin M, Marti E, González-Gordillo JI, Ubeda B, Gálvez JA (2015) Plastic accumulation in the Mediterranean Sea. PLoS One 10(4):e0121762. https://doi.org/10.1371/journal.pone.0121762
Daniel DB, Thomas SN, Thomson KT (2019) Assessment of fishing-related plastic debris along the beaches in Kerala Coast, India. Mar Pollut Bull 150:110696
Davis E, Raja S (2020) Sources and impact of microplastic pollution in Indian aquatic ecosystem: a review. Sustain Mining, Special Issue 1
De Frond HL, van Sebille E, Parnis JM, Diamond ML, Mallos N, Kingsbury T, Rochman CM (2019) Estimating the mass of chemicals associated with ocean plastic pollution to inform mitigation efforts. Integr Environ Assess Manag 15(4):596–606
de Lucia GA, Caliani I, Marra S, Camedda A, Coppa S, Alcaro L et al (2014) Amount and distribution of neustonic micro-plastic off the western Sardinian coast (Central-Western Mediterranean Sea). Mar Environ Res 100:10–16. https://doi.org/10.1016/j.marenvres.2014.03.017
Denuncio P, Bastida R, Dassis M, Giardino G, Gerpe M, Rodriguez D (2011) Plastic ingestion in Franciscana dolphins, Pontoporia blainvillei (Gervais and d'Orbigny, 1844), from Argentina. Mar Pollut Bull 62:1836–1841
Derraik JG (2002) The pollution of the marine environment by plastic debris: a review. Mar Pollut Bull 44(9):842–852
Desforges JP, Galbraith M, Ross PS (2015) Ingestion of Microplastics by Zooplankton in the Northeast Pacific Ocean. Arch Environ Contam Toxicol 69(3):320–330
Devriese LI, van der Meulen MD, Maes T, Bekaert K, Paul-Pont I, Frere L, Robbens J, Vethaak AD (2015) Microplastic contamination in brown shrimp (*Crangon crangon*, Linnaeus 1758) from coastal waters of the Southern North Sea and Channel area. Mar Pollut Bull 98:179–187. https://doi.org/10.1016/j.marpolbul.2015.06.051
Dharani G, Nazar AKA, Venkatesan R, Ravindran M (2003) Marine debris in Great Nicobar. Curr Sci 85:574–575
Di Beneditto APM, Awabdi DR (2014) How marine debris ingestion differs among megafauna species in a tropical coastal area. Mar Pollut Bull 88:86–90
Dris R, Imhof H, Sanchez W, Gasperi J, Galgani F, Tassin B, Laforsch C (2015) Beyond the ocean. Contamination of freshwater ecosystems with (micro-) plastic particles. Environ Chem 12:32

Dubaish F, Liebezeit G (2013) Suspended Microplastics and Black Carbon Particles in the Jade System, Southern North Sea. Water Air Soil Pollut 224:1352

Dutta J, Choudhury M (2018) Plastic pollution: a global problem from a local perspective. J Waste Manag Xenobio 1(1):000102. https://doi.org/10.23880/oajwx-16000102

Engler RE (2012) The complex interaction between marine debris and toxic chemicals in the ocean. Environ Sci Technol 46(22):12302–12315

Eriksen M, Lebreton LC, Carson HS, Thiel M, Moore CJ, Borerro JC (2014) Plastic pollution in the world's oceans: more than 5 trillion plastic pieces weighing over 250,000 tons afloat at sea. PLoS One 9(12):e111913

Esteves JL, Harris, G, Musmeci JM, Palla J, Sanchez JP (1997) Primer Censo de Contaminacion Costera de la Republica Argentina. Plan de Manejo Integrado de la Zona Costera Patagonica (Puerto Madryn, Argentina). Informe Tecnico N_41. http://www.patagonianatural.org/informes-tecnicos.html. Accessed 18 Jun 14

Fendall LS, Sewell MA (2009) Contributing to marine pollution by washing your face: microplastics in facial cleansers. Mar Pollut Bull 58:1225–1228

FICCI (2014) Potential of plastics industry in northern India with special focus on plasticulture and food processing—a report on plastics industry. Special Document

Flammang BE, Ebert DA, Cailliet GM (2007) Egg cases of the genus Apristurus (Chondrichthyes: Scyliorhinidae): phylogenetic and ecological implications. Zoology 110:308–317

Fossi MC, Panti C, Guerranti C, Coppola D, Giannetti M, Marsili L, Minutoli R (2012) Are baleen whales exposed to the threat of microplastics? A case study of the Mediterranean fin whale (*Balaenoptera physalus*). Mar Pollut Bull 64:2374–2379. https://doi.org/10.1016/j.marpolbul.2012.08.013

Fossi MC, Marsili L, Baini M, Giannetti M, Coppola D, Guerranti C, Caliani I, Minutoli R, Lauriano G, Finoia MG, Rubegni F, Panigada S, Bérubé M, Urbán Ramírez J, Panti C (2016) Fin whales and microplastics: the Mediterranean Sea and the Sea of Cortez scenarios. Environ Pollut 209:68–78. https://doi.org/10.1016/j.envpol.2015.11.022

Fossi GF, Weber C, Santillo R, Sousa D, Ingram JI, Nadal A, Romano D (2018) Marine litter plastics and microplastics and their toxic chemicals components: the need for urgent preventive measures. Environ Sci Eur 30:13. https://doi.org/10.1186/s12302-018-0139-z

Frias JPGL, Nash R (2019) Microplastics: finding a consensus on the definition. Mar Pollut Bull 138:145–147. https://doi.org/10.1016/j.marpolbul.2018.11.022

Galgani F, Hanke G, Werner S, Oosterbaan L, Nilsson P, Fleet D, Kinsey S, Thompson RC, Van Franeker J, Vlachogianni T (2013) Guidance on monitoring of marine litter in European seas. Publications Office of the European Union, Luxembourg

Gallo F, Fossi C, Weber R, Santillo D, Sousa J, Ingram I, Nadal A, Romano D (2018) Marine litter plastics and microplastics and their toxic chemicals components: the need for urgent preventive measures. Environ Sci Eur 30(1):13

Galloway TS, Cole M, Lewis C (2017) Interactions of microplastic debris throughout the marine ecosystem. Nat Ecol Evol 1

Gilgault J, Halle AT, Baudrimont M, Pascal PY, Gauffre F, Phi TL, El Hadri H, Grassl B, Reynaud S (2018) Current opinion: what is a nanoplastic? Environ Pollut 235:1030–1034. https://doi.org/10.1016/j.envpol.2018.01.024

Gonzalez Carman V, Alvarez K, Prosdocimi L, Inchaurraga MC, Dellacasa RF, Faiella A, Echenique C, Gonzalez R, Andrejuk J, Mianzan H, Campagna C, Albareda DA (2011) Argentinian coastal waters: a temperate habitat for three species of threatened sea turtles. Mar Biol Res 7:500–508

Gonzalez Carman V, Falabella V, Maxwell S, Albareda D, Campagna C, Mianzan H (2012) Revisiting the ontogenetic shift paradigm: the case of juvenile green turtles in the SW Atlantic. J Exp Mar Biol Ecol 429:64–72

Gonzalez Carman V, Acha EM, Maxwell SM, Albareda D, Campagna C, Mianzan H (2014) Young green turtles, Chelonia mydas, exposed to plastic in a frontal area of the SW Atlantic. Mar Pollut Bull 78:56–65

Guerra-Correa C, Valenzuela A, Retamal LM, Malinarich A (2007) "Influencia de los desechos plásticos en la sobrevivencia de tortugas: el caso de Chelonia mydas en Antofagasta," in VII Simposio Sobre Medio Ambiente: Estado. Actual y Perspectivas de la Investigación y Conservación de las Tortugas Marinas en las Costas del Pacífico Sur Oriental (Antofagasta) 96

Gundogdu S, Cevik C, Karaca S (2017) Fouling assemblage of benthic plastic debris collected from Mersin Bay, NE Levantine coast of Turkey. Mar Pollut Bull 124(1):147–154. https://doi.org/10.1016/j.marpolbul.2017.07.023

Hall NM, Berry KLE, Rintoul L, Hoogenboom MO (2015) Microplastic ingestion by scleractinian corals. Mar Biol 162:725–732. https://doi.org/10.1007/s00227-015-2619-7

Hernández S, Lamilla, J, Dupré E, Stotz W (2005) Desarrollo embrionario de la pintarroja común Schroederichthys chilensis (Guichenot, 1848) (Chondrichthyes: Scyliorhinidae). Gayana 69:184–190

Hidalgo-Ruz V, Gutow L, Thompson R, Thiel M (2012). Microplastics in the Marine Environment: A Review of the Methods Used for Identification and Quantification. Environ Sci Tech 46:3060-3075

Hinojosa I, Thiel M (2009) Floating marine debris in fjords, gulfs and channels of southern Chile. Mar Pollut Bull 58:341–350

Hirai H, Takada H, Ogata Y, Yamashita R, Mizukawa K, Saha M, Kwan C, Moore C, Gray H, Laursen D, Zettler ER (2011) Organic micropollutants in marine plastics debris from the open ocean and remote and urban beaches. Mar Pollut Bull 62(8):1683–1692

Holmes LA, Turner A, Thompson RC (2012) Adsorption of trace metals to plastic resin pellets in the marine environment. Environ Pollut 160:42–48

Horn D, Miller M, Anderson S, Steele C (2019) Microplastics are ubiquitous on California beaches and enter the coastal food web through consumption by Pacific mole crabs. Mar Pollut Bull 139:231–237. https://doi.org/10.1016/j.marpolbul.2018.12.039

Horn DA, Granek EF, Steele CL (2020) Effects of environmentally relevant concentrations of microplastic fibers on Pacific mole crab (*Emerita analoga*) mortality and reproduction. Limnol Oceanogr Lett 5:74–83. https://doi.org/10.1002/lol2.10137

IMARPE (2011) Informe Nacional sobre la Conservación de las Tortugas Marinas en el Perú. Callao: Instituto del Mar del Perú

Imran M, Das KR, Naik MM (2019) Co-selection of multi-antibiotic resistance in bacterial pathogens in metal and microplastic contaminated environments: an emerging health threat. Chemosphere 215:846–857

IUCN (2013) IUCN red list of threatened species. Version 2013.2. www.iucnredlist.org. Accessed 18 Jun 2014

Jabeen K et al (2016) Microplastics and mesoplastics in fish from coastal and fresh waters of China. Environ Pollut. https://doi.org/10.1016/j.envpol.2016.11.055

Jambeck JR, Geyer R, Wilcox C, Siegler TR, Perryman M, Andrady A, Narayan R, Law KL (2015) Plastic waste inputs from land into the ocean. Science 347:768–771

Jemec A, Horvat P, Kunej U, Bele M, Kržan A (2016) Uptake and effects of microplastic textile fibers on freshwater crustacean Daphnia magna. Environ Pollut 219:201–209. https://doi.org/10.1016/j.envpol.2016.10.037

Jiménez A, Pingo S, Alfaro-Shigueto J, Mangel JC, Hooker Y (2017) Feeding ecology of the green turtle Chelonia mydas in northern Peru. Lat Am J Aquat Res 45:585–596

Joshi R, Ahmed S (2016) Status and challenges of municipal solid waste management in India: a review. Cogent Environ Sci 2(1):1139434

Kenyon KW, Kridler E (1969) Laysan Albatrosses swallow indigestible matter. The Auk 86(2):339–343

Khan S, Dutta J, Choudhury M (2018) Searching alternatives to plastic with special reference to Indian rural sector. ESSENCE Int J Environ Rehab Conserv IX(1):162–165. https://doi.org/10.31786/09756272.18.9.1.119

Kiessling T, Gutow L, Thiel M (2015) Marine litter as a habitat and dispersal vector. In M. Bergmann, L. Gutow, & M. Klages (Eds.). Marine anthropogenic litter (pp. 141–181). Springer, Berlin

Kiessling T, Salas S, Mutafoglu K, Thiel M (2017) Who cares about dirty beaches? Evaluating environmental awareness and action on coastal litter in Chile. Ocean Coast Manag 137:82–95
Kole PJ, Lohr AJ, Van Belleghem FG, Ragas AM (2017) Wear and tear of tyres: a stealthy source of microplastics in the environment. Int J Environ Res Publ Health 14:126
Lattin GL, Moore CJ, Zellers AF, Moore SL, Weisberg SB (2004) A comparison of neustonic plastic and zooplankton at different depths near the southern California shore. Mar Pollut Bull 49:291–294. https://doi.org/10.1016/j.marpolbul
Lavers JL, Bond AL, Hutton I (2014) Plastic ingestion by flesh-footed shearwaters (*Puffinus carneipes*): implications for chick body condition and the accumulation of plastic-derived chemicals. Environ Pollut 187:124–129
Law KL (2017) Plastics in the marine environment. Annu Rev Mar Sci 9:205–229
Lei K, Qiao F, Liu Q, Wei Z, Qi H, Cui S (2017) Microplastics releasing from personal care and cosmetic products in China. Mar Pollut Bull 123:122–126
Li WC (2018) The occurrence, fate, and effects of microplastics in the marine environment. In: Microplastic contamination in aquatic environments. Elsevier, Amsterdam, pp 133–173
Li J, Yang D, Li L, Jabeen K, Shi H (2015) Microplastics in commercial bivalves from China. Environ Pollut 207:190–195
Li J, Qu X, Su L, Zhang W, Yang D, Kolandhasamy P, Li D, Shi H (2016) Microplastics in mussels along the coastal waters of China. Environ Pollut 214:177–184
Litter MTSOM (2013) Guidance on monitoring of marine litter in European seas, JRC scientific and policy reports. https://mcc.jrc.ec.europa.eu/documents/201702074014.pdf. Accessed 16 May 2020
Lusher AL (2015) Microplastics in the marine environment: distribution, interactions and effects. In: Marine anthropogenic litter. Springer, Cham. https://doi.org/10.1007/978-3-319-16510-3
Lusher AL, McHugh M, Thompson RC (2013) Occurrence of microplastics in the gastrointestinal tract of pelagic and demersal fish from the English Channel. Mar Pollut Bull 67:94–99. https://doi.org/10.1016/j.marpolbul.2012.11.028
Lusher AL, Hernandez-Milian G, O'Brien J, Berrow S, O'Connor I, Officer R (2015) Microplastic and macroplastic ingestion by a deep diving, oceanic cetacean: the True's beaked whale Mesoplodon mirus. Environ Pollut 199:185–191
Malakar K, Mishra T, Patwardhan A (2019) Factors linked with adaptation in the Indian marine fishing community. Ocean Coast Manag 171:37–46
Maximenko N, Hafner J, Niiler P (2012) Pathways of marine debris derived from trajectories of Lagrangian drifters. Mar Pollut Bull 65:51–62
Mendez M, Rosenbaum H, Bordino P (2008) Conservation genetics of the Franciscana dolphin in Northern Argentina: population structure, by-catch impacts, and management implications. Conserv Genet 9:419–435
Miranda-Urbina D, Thiel M, Luna-Jorquera G (2015) Litter and seabirds found across a longitudinal gradient in the South Pacific Ocean. Mar Pollut Bull 96:235–244
Moore CJ, Moore SL, Leecaster MK, Weisberg SB (2001) A comparison of plastic and plankton in the north Pacific central gyre. Mar Pollut Bull 42:1297–1300. https://doi.org/10.1016/S0025-326X(01)00114-X
Mugilarasan M, Venkatachalapathy R, Sharmila N, Gurumoorthi K (2017) Occurrence of microplastic resin pellets from Chennai and Tinnakkara Island: towards the establishment of background level for plastic pollution. Indian J Geomarine Sci 46(6):1210–1212
Murray F, Cowie PR (2011) Plastic contamination in the decapod crustacean *Nephrops norvegicus* (Linnaeus, 1758). Mar Pollut Bull 62(6):1207–1217. https://doi.org/10.1016/j.marpolbul.2011.03.032
Nadal MA, Alomar C, Deudero S (2016) High levels of microplastic ingestion by the semipelagic fish Bogue Boops boops (L.) around the Balearic Islands. Environ Pollut 214:517–523
Nagarajan VM, Kannappan PG, Krishnan A, Rajendran R, Krishnan A (2020) A critical review on various trophic transfer routes of microplastics in the context of the Indian coastal ecosystem. Watershed Ecology and the Environment, 2:25-41.

Naidu SA, Ranga Rao V, Ramu K (2017) Microplastics in the benthic invertebrates from the coastal waters of Kochi, Southeastern Arabian Sea. Environ Geochem Health 40:1377–1383. https://doi.org/10.1007/s10653-017-0062-z

Naik RK, Naik MM, D'Costa PM, Shaikh F (2019) Microplastics in ballast water as an emerging source and vector for harmful chemicals, antibiotics, metals, bacterial pathogens and HAB species: a potential risk to the marine environment and human health. Mar Pollut Bull 149:110525

Napper IE, Thompson RC (2016) Release of synthetic microplastic plastic fibres from domestic washing machines: effects of fabric type and washing conditions. Mar Pollut Bull 112:39–45

Nelms SE, Duncan EM, Broderick AC, Galloway TS, Godfrey MH, Hamann M, Lindeque PK, Godley BJ (2016) Plastic and marine turtles: a review and call for research. ICES J Mar Sci 73:165–181

Ng KL, Obbard JP (2006) Prevalence of microplastics in Singapore's coastal marine environment. Mar Pollut Bull 52:761–767. https://doi.org/10.1016/j.marpolbul.2005.11.017

Nigam R (1982) Plastic pellets on the Caranzalem beach sands, Goa, India. Mahasagar Bull Natl Inst Oceanogr 15:125–127

Oddone MC, Vooren CM (2002) Egg cases and size at hatching of Sympterygia acuta in the south-western Atlantic. J Fish Biol 61:858–861

Oddone MC, Vooren CM (2008) Comparative morphology and identification of egg capsules of skate species of the genera Atlantoraja Menni, 1972, Rioraja Whitley, 1939 and Sympterygia Müller & Henle, 1837. Arquivos de Ciências do Mar 41:5–13

Ogata Y, Takada H, Mizukawa K, Hirai H, Iwasa S, Endo S, Mato Y, Saha M, Okuda K, Nakashima A et al (2009) International pellet watch: global monitoring of persistent organic pollutants (POPs) in coastal waters. 1. Initial phase data on PCBs, DDTs, and HCHs. Mar Pollut Bull 58(10):1437–1446

Olguner MT, Olguner C, Mutlu E, Deval MC (2018) Distribution and composition of benthic marine litter on the shelf of Antalya in the eastern Mediterranean. Mar Pollut Bull 136:171–176. https://doi.org/10.1016/j.marpolbul.2018.09.020

Ory NC, Sobral P, Ferreira JL, Thiel M (2017) Amberstripe scad Decapterus muroadsi (Carangidae) fish ingest blue microplastics resembling their copepod prey along the coast of Rapa Nui (Easter Island) in the South Pacific subtropical gyre. Sci Total Environ 586:430–437. https://doi.org/10.1016/j.scitotenv.2017.01.175

OSPAR (2014) Imposex and TBT: status, trends and effects in marine molluscs: an improving situation? Reference Number: HASEC14/AS01

Peda C, Caccamo L, Fossi MC, Gai F, Andaloro F, Genovese L, Perdichizzi A, Romeo T, Maricchiolo G (2016) Intestinal alterations in European sea bass Dicentrarchus labrax (Linnaeus, 1758) exposed to microplastics: preliminary results. Environ Pollut 212:251–256. https://doi.org/10.1016/j.envpol.2016.01.083

Peters CA, Bratton SP (2016) Urbanization is a major influence on microplastic ingestion by sunfish in the Brazos River Basin, Central Texas, USA. Environ Pollut 210:380–387

Porecha M (2015) BMC makes strides in controlling malaria. DNA India. Available at: https://www.dnaindia.com/mumbai/report-bmc-makes-strides-in-controlling-malaria-2080481. Accessed 9 Apr 2020

Pozo K, Gomez V, Torres M, Vera L, Nunez D, Pribylova P, Audi O, Clarke B, Arias A, Tombesi N, Guida Y, Kl´anova J (2019) Presence and characterization of microplastics in fish of commercial importance from the Biobío region in Central Chile. Mar Pollut Bull 140:315–319

Pradhan U, Naik S, Begum M, Kumar SS, Panda US, Mishra P, Murthy MVR (2018) Marine litter: post-flood nuisance for Chennai beaches. Curr Sci 115:1454–1455

Raju P, Gunabal S, Santhanam P (2018) The impact of microplastics on marine copepods. In: Basic and applied zooplankton biology. Springer, Singapore, pp 429–442

Rech S, Macaya-Caquilpan V, Pantoja J, Rivadeneira M, Madariaga DJ, Thiel M (2014) Rivers as a source of marine litter-a study from the SE Pacific. Mar Pollut Bull 82(1):66–75

Rech S et al (2015) Sampling of riverine litter with citizen scientists—findings and recommendations. Environ Monit Assess 187:1–18

Reichert J, Schellenberg J, Schubert P, Wilke T (2018) Responses of reef building corals to microplastic exposure. Environ Pollut 237:955–960. https://doi.org/10.1016/j.envpol.2017.11.006
Ryan P (1988) Effects of ingested plastic on seabird feeding: evidence from chickens. Mar Pollut Bull 19:125–128
Sharma MD, Elanjickal AI, Mankar JS, Krupadam RJ (2020) Assessment of cancer risk of microplastics enriched with polycyclic aromatic hydrocarbons. J Hazard Mater 398:122994
Silva A, Retamal LM, Guerra-Correa C (2007) "Registro de tortugas marinas ingresadas al centro de rescate y rehabilitación de fauna silvestre," in VII Simposio sobre Medio Ambiente: Estado Actual y Perspectivas de la Investigación y Conservación de las Tortugas Marinas en las Costas del Pacífico Sur Oriental (Antofagasta) 96
Sridhar KR, Deviprasad B, Karamchand KS, Bhat R (2007) Plastic debris along the beaches of Karnataka, southwest coast of India. Asian J Water Environ Pollut 6(2):87–93
Sruthy S, Ramasamy EV (2017) Microplastic pollution in Vembanad Lake, Kerala, India: the first report of microplastics in lake and estuarine sediments in India. Environ Pollut 222:315–322
Su L, Xue YG, Li LY, Yang DQ, Kolandhasamy P, Li DJ et al (2016) Microplastics in Taihu Lake. China Environ Pollut 216:711–719
Syakti AD, Jaya JV, Rahman A, Hidayati NV, Raza'i TS, Idris F, Trenggono M, Doumenq P, Chou LM (2019) Bleaching and necrosis of staghorn coral (*Acropora formosa*) in laboratory assays: immediate impact of LDPE microplastics. Chemosphere 228:528–535. https://doi.org/10.1016/j.chemosphere.2019.04.156
Tang J, Ni X, Zhou Z, Wang L, Lin S (2018) Acute microplastic exposure raises stress response and suppresses detoxification and immune capacities in the scleractinian coral Pocillopora damicornis. Environ Pollut 243:66–74. https://doi.org/10.1016/j.envpol.2018.08.045
Thiel M, Hinojosa I, Miranda L, Pantoja J, Rivadeneira M, Vásquez N (2013) Anthropogenic marine debris in the coastal environment: a multi-year comparison between coastal waters and local shores. Mar Pollut Bull 71:307–316. https://doi.org/10.1016/j.marpolbul.2013.01.005
Thiel M, Luna-Jorquera G, Álvarez-Varas R, Gallardo C, Hinojosa IA, Luna N, Miranda-Urbina D, Morales N, Ory N, Pacheco AS, Portflitt-Toro M, Zavalaga C (2018) Impacts of marine plastic pollution from continental coasts to subtropical gyres—fish, seabirds, and other vertebrates in the SE Pacific. Front Mar Sci 5. https://doi.org/10.3389/fmars.2018.00238
Thompson RC (2015) Microplastics in the marine environment: sources, consequences and solutions. In: Marine anthropogenic litter. Springer International Publishing, Cham, pp 185–200
Thompson R, Olsen Y, Mitchell R, Davis A, Rowland SJ, John AWG, McGonigle D, Russell AE (2004) Lost at sea: where is all the plastic? Science 304:838
Tuncer S, Artuz OB, Demirkol M, Artuz ML (2018) First report of occurrence, distribution, and composition of microplastics in surface waters of the Sea of Marmara, Turkey. Mar Pollut Bull 135:283–289. https://doi.org/10.1016/j.marpolbul.2018.06.054
UNEP/MAP (2014) State of the Mediterranean marine and coastal environment. United Nations Environment Program—Mediterranean Action Plan, Barcelona Convention, Athens, p 201220
Veerasingam S, Mugilarasan M, Venkatachalapathy R, Vethamony P (2016) Influence of 2015 flood on the distribution and occurrence of microplastic pellets along the Chennai coast, India. Mar Pollut Bull 109(1):196–204
Veerasingam S, Ranjani M, Venkatachalapathy R, Bagaev A, Mukhanov V, Litvinyuk D, Verzhevskaia L, Guganathan L, Vethamony P (2020) Microplastics in different environmental compartments in India: analytical methods, distribution, associated contaminants and research needs. TrAC Trends Anal Chem 133:116071
Vianello A, Boldrin A, Guerriero P, Moschino V, Rella R, Sturaro A, Da Ros L (2013) Microplastic particles in sediments of lagoon of Venice, Italy: first observations on occurrence, spatial patterns and identification. Press Stress Shock Trends Estuar Ecosyst 130:54–61. https://doi.org/10.1016/j.ecss.2013.03.022
Wagner M, Scherer C, Alvarez-Munoz D, Brennholt N, Bourrain X, Buchinger S, Fries E, Grosbois C, Klasmeier J, Marti T, Rodriguez-Mozaz S, Urbatzka R, Vethaak AD, Winther-Nielsen M,

Reifferscheid G (2014) Microplastics in freshwater ecosystems: what we know and what we need to know. Environ Sci Eur 26:12

Wang HM, He Y, Sen B (2019) Research and management of plastic pollution in coastal environments of China. Environ Pollut 248:898–905.

Ward JE, Kach DJ (2009) Marine aggregates facilitate ingestion of nanoparticles by suspension-feeding bivalves. Mar Environ Res 68:137–142. https://doi.org/10.1016/j.marenvres.2009.05.002

Welden NAC, Cowie PR (2016) Environment and gut morphology influence microplastic retention in langoustine, Nephrops norvegicus. Environ Pollut 214:859–865

Wilcox C, Van Sebille E, Hardesty BD (2015) Threat of plastic pollution to seabirds is global, pervasive, and increasing. Proc Natl Acad Sci 112(38):11899–11904. Available at: https://www.pnas.org/content/pnas/early/2015/08/27/1502108112.full.pdf?with-ds=yes

Wilcox C, Mallos NJ, Leonard GH, Rodriguez A, Hardesty BD (2016) Using expert elicitation to estimate the impacts of plastic pollution on marine wildlife. Mar Policy 65:107–114

Windsor FM, Durance I, Horton AA, et al. (2019) A catchment-scale perspective of plastic pollution. Glob Change Biol 25:1207–1221

Wright SL, Rowe D, Thompson RC, Galloway TS (2013) Microplastic ingestion decreases energy reserves in marine worms. Curr Biol. https://doi.org/10.1016/j.cub.2013.10.068

Yu X, Peng J, Wang J, Wang K, Bao S (2016) Occurrence of microplastics in the beach sand of the Chinese inner sea: the Bohai Sea. Environ Pollut 214:722–730

Zhao SY, Zhu LX, Wang T, Li DJ (2014) Suspended microplastics in the surface water of the Yangtze Estuary system, China: first observations on occurrence, distribution. Mar Pollut Bull 86:562–568

Zhao SY, Zhu LX, Li DJ (2015) Microplastic in three urban estuaries, China. Environ Pollut 206:597–604

Chapter 7
Variability of Nutrients and Their Stoichiometry in Chilika Lagoon, India

Pradipta R. Muduli, Manas Barik, Prasannajit Acharya, Alaya T. Behera, and Ishan B. Sahoo

Abstract The biogeochemical process on a spatial and temporal scale can have a significant influence on the regulation of the stoichiometry of nutrients in the waters of coastal and nearshore ecosystems. Such changes may result in alteration of the plankton population and diversity and ultimately the entire food chain. Chilika, the first Ramsar site of India and largest brackish water lagoon of Asia, was investigated for 7 years (2013–2020) to understand the nutrient variability and their stoichiometry. During the study period, crucial parameters showed a significant variation spatially as well as seasonally ($p < 0.05$, $n = 2520$). Nutrient concentrations in Chilika were found to be 0.4 ± 0.3, 5 ± 4, 7 ± 4, 0.5 ± 0.6, and 71 ± 41 μM for nitrite (NO_2), nitrate (NO_3), ammonia (NH_3), phosphate (P), and silicate (Si). The lagoon maintained mesotrophic condition irrespective of seasons. Shifts in the stoichiometry of dissolved inorganic nitrogen (N) to dissolved inorganic phosphate (P) and Si (N/P/Si) were investigated and found N/P and Si/P were maintained between 0.1 and 2700 with an avg. of 61 ± 125 and 0.1 and 15,439 with an avg. of 514 ± 1049, respectively, whereas N/Si varied between 0.01 and 4 with an avg. of 0.3 ± 0.3. A significant positive correlation ($p < 0.01$) of N/Si ($r = 0.79$), N/P ($r = 0.79$), and Si/P ($r = 0.67$) with chlorophyll a (Chl-a) indicated nutrient stoichiometry is the major factor that controls the productivity of the Chilika lagoon. OC (Outer Channel) recorded the lowest N/P as compared to other sectors indicating nitrogen limitation due to the mixing of seawater with poor nitrogen level. In the present study, N and P were limiting with respect to Si, and P was limiting with respect to N as evidenced from N/Si < 1; Si/P > 16 and N/P > 16, respectively. This study suggested that the NH_3 has a major role in Chilika (along with NO_3) for the calculation of N/P and deciding the limiting factors.

Keywords Lagoon · Nutrients · Biogeochemical process · Stoichiometry · Redfield ratio

P. R. Muduli (✉) · M. Barik · P. Acharya · A. T. Behera · I. B. Sahoo
Chilika Development Authority, Wetland Research and Training Centre, Balugaon, Odisha, India

S. Madhav et al. (eds.), *Coastal Ecosystems*, Coastal Research Library 38,
https://doi.org/10.1007/978-3-030-84255-0_7

1 Introduction

A lagoon is recognized as a water body separated from a larger body of water like the ocean by barriers of marine origin and mostly parallel to the coastline. The Indian lagoon ecosystem is facing challenges from natural and anthropogenic aggravation which can significantly affect biodiversity and nutrient dynamics. Coastal lagoons are generally influenced by riverine and seawater influx simultaneously making brackish water ecosystems. Coastal lagoons cover about 13% of the total world's coastline (Kjerfve 1994). The lagoons are highly useful and productive with a wide array of biodiversity and an abundance of flora, fauna, and avifauna. This also acts as the hub of the blue economy by supporting large biodiversity of fisheries, wintering ground of migratory birds, maintenance of hydrology, climate regulation, food protection coastal tourism, and a large amount of fishing harbor (Newton et al. 2018). However, anthropogenic activity is gradually increasing and degrading the ecosystems in terms of loss of ecological characters and biodiversity.

There are 17 coastal lagoons, present on the Indian coast (8 on the east coast and 9 on the west coast; Mahapatro et al. 2013), which are shallow and well mixed by waves and currents, and an average depth of the lagoon is 2 m with a photic depth of the coastal lagoons extending up to 2 m. Lagoons not only have importance for biodiversity conservation but also support the livelihood of the local community as well as the economy on an international scale. For instance, the Chilika lagoon supports more than 0.2 million fishermen. This warrants the study of such ecosystems with respect to the environmental characteristics and factors responsible for the adverse changes. The water quality indicators and its dynamics in the systems have been proved to be a critical indicator of the anthropogenic nutrient fluxes and overall ecosystem health (Muduli et al. 2021; Mishra et al. 2020). Such studies are also helpful for policymakers for the management action plan formulation (Barik et al. 2017; Robin et al. 2016; Muduli et al. 2017; Patra et al. 2016). The nutrient concentration that is maintained in the lagoon, its variability with respect to seasons, and its stoichiometry maintained with changing environmental conditions could impact the plankton biodiversity and higher food chain. In order to predict phytoplankton species composition, the stoichiometry of ambient available nutrients, such as nitrogen in terms of nitrate, nitrite, and ammonia and phosphorus, has been used by several researchers (Tilman 1982; Sommer et al. 2007). Models are also developed and validated for species composition prediction based on the stoichiometry of nutrients and phytoplankton elemental stoichiometries. The same also has been extended recently for zooplankton (Sterner 1990; Sterner and Hessen 1994). According to Siddiqui et al. (2019), eutrophication prediction can be done considering the nutrient concentration and stoichiometry. These approaches unite predictive models of populations with environmental processes and help to determine patterns of nitrogen and phosphorus limitation.

Water bodies containing low N and P naturally are very much sensitive to the nutrient fluxes as they influence the nutrient balance in lagoon ecosystems. The

Redfield ratio is an important indicator of nutrient limitation for the phytoplankton growth, and its predictive power prompts for searching similar patterns and relationships in other ecosystems. Ecological stoichiometry helps to understand the balance between different chemical elements responsible for ecological functions. For instance, P is considered to be the limiting nutrient when the N/P ratio is >16, and N is considered to be a limitation for the phytoplankton growth if N/P ratios are <16. Similarly, when the stoichiometric ratio of N/P and Si maintains as 16:1:16, it is considered as a limitation of nutrients for diatoms. Lack of management policy formulation and actions has resulted in disproportionate nutrient loads in many aquatic systems around the globe. Such nutrients whether it may be N or P have the potential to alter the nutrient stoichiometry which may lead to either P limitation or N limitation (Jabir et al. 2020). Silicon loading is mostly controlled by natural factors and found to maintain consistency unlike N and P (Pandey et al. 2016).

Coastal lagoons are usually very unstable as the variations that occur in these water bodies are comparatively higher than in saline environments (Panigrahi et al. 2009). Study on nutrient dynamics in these ecosystems is crucial to demonstrate the health of a lagoon with many factors like an ocean tide, river and rivulet water mixing, and anthropogenic interferences. Ecological stoichiometry helps in the understanding of relationships between nutrient cycling and trophic status (Zhang et al. 2013). For the trophic chain, dissolved nutrients are considered as raw materials, and the lagoons act as the entry gate for the nutrients which come from continental drainage to the marine ecosystems. The supply of nutrients is higher in the lagoon ecosystem which is close to highly populated regions, because of input of industrial and domestic waste, agricultural effluents, and urban drainage. The nutrient budget on a global scale is not the same as it was during the preindustrial times as it has been changed from an almost balanced state to a nutrient enrichment state. Such changes in the lagoon and coastal waters are responsible for modifications in the environment such as increases in productivity, fishing yields, etc. When the circulation is restricted, the anthropogenic inputs can lead to excessive eutrophication in the ecosystem which consequently gets changed with varying water flows leading to alterations in water chemistry. Such changes result in different ecological consequences including species composition, phytoplankton blooms, and decline in DO (Martin et al. 2008) and eutrophication (Sonal and Kataria 2012). These forcing factors drive the quality of the habitat and also the change in biodiversity in different lagoon ecosystems.

Spatiotemporal variations in the nutrients in the lagoon could be attributed to the freshwater discharge through rivers and seawater intrusion from two mouths, Arakhkuda and Sanapatna. Apart from anthropogenic sources, atmospheric deposition and surface runoff also contribute nutrients to the Chilika lagoon (Muduli et al. 2013). Natural biogeochemical process such as nitrification and denitrification processes also attributes to a significant hike in N and P load into the lagoon (CPCB 1986). In Chilika the litter of birds (especially from Nalabana birds' sanctuary within the lagoon) also could be a significant contributor for nutrients and holds the capability to alter the nutrient stoichiometry in the lagoon. Surface runoff along with these robust factors altogether can be a more comprehensive predictor of the eutrophic status of the Chilika lagoon which can alter the pattern of ecological nutrient limitation. Depending on the nutrient availability (either allochthonous or autochthonous sources) in the water column and other physicochemical factors, the strophic status of the ecosystem such as eutrophic, oligotrophic, or mesotrophic is defined in terms of trophic state index (TSI). Carlson (1977) initially derived TSI by considering three parameters, i.e., Chl-a, total phosphorus (TP), and transparency, which was further modified by Burns (2005). In that, total nitrogen (TN) was added along with the three parameters used for TSI calculation, and it was renamed as Trophic Level Index (TLI).

Many aquatic systems were found to be switched over from nutrient-limiting to nutrient surplus over the past few decades due to an increase in anthropogenic waste addition to the ecosystem which leads to eutrophication. Like other aquatic ecosystems, nutrient and other physiochemical parameters are the major components of the Chilika lagoon that proved to be crucial for the sustenance of the good health of the ecosystem. There are several studies on Chilika that have focused on the water quality of the lagoon (Panigrahi et al. 2009; Ganguly et al. 2015). However, the studies specifically on nutrient stoichiometry are scanty. Hence the present assessment was taken up to (1) study the spatiotemporal variability of nutrients and their stoichiometry and (2) reveal the factors controlling the nutrient stoichiometry and its influence on the Chilika ecosystem.

2 Material and Methods

2.1 Study Area

The brackish water lagoon, Chilika (19°42′N–85°21′E), spreads over three districts of Odisha state (Puri, Khurda, and Ganjam) on the east coast of India, which flows into the Bay of Bengal. It has a watershed area of over 116,500 ha. After the New Caledonian barrier reef, Australia is the largest brackish water lagoon in the world. It is the largest coastal lagoon in India and has been listed as a tentative UNESCO World Heritage site. Chilika lagoon is designated as Ramsar Convention site no. 229 Ramsar, 1981 which is the first Indian wetland that got international importance by Ramsar Convention. Chilika lagoon is about 65 km in length and 20.1 km in breadth (northeast to southwest) and maintains an average depth of 2 m. The lagoon maintains an area of 950 and 1165 km^2 during summer and monsoon, respectively (Gupta et al. 2008). The lagoon is connected with the sea (Bay of Bengal) through an opening that was dredged in September 2000. To distribute the saline water throughout the lagoon, channels were dredged by the Chilika Development Authority (CDA) which enables saline water flow into the lagoon during summer and also helps to flush out the suspended matters into the sea which are received from the riverine discharge during the monsoon. The geological factors that impacted the coastal lagoon copiously are littoral drift, marine water intrusion, catchment influx, groundwater discharge, the coastal geomorphological process, inlet configuration and dimension, lagoon size, orientation with respect to prevailing winds, and water depth. Some studies explained as follows to support this; Chilika lagoon is lying parallel to the coast and maintaining the biogeochemistry of the lagoon by mixing the saline water from the sea near Satapada and freshwater flow from rivers draining into the lagoon (Cohen et al. 1999; Muduli et al. 2013; Barik et al. 2017). Daya, Makara, Bhargavi, and Luna are the major rivers which drain the copious amount of freshwater containing SPM (suspended particulate matter) and nutrients into the lagoon and attribute to significant annual and seasonal changes in hydrological conditions of the lagoon.

3 Methodology

Sampling from 30 prefixed locations was done (Fig. 7.1) during September 2013 and June 2020 on a monthly basis. All samples were analyzed for water quality parameters within 12 h at the shoreline laboratory facility of the Wetland Research and Training Centre (WRTC), CDA, Odisha. March to June, July to October, and November to February were considered as summer, monsoon (MON), and winter. The lagoon was considered as having four sectors named outer channel (OC), central sector (CS), southern sector (SS), and northern sector (NS) (Muduli et al. 2017; Muduli and Pattnaik 2020). 5 L Niskin sampler was used for the collection of subsurface water samples from 0.3 m depth from the surface. Photic depth or

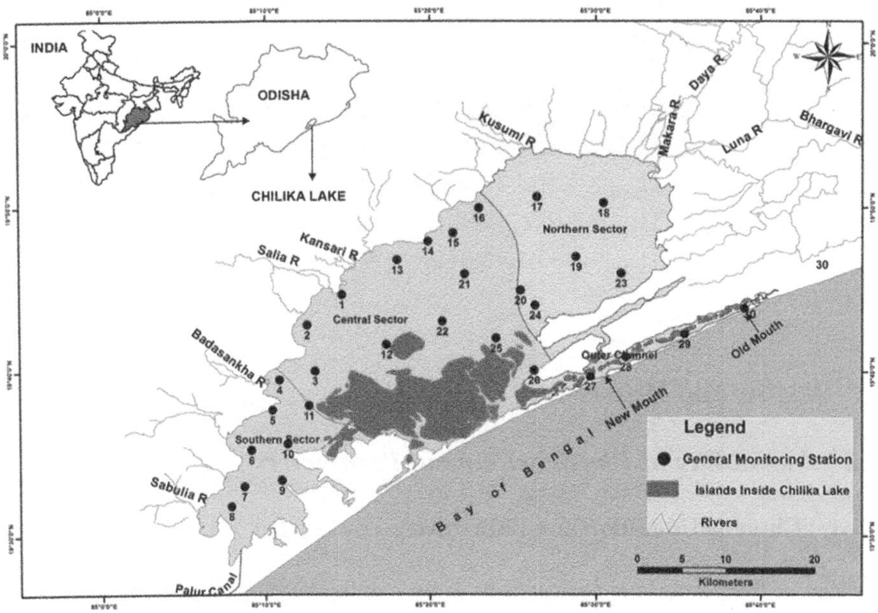

Fig. 7.1 Chilika map showing 30 sampling locations, sectoral divisions, and major rivers draining to the lagoon

transparency of the lagoon was measured as the Secchi disk depth. A Thermos meter with an accuracy of ±0.01 °C was used to measure the air temperature (AT) and water temperature (WT). Calibrated water quality Sonde (YSI, USA, V2, Model No. 6600) was used to measure pH and salinity. Water samples for NH_3, NO_2, NO_3, P, and Si were collected in HDPE bottle and filtered using membrane filter paper of 0.45 μ of size 47 mm, and finally the filtered samples were analyzed using a nutrient autoanalyzer (Make: SKALAR SANplus) with precisions of ±0.01, ±0.02, ±0.01, ±0.01, and ±0.02 μM, respectively (Grasshoff et al. 1999). $NO_2 + NO_3 + NH_3$ were represented as N. Modified Winkler's method as reported in Carrit and Carpenter (1966) was followed for dissolved oxygen (DO) analysis, and biochemical oxygen demand (BOD) was estimated by 5-day incubation of samples at 20 °C. Chlorophyll a (Chl-a) sample of 1 L was filtered (GF/F filter paper, 47 mm 0.7 μm) and extracted in the dark with 90% acetone for 24 h at 4 °C. UV-VIS spectrophotometer (Make: Thermo Scientific Evolution TM 201) was used to record the absorbance following methods described in Strickland and Parsons (1972). SPSS-18 was used for multivariate regression analyses and for deriving the Pearson correlation coefficient (*r*).

According to the availability of data, TSI and TLI were calculated for 2019 (January to December) using Chl-a, TN and TP concentration (in μg L^{-1}), and SD (in meters) as derived by Carlson (1977) and Burns (2005), respectively (El-Serehy et al. 2018), as mentioned below:

$$\mathrm{TSI} = \left[\mathrm{TSI}(\mathrm{SD}) + \mathrm{TSI}(\mathrm{TP}) + \mathrm{TSI}(\mathrm{Chl\ a})\right]/3$$
$$\mathrm{TSI}(\mathrm{SD}) = 60 - 14.42\ln(\mathrm{SD})$$
$$\mathrm{TSI}(\mathrm{TP}) = 14.42\ln(\mathrm{TP}) + 4.15$$
$$\mathrm{TSI}(\mathrm{Chl\ a}) = 9.81\ln(\mathrm{Chl\ a}) + 30.6$$

$$\mathrm{TLI} = 1/4\left(\mathrm{TL}_{\mathrm{Chla}} + \mathrm{TL}_{\mathrm{SD}} + \mathrm{TL}_{\mathrm{TP}} + \mathrm{TL}_{\mathrm{TN}}\right)$$
$$\mathrm{TL}_{\mathrm{Chla}} = 2.22 + 2.54\log(\mathrm{Chl\ a})$$
$$\mathrm{TL}_{\mathrm{SD}} = 5.10 + 2.60\log(1/\mathrm{SD} - 1/40)$$
$$\mathrm{TL}_{\mathrm{TP}} = 0.218 + 2.92\log(\mathrm{TP})$$
$$\mathrm{TL}_{\mathrm{TN}} = -3.61 + 3.01\log(\mathrm{TN})$$

4 Results and Discussion

4.1 Variability of Physicochemical Parameters

4.1.1 Climatic Condition and Bathymetry

Mixing of freshwater and seawater, flushing rate, many biophysical processes, and biotic-abiotic factors with space and time influence the physicochemical variables in the Chilika lagoon. Chilika experiences monsoon, winter, and summer from July to October, November to February, and March to June, respectively. The average rainfall in the catchment is 1238.8 mm and generally decreases from northeast to southwest. Eighty percent of the annual rainfall occurs during monsoon months varying from tentatively 39 to 200 cm. Significant annual variations in rainfall and sediment flow into the rivers are observed depending on the variations in precipitation with time.

AT recorded during the study period was ranged between 12.50 and 36.00 °C with an average of 27.42 ± 3.73 °C. The WT varied proportionately with air temperature. WT is an important component of the water chemistry in the lagoon. In the lagoon, WT is mostly influenced by solar radiation, heat transfer from the atmosphere, and turbidity. It is a very important factor that controls the pH, nutrient uptake, primary productivity, plankton diversity, rate of photosynthesis, microbial activity, degradation of organic matter, oxygen solubility, etc. It was ranged between 11.84 and 35.50 °C with an average of 27.72 ± 3.43 °C. WT showed significant variation with respect to season, the lowest recorded in winter (24.56 ± 2.49 °C) as compared to summer and monsoon (average 29.6 °C) when the WT difference was insignificant.

Depth of the Chilika lagoon in association with other environmental factors could affect the water quality. For instance, turbidity (sediment churning from benthic compartment due to wind action), water column productivity and nutrient uptake, and new production in the pelagic compartment are largely dependent on the depth of the ecosystem. In this study, the depth of the lagoon was varied from 0.06 to 6 m with an average of 1.71 ± 0.75 m. NS was the shallowest with 1.15 ± 0.38 m depth on average followed by 1.46 ± 0.43, 2.10 ± 0.66, and 2.38 ± 0.93 m in CS, SS, and OC, respectively (Table 7.1).

4.1.2 Factors Responsible for SD Variability

The photic depth usually measured as Secchi disk depth (SD) is the uppermost layer of the water body that receives the sunlight, allowing flora and fauna for photosynthesis. SD depends on physical parameters like turbidity, total suspended matter, and phytoplankton pigments (Srichandan et al. 2015b). The productivity of the lagoon mostly depends on the nutrient concentration and availability of PAR (photosynthetic active radiation) at the sub-surface water level which directly depends on the SD. It could be a critical factor for phytoplankton diversity, abundance, and spatial variation. Chilika lagoon maintained the SD of 0.68 ± 0.40 m (0–2.7 m) having the highest in SS (0.93 ± 0.37 m) followed by CS (0.72 ± 0.35 m), OC (0.68 ± 0.41 m), and NS (0.37 ± 0.25 m) (Table 7.1; Fig. 7.2a). In Chilika, the regions covered with seagrass bed and submerged macrophytes recorded SD almost the same to the depth as it contained the least turbidity due to the fact that the suspended particulate matter sticks to the surface making the water more transparent (Kim et al. 2015; Patra et al. 2016). Turbidity which is responsible for lowering the SD and light penetration to the water column ranged between 0 and 636 with an avg. of 61.08 NTU. A high amount of suspended matter occurs due to flooding and intense rainstorms mostly in the monsoon seasons that leads to decline in

Table 7.1 Descriptive statistics of physicochemical parameters of Chilika waters during 2013–2020

		Monsoon				Winter				Summer				Overall			
Parameter	Sector	Min	Max	Mean	Stdv	Min	Max	Mean	Stdv	Min	Max	Mean	Stdv	Min	Max	Mean	Stdv
AT (°C)	CS	21.22	36.00	29.15	2.46	17.50	33.10	24.11	3.30	21.30	36.00	29.21	3.04	17.50	36.00	27.35	3.83
	SS	20.10	35.50	29.36	2.78	16.00	34.50	24.60	3.21	21.47	35.90	28.69	3.14	16.00	35.90	27.44	3.73
	NS	23.00	34.00	29.04	2.22	12.50	33.34	24.29	3.47	21.76	35.40	29.81	2.73	12.50	35.40	27.55	3.80
	OC	24.00	33.00	28.99	2.05	16.18	30.30	24.28	2.80	22.46	35.00	29.14	2.63	16.18	35.00	27.31	3.41
	Overall	20.10	36.00	29.16	2.44	12.50	34.50	24.33	3.25	21.30	36.00	29.20	2.96	12.50	36.00	27.42	3.73
WT (°C)	CS	22.27	34.50	29.72	2.34	11.84	32.80	24.32	2.62	23.48	34.80	29.03	2.49	11.84	34.80	27.55	3.49
	SS	23.34	34.30	30.00	2.30	19.00	30.60	24.91	2.25	24.00	35.50	29.27	2.30	19.00	35.50	27.92	3.23
	NS	21.18	34.70	29.68	2.59	18.00	33.90	24.35	2.75	23.29	35.50	29.45	2.51	18.00	35.50	27.66	3.63
	OC	21.65	35.00	29.80	2.34	19.00	31.30	24.68	2.15	23.35	34.40	29.17	2.55	19.00	35.00	27.74	3.30
	Overall	21.18	35.00	29.80	2.39	11.84	33.90	24.56	2.49	23.29	35.50	29.23	2.45	11.84	35.50	27.72	3.43
Depth (m)	CS	0.88	3.05	1.82	0.38	0.17	2.52	1.31	0.37	0.52	2.40	1.24	0.30	0.17	3.05	1.46	0.43
	SS	0.67	3.90	2.41	0.65	0.32	3.22	1.98	0.64	0.30	2.97	1.89	0.56	0.30	3.90	2.10	0.66
	NS	0.91	2.40	1.50	0.28	0.06	1.70	0.97	0.32	0.50	1.68	0.97	0.24	0.06	2.40	1.15	0.38
	OC	0.80	6.00	2.67	0.88	0.37	5.10	2.16	0.91	0.82	4.60	2.34	0.92	0.37	6.00	2.38	0.93
	Overall	0.67	6.00	2.04	0.70	0.06	5.10	1.55	0.72	0.30	4.60	1.53	0.70	0.06	6.00	1.71	0.75
Salinity	CS	0.01	34.56	7.83	7.51	0.42	22.82	5.97	3.18	0.26	36.16	14.79	8.45	0.01	36.10	9.29	7.61
	SS	2.10	21.58	11.94	5.01	3.80	16.22	8.37	2.92	3.22	24.70	12.24	4.52	2.10	24.70	10.73	4.55
	NS	0.00	27.97	3.06	6.11	0.06	10.47	2.26	2.65	0.10	34.76	9.49	8.66	0.00	34.76	4.75	6.91
	OC	0.18	36.47	11.29	10.68	0.50	32.05	16.17	8.69	12.17	36.16	28.72	6.30	0.18	36.47	18.35	11.28
	Overall	0.00	36.47	8.31	7.93	0.06	32.05	7.24	6.14	0.10	36.16	14.63	9.44	0.00	36.10	9.87	8.49
pH	CS	6.10	9.87	8.11	0.57	6.51	10.18	8.35	0.63	6.36	9.88	8.08	0.72	6.10	10.18	8.19	0.65
	SS	6.29	9.33	8.07	0.52	6.91	9.41	8.10	0.49	7.00	9.92	8.07	0.59	6.29	9.92	8.08	0.54
	NS	6.45	9.67	7.86	0.54	5.99	9.78	8.06	0.61	6.23	10.35	7.99	0.79	5.99	10.35	7.97	0.65
	OC	7.00	9.08	7.94	0.48	6.61	8.98	8.01	0.44	7.00	9.00	7.74	0.43	6.61	9.08	7.91	0.46
	Overall	6.10	9.87	8.01	0.54	5.99	10.18	8.15	0.57	6.23	10.35	8.00	0.68	5.99	10.35	8.06	0.60

Parameter	Sector	Monsoon				Winter				Summer				Overall			
		Min	Max	Mean	Stdv	Min	Max	Mean	Stdv	Min	Max	Mean	Stdv	Min	Max	Mean	Stdv
SD (m)	CS	0.09	2.06	0.77	0.41	0.10	1.73	0.75	0.32	0.09	1.57	0.61	0.29	0.09	2.06	0.72	0.35
	SS	0.13	2.33	1.12	0.39	0.23	2.04	0.92	0.29	0.14	1.76	0.73	0.32	0.13	2.33	0.93	0.37
	NS	0.05	1.74	0.38	0.31	0.06	1.52	0.34	0.22	0.04	1.10	0.38	0.22	0.04	1.74	0.37	0.25
	OC	0.06	1.75	0.54	0.28	0.00	2.70	0.76	0.44	0.00	2.02	0.74	0.46	0.00	2.70	0.68	0.41
	Overall	0.05	2.33	0.74	0.46	0.00	2.70	0.70	0.38	0.00	2.02	0.61	0.35	0.00	2.70	0.68	0.40
Turbidity (NTU)	CS	0.00	386.00	39.97	58.76	0.00	427.20	32.86	59.49	0.00	294.00	34.45	48.79	0.00	427.20	35.51	55.94
	SS	0.40	97.70	21.76	29.64	0.60	140.00	19.18	24.67	0.10	274.00	30.94	36.96	0.10	274.00	23.91	31.10
	NS	1.72	635.70	122.76	119.33	0.00	614.70	103.61	110.64	1.06	830.20	121.50	166.30	0.00	830.20	115.19	133.65
	OC	0.90	424.00	67.70	78.67	0.20	374.40	34.53	49.12	0.20	291.20	38.98	60.44	0.20	424.00	46.16	64.41
	Overall	0.00	635.70	60.90	86.85	0.00	614.70	47.79	77.07	0.00	830.20	56.71	101.41	0.00	830.20	54.65	88.70
DO (mg L^{-1})	CS	3.18	12.66	7.24	1.43	2.51	15.59	8.33	1.85	3.05	21.40	7.43	1.92	2.51	21.40	7.69	1.81
	SS	3.60	11.30	7.03	1.25	3.55	13.38	7.97	1.37	1.42	19.40	7.58	2.03	1.42	19.40	7.55	1.62
	NS	3.50	12.29	7.41	1.49	1.51	13.15	8.52	1.96	4.85	15.93	8.18	1.92	1.51	15.93	8.05	1.87
	OC	4.10	10.69	7.38	1.17	2.89	12.66	8.07	1.49	4.54	10.20	7.51	1.28	2.89	12.66	7.68	1.36
	Overall	3.18	12.66	7.24	1.37	1.51	15.59	8.23	1.72	1.42	21.40	7.68	1.90	1.42	21.40	7.74	1.72
BOD (mg L^{-1})	CS	0.09	7.61	2.14	1.36	0.08	7.33	2.29	1.22	0.06	8.10	2.38	1.53	0.06	8.10	2.27	1.37
	SS	0.08	6.40	2.07	1.13	0.04	6.80	2.01	1.10	0.17	5.35	2.17	1.28	0.04	6.80	2.08	1.17
	NS	0.16	8.80	2.39	1.61	0.04	10.19	2.58	1.59	0.14	10.40	2.87	1.95	0.04	10.40	2.61	1.72
	OC	0.16	7.57	2.05	1.45	0.13	6.66	2.14	1.14	0.22	5.56	2.30	1.25	0.13	7.57	2.16	1.28
	Overall	0.08	8.80	2.17	1.38	0.04	10.19	2.26	1.30	0.06	10.40	2.43	1.56	0.04	10.40	2.28	1.42

(continued)

Table 7.1 (continued)

Parameter	Sector	Monsoon				Winter				Summer				Overall			
		Min	Max	Mean	Stdv	Min	Max	Mean	Stdv	Min	Max	Mean	Stdv	Min	Max	Mean	Stdv
Alkalinity (mg L^{-1})	CS	57.60	205.00	136.57	27.11	66.00	240.00	148.48	26.18	76.00	239.00	144.76	29.76	57.60	240.00	143.28	28.04
	SS	37.20	223.75	151.50	27.60	96.00	237.00	153.49	23.96	76.00	277.00	154.84	26.55	37.20	277.00	153.23	26.03
	NS	43.00	182.00	113.84	26.13	47.00	188.00	131.70	22.93	35.00	209.00	119.61	31.46	35.00	209.00	121.97	27.85
	OC	41.70	216.00	127.49	28.87	62.00	194.60	141.90	23.22	55.00	187.00	144.02	20.27	41.70	216.00	137.62	25.52
	Overall	37.20	223.75	133.71	30.76	47.00	240.00	144.59	25.67	35.00	277.00	141.03	31.14	35.00	277.00	139.81	29.52
Chl-a (μg L^{-1})	CS	0.00	49.82	5.34	6.75	0.00	38.80	3.80	5.21	0.02	39.00	4.55	5.49	0.00	49.82	4.54	5.89
	SS	0.04	95.80	4.59	9.64	0.00	48.50	3.50	5.36	0.01	66.00	5.16	9.62	0.01	95.80	4.36	8.35
	NS	0.17	34.09	6.54	6.78	0.00	75.60	6.50	9.06	0.24	58.00	6.33	8.39	0.01	75.60	6.47	8.11
	OC	0.02	35.00	5.23	6.05	0.10	48.00	4.30	6.87	0.03	27.00	4.70	4.84	0.02	48.00	4.73	6.04
	Overall	0.00	95.80	5.40	7.64	0.00	75.60	4.40	6.75	0.01	66.00	5.20	7.61	0.00	95.80	4.99	7.33

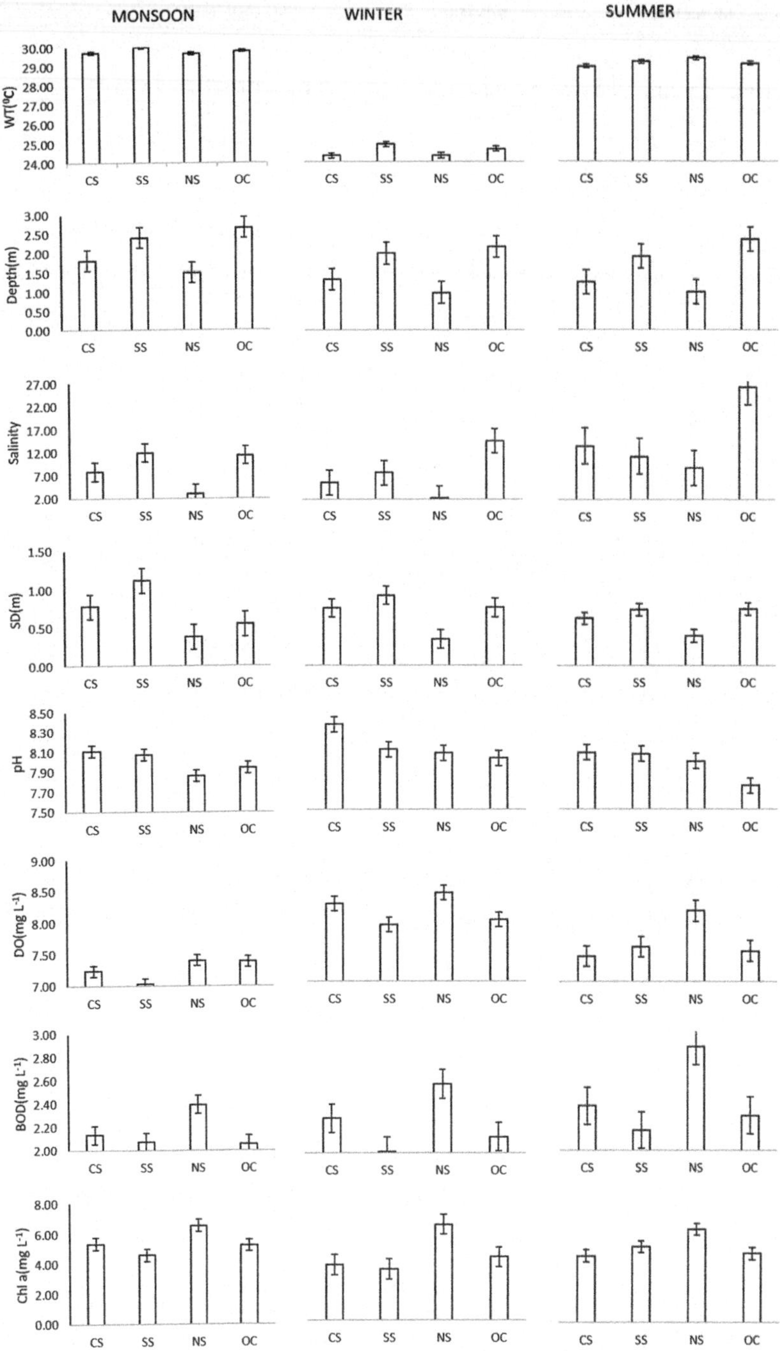

Fig. 7.2 (**a**) Sectoral variation of physicochemical parameters in monsoon, winter, and summer (CS, SS, NS, and OC represent central, southern, northern sector, and outer channel, respectively). (**b**) Sectoral variation of nutrients and their stoichiometry in monsoon, winter, and summer (CS, SS, NS, and OC represent central, southern, northern sector, and outer channel, respectively)

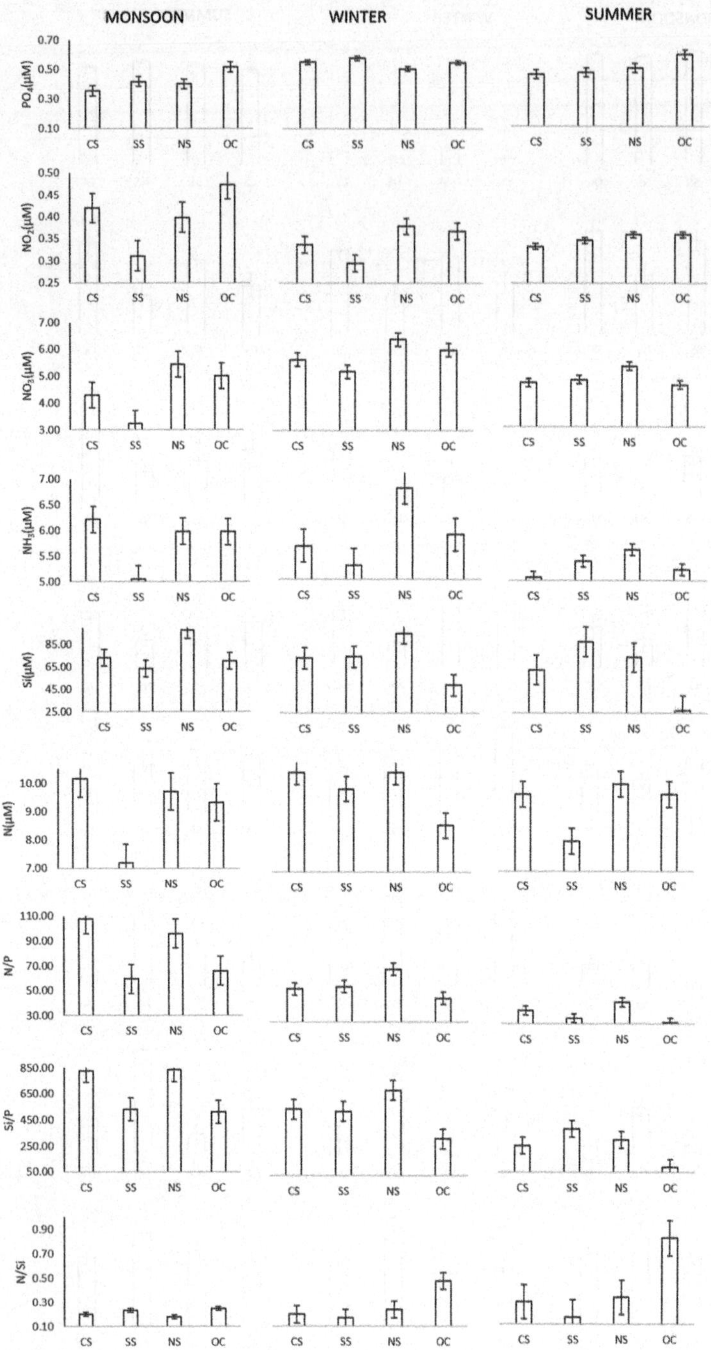

Fig. 7.2 (continued)

SD. Cyclonic events such as Phailin, Titli, and Fani also had a significant impact on the flash change in SD level along with other water quality parameters (Barik et al. 2017; Muduli et al. 2017; Mishra et al. 2021) as a result of flash flood induced by severe cyclones that had landfall proximate to Chilika lagoon. SD was significantly correlated with alkalinity ($r = 0.21$, $p < 0.01$) which could be due to the fact that lower SD supports high photosynthesis which utilizes CO_2 and increases the alkalinity level in the pelagic compartment (Muduli et al. 2012) (Table 7.2). The earlier reported values for SD and other physicochemical parameters are listed in Table 7.3.

4.1.3 pH, DO, and Salinity Variability Factors

pH is the measurement of the hydrogen and hydroxyl ion concentration in the water. It is the important component of the water that determines whether it is acidic or basic. The present study observed a pH of 8.06 ± 0.6 on average and ranged between 5.99 and 10.35 which indicates the lagoon water maintained alkaline condition. pH varied significantly with space and time. pH in the lagoons is mostly affected by the physical processes such as mixing of fresh and saline water with different pH conditions and biological processes such as respiration, photosynthetic activity of the phytoplankton, etc. (Ganguly et al. 2015; Muduli et al. 2013). According to Srichandan et al. (2015a, b), the variability of pH with respect to season can impact the assimilation of phytoplankton and macrophytes in the lagoon. The lowest pH recorded in NS could be attributed to the decomposition of freshwater vegetation observed dominantly in the NS. Usually, lagoon turns alkaline during the winter due to the seawater influence and biological activity. During summer and monsoon, it gradually decreases because of the decomposition of organic matters and freshwater influx (Upadhyay et al. 2015). pH was positively correlated with transparency and DO (overall as well as for all seasons) which reveals that the pH in Chilika is predominantly controlled by photosynthetic activity which is most favored in the higher SD region of Chilika (Tables 7.2, 7.4, 7.5, and 7.6).

Dissolved oxygen (DO) indicates the oxygen quantity in dissolved form and for which regulating authorities have fixed thresholds depending on the purpose of uses. The coastal lagoon system is receiving organic load and nutrients throughout the world due to urbanization and industrialization leading to the formation of the algal bloom causing hypoxic (low oxygen concentration) conditions in the lagoon which is a matter of concern and needs its monitoring. It is the important element of an aquatic organism for the process of respiration and is produced in the process of photosynthesis by the phytoplankton, macrophytes, and submerged vegetation. Chilika lagoon maintains itself well oxygenated throughout the year (Sundaray et al. 2006; Barik et al. 2017). The DO concentrations were within the threshold range of 4 mg L^{-1} which is suitable for the healthy aquatic life, wildlife propagation, and fisheries (CPCB 1986). The present study also recorded a fair level of DO with respect to the threshold (7.74 ± 1.72 mg L^{-1} in average and ranged between 1.42 and 21.4 mg L^{-1}). All the studies to date reported the overall DO of >5 mg L^{-1}

Table 7.2 Pearson's correlation matrix of physicochemical parameters and nutrients of Chilika lagoon during 2013–2020

	AT	WT	Depth	Salinity	pH	SD	Turbidity	DO	BOD	NO_2	NO_3	NH_3	P	SI	Alkalinity	Chl-a	N	N/P	Si/P
WT	0.799[a]																		
Depth	0.060[a]	0.148[a]																	
Salinity	0.146[a]	0.111[a]	0.213[a]																
pH	0.003	0.001	−0.135[a]	−0.198[a]															
SD	−0.009	0.039	0.383[a]	0.173[a]	0.129[a]														
Turbidity	−0.060[a]	0.008	−0.150[a]	−0.242[a]	−0.129[a]	−0.420[a]													
DO	−0.068[a]	−0.140[a]	−0.189[a]	−0.165[a]	0.310[a]	−0.091[a]	0.004												
BOD	0.045[b]	0.054[b]	−0.107[a]	−0.110[a]	0.025	−0.092[a]	0.162[a]	0.247[a]											
NO_2	−0.007	0.010	−0.073[a]	0.022	−0.020	−0.152[a]	0.009	0.024	0.002										
NO_3	−0.021	0.024	−0.014	−0.131[a]	−0.134[a]	−0.104[a]	0.185[a]	−0.132[a]	0.053[b]	0.068[a]									
NH_3	−0.032	−0.006	0.022	0.038	−0.187[a]	−0.031	0.090[a]	−0.119[a]	−0.041	−0.017	0.173[a]								
P	−0.103[a]	−0.082[a]	0.054[b]	0.032	−0.133[a]	0.005	0.150[a]	−0.098[a]	0.026	−0.045[b]	0.211[a]	0.086[a]							
SI	−0.060[a]	−0.011	−0.061[a]	−0.509[a]	0.005	−0.261[a]	0.193[a]	0.032	0.048[b]	0.142[a]	0.216[a]	−0.075[a]	−0.011						
Alkalinity	−0.057[a]	−0.102[a]	0.032	0.238[a]	0.098[a]	0.210[a]	−0.249[a]	−0.001	−0.134[a]	−0.039	−0.215[a]	−0.103[a]	−0.023	−0.202[a]					
Chl-a	−0.058[b]	0.040	0.021	−0.118[a]	−0.042	−0.063[b]	0.188[a]	−0.073[a]	0.025	−0.066[a]	0.125[a]	0.111[a]	0.093[a]	0.074[a]	−0.137[a]				
N	−0.045[b]	0.008	0.007	−0.057[a]	−0.227[a]	−0.080[a]	0.187[a]	−0.176[a]	0.011	0.084[a]	0.776[a]	0.753[a]	0.196[a]	0.100[a]	−0.215[a]	0.149[a]			
N/P	−0.045[b]	−0.033	0.013	−0.159[a]	−0.015	−0.035	−0.022	−0.025	−0.004	0.047[b]	0.056[a]	0.169[a]	−0.223[a]	0.105[a]	−0.065[a]	0.079[a]	0.148[a]		
Si/P	−0.057[a]	−0.032	0.026	−0.242[a]	0.081[a]	−0.084[a]	−0.011	0.040	0.013	0.063[a]	−0.077[a]	−0.097[a]	−0.256[a]	0.364[a]	−0.065[a]	0.067[a]	−0.114[a]	0.716[a]	
N/Si	0.022	0.038	0.045[b]	0.127[a]	−0.052[b]	0.038	0.017	0.009	0.042	−0.008	0.074[a]	0.081[a]	0.052[b]	−0.134[a]	−0.017	0.079[a]	0.100[a]	−0.010	−0.047[b]

[a].Correlation is significant at the 0.01 level (2-tailed)
[b].Correlation is significant at the 0.05 level (2-tailed)

Table 7.3 Physicochemical parameter variation in the Chilika lagoon since 1950 (as per the published literature)

Si no.	Depth (m)	SD (m)	WT (°C)	pH	Salinity	DO (mg L^{-1})	BOD (mg L^{-1})	Alkalinity (mg L^{-1})	Chl-a (μM)	Study period	Reference
1			25–33		0.96–34.9					1950–1951	Roy (1954)
2			24.44–26.05		>1.8–23.6					1960–1961	Ramanadham et al. (1964)
3			19–32	8–9.6	0.29–36	3.3–11.4		26.8–122		1957–1961	Banarjee and Roy Choudhury (1971)
4	0.2–3.0	0–2.2	28.0–36	07–10	0.55–15.83	1.9–16.9			0–13.38		Raman et al. (1990)
5	0.12–0.23		22–35	7.35–9.72	0.04–20.5	2.41–9.54				1988–1991	Tripathy (1995)
6	0.4–3.3	0.2–2	23–31.5	7–10.66	0–35.4	3–14.6		51–495		1985–1987	Siddiqui and Rao (1995)
7				7.37–10.2	0.37–31.73	3–13.2	0.5–7		0.233–54.04	1998–2001	Nayak et al. (2004)
8	0.35–2.5	0.1–1.5			0.59–32.73	3.58–9.98			0.09–48.53	2000–2003	Panigrahi et al. (2007)
9	0.28–6.82	0.05–2.89	19–35.5	6.1–10.3	0–37	0–16.36		0–304		1999–2004	*Jeong et al.* (2008)
10	0.3–3.63	0.10–0.24	18.9–330.6	6.92–10.07	0.04–36.5	0.3–10.98	0.1–13.68	22–326		2004–2007	Mohanty et al. (2009)
11			25.7–34.2	7.13–8.41	4.1–36.82	1.63–10.32	0.22–6.01			2008–2009	Patra et al. (2010)
12				7.95–8.6	0–32.5	3.6–5.22			0.3–17.68	2011	Ganguly et al. (2015)
13		0.72	28.3	8.5	13	7.1			18.4	2011–2012	Srichandan et al. (2015a)

(continued)

Table 7.3 (continued)

Si no.	Depth (m)	SD (m)	WT (°C)	pH	Salinity	DO (mg L^{-1})	BOD (mg L^{-1})	Alkalinity (mg L^{-1})	Chl-a (µM)	Study period	Reference
14		0.63–0.77	24.3–30.5	7.6–8.5	5.6–19.6	6.3–8.8			4.9–17.6	2012–2014	Srichandan et al. (2015b)
15	0.78–5.29	0.3–1.5	23.91–31.34	7.78–8.99	0.34–30.1	5.7–9.81	1.89–5.14	103.58–158.07	0.13–51	2011–2015	Barik et al. (2017)
16			27.4–30.1	7.75–8.07	5.57–22.4	214.3–231.0			3.72–5	2016	Ganguly et al. (2018)
17				7.4–10.2	0.2–32.8	2.30–17.39				2013–2014	Nazneen et al. (2019)
18		0.45–0.97	22.68–32.53			6.86–7.28			1.87	2017–2018	Srichandan et al. (2019)
19	1.83	0.74	28.16	7.88	7.64	7.32			5.91	2014–2015	Mohapatra et al. (2020)
20	0.27–7.48	0.07–4.0	18.9–35.9	6.1–10.35	0–37	0.3–14	0.04–14.52	20–304		1999–2015	Muduli and Pattnaik (2020)
21		0.71	27.97	8.12	11.96	7.32			11.46	2011–2015	Tarafdar et al. (2021)
22	**0.06–6**	**0–2.7**	**11.84–35.5**	**5.99–10.35**	**0–36**	**1.42–21.4**	**0.04–10.4**	**35–277**	**0–95.8**	**2013–2020**	**Present study**

Table 7.4 Pearson's correlation matrix of physicochemical parameters and nutrients of Chilika lagoon during monsoon (2013–2020)

	AT	WT	Depth	Salinity	pH	SD	Turbidity	DO	BOD	NO_2	NO_3	NH_3	P	SI	Alkalinity	Chl-a	N	N/P	Si/P
WT	0.596[a]																		
Depth	−0.038	0.072																	
Salinity	0.008	−0.076	0.113[a]																
pH	0.071	0.165[a]	−0.011	0.086[b]															
SD	0.128[a]	0.137[a]	0.268[a]	0.214[a]	0.179[a]														
Turbidity	−0.317[a]	−0.217[a]	−0.152[a]	−0.352[a]	−0.132[a]	−0.483[a]													
DO	0.088[b]	0.012	−0.027	−0.144[a]	0.203[a]	−0.066	0.026												
BOD	0.016	−0.034	−0.043	−0.163[a]	0.029	−0.028	0.061	0.302[a]											
NO_2	0.019	0.018	−0.104[a]	0.073	−0.055	−0.223[a]	−0.024	−0.031	−0.001										
NO_3	−0.034	0.001	−0.046	−0.272[a]	0.013	−0.302[a]	0.264[a]	0.021	−0.011	0.285[a]									
NH_3	−0.082[b]	−0.010	−0.091[b]	0.022	−0.141[a]	−0.072	0.015	−0.125[a]	−0.084[b]	0.043	0.107[a]								
P	−0.021	−0.029	0.113[a]	0.055	−0.142[a]	−0.021	0.140[a]	−0.096[b]	−0.052	0.024	−0.036	−0.006							
SI	0.036	0.038	−0.068	−0.499[a]	−0.052	−0.319[a]	0.131[a]	0.107[a]	0.058	0.179[a]	0.273[a]	−0.116[a]	−0.148[a]						
Alkalinity	0.036	0.067	0.076[b]	0.330[a]	0.130[a]	0.352[a]	−0.150[a]	−0.146[a]	−0.193[a]	−0.105[a]	−0.081[b]	0.057	0.005	−0.389[a]					
Chl-a	−0.191[a]	−0.044	0.047	−0.134[a]	−0.034	−0.067	0.113[b]	−0.115[a]	−0.049	−0.095[b]	0.074	0.096[b]	0.009	0.103[b]	−0.102[b]				
N	−0.077[b]	−0.005	−0.098[a]	−0.158[a]	−0.089[b]	−0.251[a]	0.196[a]	−0.071	−0.064	0.269[a]	0.735[a]	0.751[a]	−0.026	0.107[a]	−0.020	0.107[b]			
N/P	−0.175[a]	−0.122[a]	−0.035	−0.165[a]	−0.044	−0.102[a]	−0.007	0.021	0.026	0.053	0.172[a]	0.240[a]	−0.279[a]	0.151[a]	−0.062	0.135[a]	0.277[a]		
Si/P	−0.208[a]	−0.149[a]	0.037	−0.251[a]	0.035	−0.131[a]	−0.008	0.092[b]	0.074	−0.021	0.049	0.014	−0.330[a]	0.369[a]	−0.151[a]	0.179[a]	0.040	0.783[a]	
N/Si	−0.046	0.025	0.042	0.145[a]	0.054	0.099[a]	0.054	−0.122[a]	−0.031	0.070	0.194[a]	0.406[a]	0.030	−0.441[a]	0.177[a]	−0.005	0.405[a]	0.003	−0.152[a]

[a].Correlation is significant at the 0.01 level (2-tailed)
[b].Correlation is significant at the 0.05 level (2-tailed)

Table 7.5 Pearson's correlation matrix of physicochemical parameters and nutrients of Chilika lagoon during winter (2013–2020)

	AT	WT	Depth	Salinity	pH	SD	Turbidity	DO	BOD	NO_2	NO_3	NH_3	P	SI	Alkalinity	Chl-a	N	N/P	Si/P
WT	0.693[a]																		
Depth	−0.005	0.047																	
Salinity	0.031	0.060	0.405[a]																
pH	0.033	0.038	−0.137[a]	−0.139[a]															
SD	−0.004	0.021	0.457[a]	0.408[a]	0.106[a]														
Turbidity	−0.140[a]	−0.080[b]	−0.165[a]	−0.350[a]	−0.019	−0.452[a]													
DO	0.146[a]	0.092[b]	−0.174[a]	−0.117[a]	0.359[a]	−0.092[b]	0.092[b]												
BOD	0.088[b]	0.079[b]	−0.144[a]	−0.163[a]	0.120[a]	−0.148[a]	0.193[a]	0.300[a]											
NO_2	−0.053	0.041	−0.116[a]	0.059	−0.004	−0.098[a]	−0.026	0.042	0.015										
NO_3	−0.017	−0.050	0.075[b]	−0.083[b]	−0.293[a]	0.059	0.019	−0.207[a]	−0.079[b]	−0.010									
NH_3	−0.032	−0.028	0.101[a]	−0.006	−0.147[a]	0.043	0.028	−0.081[b]	−0.042	−0.022	0.164[a]								
P	−0.098[a]	−0.096[a]	0.067	0.048	−0.110[a]	0.019	0.037	−0.105[a]	−0.083[b]	−0.095[a]	0.224[a]	0.053							
SI	−0.046	0.029	−0.105[a]	−0.452[a]	−0.039	−0.339[a]	0.296[a]	−0.010	0.022	0.116[a]	0.256[a]	−0.018	0.018						
Alkalinity	0.078[b]	−0.022	0.048	0.265[a]	−0.010	0.212[a]	−0.299[a]	−0.067	−0.057	−0.081[b]	−0.212[a]	−0.098[a]	0.041	−0.207[a]					
Chl-a	−0.101[b]	−0.039	0.003	−0.097[b]	−0.031	−0.083[b]	0.265[a]	−0.085[b]	−0.068	−0.095[b]	0.052	0.136[a]	0.066	−0.033	−0.148[a]				
N	−0.034	−0.049	0.111[a]	−0.056	−0.289[a]	0.070	0.029	−0.187[a]	−0.079[b]	0.022	0.763[a]	0.761[a]	0.178[a]	0.161[a]	−0.207[a]	0.124[a]			
N/P	0.002	−0.082[b]	−0.025	−0.139[a]	−0.015	0.018	−0.052	−0.061	0.006	0.052	−0.063	0.162[a]	−0.238[a]	0.003	−0.030	0.036	0.068		
Si/P	0.024	0.038	−0.042	−0.206[a]	0.052	−0.093[b]	0.017	−0.015	0.013	0.146[a]	−0.130[a]	−0.166[a]	−0.225[a]	0.342[a]	−0.051	−0.020	−0.188[a]	0.609[a]	
N/Si	−0.037	−0.030	0.120[a]	0.269[a]	−0.016	0.155[a]	0.081[b]	0.020	−0.005	0.016	0.086[b]	0.298[a]	−0.004	−0.417[a]	−0.125[a]	0.113[a]	0.252[a]	0.068	−0.172[a]

[a]Correlation is significant at the 0.01 level (2-tailed)

[b].Correlation is significant at the 0.05 level (2-tailed)

Table 7.6 Pearson's correlation matrix of physicochemical parameters and nutrients of Chilika lagoon during summer (2013–2020)

	AT	WT	Depth	Salinity	pH	SD	Turbidity	DO	BOD	NO_2	NO_3	NH_3	P	SI	Alkalinity	Chl-a	N	N/P	Si/P
WT	0.657[a]																		
Depth	−0.130[a]	−0.043																	
Salinity	−0.035	−0.112[a]	0.373[a]																
pH	0.192[a]	0.154[a]	−0.216[a]	−0.398[a]															
SD	−0.111[a]	0.050	0.436[a]	0.136[a]	0.099[b]														
Turbidity	−0.009	0.049	−0.206[a]	−0.175[a]	−0.193[a]	−0.372[a]													
DO	0.045	0.012	−0.185[a]	−0.205[a]	0.302[a]	−0.112[a]	−0.036												
BOD	0.039	0.169[a]	−0.092[b]	−0.126[a]	−0.051	−0.092[b]	0.218[a]	0.167[a]											
NO_2	−0.056	−0.129[a]	−0.075	−0.014	0.005	−0.145[a]	0.063	0.095[b]	0.006										
NO_3	−0.061	0.186[a]	0.021	−0.211[a]	−0.085[b]	−0.021	0.291[a]	−0.196[a]	0.214[a]	−0.031									
NH_3	−0.175[a]	−0.158[a]	0.019	0.053	−0.261[a]	−0.060	0.203[a]	−0.140[a]	−0.001	−0.093[b]	0.253[a]								
P	−0.114[a]	0.020	0.088[b]	0.008	−0.248[a]	0.031	0.333[a]	−0.217[a]	0.322[a]	−0.012	0.460[a]	0.326[a]							
SI	0.000	0.113[a]	−0.151[a]	−0.556[a]	0.087[b]	−0.168[a]	0.183[a]	0.007	0.129[a]	0.109[a]	0.211[a]	−0.099[b]	0.128[a]						
Alkalinity	−0.034	−0.101[b]	0.125[a]	0.198[a]	0.130[a]	0.053	−0.283[a]	0.087[b]	−0.159[a]	0.105[a]	−0.401[a]	−0.286[a]	−0.248[a]	0.076					
Chl-a	−0.052	0.102[b]	−0.033	−0.181[a]	−0.041	−0.033	0.190[a]	0.028	0.207[a]	−0.013	0.249[a]	0.075	0.358[a]	0.206[a]	−0.152[a]				
N	−0.145[a]	0.040	0.031	−0.067	−0.269[a]	−0.029	0.326[a]	−0.249[a]	0.150[a]	−0.021	0.822[a]	0.755[a]	0.513[a]	0.073	−0.440[a]	0.201[a]			
N/P	0.015	0.073	−0.074	−0.120[a]	0.082[b]	0.003	−0.029	0.033	−0.072	−0.024	0.178[a]	0.049	−0.369[a]	0.009	−0.101[b]	−0.055	0.140[a]		
Si/P	0.083[b]	0.067	−0.145[a]	−0.262[a]	0.357[a]	−0.088[b]	−0.094[b]	0.209[a]	−0.092[b]	0.132[a]	−0.225[a]	−0.399[a]	−0.440[a]	0.381[a]	0.286[a]	−0.035	−0.401[a]	0.471[a]	
N/Si	0.018	0.059	0.093[b]	0.135[a]	−0.080[b]	0.066	0.013	0.025	0.064	−0.018	0.090[b]	0.075	0.137[a]	−0.162[a]	−0.036	0.129[a]	0.105[a]	−0.030	−0.076

[a]Correlation is significant at the 0.01 level (2-tailed)
[b].Correlation is significant at the 0.05 level (2-tailed)

whichvindicates Chilika is well oxygenated and maintains a healthy state irrespective of seasons.

DO recorded the lowest during the monsoon period and highest in winter followed by summer (Fig. 7.2a). Similar observations were also made by several researchers (Panda 2020; Upadhyay et al. 2015). During the winter higher DO was attributed to abundant phytoplankton growth leading to high primary productivity, whereas high turbid water hindering light penetration and productivity could be responsible for low DO during monsoon. Lower DO in summer as compared to winter could be due to utilization of DO for degradation of organic matter which is accelerated due to water volume reduction and growth of microbes in higher temperature by respiration process by the phytoplankton, microbes, macrophytes, and other living organisms in the lagoon (Robin et al. 2016). DO is influenced by salinity and temperature (Vijayakumar et al. 2000) as higher salinity tends to decrease in DO solubility (Mishra and Shaw 2003). In the present study, it was observed that DO did not correlate to WT significantly but negatively and significantly correlated with salinity (Table 7.2). However, the higher DO observed in low saline regions also could be due to the submerged macrophytes from which DO is sourced through the photosynthesis process (Muduli and Pattnaik 2020).

Salinity is an important factor for determining natural and biological processes in the lagoon. It is a strong determinant of the community composition of phytoplankton and their distribution (Huang 2004; Lueangthuwapranit et al. 2011). Salinity is a crucial parameter which determines the species distribution as the more tolerant species selects regions with brackish and higher saline waters, whereas the less tolerant species confine themselves in the freshwater area. Overall salinity of the lagoon was recorded as 9.87 ± 8.49 (mesohaline condition) and ranged between 0 and 36.1 (Table 7.1). The average salinity showed a trend of summer > winter > monsoon. During summer, the lagoon was completely saline water-dominated, whereas in monsoon it was freshwater-dominated. The salinity decreases to its minimum level during the southwest monsoon when the heavy rainfall increases the freshwater flux in the rivers (western catchment and Mahanadi). During peak monsoon, the water near the sea mouth is also found fresh because of the unidirectional flow of water which drains to the Bay of Bengal coming from the riverine system through the lead channel. This study showed a salinity gradient of OC > SS > CS > NS depending on the quantity of saline and freshwater mixing. NS during monsoon recorded the lowest due to mixing of freshwater from northeast rivers, whereas the OC in summer recorded highest which was attributed to minimum freshwater addition, high evaporation, and low precipitation (Mohanty and Mohanty 2002). The SS recorded higher salinity than the CS during monsoon and winter because of saline water intrusion from Rushikulya estuary through the Palur canal (Fig. 7.1). However, during summer, the CS recorded high salinity than SS which could be attributed to high salinity maintained in the OC and nearby regions (CS) due to the least freshwater flow and increased tidal saltwater mixing.

4.2 Nutrient Dynamics

Nutrients are the primary component of the aquatic food chain, and the key source into the lagoon environment is continental drainage from estuaries. The source of the nutrient can be both autochthonous and allochthonous. Inorganic nutrients such as N, P, and Si are very much crucial for the growth of the phytoplankton community in aquatic ecosystems. The freshwater influx and tidal condition in association with season greatly impact the nutrient distribution in the Chilika lagoon (Patra et al. 2016). The decline in nutrient level reflects along with increasing salinity; however, such phenomena interestingly not observed in Chilika could be due to the different point sources which increase the effluent load in the lagoon significantly (Sundaray et al. 2006). As per, the nutrient levels usually decrease during high tide and vice versa, as the high tide water is dominated by seawater with lower nutrient concentration than the riverine or estuarine ecosystem. Rainfall on the lagoon, on the Chilika watershed area, riverine freshwater discharge, seawater exchange, and in situ biogeochemical processes could change the nutrient stoichiometry and the concentration in water. Individual nutrient variability and the influencing factors have been discussed in the following sections.

4.2.1 Variability of Dissolved Inorganic Nitrogen Species

Nitrate in the aquatic environments is influenced by microbial oxidation of ammonia, advective transport into euphotic surface waters, and uptake by primary producers or denitrification in anoxic conditions (Grasshoff et al. 1999). Low summer NO_3-concentrations occur either due to low discharge or high biological uptake. Atmospheric input of nitrogen, in the form of N_2 gas, into aquatic systems, or associated with catchment rain events, has been recognized in the past decade as a significant allochthonous nitrogen source (Peierls et al. 2003). Loss of nitrate occurs through denitrification by microbial activity, which is the cause of the successive decrease of fixed NO_2 and NO_3 which gets converted to gaseous N_2 and N_2O. Denitrification processes within the lagoon are influenced by nitrate availability, oxygen, organic matter, temperature, and benthic in faunal activity (Nowicki et al. 1997). The major sink of nitrogen occurs through the denitrification processes in the sediments of aquatic environments, converting useable inorganic nitrogen to non-useable gaseous form, and, in the process, it alters the stoichiometric ratios of nutrients available to primary producers. NO_3 in the lagoon ranged between 0.02 and 20.79 μM with an average of 5.05 ± 4.3 μM. The highest NO_3 of 20.79 μM was observed in NS during the peak discharge period, in monsoon (Srichandan et al. 2019; Pattanaik et al. 2020). Irrespective of all the seasons, NS recorded the highest NO_3 (Fig. 7.2b; Table 7.7) which could be due to the release of NO_3 by microbial respiration of organic matter sourced from dominated vegetation in the NS. NO_3 constituted ~45% of N having the highest % in monsoon (56%) followed by winter (50%) and summer (47%).

During the present study, the intermediate species NO_2 (between NH_3 and NO_3) ranged between 0 and 3.16 μM (avg. 0.36 ± 0.32 μM). As reported by Chandran and Ramamoorthi (1984), NO_2 is sourced from planktons through metabolic activity and gets released into the water. Seasonally it followed the pattern: monsoon > winter > summer, as earlier observed by Srichandan et al. (2015b). The present study also found a significant variability of NO_2 with respect to seasons as confirmed by ANOVA ($p = 0.01$; $n = 2097$). A similar pattern was also recorded for other ecosystems (Pandey et al. 2015). Comparatively higher nitrite values have also been reported for the summer season which could be attributed to denitrification processes that occur in the sediment-water interface (Muduli and Pattnaik 2020). NO_2 constituted ~5% of total N having the highest % in summer (5.42%) followed by winter (4.85%) and monsoon (4.37%); this could be the indication of in situ biogeochemical process through which NO_2 is formed in the system (Barik et al. 2017).

Ammonium (NH_4) is generated in the pelagic or benthic compartment through the degradation of organic matter by bacterially mediated deamination (Seitzinger 1988) and animal excretion (McCarthy 1981). This has been shown to be a rapid and irreversible loss process for NH_3 (Lipschultz et al. 1986). The concentration of NH_4 relative to other nutrients may be low, and regeneration rates are variable and may be high relative to ambient concentrations (i.e., Gilbert et al. 1982), providing a source of available nutrients. NH_4 concentration can be altered through the nitrification process, i.e., ammonia oxidation to NO_3. When there enough concentration of NH_4 is available, NO_3 remains unutilized and subsequently lost through advective processes. The main input of NH_3 into the lagoon is through freshwater influx associated with local anthropogenic pollutants during monsoon which is high as compared to summer and winter (Muduli and Pattnaik 2020). In Chilika, it ranged between 0.55 and 28.99 μM with an average of 6.73 μM. NH_3 recorded highest in the NS and also during the summer period (Table 7.7, Fig. 7.2a) which could be due to release of NH_3 from macrophyte decomposition triggered by increasing salinity stroke as the freshwater weeds (such as *Potamogeton* and *Ichornia*, which are dominantly found in NS and CS) during monsoon keep on decomposing as the salinity keeps on increasing having peaked in summer. Apart from these few dominant macrophytes such as *Phragmites karka*, *Schoenoplectus* and *Salicornia* could also contribute for nutrients on decomposition according to the changing environmental characteristics. The present study showed the NH_3 had a significant negative correlation with DO ($r = -0.095$, $p < 0.05$) (Table 7.2) which indicated that NH_3 is formed by in situ process by decomposition of organic matter by utilizing DO (Robin et al. 2016). The phenomena occurred in all the seasons as supported by significant correlations (Tables 7.4, 7.5, and 7.6). As compared to NO_3 and NO_2, NH_3 species constituted the highest %, i.e., more than 50% of total N. In Chilika lagoon it is very crucial to consider NH_3 while calculating the N/P. Several studies have reported an N/P ratio considering NO_3 as N. This may not add much error for the ecosystems with a very low % of NH_3. However, in Chilika it could lead to misinterpretation as the addition of NH_3 concentration to NO_3 for N/P ratio calculation may change the ratio which decides the nutrient limitation. During the study period, the NH_3% varied significantly with respect to season, having the highest in

Table 7.7 Descriptive statistics of nutrients and their stoichiometry of Chilika waters during 2013–2020

Parameter	Sector	Monsoon				Winter				Summer				Overall			
		Min	Max	Mean	Stdv	Min	Max	Mean	Stdv	Min	Max	Mean	Stdv	Min	Max	Mean	Stdv
NO_2 (μM)	CS	0.03	1.91	0.42	0.35	0.00	2.88	0.33	0.29	0.00	1.87	0.34	0.29	0.00	2.88	0.36	0.31
	SS	0.02	1.10	0.31	0.25	0.00	2.95	0.35	0.34	0.01	2.53	0.30	0.32	0.00	2.95	0.32	0.31
	NS	0.00	2.01	0.40	0.36	0.01	1.33	0.36	0.24	0.00	3.16	0.38	0.38	0.00	3.16	0.38	0.33
	OC	0.00	1.71	0.47	0.34	0.00	1.58	0.36	0.30	0.01	1.48	0.37	0.26	0.00	1.71	0.40	0.30
	Overall	0.00	2.01	0.39	0.33	0.00	2.95	0.35	0.30	0.00	3.16	0.34	0.32	0.00	3.16	0.36	0.32
NO_3 (μM)	CS	0.10	19.47	4.30	3.66	0.12	19.54	4.80	4.52	0.08	19.10	5.71	4.44	0.08	19.54	4.91	4.25
	SS	0.12	19.45	3.26	2.98	0.24	19.88	4.92	4.55	0.06	19.50	5.25	4.17	0.06	19.88	4.46	4.05
	NS	0.17	20.79	5.45	4.38	0.02	18.68	5.47	4.65	0.12	18.33	6.48	4.88	0.02	20.79	5.77	4.65
	OC	0.36	17.21	5.01	4.00	0.22	19.13	4.71	3.68	0.03	16.10	6.08	4.39	0.03	19.13	5.22	4.04
	Overall	0.10	20.79	4.40	3.82	0.02	19.88	5.00	4.45	0.03	19.50	5.83	4.49	0.02	20.79	5.05	4.30
NH_3 (μM)	CS	0.63	25.42	6.22	4.38	0.11	39.97	5.07	4.46	0.02	19.60	5.63	3.92	0.02	39.97	5.62	4.30
	SS	0.55	16.92	5.05	3.23	0.30	49.30	5.37	4.51	0.06	26.38	5.27	3.81	0.06	49.30	5.23	3.90
	NS	0.52	23.74	5.98	4.15	0.41	28.23	5.60	4.16	0.06	19.62	6.72	3.92	0.06	28.23	6.05	4.11
	OC	0.30	28.99	5.97	4.22	0.11	41.42	5.21	4.78	0.11	18.45	5.84	3.70	0.11	41.42	5.66	4.29
	Overall	0.30	28.99	5.77	4.00	0.11	49.30	5.32	4.44	0.02	26.38	5.84	3.89	0.02	49.30	5.63	4.14
P (μM)	CS	0.00	2.20	0.36	0.36	0.01	7.43	0.53	0.98	0.03	2.00	0.44	0.39	0.00	7.43	0.44	0.66
	SS	0.00	7.01	0.42	0.60	0.02	10.19	0.56	1.12	0.03	2.46	0.45	0.39	0.00	10.19	0.48	0.79
	NS	0.00	1.58	0.40	0.36	0.01	2.87	0.48	0.56	0.03	2.90	0.49	0.50	0.00	2.90	0.46	0.48
	OC	0.00	4.11	0.52	0.52	0.01	2.04	0.53	0.48	0.07	2.51	0.60	0.49	0.00	4.11	0.55	0.50
	Overall	0.00	7.01	0.41	0.47	0.01	10.19	0.53	0.88	0.03	2.90	0.48	0.44	0.00	10.19	0.47	0.64

(continued)

Table 7.7 (continued)

Parameter	Sector	Monsoon				Winter				Summer				Overall			
		Min	Max	Mean	Stdv	Min	Max	Mean	Stdv	Min	Max	Mean	Stdv	Min	Max	Mean	Stdv
SI (μM)	CS	9.77	174.13	73.65	38.69	2.83	282.36	72.74	44.46	7.80	151.00	54.71	30.66	2.83	282.36	67.66	39.65
	SS	4.80	173.70	63.48	29.05	17.10	177.90	73.82	30.74	15.68	137.10	74.19	24.62	4.80	177.90	70.47	28.83
	NS	13.96	214.21	97.99	51.92	2.40	211.28	94.25	51.36	2.63	180.16	63.41	33.58	2.40	214.21	86.19	49.20
	OC	7.50	180.60	70.45	45.44	1.79	142.52	49.47	33.24	1.90	90.70	26.89	19.64	1.79	180.60	49.94	38.87
	Overall	4.80	214.21	76.51	43.26	1.79	282.36	75.22	43.63	1.90	180.16	58.80	32.17	1.79	282.36	70.73	41.13
N (μM)	CS	0.00	31.20	10.16	6.46	0.90	40.80	10.20	6.99	0.00	39.00	9.50	7.43	0.00	40.80	9.95	6.97
	SS	0.00	24.80	7.21	5.66	0.00	52.50	9.66	7.19	0.00	33.40	7.97	7.09	0.00	52.50	8.28	6.75
	NS	0.00	33.30	9.72	7.26	0.00	32.60	10.20	7.30	0.00	36.10	9.85	8.45	0.00	36.10	9.92	7.68
	OC	0.00	30.90	9.33	6.75	0.00	42.40	8.50	6.96	0.00	27.30	9.50	7.82	0.00	42.40	9.09	7.16
	Overall	0.00	33.30	9.02	6.62	0.00	52.50	9.76	7.14	0.00	39.00	9.13	7.71	0.00	52.50	9.30	7.17
N/P	CS	2.20	2700.00	107.76	266.65	0.10	699.50	55.32	80.56	5.20	321.60	39.64	34.12	0.10	2700.00	68.41	165.96
	SS	0.70	535.50	59.44	86.06	0.20	459.50	57.04	67.78	2.20	485.40	34.29	39.62	0.20	535.50	50.99	68.79
	NS	4.50	1305.00	96.13	190.69	2.00	1474.00	70.12	132.32	1.50	515.00	46.04	50.07	1.50	1474.00	71.52	140.02
	OC	4.10	470.70	65.96	93.98	2.20	690.00	48.24	96.46	3.90	237.90	30.78	30.74	2.20	690.00	49.06	82.58
	Overall	0.70	2700.00	84.16	184.88	0.10	1474.00	58.61	95.96	1.50	515.00	38.43	40.38	0.10	2700.00	61.15	125.16
Si/P	CS	17.70	15439.00	828.79	1903.85	4.70	9412.00	541.28	1077.61	11.00	2798.00	252.36	324.10	4.70	15439.00	552.63	1314.83
	SS	6.20	6459.00	531.44	912.40	3.50	6102.00	524.72	819.15	19.70	2775.70	380.56	459.15	3.50	6459.00	483.54	767.60
	NS	12.10	6893.50	834.83	1275.35	13.00	14819.00	683.18	1328.45	3.80	3082.40	298.10	416.41	3.80	14819.00	616.81	1130.46
	OC	14.90	4481.30	511.76	899.62	3.40	5680.00	322.62	774.83	0.10	425.10	84.91	98.00	0.10	5680.00	316.14	722.33
	Overall	6.20	15439.00	694.80	1368.42	3.40	14819.00	540.41	1046.52	0.10	3082.40	278.30	386.19	0.10	15439.00	513.66	1048.70
N/Si	CS	0.00	0.90	0.20	0.16	0.00	2.60	0.20	0.26	0.00	1.60	0.29	0.26	0.00	2.60	0.23	0.24
	SS	0.00	4.20	0.23	0.44	0.00	1.20	0.17	0.14	0.00	0.50	0.16	0.11	0.00	4.20	0.19	0.28
	NS	0.00	1.20	0.18	0.17	0.00	4.40	0.24	0.47	0.00	6.00	0.33	0.54	0.00	6.00	0.25	0.43
	OC	0.00	1.40	0.25	0.25	0.00	4.80	0.48	0.76	0.00	8.90	0.84	1.10	0.00	8.90	0.51	0.80
	Overall	0.00	4.20	0.21	0.29	0.00	4.80	0.24	0.42	0.00	8.90	0.34	0.57	0.00	8.90	0.26	0.44

monsoon (56.47%) followed by winter (50.19%) and summer (46.94%) which could be the indication of the source of NH_3 from freshwater discharge from rivers which diminishes as from monsoon to summer (Ganguly et al. 2015).

4.2.2 Variability of Dissolved Inorganic Phosphate

Dissolved inorganic phosphate is one of the micronutrients which controls the trophic status depending on the availability of N. As per De Busk (1999), the P in the water body could be organic or inorganic compounds either in the form of dissolved or particulate matter. In the surface water, the P is sourced from rock weathering and organic matter decomposition. Under favorable environmental conditions such as light and temperature, the P gets assimilated by phytoplankton and bacteria. As per Sobehrad 1997), the organic P available in SPM and on the surface of organic detritus is consumed by filter feeders and released as inorganic P. Lagoon water with low P declines the productivity of water as the phytoplankton growth gets hindered, whereas excess P can be the cause of the eutrophic condition which may lead to a bloom of some dominating species. To date, no studies have reported bloom caused by high P content in Chilika which could be due to the fact that the water column transparency gets declined by turbid water input from rivers which hinders photosynthesis leading to lower primary productivity (Srichandan et al. 2015a). This study showed Chilika maintains a very low concentration of P, varying from 0 to 10.19 μM with an average of 0.47 ± 0.64 μM. In Chilika lagoon, P from a point source has not been reported as there is no such industry situated in proximity to the lagoon, and whatever input that comes through the river has a minimal impact on the variability of P concentration in Chilika lagoon. As reported in DWAF (1995), high levels of P are originated from industrial effluents, domestic discharge, drainage from agricultural land, urban runoff, and atmospheric precipitation. Chilika lagoon is safe from such effluents which could influence the P level in monsoon and subsequent seasons.

The sector-wise variation in summer and winter showed a similar trend; however, its difference from monsoon might be due to the impact of abundant riverine discharge from northeast rivers. Lower P level recorded during the monsoon could be attributed to adsorption to SPM (Sobehrad 1997) and dilution effect whose factor is also earlier reported by Muduli et al. (2017). However, a couple of studies also recorded comparatively higher P level in monsoon attributing to freshwater discharge with fertilizer content and weathering to the spike (Srichandan et al. 2019). In the present study, during none of the seasons, P showed a significant correlation with salinity (Tables 7.2, 7.4, 7.5, and 7.6) which indicated it is not controlled by either freshwater input during monsoon or seawater exchange in summer. Rather, the in situ biogeochemical processes controlled the P level in Chilika. Supporting the same, P showed a significant positive correlation with turbidity indicating the P release from sediment by the churning effect.

4.2.3 Variability of Silicate

Silicate (Si) is a bio-limiting nutrient and a major constituent of diatoms. As per DWAF (1995), diatoms use Si to encase their cells. Along with N and P, this is also required for primary production. This is sourced from the terrestrial system where erosion of adjacent land takes place and also from the anthropogenic activity-influenced areas. In the Chilika lagoon, this process was evidenced by higher Si values during the monsoon. In the present study, the Si concentration in Chilika ranged from 5 to 214 μM. A gradual reduction in Si level recorded from winter season to summer (Table 7.7) could be attributed to the removal of dissolved silicate by two processes: (1) uptake of Si by diatoms for shell formation and (2) absorption on SPM which gets triggered in summer due to increase in salinity. As per Borole (1993), Si behaves conservatively, and during summer periods, silicate behaves non-conservatively. The Si observed in the OC during different seasons showed a trend of monsoon > winter > summer which indicated that the Si transfer to the Bay of Bengal through OC faces a decline from monsoon to the summer season (Table 7.7). Si is a critical factor that decides the plankton biodiversity especially the diatoms in the Chilika lagoon (Srichandan et al. 2015b). In the Veli lagoon, also a clear inverse relationship of silicate with Bacillariophyta showed the dependency of the phytoplanktons on respective nutrients (Mathew and Nair 1981). Overall as well as in respect to seasons, Si was negatively correlated with salinity ($r = -0.499$, $p < 0.01$) and SD ($r = -0.320$, $p < 0.01$) which indicated the source of Si was from riverine freshwater input and high silicate maintained in the low transparent water (Tables 7.2, 7.4, 7.5 and 7.6).

4.3 *Spatiotemporal Variability in Trophic Index*

During 2019, the Chilika lagoon maintained mesotrophic status as evidenced by a TSI value of 45.92. As recommended by Carlson (1977), the ecosystem is considered as eutrophic, mesotrophic, and oligotrophic if the calculated TSI value is >50, 40–50, and <40, respectively. There was no difference in trophic status with respect to the season as the TSI recorded for the individual season was as follows: summer (46.8), winter (46.58), and monsoon (44.35). TSI values indicated irrespective of seasons the lagoon maintained mesotrophic nature. Similar to seasons, all the sectors were also found to maintain mesotrophic status (TSI of 46.12, 45.23, and 42.57 for CS, NS, and SS, respectively) with the exception of the OC. The TSI of OC was calculated to be 52.75 which is very close to the eutrophic boundary. The higher TSI of OC was due to the lower transparency recorded in the OC which was attributed to surfing of water by frequent movement of motorized boats operated in the OC for tourism activities (Mohanty et al. 2016). Apart from this, the tidal fluctuations in the OC could be another factor for lower SD (Muduli and Pattnaik 2020). The present study indicates SD is the most contributing factor for TSI status, and it was ranged between 69.71 and 80.61. Such lower values of TSI_{Chl} as compared to TSI_{SD} showed, along with algae there are some other factors such as sediment particle or color which could be responsible for the light attenuation.

TLI calculated during 2019 also indicated the same trophic status (showed by TSI) of Chilika, i.e., mesotrophic. As suggested by Burns (2005), the trophic status is considered as eutrophic, mesotrophic, and oligotrophic for the TLI of >4, 3–4, and <3, respectively. This study recorded TLI values varying from 2.76 to 4.48 with an average of 3.62. Similar to TSI, the TLI also indicated mesotrophic status for individual seasons as well as sectors except the OC (TLI of 4.04, 3.55, and 3.25 in winter, summer, and monsoon, respectively; 4.17, 3.65, 3.49, and 3.30 for OC, CS, NS, and SS, respectively). The TLI recorded for OC was 4.17 which is beyond the boundary of the mesotrophic status (Burns 2005) as also revealed from TSI. The exceptional trophic status in the OC could be attributed to the factors as explained for TSI. Since there is no significant difference in the trophic status of the lagoon in different seasons and sectors explained through TSI and TLI, either of these indexes could be used for deriving the trophic status of the Chilika lagoon.

4.4 Nutrient Stoichiometry and Influencing Factors

4.4.1 N/P

The concentration of nutrients in lagoon water with specific stoichiometry plays a critical role in phytoplankton growth rate, and the ecosystems are considered to be nutrient-limited if the balance of carbon, N, and P in the environment varies from the Redfield ratio for DIC:DIN:DIP of 106:16:1 (Redfield 1958). As reported by Correll (1998), freshwater ecosystems are typically P-limited as the incorporation rate of nitrogen into plant tissue is usually controlled by P availability. NS of Chilika lagoon maintains fresh to brackish nature throughout the year, and it also showed P limiting with N/P (71.52 ± 140.02). Most of the studies also reported P limiting except few studies which reported N limiting (Table 7.8). These discrepancies could be attributed to change in sample numbers, sampling period, and sampling locations. P limitation is also reported for other ecosystems such as Apalachicola Bay, Chesapeake Bay, Hudson River, and Peel Harvey estuaries (Boynton et al. 1982; McComb et al. 1981; Myers and Iverson 1981). All the individual sectors recorded the highest N/P during the monsoon period followed by winter and summer (Table 7.7). This declining trend from monsoon to summer could be attributed to the addition of N load in monsoon from riverine freshwater discharge which gradually decreases from monsoon to summer. In the present study, N/P varied between 0.1 and 2700 μM with an average of 61.15 ± 125.16. The N/P ratio was found to be negatively correlated with salinity ($r = -0.159$, $p < 0.01$) (Table 7.2). Such an inverse relationship is also recorded for all the seasons (Tables 7.4, 7.5, and 7.6). This relationship is justified with higher N content as compared to P in the lower saline region, whereas higher saline water mostly in the OC and SS contained less concentration of N (Barik et al. 2017). The change in ratio and limitation of

Table 7.8 Variability of nutrient concentration in Chilika lagoon as per the published literature since 1990

Si No.	NO_2 (μM)	NO_3 (μM)	P (μM)	Si (μM)	NH_3 (μM)	N/P	Si/P	N/Si	Study period	Reference
1	0.28	0.26	0.2	0.39–5.34					1990	Raman et al. (1990)
2	0–4.53	0–7.61	0–0.07	0–9.14					1988–1991	Tripathy (1995)
3		0.42–50.4	1.24–8.68	11.2–44.8					1985–1987	Siddiqui and Rao (1995)
4		0–28	0.12–5.32						1998–2001	Nayak et al. (2004)
5		<0.01–4.60	0.07–1.83						2000–2003	Panigrahi et al. (2007)
6		0–34.11	0–20.43						1999–2004	Jeong et al. (2008)
7	0.13–42.88	0.27–87.6	0.04–4.14						2004–2007	Mohanty et al. (2009)
8		1.55–117.4	0.17–5.4	3.55–156.25		10.39	133.95		2008–2009	Patra et al. (2010)
9	0.01–0.55	1–35	0.4–1.3	20–105		<16			2011	Ganguly et al. (2015)
10		3.13	1.1	70					2011–2012	Srichandan et al. (2015a)
11		1.7–16.2	0.37–1.46	60.7–88.8		5.1	16	<1	2012–2014	Srichandan et al. (2015b)
12	0.19–0.87	2.21–9.44	0.32–1.14	34.07–115.11		8.49–16.8	4.22–168.14	0.08–0.21	2011–2015	Barik et al. (2017)
13	0–2.35	0.2–9.6	0–0.09						2013–2014	Nazneen et al. (2019)
14	0.17–0.87	0–10.81	0.75–1.66	64.74	9				2017–2018	Srichandan et al. (2019)
15	0.26	8.97	0.82	66.14	7.52				2014–2015	Mohapatra et al. (2020)
16	0.01–2.01	0.12–19.88	0.01–2.85	0.1–363	38–1629	30.6	>16	<1	1999–2015	Muduli and Pattnaik (2020)
17	0.38	5.69	0.81	68.93		16.84	233.25	0.28	2011–2015	Tarafdar et al. (2021)
18	**0–3.16**	**0.02–20.79**	**0–10.19**	**1.8–282**	**0.02–49.3**	**0.1–2700**	**0.1–15,439**	**0–8.90**	**2013–2020**	**Present study**

nutrients in relation to salinity also has been reported in several ecosystems overseas. For instance, Sakshaug and Olsen (1986) recorded P limitation in fresh and brackish waters during phytoplankton blooms in Norwegian waters, and Paasche and Erga (1988) found N limitation in marine waters. Estuaries are typically nitrogen-limited, with some variation in nutrient limitation in brackish waters observed like in Chilika ecosystems (Correll 1998). P limitation in Chilika has been reported by earlier studies (Panigrahi et al. 2009; Sarma et al. 2010; Barik et al. 2017) supporting the observations in the present study.

Nitrogen limitation is recorded for several ecosystems due to anthropogenic material influx from rivers which shifts the nutrient stoichiometry leading to N limitation (Siddiqui et al. 2019). Studies also reported N/P very close to 16, for instance, Martin et al. (2008) reported N limitation over P due to hike in P influx in the southwest coast of India. As per Klug (2006), a decline in the N/P also has the potential to alter the phytoplankton species composition. In coastal ecosystems, the phytoplankton productivity under favorable environmental conditions including nutrient stoichiometry is the cause of increasing toxic blooms. The stoichiometry study for ten large world rivers and two river-dominated coastal ecosystems similar to the Chilika lagoon ecosystem was found to be in a eutrophic state (Justic et al. 1995), and the study revealed that the nutrient stoichiometry of the river waters strongly altered the stoichiometry in the coastal waters.

Almost all studies where the nutrient stoichiometry has been reported used dissolved inorganic nitrogen (N) as NO_3 or $NO_3 + NO_2$ excluding NH_3. In the present study, NH_3 was considered to calculate N/P (61.1) and it was found that ~50% of the decrease in N/P (30.4) was without consideration of NH_3. It is noteworthy to mention that there was no change in "nutrient limitation" and it remained as P limiting in both cases. In the case of the Chilika lagoon, the inclusion of NH_3 to calculate N/P did not make any difference in deciding the nutrient-limiting factor. However, in case of change in the environmental condition of Chilika in the future and any other ecosystem, maintaining relatively higher NH_3 could decrease the N/P to <16 and alter the nutrient-limiting factor to N limiting. Hence is it recommended to include NH_3 along with $NO_2 + NO_3$ for the N/P interpretation in further studies on the Chilika lagoon.

4.4.2 Si/P

Si/P could be used as an indicator to understand the nutrient dynamics and impact of riverine discharge on ecosystems like Chilika (Paul et al. 2008). The Si/P status maintained in the lagoon indicates the weathering forms around the lagoon which is mostly dependent on climate conditions (Turner and Rabalais 2003). This study showed a Si/P varying between 0.1 and 15,439 μM with an average of 513.66 ± 1048 μM. Similar to previous studies (Table 7.8), the present study also recorded Si/P > 16 which indicated P was limiting, making the Chilika water favorable for diatom growth (Panigrahi et al. 2009; Srichandan et al. 2015b). The abundance of diatoms due to such factors is also registered by Domingues (2007). There are examples of an ecosystem where Si limitation is encountered which leads to the dominance of phytoplankton which is not siliceous. Pereira et al. (2009) recorded such observation in Obidos lagoon, Portugal. In all the seasons (except little deviation in summer), Chilika Si/P followed a sectoral trend NS > CS > SS > OC which was in parallel with reverse salinity gradient that indicated the freshwater in the lagoon with high Si controls the Si/P (Table 7.7, Fig. 7.2b). This fact is also supported by the seasonal trend observed for Si/P as monsoon > winter > summer (lagoon dominates with freshwater and Si in monsoon which keeps on decreasing till the end of summer). Similar to N/P, Si/P correlated significantly with salinity ($r = -0.242$, $p < 0.01$) in all the seasons which also revealed the abovementioned facts (Tables 7.2, 7.4, 7.5, and 7.6).

4.4.3 N/Si

N to silicate ratio (N/Si) is a major indicator to quantify the health of a lagoon ecosystem which gives information of acute phosphorus depletion (Lucea et al. 2005). As per the Redfield ratio, the stoichiometry should be maintained as 1 (N/P/Si = 16:1:16). In the present investigation, the ratio ranged between 0 and 8.9 (avg. 0.26 ± 0.43) which is <1 indicating the N limitation over Si (Table 7.7; Srichandan et al. 2015a) similar to observations made earlier (Table 7.8). Sector-wise the trend followed as SS < CS < NS < OC, and the highest N/Si observed in the OC was attributed to the least Si recorded in the OC as compared to other sectors. On a temporal scale, the variability was also significant as the maximum ratio found in summer (0.54 ± 3.87) followed by winter (0.24 ± 0.42) and monsoon (0.21 ± 0.29) (Table 7.7). N/Si nearing 1 (Redfield ratio) in OC, i.e., 0.91, makes a most favorable condition for primary productivity as suggested by Redfield (Brzezinski 1985). Pearson correlation showed that the salinity maintained a significant correlation with N/Si ($r = 0.127$, $p < 0.01$) (Table 7.2) in all the seasons (Tables 7.4, 7.5, and 7.6) as the high saline waters contained low silicate and supported by a significant negative correlation of Si to salinity ($r = -0.509$, $p < 0.01$; Table 7.2). Hence the occurrence of silica-enriched conditions could ensure abundant Si availability to phytoplankton. Chlorophyll-a positively correlated with N/Si ($r = 0.079$, $p < 0.01$; Table 7.2) indicated the uptake of Si by phytoplankton. Low N/Si ratio in monsoon as compared to other seasons is also observed by other ecosystems when the productivity is fuelled by the supply of nutrients (Yadav and Pandey 2018). Only few studies have reported Si limitation over N (N/Si > 1) (Siddiqui et al. 2019).

5 Conclusion

This study highlighted the influencing factors of nutrient variability and the role of nutrient stoichiometry on lagoon productivity. The water quality changes due to seasons were found to be crucial for the nutrient biogeochemistry and other physicochemical parameters of the lagoon. N and P recorded in the study period along with other physicochemical parameters indicated good health of Chilika lagoon. However, on seasonal scale winter scores the best and monsoon least owing to the least transparency nutrient load from northeast rivers. The study indicated the NH_3

species constituted more than 50% of total N, and it is a vital parameter to consider while calculating the N/P, and the N must include NH_3. It could be misleading consideration of the only NO_3 as N for the N/P stoichiometry as the addition of NH_3 concentration to NO_3 may shift the stoichiometry either close to Redfield ratio or far from it and may lead to misinterpretation on the nutrient-limiting factors. TSI and TLI index indicated the lagoon maintains mesotrophic condition. However, long-term monitoring of TN along with TP and other physicochemical parameters is needed for a more appropriate representation of the trophic status of the Chilika lagoon. Including the present study, Chilika has been studied in major aspects of nutrient variability such as nutrient uptake by the plankton community of Chilika, nutrient flux from riverine input, spatiotemporal variation of nutrients, etc. However, there is still some critically important figure related to nutrient dynamics yet to be studied for which it is recommended to (1) quantify the N exchange through nitrification and denitrification process, (2) estimate the nutrient exchange from sea, (3) quantify the nutrient exchange from benthic compartment, (4) estimate the nutrient uptake by macrophytes of Chilika especially the *Phragmites karka* spread over the NS of the lagoon, and (5) long-term monitoring of TN and TP.

Acknowledgments The authors are thankful to the Chief executive, CDA, for providing the data and encouragement to carry out the research. Funding from the World Bank (Credit no. 4765-IN) is acknowledged for its support to develop laboratory facilities at WRTC which was utilized to carry out the research. Thanks are also due to the research staffs of CDA who helped during the sampling and laboratory works to carry out the study.

References

Banarjee AC, Roy Choudhury NC (1971) Observation on some physicochemical features of Chilika Lake. Indian J Fish 13(1–2):395–429

Barik SK, Muduli PR, Mohanty B, Behera AT, Mallick S, Das A, Samal RN, Rastogi G, Pattnaik AK (2017) Spatio-temporal variability and the impact of Phailin on water quality of Chilika lagoon. Cont Shelf Res 136:39–56

Borole DV (1993) Late Pleistocene sedimentation: a case study of the Central Indian Ocean Basin. Deep-Sea Res I Oceanogr Res Pap 40(4):761–775

Boynton WR, Kemp WM, Keefe CW (1982) A comparative analysis of nutrients and other factors influencing estuarine phytoplankton production. In: Estuarine comparisons. Academic Press, New York, pp 69–90

Bita MPR, Muduli AT, Behera D, Mahapatro SK, Barik SK, Nag RN, Samal AK, Pattnaik (2016) (2016) Assessment of petroleum hydrocarbon in a tropical brackish water lagoon: Chilika India. Chemistry and Ecology 32(7):653–668. https://doi.org/10.1080/02757540.2016.1177521

Brzezinski MA (1985) The si:c:n ratio of marine diatoms: interspecific variability and the effect of some environmental variables1. Journal of Phycology 21(3):347–357. https://doi.org/10.1111/j.0022-3646.1985.00347.x

Burns N, McIntosh J, Scholes P (2005) Strategies for managing the lakes of the rotorua district new zealand. Lake and Reservoir Management 21(1):61–72. https://doi.org/10.1080/07438140509354413

Carlson RE (1977) A trophic state index for lakes 1. Limnol Oceanogr 22(2):361–369

Carrit DE, Carpenter JH (1966) Recommendation procedure for Winkler analyses of sea water for dissolved oxygen. J Mar Res 24:313–318

Chandran R, Ramamoorthi K (1984) Hydrobiological studies in the gradient zone of the vellar estuary: III. Heterotrophic bacteria. Mahasagar Bull Natl Inst Oceanogr 17(4):197–201

Cohen MCL, Lara RJ, Ramos JDF, Dittmar T (1999) Factors influencing the variability of Mg, Ca and K in waters of a mangrove creek in Bragança, North Brazil. Mangrove Salt Marshes 3(1):9–15

Correll DL (1998) The role of phosphorus in the eutrophication of receiving waters: a review. J Environ Qual 27(2):261–266

CPCB, New Delhi (1986). https://www.cpcb.nic.in/Waterquality_Criteria.php. Accessed 5.1.16

De Busk WF (1999) Document SL170. Soil and Water Science Department, Florida Cooperative Extension Service, Institute of Food and Agricultural Sciences, University of Florida

Ganguly D, Singh G, Purvaja R, Bhatta R, Paneer Selvam A, Banerjee K, Ramesh R (2018) Valuing the carbon sequestration regulation service by seagrass ecosystems of Palk Bay and Chilika India. Ocean & Coastal Management 15926–33. https://doi.org/10.1016/j.ocecoaman.2017.11.009

Domingues RB (2007) Phytoplankton succession in the Guadiana upper estuary: before and after the Alqueva dam phytoplankton succession in the Guadiana upper estuary: before and after the Alqueva dam

DWAF (1995) South African water quality guidelines for coastal marine waters, Natural environment, vol 1. DWAF, Pretoria

El-Serehy HA, Abdallah HS, Al-Misned FA, Irshad R, Al-Farraj SA, Almalki ES (2018) Aquatic ecosystem health and trophic status classification of the Bitter Lakes along the main connecting link between the Red Sea and the Mediterranean. Saudi J Biol Sci 25(2):204–212

Ganguly D, Patra S, Muduli PR, Vardhan KV, Abhilash KR, Robin RS, Subramanian BR (2015) Influence of nutrient input on the trophic state of a tropical brackish water lagoon. J Earth Syst Sci 124(5):1005–1017

Gilbert PM, Lipschultz F, McCarthy JJ, Altabet MA (1982) Isotope dilution models of uptake and remineralization of ammonium by marine plankton 1. Limnol Oceanogr 27(4):639–650

Grasshoff K, Kremling K, Ehrhardt M (1999) Methods of seawater analysis, third completely revised and extended edn. Verlag Chemie, Weinheim

Gupta GVM, Sarma VVSS, Robin RS, Raman AV, Kumar MJ, Rakesh M, Subramanian BR (2008) Influence of net ecosystem metabolism in transferring riverine organic carbon to atmospheric CO2 in a tropical coastal lagoon (Chilika Lake, India). Biogeochemistry 87(3):265–285

Huang L (2004) Species diversity and distribution for phytoplankton of the Pearl River estuary during rainy and dry seasons. Mar Pollut Bull 49(7–8):588–596. https://doi.org/10.1016/j.marpolbul.2004.03.015

Jabir T, Vipindas PV, Jesmi Y, Valliyodan S, Parambath PM, Singh A, Abdulla MH (2020) Nutrient stoichiometry (N: P) controls nitrogen fixation and distribution of diazotrophs in a tropical eutrophic estuary. Mar Pollut Bull 151:110799

Jeong KS, Kim DK, Pattnaik A, Bhatta K, Bhandari B, Joo GJ (2008) Patterning limnological characteristics of the Chilika lagoon (India) using a self-organizing map. Limnology 9(3):231–242

Justic D, Rabalais NN, Turner RE (1995) Stoichiometric nutrient balance and origin of coastal eutrophication. Mar Pollut Bull 30(1):41–46. https://doi.org/10.1016/S0272-7714(05)80014-9

Kim JY, Rastogi G, Do Y, Kim DK, Muduli PR, Samal RN, Pattnaik AK, Joo GJ (2015) Trends in a satellite-derived vegetation index and environmental variables in a restored brackish lagoon. Glob Ecol Conserv 4:614–624

Kjerfve B (1994) Coastal lagoons. Coast Lagoon Proc 1952:1–8. https://doi.org/10.2307/634562

Klug JL (2006) Nutrient limitation in the lower Housatonic River estuary. Estuar Coasts 29(5):831–840

Lucea A, Duarte CM, Agustí S, Kennedy H (2005) Nutrient dynamics and ecosystem metabolism in the Bay of Blanes (NW Mediterranean). Biogeochemistry 73(2):303–323

Lueangthuwapranit C, Sampantarak U, Wongsai S (2011) Distribution and abundance of phytoplankton: influence of salinity and turbidity gradients in the Na Thap River, Songkhla Province, Thailand. J Coast Res 27(3):585–594

Lipschultz F, Wofsy SC, Fox LE (1986) Nitrogen metabolism of the eutrophic Delaware River ecosystem 1. Limnology and Oceanography 31(4):701–716

Mahapatro D, Panigrahy RC, Panda S (2013) Coastal lagoon: present status and future challenges. Int J Mar Sci 3:178

Martin GD, Vijay JG, Laluraj CM, Madhu NV, Joseph T, Nair M, Gupta GVM, Balachandran KK (2008) Fresh water influence on nutrient stoichiometry in a tropical estuary, southwest coast of India. Appl Ecol Environ Res 6:57–64

Mathew T, Nair NB (1981) Phytoplankton of the Veli Lake, a lagoon on the south-west coast of India. Mahasagar Bull Natl Inst Oceanogr 14(1):45–54

McCarthy JJ (1981) Uptake of major nutrients by estuarine plants. In: Estuaries and nutrients. Humana Press, Totowa, pp 139–163

McComb AJ, Atkins RP, Birch PB, Gordon DM, Lukatelich RJ (1981) Eutrophication in the Peel-Harvey estuarine system, western Australia. In: Estuaries and nutrients. Humana Press, Totowa, pp 323–342

Mishra RK, Shaw BP (2003) Spatio-temporal variation of optically active substances in the coastal waters off Orissa from Rushikulya to Dhamra (east coast of India). Indian J Mar Sci 32:133–140

Mishra DR, Kumar A, Muduli PR, Equeenuddin SKM, Rastogi G, Acharyya T, Swain D (2020) Decline in phytoplankton biomass along Indian coastal waters due to COVID-19 lockdown. Remote Sens 12(16):2584. https://doi.org/10.3390/rs12162584

Mishra DR, Kumar A, Muduli P, Acharyya T, Acharya P, Singh S, Rastogi G (2021) Landfall season is critical to the impact of a cyclone on a monsoon-regulated tropical coastal lagoon. Sci Total Environ 770:145235

Mohanty RK, Mohanty SK (2002) Factors affecting finfish and shellfish biodiversity in Chilika lagoon. Fish Chimes 22(5):42–43

Mohanty RK, Mohapatra A, Mohanty SK (2009) Assessment of the impacts of a new artificial lake mouth on the hydrobiology and fisheries of Chilika Lake, India. Lakes Reserv Res Manag 14(3):231–245

Mohapatra M, Behera P, Kim JY, Rastogi G (2020) Seasonal and spatial dynamics of bacterioplankton communities in a brackish water coastal lagoon. Sci Total Environ 705:134729

Muduli PR, Pattnaik AK (2020) Spatio-temporal variation in physicochemical parameters of water in the Chilika Lagoon. In: Ecology, conservation, and restoration of Chilika Lagoon, India. Springer, New York, pp 203–229

Muduli PR, Kanuri VV, Robin RS, Kumar BC, Patra S, Raman AV, Rao GN, Subramanian BR (2012) Spatio-temporal variation of CO2 emission from Chilika Lake, a tropical coastal lagoon, on the east coast of India. Estuar Coast Shelf Sci 113:305–313

Muduli PR, Kanuri VV, Robin RS, Kumar BC, Patra S, Raman AV, Rao GN, Subramanian BR (2013) Distribution of dissolved inorganic carbon and net ecosystem production in a tropical brackish water lagoon, India. Cont Shelf Res 64:75–87

Muduli PR, Barik SK, Mohapatra D, Samal RN, Rastogi G (2017) The impact of tropical cyclone 'Phailin' on the hydrology of Chilika lagoon, India. Int J Environ Sci Nat Res 4(2):555632

Muduli PR, Kumar A, Kanuri VV et al (2021) Water quality assessment of the Ganges River during COVID-19 lockdown. Int J Environ Sci Technol. https://doi.org/10.1007/s13762-021-03245-x

Myers VB, Iverson RI (1981) Phosphorus and nitrogen limited phytoplankton productivity in northeastern Gulf of Mexico coastal estuaries. In: Estuaries and nutrients. Humana Press, Totowa, pp 569–582

Nayak BK, Acharya BC, Panda UC, Nayak BB, Acharya SK (2004) Variation of water quality in Chilika lake, Orissa. Indian J Mar Sci 33:164–169

Nazneen S, Raju NJ, Madhav S, Ahamad A (2019) Spatial and temporal dynamics of dissolved nutrients and factors affecting water quality of Chilika lagoon. Arab J Geosci 12(7):1–23

Newton A, Brito AC, Icely JD, Derolez V, Clara I, Angus S, Schernewski G, Inácio M, Lillebø AI, Sousa AI, Béjaoui B (2018) Assessing, quantifying and valuing the ecosystem services of coastal lagoons. J Nat Conserv 44:50–65

Nowicki BL, Kelly JR, Requintina E, Van Keuren D (1997) Nitrogen losses through sediment denitrification in Boston Harbor and Massachusetts Bay. Estuaries 20(3):626–639

Paasche E, Erga SR (1988) Phosphorus and nitrogen limitation of phytoplankton in the inner Oslofjord (Norway). Sarsia 73(3):229–243

Panda BP (2020) Relationship among the physico-chemical parameters of soil and water in different wetland ecosystems. Asian J Chem 32(7):1681–1690. https://doi.org/10.14233/ajchem.2020.22616

Pandey J, Yadav A, Singh R (2015) The N: P: Si stoichiometry and relative abundance of diatoms in Ganga River. Curr J Appl Sci Technol 9:137–147

Pandey J, Tripathi S, Pandey U (2016) Anthropogenically induced shifts in N: P: Si stoichiometry and implications in Ganga River. Air Soil Water Res 9:35–43

Panigrahi S, Acharya BC, Panigrahy RC, Nayak BK, Banarjee K, Sarkar SK (2007) Anthropogenic impact on water quality of Chilika lagoon RAMSAR site: a statistical approach. Wetl Ecol Manag 15(2):113–126

Panigrahi S, Wikner J, Panigrahy RC, Satapathy KK, Acharya BC (2009) Variability of nutrients and phytoplankton biomass in a shallow brackish water ecosystem (Chilika lagoon, India). Limnology 10(2):73–85

Patra S, Raman AV, Ganguly D, Robin RS, Muduli PR, Kanuri V, Abhilash KR, Kumar BC, Subramanian BR (2016) Influence of suspended particulate matter on nutrient biogeochemistry of a tropical shallow lagoon, Chilika, India. Limnology 17(3):223–238

Patra AP, Patra JK, Mahapatra NK, Das S, Swain GC (2010) Seasonal Variation in Physicochemical Parameters of Chilika Lake after Opening of New Mouth near Gabakunda, Orissa, India. World Journal of Fish and Marine Sciences 2:109–117.

Pattanaik S, Chanda A, Sahoo RK, Swain S, Satapathy DR, Panda CR, Choudhury SB, Mohapatra PK (2020) Contrasting intra-annl inorganic carbon dynamics and air–water CO2 exchange in Dhamra and Mahanadi Estuaries of northern Bay of Bengal, India. Limnology 21(1):129–138

Paul JT, Ramaiah N, Sardessai S (2008) Nutrient regimes and their effect on distribution of phytoplankton in the Bay of Bengal. Mar Environ Res 66(3):337–344

Pereira P, de Pablo H, Vale C, Franco V, Nogueira M (2009) Spatial and seasonal variation of water quality in an impacted coastal lagoon (Óbidos Lagoon Portugal). Environmental Monitoring and Assessment 153(1-4):281–292. https://doi.org/10.1007/s10661-008-0355-x

Peierls BL, Christian RR, Paerl HW (2003) Water quality and phytoplankton as indicators of hurricane impacts on a large estuarine ecosystem. Estuaries 26(5):1329–1343

Raman A, Satyanarayana C, Adiseshasai K, Prakash K (1990) Phytoplankton characteristics of Chilka lake, a brackish water lagoon along east coast of India. Indian Journal of Marine Sciences 19:274–277

Ramanadham R, Reddy MPM, Murty AVS (1964) Physico-chemical characteristics of the Vellar estuary in relation to shrimp farming. J Mar Biol Assoc India 6(2):183–201

Ramsar (1981). https://rsis.ramsar.org/ris/229 accessed on 18th September 2021

Redfield AM (1958) The biological control of chemical factors in the environment. Am Sci 46:205–222

Robin RS, Kanuri VV, Muduli PR, Ganguly D, Patra S, Hariharan G, Subramanian BR (2016) CO2 saturation and trophic shift induced by microbial metabolic processes in a river-dominated ocean margin (tropical shallow lagoon, Chilika, India). Geomicrobiol J 33(6):513–529

Roy JC (1954) Periodicity of the plankton diatoms of the Chilika lake for the year 1950. J Bombay Nat Hist Soc 52:112–123

Sterner RW, Hessen DO (1994) Algal nutrient limitation and the nutrition of aquatic herbivores. Annual review of ecology and systematics 25(1):1–29

Sakshaug E, Olsen Y (1986) Nutrient status of phytoplankton blooms in Norwegian waters and algal strategies for nutrient competition. Can J Fish Aquat Sci 43(2):389–396

Sarma VVSS, Prasad VR, Kumar BSK, Rajeev K, Devi BMM, Reddy NPC, Sarma VV, Kumar MD (2010) Intra-annual variability in nutrients in the Godavari estuary, India. Cont Shelf Res 30(19):2005–2014

Seitzinger SP (1988) Denitrification in freshwater and coastal marine ecosystems: ecological and geochemical significance. Limnol Oceanogr 33(4 part 2):702–724

Siddiqui SZ, Rao KVR (1995) Limnology of Chilika Lake. In: Director of Zoological survey of India (Calcutta), Fauna of Chilika Lake (wetland ecosystem series I). Zoological Survey of India, Calcutta

Siddiqui E, Pandey J, Pandey U (2019) The N:P:Si stoichiometry as a predictor of ecosystem health: a watershed scale study with Ganga River, India. Int J River Basin Manag 17(2):199–207. https://doi.org/10.1080/15715124.2018.1476370

Sobehrad S (1997) Phosphorus cycling in the estuarine environment. http://www.bellnetweb.brc.tamus.edu/phosphor.htm

Sommer T, Armor C, Baxter R, Breuer R, Brown L, Chotkowski M, Culberson S, Feyrer F, Gingras M, Herbold B, Kimmerer W (2007) The collapse of pelagic fishes in the upper San Francisco Estuary: El colapso de los peces pelagicos en la cabecera del Estuario San Francisco. Fisheries 32(6):270–277

Sonal T, Kataria HC (2012) Physico-chemical studies of water quality of Shahpura lake, Bhopal (MP) with special reference to pollution effects on ground water of its fringe areas. Curr World Environ 7(1):139–144

Srichandan S, Kim JY, Kumar A, Mishra DR, Bhadury P, Muduli PR, Pattnaik AK, Rastogi G (2015a) Interannual and cyclone-driven variability in phytoplankton communities of a tropical coastal lagoon. Mar Pollut Bull 101(1):39–52

Srichandan S, Kim JY, Bhadury P, Barik SK, Muduli PR, Samal RN, Pattnaik AK, Rastogi G (2015b) Spatiotemporal distribution and composition of phytoplankton assemblages in a coastal tropical lagoon: Chilika, India. Environ Monit Assess 187(2):1–17

Srichandan S, Baliarsingh SK, Prakash S, Lotliker AA, Parida C, Sahu KC (2019) Seasonal dynamics of phytoplankton in response to environmental variables in contrasting coastal ecosystems. Environ Sci Pollut Res 26(12):12025–12041

Strickland JDH, Parsons TR (1972) A practical handbook of seawater analysis. Fisheries Research Board of Canada, Ottawa. https://doi.org/10.1007/978-1-4615-5439-4_19

Sterner, R. W. (1990). The ratio of nitrogen to phosphorus resupplied by herbivores: zooplankton and the algal competitive arena. *The American Naturalist*, 136(2), 209–229v

Sundaray SK, Panda UC, Nayak BB, Bhatta D (2006) Multivariate statistical techniques for the evaluation of spatial and temporal variations in water quality of the Mahanadi river–estuarine system (India)—a case study. Environ Geochem Health 28(4):317–330

Tarafdar L, Kim JY, Srichandan S, Mohapatra M, Muduli PR, Kumar A, Mishra DR, Rastogi G (2021) Responses of phytoplankton community structure and association to variability in environmental drivers in a tropical coastal lagoon. Sci Total Environ 783:146873

Tilman D (1982) Review: structure and community competition. J Ecol 71(2):663–664

Tripathy SK (1995) Studies on the chemical properties of the Chilika lagoon, east coast of India. Doctoral dissertation, PhD thesis, Berhampur University, Odisha, India, pp 1–193

Turner RE, Rabalais NN (2003) Linking landscape and water quality in the Mississippi River Basin for 200 years. Bioscience 53(6):563–572. https://doi.org/10.1641/0006-3568(2003)053[0563:LLAWQI]2.0.CO;2

Upadhyay RK, Tiwari SK, Roy P (2015) Complex dynamics of wetland ecosystem with nonlinear harvesting: application to Chilika Lake in Odisha, India. Int J Bifurc Chaos 25(7):1540016

Vijayakumar S, Rajesh KM, Mendon MR, Hariharan V (2000) Seasonal distribution and behaviour of nutrients with reference to tidal rhythm in the Mulki estuary, southwest coast of India. J Mar Biol Assoc India 42(182):21–31

Yadav A, Pandey J (2018) The pattern of N/P/Si stoichiometry and ecological nutrient limitation in Ganga River: up-and downstream urban influences. Appl Water Sci 8(3):1–12

Zhang Z, Song X, Lu X, Xue Z (2013) Ecological stoichiometry of carbon, nitrogen, and phosphorus in estuarine wetland soils: influences of vegetation coverage, plant communities, geomorphology, and seawalls. J Soils Sediments 13(6):1043–1051

Chapter 8
A Systematic Review on the Impact of Urbanization and Industrialization on Indian Coastal Mangrove Ecosystem

Deepika Sharma, Karuna Rao, and AL. Ramanathan

Abstract The history of human civilization has witnessed a strong and rapid transformation pattern in the coastal environment. It harbors a prominent transition zone of land and sea that plays a significant part in the socioeconomic and environmental aspects. Due to tremendous pressure from anthropogenic perturbations manifested by coastal squeeze, it's protection and conservation become substantial. 5.04% of the mangrove land has been converted to aquaculture land between 1988 and 2013. Present mangrove loss is 35% which is supposed to reach 60% by 2030. Human activities increase the chances of exposure of coastal waters to effluents (organic and inorganic) released from the industrial and urban components which accelerate the metals and nutrient pollution, eutrophication, and oxygen depletion. This tends to alter ecosystem dynamics and biogeochemical processes with serious impacts on the biota. Pichavaram shows an increase in nitrate from 5.9 mg/l in 1995 to 29.9 mg/l in 2006–2007. In Sundarbans it increases from 1.14 mg/l in 2001 to 3.69 mg/l in 2006 and in Godavari from 0.61 mg/l in 2001 to 2.25 mg/l in 2016. The phosphate values increase from 0.28 mg/l in 1995 to 6.6 mg/l in 2006 in Pichavaram mangroves. Manori creek, Mumbai, shows hike in phosphate in past 25 years. The value increases from 0.06 mg/l in 1982 to 2.19 mg/l in 2007. A consistent increase in heavy metal content has been observed in Sundarban, Pichavaram, and Goa mangroves. Thus, the resultant surge of heavy metals and nutrient pollutants indicates growth of fallow land, agricultural, and aquaculture activities and industrial pollution. This chapter has been constructed to discuss a holistic view of the major drivers of coastal mangrove ecosystem degradation by reviewing the case studies to highlight the past changes and present trends of human activities through industrialization and urbanization. We evaluate the impact of these human influences on the mangrove ecosystem, with an approach to emphasize the crucial role of mangroves, both in terms of quality and quantity, and the absolute need to conserve their future.

Keywords Mangrove · Coastal ecosystem · Anthropogenic perturbation · Pollution status · Biogeochemical processes

---

D. Sharma · K. Rao · AL. Ramanathan (✉)
School of Environmental Sciences, Jawaharlal Nehru University, New Delhi, India

S. Madhav et al. (eds.), *Coastal Ecosystems*, Coastal Research Library 38,
https://doi.org/10.1007/978-3-030-84255-0_8

Abbreviations

μM	Micromolar
ADB	Asian Development Bank
As	Arsenic
BOD	Biochemical oxygen demand
cal/g	Calorie per gram
Cd	Cadmium
COD	Chemical oxygen demand
Cr	Chromium
CRZ	Coastal Regulation Zone
Cu	Copper
DAP	Diammonium phosphate
DIP	Dissolved inorganic phosphate
DO	Dissolved oxygen
E	East
EDC	Endocrine disrupter compounds (EDCs)
FAO	Food and Agricultural Organization
Fe	Iron
FSI	Forest Survey of India
GIS	Geographic Information System
ha	Hectare
JMM	Joint Mangrove Management
km^2	Square kilometer
MAP	Management Action Plan
mg/kg	Milligram per kilogram
mg/l	Milligram per liter
MMR	Mumbai Metropolitan Region
Mn	Manganese
MSL	Mean sea level
N	Nitrogen
N	North
NACA	Network of Aquaculture Centres in Asia-Pacific
NGO	Nongovernmental organization
Ni	Nickel
NOAA	National Oceanic and Atmospheric Administration
P	Phosphorus
PAH	Polycyclic aromatic hydrocarbons
Pb	Lead
POP	Persistent organic pollutants (POPs)
PPCP	Pharmaceuticals and personal care products
ppm	Parts per million
ppmv	Parts per million by volume
S	South
SEZ	Special Economic Zone

Si	Silicon
SPM	Suspended particulate matter
TSS	Total suspended solid
W	West
Zn	Zinc

1 Introduction

Mangroves are the intertidal forest ecosystems that dominate 75% of the world's shoreline (Ranjan et al. 2008) between 25° N and 25° S with a projected area of 1.7 to 2.0 × 10^5 km^2 (Borges 2003). These woody halophytes occupy severe place exposed to whims of both land-dwelling and the oceanic, hit by prevailing heavy rain and storm, with high salinities and droughts, shifting sediments, inundation, and exposure. But this unbending nature to colonize provides many rewards too. Mangrove ecosystems have some benefits over other ecosystems which include adaptations like aerial breathing roots called "pneumatophores," succulent leaves, sunken stomata, vivipary, stilt roots, and buttresses that are mainly exhibited by these salt-tolerant plant community. The crustaceans and fish move in with every tidal inflow to feed in the spaces that are shared by the insects and birds as well. The interaction of the species adds up to the rich diversity observed in these nooks, hence providing significant grounds for nursery and sites for breeding (Spalding 2010). These structures are the basis of renewable logs, locations for sediment accumulation, impurities, carbon, and nutrients which also guard against coastal erosion (Alongi 2002). These myriad patches act as major channels for the exchange of tides of dissolved and particulate matter as well as organic matter exportation and nutrients to the ocean, caused majorly by biological and physical processes within the forest ecosystem (Singh et al. 2005). Hence mangrove ecosystems play an important role in the biogeochemical cycling of these materials. However, the Earth's ecosystems have always been subjected to a persistent change through which the organisms tend to respond and adapt, thereby adjusting to climate change and other physical attributes. The biological and ecological alterations in the ecosystems are the result of both disturbances created by nature and man-made factors that can change in their period, occurrence, magnitude, and power and facilitate adaptive changes (Alongi 2008). Mangrove forests, like other ecosystems, face similar disorders which can change in their fundamental nature in time and space. Bridging the gap in the terrestrial and marine over small latitudes, they are true ecotones that fuse the components of both sea and land biomes, along with the development of unique structural and functional adaptations (Alongi 2012).

The objective of this chapter is to critically assess the impact of urbanization and industrialization on the coastal mangrove ecosystem. It deals with an approach to emphasize the crucial role of mangroves, both in terms of quality and quantity with the absolute need to conserve their future.

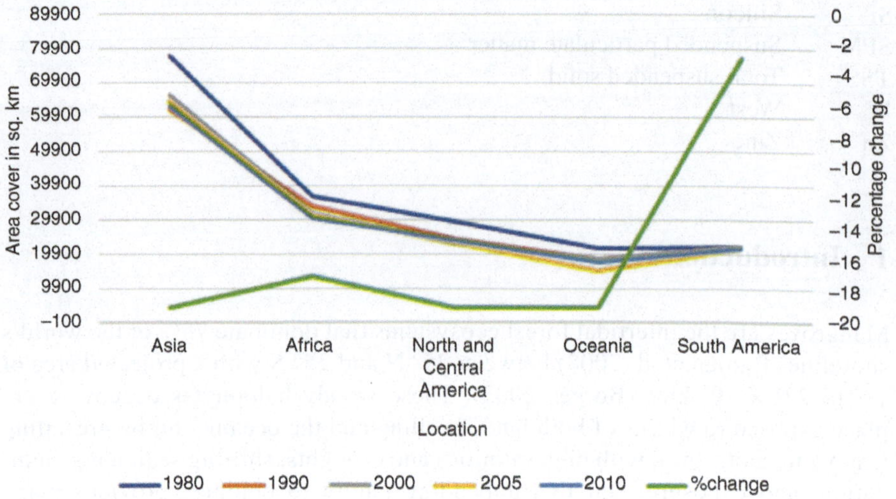

Fig. 8.1 Percentage change in the global mangrove area cover over the years around the globe. (Source: Gurjar et al. 2019)

1.1 Global Mangrove Cover

The distribution of mangroves in over 123 countries and territories makes up 15.2 million ha of the total world mangrove cover which is less than 1% of the tropical forests of the world and less than 0.4 % of the total global forest domain. The largest share of mangroves occurs in Asia where around 33.5% is found in Southeast Asia and 6.8 % in South Asia. The regions of North Central America and South America followed by West and Central Asia anchor the rest of the mangrove cover. India holds 45.8% of the total South Asian mangrove cover. Figure 8.1 shows the percentage changes in the global mangrove cover by the Food and Agricultural Organization (FAO) of the United Nations which shows a decrease of about 18% in global mangrove cover. It is observed that there is a decrease in mangrove cover of Asia, Oceania, and North and Central America by 19%, while the mangroves of South America remain relatively unchanged and show a minor decrease of 2.8%. Further it is noticed that the mangrove cover of Africa decreases by 17% Food and Agricultural Organization (FAO 2010).

1.2 Indian Coastal Mangrove Systems

Sheltering the widespread and diverse mangrove wetlands, the estuaries, and the "coasts of nine maritime states and four union territories" of the Indian peninsula, bounded by the western Arabian sea, southern Indian Ocean, and the Bay of Bengal on the east, it runs over a distance of 7516.6 km including the coastline of Lakshadweep islands and Andaman and Nicobar Islands along with the mainland. Therefore, three major classifications of the mangrove habitat arise as "deltaic (Eastern coast mangroves); estuarine and backwater (Western coast Mangroves), and Insular (Andaman and Nicobar Islands)" (George et al. 2019). The species which dominate the Indian mangrove ecosystems are *Rhizophora mucronata, Sonneratia alba, Avicennia alba, Avicennia officinalis, Morinda citrifolia, Heritiera littoralis, Phoenix paludosa, Ceriops tagal*, and *Bruguiera cylindrica*. According to Forest Survey of India (FSI), State of Forest Report, Dehra Dun, 1999, the majority of mangrove wetlands (487,100 ha) in India occurs on the east coast, which is nearly 56.7% (275,800 ha) and 23.5% (114,700 ha) along the west coast, while Andaman and Nicobar Islands accommodate the remaining 19.8%. Figure 8.2 here depicts a map showing the major mangrove forest locations in India. There is a difference in the geomorphic settings of the

Fig. 8.2 Map locating the mangrove forest in India. (Source: http://www.casmbenvis.nic.in/database/Mangroves_3893.aspx?format=Print)

mangrove wetlands of the east coast and the west of the Indian coast. On the western coast of India, mangrove wetlands are smaller in size, lesser in diversity, and lesser complicated in terms of tidal creek network which is due to the coastal zone being "narrow and steep in slope due to the presence of Western Ghats" with the absence of major west-flowing river. However, east onwards, the mangrove wetlands are larger (~90% of the total mangroves forest cover for the whole country) and comprise higher diversity, and water bodies connected with mangroves are delineated by the occurrence of larger brackish water bodies and an intricate network of tidal creeks and canals. The greater delta created by the presence of east-flowing rivers and the gentle slope of the coast are the two features that contribute to abundance (Selvam 2003).

Sundarbans (West Bengal): The Sundarbans mangrove forests, the world's largest coastal wetland, are found in the delta created by the rivers Ganga, Brahmaputra, and Meghana, with a cover of about 1 million ha; the forests get 60% shared with Bangladesh and 40% lie within India. They are found on the upper side of the Bay of Bengal between 21°40′ N and 22°40′ N latitude and 88°03′ E and 89°07′ E longitude. They are influenced by the enormous amounts of sediments carried by the rivers which lead to the expansion and nutrient dynamics along with the impact of subtropical monsoon climate (annual rainfall: 1600–1800 mm) and extreme cyclonic events (Gopal and Chauhan 2006; Prasad et al. 2017). This mangrove estuary has large tidal flats which is a common characteristic in mangrove estuaries dominated by tides and suitable microenvironment. Microenvironments are provided for mangrove plant colonization, which produces communities of dense and tall mangrove plants.

Bhitarkanika (Odisha): The "second largest" Indian mangrove ecosystem comprises the mangrove forests, estuary, creeks, rivers, backwater, accreted land, and mudflats, flourishing the delta region of "Brahmani and Baitarani rivers." Geographically this ecosystem is located in the Kendrapara district of Orissa between 20°4′–20°8′ N latitudes and 86°45′–87°50′ E longitudes. It has been declared as a Wildlife Sanctuary covering an estimated area of 672 km^2 in 1975. It is a tide-dominated mangrove with a mean tide level of 1.5–3.4 m that consists of widespread low gradient intertidal zones available for colonization of mangroves supporting a rich floral diversity (Chauhan and Ramanathan 2008).

Coringa and Gaderu (Andhra Pradesh): Located in Andhra Pradesh between 16°51′–17°00′ N latitudes and 82°14′–82°22′ N longitude, it occurs over the delta formed by the second largest river in India, Godavari, that before discharging into the Bay of Bengal southwest of Visakhapatnam branches into Vasishta Godavari and Gautami Godavari. The region between Kakinada Bay and Gautami Godavari is characterized by condensed vegetation and that belongs to Coringa wildlife sanctuary (Dehairs et al. 2000).

Pichavaram (Tamil Nadu): Situated between the estuaries of Vellar and Coleroon (Lat. 11°2′; Long. 79°47′ E), the forest occupies 51 islets that range from 10 m^2 to 2 km^2, covering an area of about 1100 ha, which are characterized by complicated waterways' separation, that connect the Coleroon and Vellar estuaries. The southern region near the Coleroon estuary is dominated by mangrove vegetation, while the northern part that resides close to the Vellar estuary is characterized by maximum mudflats. This region is influenced by three types of waters that include neritic water (Bay of Bengal), brackish water (Vellar and Coleroon estuaries), and freshwater (irrigation channel as well as the main channel of Coleroon river). The majority of the area is covered by the forest, i.e., 50% and 40% by the waterways out of which remains for the sand-flats and mudflats (Kathiresan 2000).

Kerala mangroves: The mangrove vegetation occupies the estuarine water body banks and as narrow continuous belt or patches, adjacent to the backwater channels (Lat. 9°28′ and 10°10′ N; Long. 76°13′ E). They are influenced by the tidal flooding and 41 perennial rivers that supply fresh water to create an extensive expansion of fringing mangroves of backwaters, estuaries, and creeks. The major districts with the mangroves are Kannur and Kasaragod followed by Kollam, Trivandrum, Alappuzha, Kottayam, Thrissur, Ernakulam, Kozhikode, and Malappuram along with the three Ramsar sites, namely, Ashtami, Sasthamkotta, and Vembanad (George et al. 2019).

Goa mangroves: They are located (Lat. 14°53′–15°48′ N; Long. 73°40′–74°20′ E) along with Mandovi-Zuari estuary complex with an area of 12,000 ha on the central west coast of India (Attri and Kerkar 2011).

Mumbai mangroves (Maharashtra): The coastline of Mumbai is cushioned by a mangrove cover of 66 sq. km (Lat. 18°55′–19°20′ N; Long. 72°45′–73°00′ E) with its extensive network of creeks fringed with mangroves along both the banks.

Gulf of Kachchh mangrove (Gujarat): The state of Gujarat has four regions of mangrove cover, i.e., Kachchh, Gulf of Kachchh, Saurashtra, and South Gujarat that constitute the mangrove coastline (1048 km^2) (Pandey and Pandey 2013). Twenty percent of the total area is occupied by the dense vegetation, and the remaining area is constituted by degraded mangroves and saline-encrusted mudflats.

Andaman and Nicobar mangrove: It contributes 13% of the entire Indian mangrove area where the diversity is found to be similar to the Southeast Asian mangroves. They are recognized to be best in terms of density and growth in the country with a relative mangrove density of 76.5% (Goutham-Bharathi et al. 2014). It is located in an extensive group of 572 islands (8249 km^2) that lie in the Bay of Bengal (Lat. 6°45′–13°41′ N; Long. 92°12′–93°57′ E) on the eastern side of India. The mangroves originate along tidal creeks, bays, and lagoons where the creeks form the outlets to the rain-fed streams that bring silt from the interior to the shore for the formation of muddy plains facilitating the spread and regeneration of mangroves (Selvam 2003).

1.3 Threat to Mangrove Ecosystems

Despite their ecological and economic importance, the mangroves are still facing destruction, majorly related to the density of the human population. The degradation and devastation of mangrove ecosystems come under both natural and anthropogenic influences.

- Forest clearing
- Overharvesting
- River changes
- Overfishing
- Pollution
- Climate change

The key explanations for devastation can be depicted through urbanization, industrialization, shrimp aquaculture, mining activity, and overexploitation of resources like wood and fisheries. The restoration and rehabilitation projects are increasing all over the world with few country areas showing an increase in mangrove area. Till 2025, the exploitation is expected to continue unless they are valued for the services they provide in a sustainable manner with their greatest future hope in reduced human population growth (Alongi 2002).

1.3.1 Natural Influences

The impact of natural factors on the structure and function of the mangrove coastal ecosystem can be seen on the spatial and temporal scales. Natural disturbances such as cyclones and other storms, lightning, tsunami, and floods adversely affect the mangroves. On the Indian coast, recurrent tropical cyclones, storms, and tsunamis have damaged the mangrove forests. For example, during 1999 in Odisha, a major super cyclone devastated a large area of mangroves with an estimated loss from 307.66 to 179 km^2. Similarly, loss of mangrove forest was observed during the tsunami in 2004 in the south coast and Andaman and Nicobar Islands (Suresh and Sahu 2015). After 1999, the most devastating tropical storm reported in the region was the Amphan cyclone which impacted the Sundarbans, Bengal's first line of defense from the violent storms that periodically arise in the Bay of Bengal (Sen 2020). Other natural factors include pests and invasive species which show a severe impact on the mangrove forest. Twenty percent of a species, *Heritiera fomes*, of the trees have been harshly affected by the "top dying" disease in the Sundarbans of Bangladesh.

1.3.2 Anthropogenic Influences

Anthropogenic activities include not only the activities done to meet the food, clothing, housing, and energy, but they also include the developmental activities like dam construction, mining, etc. where these human activities affect the mangrove ecosystem directly as well as indirectly. Previously, the flawed picture of mangroves being categorized as "waste lands" led to their conversion to agricultural, industrial, and residential uses (Hema and Devi 2015). The major impacts of the human influence on the coastal mangrove ecosystem are given below.

Agricultural Activities

Farming has affected a large fraction of the mangrove forest in India which aligns with the two main causes of this decline, i.e., destruction of the habitat and its alteration. The expansion in agriculture during the past 100 years in India and Bangladesh has destroyed an estimated area of 150,000 ha of mangroves (Dhargalkar et al. 2014). In the states of Goa, Karnataka, and Andhra Pradesh, plantations of coconut and paddy are commonly carried out. The salinity of the soil is reduced using rainwater after destroying the mangrove patches. Further, these areas are protected from soil water intrusion by forming embankments which makes these areas suitable for plantation.

Industrial Development

One of the victims of rapid industrial development are the mangrove belts present across those regions. The escalated industrialization and its uncontrolled pressure has increased in the last few decades. The industrial waste discharge introduces heavy metals to the system that remains the major reason to impact the health of mangroves of a region. These virgin mangroves receive various chemical contaminants like heavy metals, inorganic nutrients, organic contaminants, hydrocarbons, etc., from the effluents of the industries (Maiti and Chowdhury 2013). When heavy metals are introduced, the mangroves absorb them mainly through roots and transport a part upward into the sensitive tissues; therefore, the concentration of heavy metals is found to be more in the roots than in the shoots. This introduction can cause changes in metabolic activities, cell structure, and plant growth (He et al. 2014). A study was done on seven different estuarine regions on the South Gujarat coast consisting of seven different rivers: Ambica, Purna, Par, Varoli, Damanganga, Kolak, and Auranga. It revealed the accumulation of eight different heavy metals, Pb, Ni, Cr, Cd, Zn, Cu, Fe, and Hg, in the mangrove plant tissues as well as mangrove sediments from the surrounding industrial areas. Here the industries majorly include manufacturing and engineering, papers, dyes, textiles, chemicals and petrochemicals, pharmaceuticals, shipbuilding, diamond processing, etc. Another coastal region which resides along the Bay of Bengal, Visakhapatnam, here the sum up of decades of industrialization and urban development projects along the wetlands of Meghadrigedda creek has reduced the extent of 400 acres to less than 40. The richness of the ecosystem and home to many birds and endangered species has been doomed with the drastic beach erosion (The Hindu, 2020). Further, the mangroves of Mumbai region are also impacted by the resultant of around 9000 industries of chemicals, fertilizers, iron and steel, oil refineries, and thermal power which give a huge output of emissions of gas, solid and liquid wastes, and toxic and hazardous wastes, thus resulting in the degradation. The wastes are being discarded into the creeks which leads to the deterioration of water quality with heavy siltation (Harun et al. 2015). These recent industrial and domestic activities have transformed the once lush flourishing mangrove areas. With great fisheries and oyster beds, Mahim bay and Thane creek were affected with high concentration polluted areas that led to nonexistent fisheries and lower dissolved oxygen. These important areas for the spawning of fishes and other marine flora and fauna have been impacted due to the anthropogenic construction and mixing of effluents. As reported by a local fisherman, instead of fish, the grounds are being occupied by the multiplying mosquitoes.

Heavy Metals

Heavy metals occur naturally in the environment, but due to interference of human activities, their background concentration has increased dramatically. They come under the category of serious pollutants as they are nonbiodegradable and remain persistent for years. The source of heavy metals is rivers, and they make their way to coastal and mangrove environments. Other sources include rainfall, tidal activities, and land runoff (Nriagu and Pacyna 1998). Upon entering the coastal ecosystem, they get absorbed by the sediments by the processes like co-precipitation and adsorption on the solid particles (Santschi et al. 1990). After some time these adsorbed metals become remobilized and available to the water column when the soil gets saturated or when there is a change in the environmental conditions (Tam and Wong 1993).

A case study on the Sundarbans upon the occurrence of elevated levels of heavy metals in India and Bangladesh wetlands has been done. Trace metals like As, Cd, Co, Cu, Fe, and Ni show an increase from 2004 to 2012 as shown in Table 8.1. A consistent increase in Cu, Ni, and Zn has been observed in Goa mangroves from 2011 to 2016. High values of heavy metals in Pichavaram sediments, then other mangroves, might be due to the higher presence of trace metals in the Vellar and Coleroon rivers and display anthropogenic influence through domestic sewage and agricultural runoff (Ramanathan et al. 1999). These two rivers pass through the densely populated industrial city along with the addition of fertilizers, pesticides, and heavy metals from the upstream region (Prasad 2012). By analyzing the course of industrial influence that has shaped up the mangroves towards new adaptations, a long-term sustainable functioning of the estuarine ecosystem can be deduced by introducing the tolerant species that can help in sustaining high heavy metal content in the sediments. While in Pichavaram, tsunamigenic sediments show the highest trace metal concentration (R. K. Ranjan et al. 2008) for all the trace metals which might be due to higher discharge of wastewater at that time along with the tsunami-driven sediments derived from the deep sea. The heavy metals like Cu, Cd, Pb, and Zn have anthropogenic origin, for example, they are derived from the untreated wastewater discharge from several industries located on the coastal shorelines. Runoff from agriculture, sewage, and other effluents are other anthropogenic sources.

The concentration of metal in sediment can sometimes depend on the variation in the geography for the same trace metal. The studies have suggested the effect of

Table 8.1 Trace metal concentration in the mangrove ecosystems of Goa, Pichavaram, and Sundarbans (Ranjan et al. 2017)

Site	Year	As (ppm)	Cd (ppm)	Co (ppm)	Cr (ppm)	Cu (ppm)	Fe (%)	Mn (%)	Ni (ppm)	Pb (ppm)	Zn (ppm)
Goa	2011			34.36	17.32	45.34	12.18	0.16	–	22.51	72.853
Goa	2015			22.85	271.18	36.26	14.53	0.512	47.15	–	104.34
Goa	2016			114.35	–	64.99	8.72	0.25	118.565	–	277.9
Pichavaram	1999		6.6	35.3	141.2	43.4	3.25	0.09	62	11.2	93
Pichavaram	2008		34.74	–	6200	132.3	2.5	0.08	252.1	–	106
Pichavaram	2013		23	–	152	34	3.8	0.033	51	21	16
Sundarbans	2004	3.5	0.1	12.46	36.44	35.47	3.08	0.14	33.46	17.2	74.18
Sundarbans	2008	–	–	–	–	90.75	–	–	–	38.175	303.75
Sundarbans	2009	8.3	0.15	10.41	55.98	25.74	3.12	0.58	30.17	65.59	22.8
Sundarbans	2010	8.09	0.18	–	99.01	28.94	–	–	51.86	23.01	–
Sundarbans	2012	–	1.88	23.48	44.13	38.47	3.75	0.0574	50.35	30.28	75.87
Sundarbans	2015	3.82	0.21	7.67	28.3	38.29	0.29	0.06	34.5	15.8	34.42

pollution on the plants with the use of response of biological phenomena like endurance, production of biomass, defoliation, photosynthetic effects, metallothionein expression, and enzymes. At the coast of Bhitarkanika, an investigation revealed a mangrove species, *A. officinalis*, accumulates the highest concentration of Fe, Cu, Mn, and Zn among other species, namely, *Xylocarpus granatum*, *Bruguiera cylindrica*, *Rhizophora mucronata*, and *Ceriops decandra.* Most investigations have revealed the *Avicennia* sp. has the highest tolerance in respect of heavy metals among mangroves and similarly in India, *A. marina*, in different mangrove patches. These studies show the importance of mangroves in sequestering heavy metal pollutants. Destruction and degradation of mangroves will lead to the loss of plant species and subsequently their potential to store heavy metals; hence, the restoration and management of the mangrove ecosystem are necessary. More pollution will lead to more proliferating pollution-tolerant mangroves which will lead to replacing other species and, hence, would result in deterioration of mangrove biodiversity.

Nutrients

Mangrove areas are highly productive forests which are rich in carbon but poor in nutrients. Some studies were done revealing mangroves maintaining high productivity despite facing nutrient limitation (Reef et al. 2010). This is possible only when the nutrients limit growth via processes like nutrient cycling and nutrient retention mechanisms (Ball 1988). Mangroves are benefitted from their location between land and sea and hence are generally not limited to the elements like magnesium, sulfur, boron, sodium, and potassium, but they are frequently limited by the nutrients like nitrate and phosphate. Fertilization studies (Lovelock et al. 2006) reveal that nutrient limitation of either nitrate or phosphate or both depends on several factors. These factors include the amount of terrigenous input, species composition, texture and fertility of the soil, redox status of soil, salinity, and tidal inundation.

Several studies on nutrient dynamics have been carried out across various Indian mangroves, and a considerable hike in nitrate and phosphate has been observed. A study by Prasad et al. (2006) shows a decadal increase in the value of nitrate and phosphate in mangrove water from 1987–1989 till 1998–1999. They reported that a significant decadal increase in nutrients owes to the rapid degradation and conversion of mangroves to aquaculture ponds. Further studies in the same region show the mean value of nitrate to be 5.9 mg/l in 1995 which increases to 10.64 mg/l in 2003–2004 (Prasad et al. 2006). These values further increase to 34.6 mg/l in 2005 (Ranjan et al. 2008), and 29.9 mg/l of nitrate has been observed by Kumar et al. (2015). These hikes in nitrate might be due to land use and land cover changes including an increase in fallow land and aquacultural activities which discharged their effluents to this mangrove water. In Pichavaram, a tremendous increase in the values of nitrate can be observed after the tsunami (December 2004) which reaches the level of 34.6 mg/l in 2005 and 29.9 mg/l in 2006 (Fig. 8.3). Pichavaram shows an increase in nitrate from 5.9 mg/l in 1995 to 29.9 mg/l in 2006–2007. The sudden increase in nitrate values after the tsunami may be due to the retreating water, which carries the waste from agricultural and aquacultural fields to this mangrove ecosystem (Krithika et al. 2008). In Sundarbans it increases from 1.14 mg/l in 2001 to 3.69 mg/l in 2006 and in Godavari from 0.61 mg/l in 2001 to 2.25 mg/l in 2016.

In Fig. 8.3a, the value of phosphate shows a considerable hike in creek water in the past 25 years (Kulkarni et al. 2010). The value of phosphate shows consistent increase from 0.06 mg/l in 1982, 0.32 mg/l in 1989, 1.01 mg/l in 2000, and 2.19 mg/l in 2007. High phosphate concentration in Manori creek indicates a high pollution level as this creek is located in the close vicinity of Mumbai City, which is under high stress due to increasing anthropogenic activities and receives effluents from municipal and industrial wastes. Other human activities also contribute which include barrels and oil drums washing, boats manufacturing and unauthorized discharge of hazardous waste into the mangrove creeks (Zingde and Desai 1980).

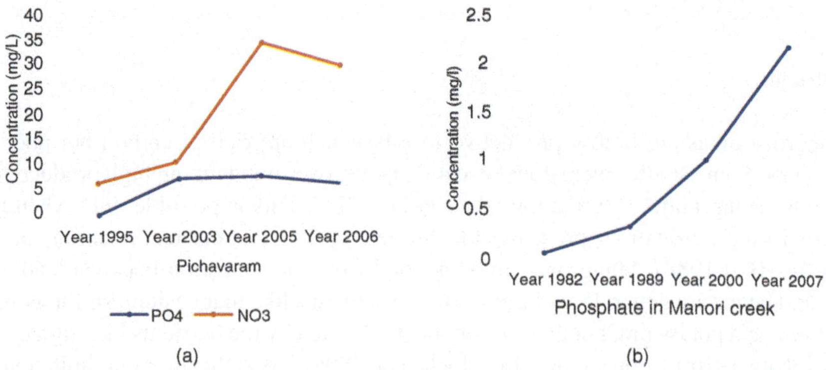

Fig. 8.3 Change in the concentration of phosphate: (**a**) Pichavaram mangroves, (**b**) Manori creek

Similarly, nitrate concentration in Godavari mangroves is 0.61 mg/l in 2001 (Tripathy et al. 2001) and increases to 2.25 mg/l in 2016 (Rao et al. 2018) which reveals human pressure mainly from effluents coming from aquacultural and agricultural runoff. Bhitarkanika has a value of 1.26 mg/l sourced by Dhamra port activities and agricultural runoff from the nearby villages (Chauhan and Ramanathan 2008), while Bhitarkanika aquacultural ponds have a value of 3.48 mg/l due to the use of diammonium phosphate (DAP) fertilizer.

(a) Nutrient ratio

Analysis of five major mangrove regions according to Table 8.2 was conducted to designate the ecological and nutrient status to study the influence of human perturbations. A standard Redfield ratio Si:N:P = 16:16:1 that defines the stoichiometric proportions of dissolved nutrients in the mangroves was found to be deviated due to the anthropogenic pressure. High input of silica to the mangrove waters due to terrestrial weathering is observed due to >1 Si:N ratios. Since the 1980s, there has been a significant increase in the dissolved nutrients mainly through sources like agriculture, aquaculture, etc. which is also depicted through decadal changes in the concentration of phosphate and nitrate in another study conducted in mangroves of Pichavaram. The presence of high DIP levels contributes to the deterioration of the water quality of Pichavaram through algal blooms, organic matter sedimentation, and depletion of oxygen. Further, the same fate has been followed by the rest of the areas. The nutrient levels in Coringa mangroves are influenced by the fluvial loads carried down from the river where allochthonous inputs are driven by agriculture and aquaculture practices, thus increasing BOD and algal blooms.

(b) Eutrophication

The elevated concentration of nutrients in coastal waters due to increasing anthropogenic activities leads to eutrophication. The increase in the nutrient loading causes increase in the harmful algal blooms which depletes the dissolved oxygen and induces toxicity leading to negligent quantities of marine fauna and disappearance. This leads to another change in the livelihood of villagers, extreme depletion in a fish catch caught over the past three decades due to effluent discharge and dumping of hazardous wastes. By enhancing the eutrophication, a shift in the phytoplankton is observed, that gives rise to blooms and further increased oxygen demands. India has the second largest population

Table 8.2 Dissolved nutrients (μM) and atomic ratios in the Indian mangrove ecosystems (Prasad 2012)

Mangrove	N:P ratio	Si:N ratio	Si:P ratio
Sundarban	11.43	53.01	4.64
Bhitarkanika	6.48	147.47	22.75
Coringa	5.46	25.87	4.74
Pichavaram	7.31	1.53	0.21
Mangalavanam	4.64	2.76	0.6

and is the fastest-growing country, so the coastal ecosystem of India could be extremely vulnerable to anthropogenically induced eutrophication.

Aquaculture

The tropical regions, especially Asia, have seen major losses due to aquaculture. Around 1 million ha of a coastal ecosystem has been converted to shrimp aquaculture. The shrimp sector development converts the flat, coastal lands to aquaculture ponds where a survey found about 5% of the shrimp aquaculture farms in India were constructed from past mangrove areas (ADB/NACA 1998). A study supported this by finding that in the Godavari delta in Andhra Pradesh around 14% of the aquaculture farms have been constructed on mangrove lands where the investigation was carried out by Andhra Pradesh Remote Sensing Application Centre. Around 80% of the mangrove land conversion occurs for shrimp aquaculture. It was observed that the rate of conversion increased from 1997 to 1999 implying that policy regulations could not prevent exploitation. From 1988 to 2013, the area under aquaculture has expanded which led to 5.04% of mangrove land being used up. According to the researchers, 35% of the worldwide mangrove loss (one-third) was due to the aquaculture that is assumed to reach 60% by 2030 (Lee et al. 2006). The modifications bring about massive landscape changes: a more urban infrastructure gets constructed that leads to change in geomorphology in the wetlands as well as the catchment area. Construction of dams and increase in water extraction demands to meet the growing population affect the coastal waters. Urban development converts the natural habitats to landscapes with impermeable surfaces which block the percolation of rainwater, thus changing the hydrology which degrades the downstream ecosystems and impacts the drainage networks which leads to more stormwater discharge in the receiving habitats (Singh et al. 2014). When we compare to other countries on the subcontinent, Bangladesh consists of the "Chakaria Sundarbans" with an area of 6020 ha, one of the oldest mangrove forests. It was the victim of shrimp aquaculture and salt production during the period of 1972 to 1989. Other countries like Thailand faced the same fate from 1918 to 1987 to accommodate the same practices. This study was directed to assess the trend of mangrove cover changes in five major mangrove-dominated countries of the subcontinent. Findings have suggested that the rate of loss of mangroves has fallen in India and Indonesia, whereas in Malaysia and Myanmar, the rate of loss amplified. The common cause of this loss was aquaculture for mangrove land conversion.

The adverse effect of aquaculture can be seen in groundwater which is an important and major source for drinking and other household purposes. Aquaculture affects the quality of water by changing its physicochemical aspects and biological activities. The physical aspect includes the pressure load on water. Chemical aspects include the extent to which it is polluting the water, and biological aspect includes the introduction of pathogens, microorganisms, exotic species, and several kinds of water-borne disease bacteria. It also causes the phenomenon of saltwater intrusion which is caused by the overuse of groundwater and its conversion to aquaculture ponds.

Oil Spills

Mangrove ecosystems are highly vulnerable to various anthropogenic activities; sometimes these activities lead to the accident like oil spills. They can happen during the extraction and transportation of oil across the world through the sea and the ocean. One of the greatest disadvantages of oil spills is that they remain in the environment for decades as they are not biodegradable and persist for a longer period. They affect the mangrove as they get deposited on the surface of the plants, roots, and soils and affect the marine life which depends on these plants and sediments (Duke and Burns 2003; NOAA 2014). Once deposited, oils get adsorbed to the oleophilic surface of both plants and animals except in the incidents where a large amount of oil have been spilled. In such cases, the oil does refloat and spread in a significant way with tidal flushing. Oil spillage causes the death of shorter plants and animals in few days, but the larger and mature may survive up to 6 months or more as oil coats the breathing surface of plant root, seedling, stems, and sediments. They also affect the fauna present in the burros and root hollows. Oil spills cause excessive harm to the aquatic fauna and seabirds. Globally, till now, around 238 notable incidents of the oil spill have been reported along the mangrove shorelines releasing a total of about 5.5 million tonnes of oil (Duke 2016). The oil spill, globally, has oiled around 1.94 million ha of mangrove ecosystems since 1958. This causes the death and decay of about 126,000 ha of mangrove vegetation.

A case study conducted in the Sundarban mangroves, Bangladesh, found direct influences on the ecosystem after the oil spill that occurred in December 2014 as shown in Fig. 8.4. The high content of oil (995 ± 429 mg/l) and high values of TSS (999 ± 447 mg/l), total hardness values (2156 ± 132 mg/l), and COD (377 ± 104 mg/l) were found in the region after the contamination. There are low transparency (12 ± 2 cm) and productivity (12 ± 2 cm) values with poor phytoplankton (32 ± 19 units/l) and zooplankton (7 ± 1.5 units/l) growth in the oil-contaminated

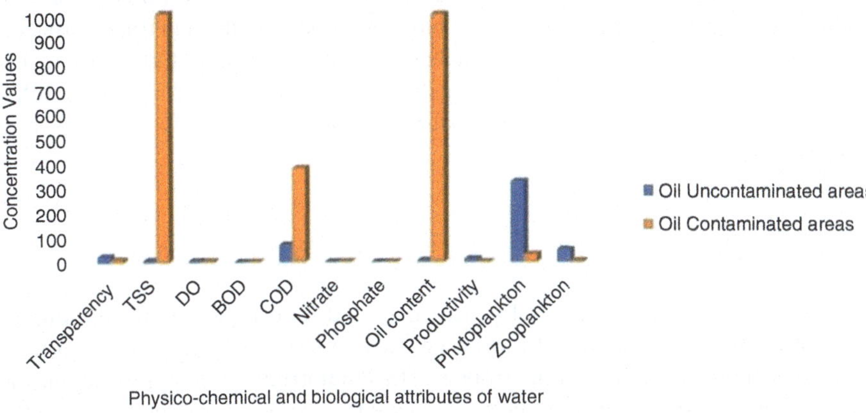

Fig. 8.4 Physicochemical and biological attributes of water of contaminated and un-contaminated areas. (Source: Harun et al. 2015)

areas. The biodiversity of mangroves gets affected by oil pollution with the Sundri plant getting covered with oil. On the other hand, the area having no contamination has better conditions. That comprises the western region which showed lower values of TSS (9.5 ± 1.8 mg/l), total hardness (965 ± 41 mg/l), and lower COD (69 ± 8). The oil content of more than 10 mg/l in aquatic habitat can also become lethal for aquatic lives (Lavate 2013). The soil of the intertidal zone in the oil-contaminated region showed higher oil content (1080 ± 420 mg/kg of 2 2-in. surface soil) than of the uncontaminated zones (5.5 ± 0.6 mg/kg of 2 2-in. surface soil). These findings through the study have shown how much oil spills can affect the coastal mangrove ecosystem (Harun et al. 2015).

Sand Mining

Another blow to the coastal habitat comes from the practice of "sand mining" that includes sand extraction from various environments such as beaches and inland dunes and dredging from ocean beds and riverbeds of deltaic regions (Pitchaiah 2017). Dried mangroves and red-colored ponds along the coastline of Kollam district of Kerala explain the widespread mining of beach sand mineral happening since the 1960s. Formation of sand bars and other interferences that include sand mining, oyster and mussel collection, and excess fishing in the Kozhikode and Malappuram districts of Kerala have witnessed the massive loss of mangrove vegetation (Bindu and Jayapal 2016). Along the Central Western Coast of India, a survey was undertaken along the two major estuaries Kundalika and Vasishthi in Maharashtra. The deterioration of these habitats gives rise to the demarcation of anthropogenic failures. Sand mining also contributes to the collapse of estuarine ecosystems. An increase in the suspended particulate levels and turbidity is followed by the input of oil and grease through vehicles used in sand removal accompanied by the changes in fish breeding. The fringing mangroves have been reclaimed to create human-enforced platforms which include landing stage for dredge vessels, loading trucks, stacking of sandbags and huts for laborer involved in trade with jetty kind of structures for cranes and winches, etc. With these perturbations to the extreme, the resulting habitat loss and modification are evident through the conflict between the livelihood crisis of the local fisherman community in comparison to bigger boats.

Resource Exploitation

Ethically very popular, the three Fs, fish, fuel, and fodder, summarize the importance of mangroves as a crucial form especially in terms of energy source for the local communities residing in the tropics. The chief origin of energy for domestic purposes in cooking and heating in the rural areas is derived from the forests in the form of firewood and charcoal. A study conducted in the Konkan region, Kolamb,

Tarkarli, Sarjekot-Kalanwali, and Achara (Sindhudurg district) carried out the evaluation of calorific values and charcoal formation in the mangrove plant species which differs from species to species. It was found that Maharashtra charcoal is produced from mangroves illegally. Although Malaya is a major exporter of charcoal from the mangroves, in India this is limited. The sequential order obtained from the local information of best charcoal for burning within different species follows *Rhizophora* > *Avicennia* > *Sonneratia*, whereas laboratory experiments found a higher percentage of the coal from *Sonneratia* (54.48% charcoal) followed by *Rhizophora* (*R. apiculata*: 53.04% charcoal). The best fuel is represented by the calorific value, and in the present study, wood logs of *Avicennia officinalis* (5922.12 cal/g), *Rhizophora mucronata* (6739.95 cal/g), and *Sonneratia alba* (4062.28 cal/g) are continuously destructed for fuel purpose (Lavate 2013). Hence, the exploitation continues because of the high calorific value of the wood and high strength. Other activities including the chipboard and paper industry also influence the clearing of forests (Rasquinha and Mishra 2020).

The consequence of extracting the fuelwood on the mangrove ecosystem has received very little consideration. A study along the east coast, Bhitarkanika mangroves, investigated this impact upon the structure, arrangement, rejuvenation, and biomass and carbon stocks. This region comes second in species richness, sheltering the maximum diversity of true mangroves species in the country. Maximum harvested communities were mainly composed of mixed-species types dominated by the presence of *A. officinalis* and *Sonneratia* species, whereas non-harvested areas exhibited (58% of the sampled species) the presence of *Heritiera fomes* and *Excoecaria agallocha*, locally called as *sundari* and *guan*. Both of them are considered to be a rich source of timber and fuelwood locally.

Historical practices of chopping these trees for building and construction have been swapped by traded materials. The impact of frequent cutting can be reflected through the bushy and scrubby mangrove patches with abundant coppicing owing to frequent cutting revealing reduction (75%) in the species of *Heritiera* and *Avicennia* since 1970. However, another species *Phoenix paludosa* is used as thatching material for house and basket making, which has increased in extent due to plantation and restricted firewood cutting along with it serving as a nesting ground for estuarine crocodile, *C. porosus*. Further, the parameters designated for the investigation were lower in areas harvested for fuelwood with a species-specific difference across both forest types where it is observed that continuous harvesting can also drive the rare species to local extinction where much needed long-term research is required. Another factor that was brought to notice is shrimp cultivation where clearing of these forest patches to include aquaculture ponds is another pressing concern for this area (Jayanthi et al. 2018).

Acknowledging the socio-cultural needs of the local people and guiding the community management initiative hold significant potential for these regions to reduce exploitation and promote sustainable methods.

Other Pollutants

The rest of the contaminants in the water, sediments, and biota of the mangrove ecosystem include:

- Pharmaceuticals and personal care products (PPCPs)
- Polycyclic aromatic hydrocarbons (PAHs)
- Endocrine disrupter compounds (EDCs)
- Persistent organic pollutants (POPs)

In a study carried out on the Thane Creek of Mumbai, west coast, India, the sediments showed 15 PAHs ranging from 902.58 to 1643.60 and 930.69 to 1158.30 ng/g in Trombay and Vashi, respectively. The four major concentrations of carcinogenic PAHs obtained from pyrogenic and petrogenic sources were benzo(b)fluoranthene, benzo(k)fluoranthene, indeno(1,2,3-cd)pyrene, and dibenz(a,h)anthracene. These concentrations are higher in the Trombay region due to the leakage of petroleum products and boat engine oil due to the fishing activities and sailing of crude oil than in the Vashi area.

2 Importance of Mangroves in Controlling Pollution

From being considered as wastelands in the past to their present role as a natural sink of pollutants and carbon capture along with the buildup of heavy metals and biomagnification, these sheltered estuarine species have become a natural fighter against pollution, hence preventing seawater pollution. They are constantly creating a balance with the nutrient cycling in the coastal and estuarine ecosystems. They decrease the water flows and enhance sediment deposits, thereby arresting coastal erosion. Their surroundings become a land accretion zone as the sediments trap heavy metal contaminants. The inundation of mangroves results in lesser oxygen in the organic-rich sediments where the sulfate ions create sulfidic conditions that will also arise that leads to immobilization of metals in the mangroves where physicochemical changes are also seen in the rhizosphere (Sukhdhane et al. 2015). The specificity of mangrove remains constant in terms of carbon and nutrient cycles and sediment characteristics which can affect the bioavailability of contaminants by not only acting as a sink or transferring but also oxidizing the metals present in the sediments. They have chemical contaminants within pore water, overlying water, and sediment, SPM, and biota. The path of human history leaves a trail of major concentrations of nutrient and organic matter behind that can be seen through a budget created upon extensive study of the coastal wetlands.

3 Current and Future Threats

The continuous degradation over the coming years due to human development will bring about a change in the global climate patterns that includes the atmosphere and oceanic processes too. The rise in sea level, global warming, and change in weather pattern manifested by hurricanes and rainfall is expected to be faced by the mangrove ecosystems that will further test the persistence of these sentinel species. The major contradiction occurs when it is found that no sound study has been carried out till now about it on the Indian mangroves.

3.1 *Global Warming*

Elevation in the levels of greenhouse gases has led to a significant rise in the mean temperature, especially carbon dioxide that will align both physical and chemical changes in marine regions. Ever since industrialization, the concentration of CO_2 has increased from 280 ppmv in 1880 to 409.8 ppmv in 2019. In India, around nine tonnes of CO_2 is removed by the mangrove forests that is approximately equivalent to 270 million US dollars in the international market. Although the mangroves are not expected to suffer from sea surface temperatures, the effects can be related to the location and species-specific occurrences depending upon the local conditions. Although the increase in temperature shows higher productivity around the temperature of 25 °C that is ideal for photosynthesis if increased, the result will affect net productivity along with potential risk to the other communities being harbored by the mangroves along with a much-emphasized change in flowering and fruiting periods that will depend upon species to species. The rise in temperatures can lead to sediment oxygen demand which can worsen hypoxia and anoxia in the aquatic region. It was stated that the water temperatures increased at the rate of 0.05 °C/year while the DO reduced at the rate of 0.4 mg/l/decade over 27 years which is mainly attributed to the climate change (Sandilyan 2014).

3.2 *Sea-Level Rise*

The global rise of sea-level is one of major consequence of global warming that is already taking place and also recorded during the 20th century (12–22 cm). The most evident outcome of the sea level rise is characterized by an upward shift in species distribution as well as ecosystem mortality that increases towards the sea along with the export and accumulation of C, N, and P nutrients. The increase in temperatures causes thermal expansion of ocean water, and melting of polar and land ice will occur. The climate change-induced sea-level rise is increasing at a rate of 9–12 cm^{-1} where the current projections have been reported to be about 0.4–0.9.

In India, a recent study was conducted in River Hooghly which was a first report on the migration of the mangroves upstream in the river that was absent before 1995 (Ghosh et al. 2020). The species of *Sonneratia, Derris, Hibiscus*, and *Thespesia* were observed which were affected by the increase in pollution, increase in the COD, sea-level rise, etc. The river has been facing a rise in the mean sea level (MSL) and toxicity from the pollution load which ultimately gets discharged into Bay of Bengal, hence influencing the regional biogeochemical aspects of the sediments and response of mangroves which act as bioindicators. The growth of *Sonneratia* species along the upstream zones of the river is much faster compared to any other mangrove species. These kinds of variation in the micro-level environment accompanied by the human-induced threats will increase the frequency of coastal hazards, hence redirecting the threats to the human population (Sandilyan 2014).

3.3 *Weather Events*

These woody halophytes take their major reputation for being the natural saviors after the devastating tsunami of 2004. The presence of a complex root system in the mangroves dissipates the sea wave energy which ultimately prevents the coastal areas from the negative impact of the weather events taking place. The force and rate of recurrence of the tropical cyclones have a big role in damaging the mangroves through uprooting, defoliation, and tree mortality that will further lead to the ecosystem conversion. Moreover, the cleared mangrove area has not been able to be revived due to the change in the hydrodynamics, low nutrients, salinity, and acidity as well as deficiency in substrates.

4 Management: Restoration and Resilience

To preserve the mangrove forest, their management plays a vital role where strategies are needed to be adopted to regulate the pressures from human development. Although they provide higher ecological services, still their destruction through the 1960s, along with the Southeast Asian countries, tells a different story where more than half of the mangrove's forests were removed for the developmental activities (Ranjan et al. 2018). The notification of 1991 regarding the Coastal Regulation Zone (CRZ) has been leading the protection of coastal environments. A study conducted in Mumbai Metropolitan Region (MMR) that has witnessed a tremendous increase in industrialization and urbanization evaluated change in the ecology and biodiversity with actions taken by the government body in following rules and regulations for the betterment of the marine environment. These actions are responsible for investigating the sustenance of long-term environmental and socioeconomic aids. Due to a project in the Gorai village, around 700 acres of mangroves field was destroyed by spraying chemicals. Further, the Mulund-Thane belt saw a dispute arise around 2005 where the special economic zone (SEZ) was reserved on 134 acres of mangroves.

Although imbibed right from the seventeenth century, their official report on "Status Report of Mangroves of India" was sent in 1987. Previously, the importance was not recognized that led to exploitation on the rise and hence huge losses in almost every country where it was also reported that the loss was even faster than coral reefs and tropical forests.

In India, the management of coastal woodlands is carried out using three major strategies: promotory, regulatory, and participatory.

- Promotory approach: Implementation by the Government of India, Management Action Plan (MAP) within 38 mangrove regions.
- Regulatory approach: The protection of mangroves is supported by policy methods and legal support through parks, sanctuary, reserved forests, and effective legislations which are often challenged by lack of financial support, poor infrastructure, and other socio-political demands. An example that can be carried forward is the condition of mangroves in Myanmar where the human pressure as well as political instability led to additional environmental degradation along with a poor economy that threatened the recovery.
- Participatory management: Stakeholders from the industrial sectors are an essential factor for this kind of approach.

4.1 Management Activities on Regional Scale

This can be explained by the following examples.

In India, along the east coast within the states of Tamil Nadu and Andhra Pradesh, a technique referred to as "Fish Bone" design was adopted for restoration of mangroves demonstrated by the M.S. Swaminathan Research Foundation in Chennai and Forest department as "Canal Bank Plantation" that increased the tidal inundation and made the soil suitable for growth by decreasing the amount of salt concentration. This technique was helpful in the restoration of forest cover by about 90% in the Pichavaram mangroves (1986–2002) along with the support of local communities. Another conservation model leads us to Maharashtra "Mangrove Cell" that was set up in January 2012, which till today has led to Maharashtra becoming the first Indian state to declare the state mangrove tree, *Sonneratia alba*, as the symbol of conservation (News report, Hindustan Times, 2020). Following the protocols, more plantations in the degraded areas and other marine-based projects have been undertaken to follow the conservation of biodiversity that successfully projects a holistic approach towards coastal ecosystem as well as marine region conservation in association with many several institutes, agencies, as well as NGOs. In May 2014, the Kannur region of Kerala undertook safety of its mangrove population by conducting a survey and declaring 236 ha mangroves as "Reserved Forest" and further acquiring them from private owners (News report, The Hindu, 2015).

Community-based co-management of mangroves in India occurs mainly in the states of West Bengal, Andhra Pradesh, Odisha, Tamil Nadu, and Gujarat, for example, restoration of 1475 ha mangroves by plantation of 6.8 million mangrove saplings by the JMM project that involved 5240 families from 28 villages on the east coast of India. Similarly, along the Gujarat coast, a project from 2001 to 2006 was managed around 5000 ha of mangroves and regeneration (Kathiresan 2018). The application of remote sensing becomes an efficient method of mapping and monitoring the mangroves. Their occurrence in inaccessible areas can be easily differentiated through the presence of conspicuous signatures in the satellite images. According to the India State of Forest Report, the analysis of satellite data along with the Geographical Information System (GIS) is the most effective way to monitor the mangrove ecosystems. The states which have witnessed a major degradation in the mangrove area are executing various measures for the conservation and management of these species.

5 Conclusion

The mangroves have been surviving the anthropogenic changes since the onset of urbanization and industrialization along the coast and adapting to the negative implications. This review of case studies across the coastal regions of India highlights various pathways through which the degradation of mangroves is being accelerated. The result of continuous mangrove degradation has been discussed with a major focus on the input of nutrients, heavy metals, oil spills, etc. contaminating the water and sediment. A consistent increase in the trace metals has been observed in the Sundarbans, Pichavaram, and Goa mangroves. The nutrient content shows a considerable increase hike in almost all the studied mangroves. Pichavaram shows an increase in nitrate from 5.9 mg/l in 1995 to 29.9 mg/l in 2006–2007. In Sundarbans it increased from 1.14 mg/l in 2001 to 3.69 mg/l in 2006 and in Godavari from 0.61 mg/l in 2001 to 2.25 mg/l in 2016. The phosphate values increase from 0.28 mg/l in 1995 to 6.6 mg/l in 2006 in Pichavaram mangroves. Manori creek, Mumbai, shows an increased hike in phosphate in the past 25 years. The value increases from 0.06 mg/l in 1982 to 2.19 mg/l in 2007. Further, in the case of oil spills, the oil content can become lethal for the aquatic species causing a reduction in biodiversity, destabilizing coastal habitat. In pristine conditions, the ecosystem can absorb disturbance and regenerate while facing alterations, also retaining similar controls on the structure and functioning. Due to the continuous damage being implicated on the ecosystems, these species can succumb to the negative pressure and become more vulnerable to the changes from being in reversible to irreversible states. Thus, we need to conserve, protect, and restore the mangrove ecosystem for the value they have and prevent degradation from reaching a threshold that can collapse their resilience (Begam et al. 2020). Restoration of mangroves can become a unique approach to decrease the emission of carbon dioxide to the atmosphere, thus mitigating climate change due to global warming.

References

ADB/NACA (1998) Report on a regional study and workshop – Aquaculture sustainability and the environment. Bangkok, Thailand. Asian Development Bank and Network of Aquaculture Centres in Asia-Pacific.

Alongi DM (2002) Present state and future of the world's mangrove forests. Environ Conserv 29(3):331–349. https://doi.org/10.1017/S0376892902000231

Alongi DM (2008) Mangrove forests: resilience, protection from tsunamis, and responses to global climate change. Estuar Coast Shelf Sci 76(1):1–13. https://doi.org/10.1016/j.ecss.2007.08.024

Alongi DM (2012) Carbon sequestration in mangrove forests. Carbon Manag 3(3):313–322. https://doi.org/10.4155/cmt.12.20

Attri K, Kerkar S (2011) Seasonal assessment of heavy metal pollution in tropical mangrove sediments (Goa, India). J Ecobiotechnol 3(8):9–15. www.scholarjournals.orgwww.journal-ecobiotechnology.com

Ball MC (1988) Ecophysiology of mangroves. Trees 2(3). https://doi.org/10.1007/BF00196018

Borges AV (2003) Atmospheric CO_2 flux from mangrove surrounding waters. Geophys Res Lett 30(11). https://doi.org/10.1029/2003GL017143

Begam MM, Chowdhury R, Sutradhar T et al (2020) Forecasting mangrove ecosystem degradation utilizing quantifiable eco-physiological resilience—a study from Indian Sundarbans. Sci Rep 10:6683. https://doi.org/10.1038/s41598-020-63586-4

Bindu KB, Jayapal G (2016) The sand bar formation and its impact on the mangrove ecosystem: a case study of Kadalundi Estuary of Kadalundi River Basin in Kerala, India. Curr World Environ 11(1):65–71. https://doi.org/10.12944/cwe.11.1.08

Chauhan R, Ramanathan AL (2008) Evaluation of water quality of Bhitarkanika mangrove system, Orissa, east coast of India. Indian J Mar Sci 37(2):153–158

Dehairs F, Rao RG, Chandra Mohan P, Raman AV, Marguillier S, Hellings L (2000) Tracing mangrove carbon in suspended matter and aquatic fauna of the Gautami-Godavari Delta, Bay of Bengal (India). Hydrobiologia 431:225–241

Dhargalkar VK, D'Souza R, Kavlekar D, Untawale AG (2014) Mangroves of Goa

Duke NC, Burns KA (2003) Fate and effects of oil and dispersed oil on mangrove ecosystems inAustralia. Environmental implications of offshore oil and gas development in Australia:further research. A compilation of three scientific marine studies. APPEA, Canberra

Duke NC (2016) Oil spill impacts on mangroves: recommendations for operational planning and action based on a global review. Mar Pollut Bull 109(2). https://doi.org/10.1016/j.marpolbul.2016.06.082

FAO (2010) Global forest resources assessment 2010 country reports. http://www.fao.org/forestry/62318/en/

George G, Krishnan P, Mini KG, Salim SS, Ragavan P, Tenjing SY, Muruganandam R, Dubey SK, Gopalakrishnan A, Purvaja R, Ramesh R (2019) Structure and regeneration status of mangrove patches along the estuarine and coastal stretches of Kerala, India. J For Res 30(2):507–518. https://doi.org/10.1007/s11676-018-0600-2

Ghosh S, Bakshi M, Gupta K, Mahanty S, Bhattacharyya S, Chaudhuri P (2020) A preliminary study on upstream migration of mangroves in response to changing environment along River Hooghly, India, Mar Pollut Bull 151:110840. ISSN 0025–326X, https://doi.org/10.1016/j.marpolbul.2019.110840. (https://www.sciencedirect.com/science/article/pii/S0025326X19309968)

Gopal B, Chauhan M (2006) Biodiversity and its conservation in the Sundarban mangrove ecosystem. Aquat Sci 68(3):338–354. https://doi.org/10.1007/s00027-006-0868-8

Goutham-Bharathi MP, Roy SD, Krishnan P, Kaliyamoorthy M, Immanuel T (2014) Species diversity and distribution of mangroves in Andaman and Nicobar Islands, India. Bot Mar 57(6):421–432. https://doi.org/10.1515/bot-2014-0033

Gurjar UR, Bunkar K, Pathak V, Takar S, Singh J (2019) Present status on distribution, biodiversity and management of Indian mangroves. Res Biotica 1:16–22

Harun A, Md A, Akber A (2015) Study of impacts of oil spill on the sundarbans mangrove forest of Bangladcsh. J Asiat Soc Bangladesh Sci 41(1):75–94

He B, Li R, Chai M, Qiu G (2014) Threat of heavy metal contamination in eight mangrove plants from the Futian mangrove forest, China. Environ Geochem Health 36(3):467–476. https://doi.org/10.1007/s10653-013-9574-3

Hema M, Devi PI (2015) Economic valuation of mangrove ecosystems of Kerala, India. J Environ Prof Sri Lanka 4(1):1

Jayanthi M, Thirumurthy S, Muralidhar M, Ravichandran P (2018) Impact of shrimp aquaculture development on important ecosystems in India. Glob Environ Chang 52:10–21. https://doi.org/10.1016/j.gloenvcha.2018.05.005

Kathiresan K (2000) A review of studies on Pichavaram mangrove, southeast India. Hydrobiologia 430:185–205

Kathiresan K (2018) Mangrove Forests of India. Curr Sci 114:976–981. https://doi.org/10.18520/cs/v114/i05/976-981

Krithika K, Purvaja R, Ramesh R (2008) Fluxes of methane and nitrous oxide from an Indian mangrove. Curr Sci 94(2):218–224

Kulkarni VA, Jagtap TG, Mhalsekar NM, Naik AN (2010) Biological and environmental characteristics of mangrove habitats from Manori creek, West Coast, India. Environ Monit Assess 168(1–4):587–596. https://doi.org/10.1007/s10661-009-1136-x

Kumar G, Kumar M, Ramanathan AL (2015) Assessment of heavy metal contamination in the surface sediments in the mangrove ecosystem of Gulf of Kachchh, West Coast of India. Environ Earth Sci 74(1):545–556. https://doi.org/10.1007/s12665-015-4062-y

Lavate R (2013) Mangrove as source of energy for rural development with special reference to Ratnagiri and Sindhudurg district (MS) India Flora of Jath Tahsil View project Water Analysis of Wetlands of Jath View project. http://biosciencediscovery.com

Lee SY, Dunn RJK, Young RA, Connolly RM, Dale PER, Dehayr R, Lemckert CJ, McKinnon S, Powell B, Teasdale PR, Welsh DT (2006) Impact of urbanization on coastal wetland structure and function. Austral Ecol 31(2):149–163. https://doi.org/10.1111/j.1442-9993.2006.01581.x

Lovelock C, Feller I, Ball M, Engelbrecht B, Ewe M (2006) Differences in plant function in phosphorus- and nitrogen-limited mangrove ecosystems. New Phytol 172: 514–22. https://doi.org/10.1111/j.1469-8137.2006.01851.x

Maiti SK, Chowdhury A (2013) Effects of anthropogenic pollution on mangrove biodiversity: a review. J Environ Prot 04(12):1428–1434. https://doi.org/10.4236/jep.2013.412163

Nriagu JO, Pacyna JM (1998) Quantitative assessment of worldwide contamination of air, water and soils by trace metals. Nature 12;333(6169):134–139. https://doi.org/10.1038/333134a0. PMID: 3285219

NOAA (2014) Oil Spills in Mangroves. Planning & Response Considerations. US Department of Commerce, National Oceanic and Atmospheric Administration (NOAA), National Ocean Service, Office of Response and Restoration, Seattle, Washington (96pp.)

Pandey CN, Pandey R (2013) Carbon sequestration by mangroves of Gujarat, India. GEER Foundation, Gandhinagar

Pitchaiah S (2017) Impacts of sand mining on environment—a review. SSRG Int J Geoinform Geol Sci (SSRG-IJGGS) 4(1). www.internationaljournalssrg.org

Prasad MBK (2012) Nutrient stoichiometry and eutrophication in Indian mangroves. Environ Earth Sci 67(1):293–299. https://doi.org/10.1007/s12665-011-1508-8

Prasad MBK, Kumar A, Ramanathan AL, Datta DK (2017) Sources and dynamics of sedimentary organic matter in Sundarban mangrove estuary from indo-Gangetic delta. Ecol Process 6(1). https://doi.org/10.1186/s13717-017-0076-6

Prasad MBK, Ramanathan AL, Alongi DM, Kannan L (2006) Seasonal variations and decadal trends in concentrations of dissolved inorganic nutrients in Pichavaram mangrove waters, Southeast India. Bull Mar Sci 79(2):287–300

Ramanathan AL, Chidambaram S, Subramanian V, Ramesh R, James A (1999) Environmental geochemistry of the Pichavaram mangrove ecosystem (tropical), southeast coast of India. In: Cases and solutions environmental geology, vol 37(3). Springer-Verlag, Berlin, p 223

Ranjan RK, Ramanathan A, Singh G, Chidambaram S (2008) Assessment of metal enrichments in tsunamigenic sediments of Pichavaram mangroves, southeast coast of India. Environ Monit Assess 147(1–3):389–411. https://doi.org/10.1007/s10661-007-0128-y

Ranjan P, Priya N, Kumar A, Ramanathan AL (2017) Distribution of trace metals in the sediments of estuarine-mangrove complex across the Indian coast. In: Wetland science: perspectives from South Asia. Springer, New Delhi, pp 163–186. https://doi.org/10.1007/978-81-322-3715-0_9

Ranjan P, Rao K, Kumar A, Ramanathan AL (2018) Heavy metal distribution and accumulation from natural and anthropogenic sources in tropical mangroves of India and Bangladesh. http://www.springer.com/series/8795

Rao K, Priya N, Ramanathan AL (2018) Impact of seasonality on the nutrient concentrations in Gautami-Godavari estuarine mangrove complex, Andhra Pradesh, India. Mar Pollut Bull 129(1):329–335. https://doi.org/10.1016/j.marpolbul.2018.02.052

Rasquinha DN, Mishra DR (2020) Impact of wood harvesting on mangrove forest structure, composition and biomass dynamics in India. Estuar Coast Shelf Sci. https://doi.org/10.1016/j.ecss.2020.106974

Reef R, Feller IC, Lovelock CE (2010) Nutrition of mangroves. Tree Physiol 30(9). https://doi.org/10.1093/treephys/tpq048

Sandilyan S (2014) Impacts of climate change on Indian mangroves: a review paper. Glob J Environ Res 8(1):1–10. https://doi.org/10.5829/idosi.gjer.2014.8.1.1104

Santschi P, Höhener P, Benoit G, Buchholtz-ten Brink M (1990) Chemical processes at the sediment-water interface. Mar Chem 30. https://doi.org/10.1016/0304-4203(90)90076-O

Selvam V (2003) Environmental classification of mangrove wetlands of India. Curr Sci 84(6):757–765

Sen S (2020) Sunderban mangroves, post Amphan: an overview, vol 8(6). www.ijcrt.org

Spalding M (2010). World Atlas of Mangroves (1st ed.). Routledge. https://doi.org/10.4324/9781849776608

Singh SK, Srivastava PK, Gupta M, Thakur JK, Mukherjee S (2014) Appraisal of land use/land cover of mangrove forest ecosystem using support vector machine. Environ Earth Sci 71(5):2245–2255. https://doi.org/10.1007/s12665-013-2628-0

Singh G, Ramanathan Al, Prasad MBK (2005) Nutrient cycling in Mangrove ecosystem: A Brief Overview. International Journal of Ecology and Environmental Sciences 31:231–244

Sukhdhane KS, Pandey PK, Vennila A, Purushothaman CS, Ajima MNO (2015) Sources, distribution and risk assessment of polycyclic aromatic hydrocarbons in the mangrove sediments of Thane Creek, Maharashtra, India. Environ Monit Assess 187(5). https://doi.org/10.1007/s10661-015-4470-1

Suresh HS, Sahu S (2015) Mangrove area assessment in India: implications of loss of mangroves. J Earth Sci Clim Change 6(5). https://doi.org/10.4172/2157-7617.1000280

Tam NFY, Wong YS (1993) Retention of nutrients and heavy metals in mangrove sediment receiving wastewater of different strengths. Environ Technol 14(8). https://doi.org/10.1080/09593339309385343

Tripathy SC, Ray AK, Patra S, Sarma VV (2001) Water quality assessment of Gautami-Godavari mangrove estuarine ecosystem of Andhra Pradesh, India during September 2001. J Earth Syst Sci 114(2):185–190

Zingde M, Desai B (1980) Waste water discharge and its effect on the quality of water of Mahim Creek and Bay. Mahasagar Bull Natl Inst Oceanogr 13:205–213

Chapter 9
Zooplankton Diversity and Their Spatiotemporal Distribution: An Ecological Assessment from a Brackish Coastal Lagoon, Chilika, Odisha

Suchismita Srichandan and Gurdeep Rastogi

Abstract Zooplankton constitutes a pivotal component in the pelagic food webs and serves as the major source of fish diet, thereby determining the productivity of coastal fisheries. Therefore, understanding zooplankton diversity and their ecology in coastal lagoon settings is a high priority research area. We examined the spatiotemporal distribution of zooplankton diversity (size >120 μm) in relation to environmental variables in Chilika lagoon. The sampling was conducted on the monthly frequency from July 2012 to June 2016 from 13 locations and identified a total of 186 zooplankton taxa which included 131 as first record from the Chilika lagoon. To date, a total inventory of 263 species of holoplankton represented by 16 diverse categories of organisms, namely, Ciliophora (51), Foraminifera (13), Tubulinea (5), Rotifera (42), Hydrozoa (1), Ctenophora (1), Nematoda (1), Polychaeta (3), Gastropoda (12), Bivalvia (5), Cladocera (13), Copepoda (95), Ostracoda (4), Malacostraca (13), Chaetognatha (2), Chordata (2), and 23 types of meroplankton were identified. Chilika lagoon exhibited a significant variation in salinity (0–35.5) at spatiotemporal scale and consisted of marine, brackish, and freshwater zooplankton along the estuarine salinity gradient. Copepods emerged as one of the most dominant and diverse zooplankton group in terms of species richness, abundance, and widespread distribution. Among the four orders of Copepoda (i.e., Calanoida, Cyclopoida, Harpacticoida, and Poecilostomatoida), Calanoida was the most abundant one. An important component of total zooplankton pool, i.e., microzooplankton (20–200 μm), was also examined in relation to environmental variables. Ciliophora dominated the microzooplankton community followed by copepod

S. Srichandan (✉)
Wetland Research and Training Centre, Chilika Development Authority, Balugaon, Odisha, India

Centurion University of Technology and Management, Bhubaneswar, Odisha, India

G. Rastogi
Wetland Research and Training Centre, Chilika Development Authority, Balugaon, Odisha, India

S. Madhav et al. (eds.), *Coastal Ecosystems*, Coastal Research Library 38,
https://doi.org/10.1007/978-3-030-84255-0_9

nauplii and Rotifera, except in the freshwater zone of the lagoon. Foraminifera, cirripede nauplii, gastropod veliger, and bivalve veliger were minor contributors in microzooplankton. Salinity and phytoplankton abundances were the major factors influencing microzooplankton community composition. The present study highlighted the necessity of a long-term systematic monitoring of zooplankton diversity and composition in Chilika lagoon.

Keywords Zooplankton · Copepoda · Salinity · Chilika · Coastal lagoon

1 Background

Coastal lagoons are highly productive and economically important aquatic environment which constitute ~13% of the world's coastline. Coastal lagoons are separated from the adjoining sea by a barrier or communicate with the sea through inlets (mouths) (Perez-Ruzafa et al. 2011). The lagoons are highly dynamic ecosystems due to continuous material influxes (dissolved and particulate) from both marine and terrestrial environments (Mitsch and Gosselink 1993). In global context, the lagoons are stressed by both natural (e.g., extreme climatic events) and anthropogenic pressures (e.g., eutrophication, sewage discharge, and overfishing) (Kumar et al. 2016; Arreola-Lizarraga et al. 2016). These natural and anthropogenic pressures on coastal lagoons along with mixing of water from riverine and marine sources yield a sharp gradient in the physicochemical factors which determine the zooplankton community composition and distribution over the spatiotemporal scales.

Zooplankton modulate carbon flow in the food chain through their trophic interactions with lower as well as higher consumers (Isari et al. 2007). They also act as a recycler and transform particulate organic matter and nutrients into dissolved organic matter (Steinberg and Landry 2017). Generally, in an aquatic ecosystem, the fishery yields are highly dependent on the availability of zooplankton standing stocks. For instance, a high quantum of fishery (e.g., sardines and anchovies) in areas with high zooplankton (e.g., *Calanus sinicus*) production has been reported in Changjiang River estuary (China) (Gao et al. 2011). In general, zooplankton feed on phytoplankton and detritus and put a higher predation pressure on the algal standing stock. For instance, an experimental study from the Zuari and Mandovi estuaries (India) revealed a significant (>60%) grazing of phytoplankton (pico and nano) standing stock by the microzooplankton (Gauns et al. 2015).

Zooplankton are also considered as bioindicator of climate change in lagoonal and marine environments (Molinero et al. 2005). Zooplankton, due to their characteristic life processes, provide an excellent proxy to track changing climatic conditions (Carter et al. 2017). Climate change influences not only zooplankton dynamics but also their phenotype, physiology, and community composition (Dam 2013). For instance, a reduction in the size of ectotherms due to long-term warming is a common prediction on the effect of changing climate on zooplankton (Rice et al. 2015).

The zooplankton communities are quite diverse in their morphology, physiology, reproductive biology, trophic status, mode of life, and responses to different environmental stimuli. In general, zooplankton range from tiny protozoan to gigantic jellyfishes and are divided into several size classes, such as microzooplankton (20–200 μm), mesozooplankton (200 μm–2 mm), macrozooplankton (2–20 mm), and megazooplankton (>20 mm). Some of the microzooplanktonic forms are tintinnids, foraminifers, radiolarians, trochophore larvae of polychaetes, copepod nauplii, gastropod veligers, and barnacle nauplii. Cladocerans, copepods, ostracods, and amphipods are the ideal examples of mesozooplankton. Some examples of macrozooplankton are the pteropods, mysids, chaetognaths, lucifers, dolioloids, and salps. The megazooplankton are only few in numbers and are mostly represented by siphonophores.

In recent past, investigations on zooplankton have targeted the taxonomic diversity, abundances, and environmental drivers in coastal lagoons (Ziadi et al. 2015; Varghese et al. 2018; Gutierrez et al. 2018). Spatiotemporal variations in zooplankton are regulated by multitude of environmental factors such as trophic state, food availability, and predation pressure (Souza et al. 2011; Miron et al. 2014). Among physical forcing, salinity has been recognized as one of the crucial factors in controlling the spatiotemporal distribution of zooplankton (Santangelo et al. 2007; Etile et al. 2009; Antony et al. 2020). Zooplankton also respond to variations in hydrobiological factors such as temperature, pH, transparency, and food availability. For instance, temperature, pH, transparency, and chlorophyll were the primary environmental variables that regulated the zooplankton ecology in Sontecomapan Lagoon (Mexico) (Miron et al. 2014). Further, zooplankton communities also respond to the trophic variations in estuarine ecosystems (Park and Marshall 2000; Gopko and Telesh 2013). For example, higher relative abundances of rotifers (*Keratella* sp.) were indicative of trophic status of Neva Estuary (Finland) (Gopko and Telesh 2013).

Chilika lagoon (hereafter Chilika), a Ramsar site (no. 229), located on the east coast of India is an ideal ecosystem to examine zooplankton communities and their response to contrasting physicochemical regimes. Considering this, several studies have targeted zooplankton to decipher their community composition, variability, and ecological preferences from this lagoon (Devasundaram and Roy 1954; Patnaik 1973; Pattanaik and Sarma 1997; Naik et al. 2008; Mukherjee et al. 2014, 2015, 2018; Rakhesh et al. 2015; Sahu et al. 2016). Most of these studies have either focused on a particular zooplankton group (Mukherjee et al. 2014, 2015) or examined the community composition only up to the order level based on seasonal and monthly surveys (Patnaik 1973; Pattanaik and Sarma 1997; Naik et al. 2008). Importantly, species-level zooplankton community structure with detailed quantitative accounts has been investigated only in few studies (Devasundaram and Roy 1954; Rakhesh et al. 2015; Sahu et al. 2016; Mukherjee et al. 2018). The present chapter deals with long-term spatiotemporal patterns of zooplankton communities and their environmental controlling factors from Chilika based on systemic field surveys. The comprehensive dataset generated with current study was integrated with existing literature to synthesize the present status of the spatiotemporal distribution of the zooplankton from this lagoon.

2 Materials and Methods

2.1 Study Area

Chilika is connected to the northwestern Bay of Bengal (BoB) on the east coast of India (19°28′–19°54′ N and 85°06′–85°35′ E). Chilika spans over an area of 906 km^2 during summer and 1165 km^2 during monsoon (Srichandan and Rastogi 2020). The lagoon is connected to the BoB through outer channel as well as through Palur Canal at the southern end (Fig. 9.1). The hydrology of Chilika is strongly influenced by the tropical southwest monsoon (July–October). Chilika receives freshwater discharge from 52 rivers and rivulets; however, 19 of them are major contributors (Ganguly et al. 2015). The freshwater influx into the lagoon occurs in the upper reaches of northern sector mainly from the distributaries of Mahanadi delta, while seawater influx mostly occurs through inlets located at the outer channel. Chilika is spatially categorized into four ecological sectors, namely, southern sector (SS), central sector (CS), northern sector (NS), and outer channel (OC), based on the salinity gradient (Srichandan et al. 2015a). Chilika also experiences different salinity regimes in different sectors such as oligohaline (NS: 0.5–5), mesohaline (CS and SS: 5–18), and polyhaline (OC: 18–30) (Muduli and Pattnaik 2020). In addition, extreme weather events such as *Phailin* (October 12, 2013) and *Hudhud* (October

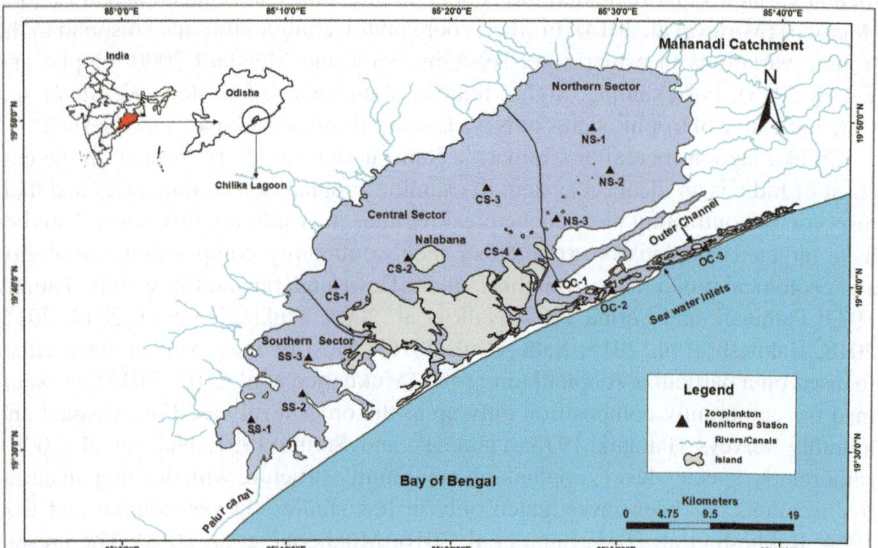

Fig. 9.1 Geographical map of Chilika lagoon showing 13 sampling stations used in zooplankton sampling. Physical boundaries are hypothetical to demonstrate the *SS* southern sector, *CS* central sector, *NS* northern sector, *OC* outer channel of the lagoon

12, 2014) have been shown to cause variability in nutrient molar ratios and phytoplankton biomass leading to proliferation of blooms (Kumar et al. 2016; Srichandan et al. 2015b).

2.2 *Sampling and Analysis*

2.2.1 Zooplankton

Microzooplankton (20–200 μm) were examined from July 2011 to June 2012; thereafter, zooplankton (>120 μm) were examined from July 2012 to June 2016. Thus, the study period for zooplankton included a total of 4 years which were referred as Y–1 (July 2012–June 2013), Y–2 (July 2013–June 2014), Y–3 (July 2014–June 2015), and Y–4 (July 2015–June 2016) throughout this chapter. Field surveys were carried out at a monthly frequency from 13 selected stations across 4 sectors and 3 distinct seasons, i.e., monsoon (July–October), post-monsoon (November–February), and pre-monsoon (March–June).

Microzooplankton were sampled by filtering ~100 l of water through 20 μm plankton net (make: KC Denmark; mouth diameter: 25 cm; length: 40 cm) which were subsequently passed through a 200 μm mesh to exclude large size zooplankton. Lugol's iodine solution (final concentration 1%) and formaldehyde (final concentration 2%) were added to the sample for preservation. Samples were concentrated by the gravimetric sedimentation technique. Subsequently, the supernatant was siphoned out leaving 100 ml as the final volume. One milliliter of concentrated sample was transferred to a Sedgewick Rafter counting chamber. The qualitative and quantitative analysis of microzooplankton was carried out using an inverted microscope (make: Olympus; model: IX73) following the standard taxonomic keys of Kofoid and Campbell (1929), Maeda (1986), Altaff (2004), Al-Yamani et al. (2011), and Gao et al. (2016).

Water samples for zooplankton (>120 μm) were collected with a plankton net (make: KC Denmark; mouth diameter: 25 cm; length: 48 cm) which was towed horizontally for 5–10 min. The amount of water passed through the net was quantified using a digital flow meter fitted with the net. Samples were preserved with 5% formaldehyde and subsampled using a plankton splitter (make: KC Denmark). A subsample (45 ml) was withdrawn from each sample, dispensed on the zooplankton counting chamber (dimensions 220 × 100 mm, inner diameter 76 mm, make: KC Denmark) and enumerated using an inverted microscope (make: Olympus; model: IX73). Zooplankton were identified up to the genus/species level based on standard literature (Kasturirangan 1963; Battish 1992; Conway et al. 2003). For compilation of zooplankton species checklist, classification system and updated scientific names as per WoRMS (World Register of Marine Species, http://www.marinespecies.org/) were referred.

2.2.2 Physicochemical Parameters and Phytoplankton Enumeration

At each sampling station, in situ measurement of water temperature, pH, salinity, and turbidity (nephelometric turbidity units (NTU)) was carried out by water quality Sonde (YSI, Model No. 6600, V2) throughout the study period. The detailed procedure for collection and analysis of dissolved oxygen (DO) and dissolved nutrients (nitrate, NO_3^-; phosphate, PO_4^{3-}; and silicate, SiO_4^{4-}) is described in Srichandan et al. (2015a).

Phytoplankton samples from each station were collected by filtering ~100 l of water through a plankton net (make: KC Denmark; mesh size: 10 μm; mouth diameter: 25 cm) and preserved with 2% neutralized formaldehyde and 1% Lugol's iodine solution. The phytoplankton cells were enumerated and identified as described earlier (Srichandan et al. 2015a). Total chlorophyll *a* (Chl *a*) was estimated by filtering 1 l of water through Whatman GF/F filters (pore size: 0.7 μm) using 90% acetone extraction method, and optical density was measured using a UV–Visible Spectrophotometer (Thermo Scientific™ Evolution 201).

2.3 *Statistical Analysis*

Canonical correspondence analysis (CCA) was applied to identify major environmental drivers of the dominant zooplankton groups. CCA was performed using CANOCO (version 4.5), and CCA biplots were generated based on the statistical significance of the environmental variables evaluated through Monte Carlo permutation (number of permutation: 499). Pearson's correlation coefficient (*r*) between environmental variables and zooplankton groups was computed using SPSS (v. 20).

3 Results and Discussion

3.1 *Zooplankton Diversity*

Zooplankton communities of the Chilika represented almost all animal phyla either as holoplankton or meroplankton. Zooplankton can be permanent forms (holoplankton) or temporary forms (meroplankton). Holoplankton include different groups such as Ciliophora, Foraminifera, Tubulinea, Rotifera, Hydrozoa, Ctenophora, Nematoda, Polychaeta, Gastropoda, Bivalvia, Cladocera, Copepoda, Ostracoda, Malacostraca, Chaetognatha, and Chordata. On the other hand, meroplankton includes the larvae of certain invertebrates and vertebrates.

Based on past and present studies, so far, a total of 263 species of holoplankton represented by 16 diverse categories of organisms, namely, Ciliophora (51), Foraminifera (13), Tubulinea (5), Rotifera (42), Hydrozoa (1), Ctenophora (1),

Nematoda (1), Polychaeta (3), Gastropoda (12), Bivalvia (5), Cladocera (13), Copepoda (95), Ostracoda (4), Malacostraca (13), Chaetognatha (2), and Chordata (2), and 23 types of meroplankton have been catalogued from Chilika (Table 9.1). The photomicrographs of some newly recorded zooplankton taxa in Chilika are presented in Plate 9.1. Importantly, earlier studies have adopted various different methods for collection, preservation, concentration, and microscopy of zooplankton in Chilika. For instance, some earlier studies have used plankton nets of 74 μm for microzooplankton collection (Mukherjee et al. 2018), while others have used sedimentation technique without plankton net (Sahu et al. 2016). Sampling frequency

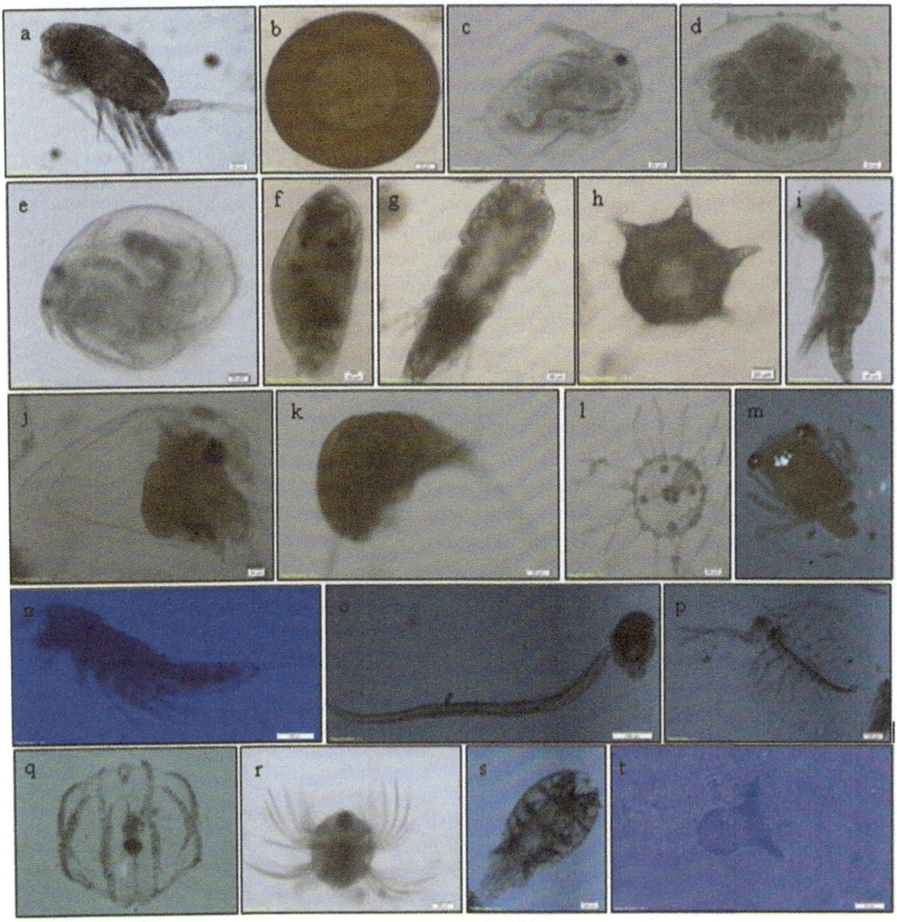

Plate 9.1 Photographs of some newly reported zooplankton taxa (**a**) *Acrocalanus gibber*; (**b**) *Arcella discoides*; (**c**) *Bosminopsis deitersi*; (**d**) brachiopod larva; (**e**) *Chydorus* sp.; (**f**) cirripede cypris larva; (**g**) *Clytemnestra scutellata*; (**h**) *Difflugia corona*; (**i**) *Euterpina acutifrons*; (**j**) *Pseudevadne tergestina*; (**k**) *Metis* sp.; (**l**) *Obelia* sp.; (**m**) brachyuran megalopa larva; (**n**) *Microsetella norvegica*; (**o**) *Oikopleura dioica*; (**p**) *Penilia avirostris*; (**q**) *Pleurobrachia pileus*; (**r**) polychaete larva; (**s**) *Sapphirina* sp.; (**t**) *Tintinnopsis mortensenii*

Table 9.1 List of zooplankton taxa from Chilika

Phylum	Ciliophora
Class	Oligotrichea
Subclass	Oligotrichia
Order	Choreotrichida
Family	Tintinnidiidae
	Leprotintinnus nordqvistii (Brandt 1906) Kofoid and Campbell 1929[a–c], *Leprotintinnus simplex* Schmidt 1902[a,c]
Family	Codonellidae
	Tintinnopsis beroidea Stein 1867[b–d], *Tintinnopsis cylindrica* Daday 1887[a–e], *Tintinnopsis mortensenii* Schmidt 1902[b,c,f], *Tintinnopsis tocantinensis* Kofoid and Campbell 1929[a–e], *Tintinnopsis tubulosa* Levander 1900[a–d], *Tintinnopsis uruguayensis* Balech 1948[b–d], *Tintinnopsis bermudensis* Brandt 1906[b–d], *Tintinnopsis buetschlii* Daday 1887[b–d], *Tintinnopsis tenuis* Hada 1932[b,c,f], *Tintinnopsis acuminata* Daday 1887[b,c,f], *Tintinnopsis dadayi* Kofoid 1905[b,c,f], *Tintinnopsis gracilis* Kofoid and Campbell 1929[a–e], *Tintinnopsis sacculus* Brandt 1896[b–d], *Tintinnopsis fimbriata* Meunier 1919[a,c,e], *Tintinnopsis directa* Hada 1932[a,c–e], *Tintinnopsis compressa* Daday 1887[a,c], *Tintinnopsis rotundata* Kofoid and Campbell 1929[a,c], *Tintinnopsis radix* Imhof 1886[a,c–e], *Tintinnopsis nucula* Fol 1884[a,c], *Tintinnopsis parvula* Jorgensen 1912[a,c,e], *Tintinnopsis spiralis* Kofoid and Campbell 1929[a,c,e], *Tintinnopsis filakinensis* Al-Yamani et al. 2011[a,c], *Tintinnopsis lohmanni* Laackmann 1906[c,d], *Tintinnopsis nana* Lohmann 1908[c,d], *Tintinnopsis karajacensis* Brandt 1896[a,c,e], *Tintinnopsis* sp. Stein 1867[b–e,g], *Codonella* sp. Haeckel 1873[c,g]
Family	Tintinnidae
	Dadayiella bulbosa Brandt 1906[a,c], *Eutintinnus fraknoii* Daday 1887[a,c], *Eutintinnus apertus* Kofoid and Campbell 1929[a,c], *Eutintinnus elongatus* Jorgensen 1924[a,c], *Eutintinnus* sp. Kofoid and Campbell 1939[c,e], *Amphorellopsis acuta* Schmidt 1902[c,d]
Family	Codonellopsidae
	Stenosemella nivalis Meunier 1910[a,c], *Stenosemella ventricosa* (Claparede and Lachmann 1858) Jorgensen 1924[a,c], *Stenosemella* sp. Jorgensen[c,e], *Codonellopsis ostenfeldi* (Schmidt 1902) Kofoid and Campbell 1929[a–d]
Family	Ptychocylididae
	Favella philippinensis Roxas 1941[b–d], *Favella brevis* Kofoid and Campbell 1929[b,c,f], *Favella adriatica* (Imhof 1886) Jorgensen 1924[a,c,e], *Favella campanula* (Schmidt 1902) Jorgensen 1924[a,c,e], *Favella ehrenbergii* (Claparede and Lachmann 1858) Jorgensen 1924[a,c,e], *Favella* sp. Jorgensen 1924[b,c,f]
Family	Metacylididae
	Metacylis tropica Duran 1957[a,c], *Metacylis jorgensenii* Cleve 1902[c,d]
Family	Dictyocystidae
	Dictyocysta seshaiyai Krishnamurthy and Santhanam 1975[b,c,f], *Dictyocysta* sp. Ehrenberg 1854[b–d], *Luminella* sp. Kofoid and Campbell 1939[b,c,f]
Family	Cyttarocylididae
	Cyttarocylis sp. Fol 1881[c,g]
Phylum	Foraminifera
Class	Polythalamea
Order	Globigerinida
Family	Globigerinidae

(continued)

Table 9.1 (continued)

	Globigerina bulloides d'Orbigny 1826[b,c,f], *Globigerina* sp. d'Orbigny 1826[b,c,f,h]
Class	Globothalamea
Order	Rotaliida
Family	Ammoniidae
	Ammonia sp. Brünnich 1771[b,c,f]
Family	Bolivinitidae
	Bolivina sp. d'Orbigny 1839[b,c,f]
Family	Discorbidae
	Discorbis sp. Lamarck 1804[b,c,f,h]
Family	Nonionidae
	Nonionella sp. Cushman 1926[b,c,f]
Family	Elphidiidae
	Elphidium sp. Montfort 1808[c,i]
Order	Lituolida
Family	Lituolidae
	Flabellammina sp. Cushman 1928[b,f]
Order	Textulariida
Family	Textulariidae
	Textularia sp. Defrance 1824[b,c,f,j]
Class	Tubothalamea
Order	Miliolida
Family	Spiroloculinidae
	Spiroloculina sp. d'Orbigny 1826[b,c,f,h]
Family	Hauerinidae
	Quinqueloculina sp. d'Orbigny 1826[b,c,f], *Triloculina* sp. d'Orbigny 1826[b,c,f,h]
Order	Spirillinida
Family	Ammodiscidae
	Ammodiscus sp. Reuss 1862[b,c,f]
Phylum	Amoebozoa
Class	Tubulinea
Order	Arcellinida
Family	Arcellidae
	Arcella discoides Ehrenberg 1843[b,f,h], *Arcella* sp. Ehrenberg 1832[b,f,h]
Family	Centropyxidae
	Centropyxis sp.
Family	Difflugiidae
	Difflugia corona Wallich 1864[b,f,h], *Difflugia* sp. Leclerc 1815[b,c,h,i]
Phylum	Rotifera
Class	Eurotatoria
Subclass	Monogononta
Order	Ploima
Family	Brachionidae

(continued)

Table 9.1 (continued)

	Anuraeopsis fissa Gosse 1851[b,c,f], *Anuraeopsis* sp. Lauterborn 1900[b,c,f], *Brachionus dichotomus reductus* Koste and Shiel 1980[b,f,h], *Brachionus falcatus* Zacharias 1898[b,e,h], *Brachionus quadridentatus* Hermann 1783[b,h,k], *Brachionus rubens* Ehrenberg 1838[b,d,h], *Brachionus* sp. Pallas 1766[b–d,g,h,j], *Brachionus angularis angularis* Gosse 1851[c,k], *Brachionus plicatilis* Muller 1786[c,e,j,k], *Brachionus bidentata* Anderson 1889[c,e,k], *Brachionus urceolaris* Müller 1773[k], *Brachionus calyciflorus* Pallas 1766[k], *Kellicottia longispina* Kellicott 1879[d,h], *Keratella tropica* Apstein 1907[b,h,j,k], *Keratella tecta* Gosse 1851[b,c,f], *Keratella* sp. Bory de St. Vincent 1822[c,d,g,h,j], *Plationus patulus* Müller 1786[b,h,k]
Family	Asplanchnidae
	Asplanchna brightwellii Gosse 1850[b,f,h], *Asplanchna* sp. Gosse 1850[b–d,h]
Family	Dicranophoridae
	Dicranophorus sp. Nitzsch 1827[b,f,h]
Family	Lecanidae
	Lecane batillifer Murray 1913[e,h,k], *Lecane crepida* Harring 1914[h,k], *Lecane inopinata* Harring and Myers 1926[e,k], *Lecane leontina* Turner 1892[k], *Lecane styrax* Harring and Myers 1926[e,k], *Lecane ungulata* Gosse 1887[k], *Monostyla bulla* Gosse 1851[d,e,h,k], *Monostyla luna* Muller 1776[e,h,k], *Monostyla* sp. Ehrenberg 1930[d,h], *Lecane* sp. Nitzsch 1827[b–d,h,j]
Family	Lepadellidae
	Lepadella sp. Bory de St. Vincent 1826[b–d,h,j]
Family	Synchaetidae
	Polyarthra vulgaris Carlin 1943[b,f,h], *Polyarthra* sp. Ehrenberg 1834[d,h]
Order	Flosculariaceae
Family	Hexarthridae
	Hexarthra sp. Schmarda 1854[b–e,h,j,k]
Family	Conochilidae
	Conochilus dossuarius Hudson 1885[c,k]
Family	Filiniidae
	Filinia longiseta Ehrenberg 1834[h,k], *Filinia opoliensis* Zacharias 1898[h,k], *Filinia* sp. Bory de St. Vincent 1824[e,g,h]
Family	Testudinellidae
	Pompholyx sulcata Hudson 1885[h,k], *Testudinella patina* Hermann 1783[c,h,j,k], *Testudinella* sp. Bory de St. Vincent 1826[c,d,h,j]
Family	Trichocercidae
	Trichocerca sp. Lamarck 1801[c,d,h,j]
Phylum	Cnidaria
Class	Hydrozoa
Subclass	Hydroidolina
Order	Leptothecata
Family	Campanulariidae
	Obelia sp. Peron and Lesueur 1810[b,c,f]
Phylum	Ctenophora
Class	Tentaculata
Order	Cydippida

(continued)

Table 9.1 (continued)

Family	Pleurobrachiidae
	Pleurobrachia pileus O.F. Müller 1776[b,c,f,j]
Phylum	Nematoda
Class	Enoplea
Subclass	Enoplia
Order	Enoplida
Family	Enchelidiidae
	Belbolla sp. Andrassy 1973[b,c,f,j]
Phylum	Annelida
Class	Polychaeta
Subclass	Errantia
Order	Phyllodocida
Family	Nereididae
	Nereis chilkaensis Southern 1921[c,i], *Neanthes glandicincta* Southern 1921[c,h,i], *Perinereis marjorii* Southern 1921[c,h,i]
Phylum	Mollusca
Class	Gastropoda
Subclass	Heterobranchia
Order	Pteropoda
Family	Creseidae
	Creseis acicula Rang 1828[b,c,f]
Family	Heliconoididae
	Heliconoides inflatus d'Orbigny 1835[b,c,f]
Order	Pylopulmonata
Family	Pyramidellidae
	Quirella humilis Preston 1905[c,i]
Order	Cephalaspidea
Family	Tornatinidae
	Acteocina estriata Preston 1914[c,i]
Subclass	Caenogastropoda
Order	Littorinimorpha
Family	Atlantidae
	Atlanta sp. Lesueur 1817[b,c,f]
Family	Stenothyridae
	Stenothyra sp. Benson 1856[h,i]
Order	Caenogastropoda
Family	Epitoniidae
	Janthina sp. Roding 1798[b,c,f]
Family	Potamididae
	Pirenella cingulata Gmelin 1791[c,i,j]
Family	Litiopidae
	Litiopa copiosa Preston 1915[c,i]
Order	Neogastropoda

(continued)

Table 9.1 (continued)

Family	Nassariidae
	Nassa denegabilis Preston 1914[c,i], *Nassarius orissaensis* Preston 1914[c,i], *Tritia burchardi* Dunker 1849[c,i]
Class	Bivalvia
Subclass	Autobranchia
Order	Mytilida
Family	Mytilidae
	Modiola undulatus var. *crassicostata* Preston 1914[c,i]
Order	Veneroida
Family	Veneridae
	Clementia annandalei Preston 1914[c,i], *Meretrix casta* Gmelin 1791[c,i], *Marcia opima* Gmelin 1791[c,i]
Order	Cardiida
Family	Semelidae
	Theora opalina Hinds 1843[c,i]
Phylum	Arthropoda
Class	Branchiopoda
Subclass	Diplostraca
Order	Onychopoda
Family	Podonidae
	Pseudevadne tergestina Claus 1877[b,c,f], *Evadne nordmanni* Loven 1836[b,c,f], *Evadne* sp. Loven 1836[c,h–j]
Order	Ctenopoda
Family	Sididae
	Penilia avirostris Dana 1849[b,c,f,h,j], *Diaphanosoma excisum* G.O. Sars 1885[b,c,f,h,j], *Diaphanosoma* sp. Fischer 1850[b,c,f,h,j]
Order	Anomopoda
Family	Chydoridae
	Chydorus sphaericus O.F. Müller 1776[b,c,f,h,j], *Chydorus* sp. Leach 1816[b,c,f,h,j], *Alona* sp. Baird 1843[b,c,f,h,j]
Family	Bosminidae
	Bosminopsis deitersi Richard 1895[b,f,h]
Family	Macrothricidae
	Macrothrix sp. Baird 1843[b,f,h]
Family	Moinidae
	Moina micrura Kurz 1875[b,f,h,j], *Moina* sp. Baird 1850[b,g,h,j]
Class	Hexanauplia
Subclass	Copepoda
Order	Calanoida
Family	Acartiidae
	Acartia centrura Giesbrecht 1889[b,c,i], *Acartia danae* Giesbrecht 1889[b,c,f], *Acartia erythraea* Giesbrecht 1889[b,c,f], *Acartia negligens* Dana 1849[b,c,f], *Acartia southwelli* Sewell 1914[b,c,f], *Acartia spinicauda* Giesbrecht 1889[b,c,f], *Acartiella major* Sewell 1919[c,i,j,l], *Acartiella minor* Sewell 1919[c,i,j], *Acartia chilkaensis* Sewell 1919[c,i,l], *Acartia* sp. Dana 1846[b,c,g]

(continued)

Table 9.1 (continued)

Family	Candaciidae
	Candacia discaudata Scott A. 1909[b,c,f]
Family	Centropagidae
	Centropages furcatus Dana 1849[b,c,f], *Centropages orsinii* Giesbrecht 1889[b,c,f], *Centropages tenuiremis* Thompson I.C. and Scott A. 1903[b,c,f], *Centropages calaninus* Dana 1849[b,c,f], *Centropages* sp. Kroyer 1849[b,c,f]
Family	Pontellidae
	Calanopia minor Scott A. 1902[b,c,f], *Calanopia* sp. Dana 1852[b,c,f], *Labidocera acuta* Dana 1849[b,c,f], *Labidocera pectinata* Thompson I.C. and Scott A. 1903[b,c,f], *Labidocera pavo* Giesbrecht 1889[b,c,i,l], *Labidocera* sp. Lubbock 1853[b,c,g], *Pontella spinipes* Giesbrecht 1889[b,c,f], *Pontella danae* Giesbrecht 1889[b,c,f], *Pontella securifer* Brady 1883[b,c,f]
Family	Temoridae
	Temora discaudata Giesbrecht 1889[b,c,f], *Temora turbinata* Dana 1849[b,c,l], *Temora* sp. Baird 1850[c,g]
Family	Tortanidae
	Tortanus forcipatus Giesbrecht 1889[b,c,f]
Family	Calanidae
	Mesocalanus tenuicornis Dana 1849[b,c,f], *Canthocalanus pauper* Giesbrecht 1888[b,c,f], *Nannocalanus minor* Claus 1863[b,c,f]
Family	Paracalanidae
	Acrocalanus gibber Giesbrecht 1888[b,c,f], *Acrocalanus gracilis* Giesbrecht 1888[b,c,f], *Acrocalanus longicornis* Giesbrecht 1888[b,c,f], *Acrocalanus monachus* Giesbrecht 1888[b,c,f], *Acrocalanus* sp. Giesbrecht 1888[b,c,f], *Paracalanus aculeatus* Giesbrecht 1888[b,c,f], *Paracalanus parvus* Claus 1863[b,c,f,j], *Paracalanus crassirostris* Dahl F. 1894[c,i], *Paracalanus* sp. Boeck 1865[b,c,f], *Bestiolina similis* Sewell 1914[c,l]
Family	Eucalanidae
	Eucalanus sp. Dana 1852[b,c,g], *Subeucalanus subcrassus* Giesbrecht 1888[b,c,f], *Subeucalanus monachus* Giesbrecht 1888[b,c,f], *Subeucalanus* sp. Geletin 1976[b,c,f], *Pareucalanus* sp. Geletin 1976[b,c,f]
Family	Pseudodiaptomidae
	Pseudodiaptomus annandalei Sewell 1919[b,c,h–j,l], *Pseudodiaptomus aurivilli* Cleve 1901[b,c,f,j], *Pseudodiaptomus serricaudatus* Scott T. 1894[b,c,f,j], *Pseudodiaptomus binghami* Sewell 1912[c,i,j], *Pseudodiaptomus hickmani* Sewell 1912[c,h–j], *Pseudodiaptomus* sp. Herrick 1884[b,c,g,h,j]
Family	Diaptomidae
	Heliodiaptomus sp. Kiefer 1932[b,f,h], *Diaptomus* sp. Westwood 1836[g,h]
Order	Cyclopoida
Family	Oithonidae
	Oithona attenuata Farran 1913[b,c,f,h,j], *Oithona brevicornis* Giesbrecht 1891[b,c,i,j], *Oithona setigera* Dana 1849[b,c,f], *Oithona similis* Claus 1866[b,c,f,h,j], *Oithona nana* Giesbrecht 1893[c,h–j], *Oithona hebes* Giesbrecht 1891[c,h,j,l], *Oithona* sp. Baird 1843[b,c,g,h,j]
Family	Cyclopidae
	Mesocyclops sp. Sars G.O. 1914[b,g,h], *Thermocyclops* sp. Kiefer 1927[b,f,h], *Microcyclops* sp. Claus 1893[b,g,h], *Cyclops buxtoni* Gurney 1921[h,l], *Cyclops* sp. Müller O.F. 1785[g,h]
Order	Harpacticoida

(continued)

Table 9.1 (continued)

Family	Miraciidae
	Miracia efferata Dana 1849[b,c,f], *Distioculus minor* Scott T. 1894[b,c,f], *Macrosetella gracilis* Dana 1846[b,c,f], *Macrosetella oculata* Sars G.O. 1916[b,c,f]
Family	Ectinosomatidae
	Microsetella rosea Dana 1847[b,c,f], *Microsetella norvegica* Boeck 1865[b,c,f,j]
Family	Peltidiidae
	Clytemnestra scutellata Dana 1847[b,c,f]
Family	Tachidiidae
	Euterpina acutifrons Dana 1847[b,c,f,j]
Family	Longipediidae
	Longipedia weberi Scott A. 1909[b,c,f]
Family	Metidae
	Metis jousseaumei Richard 1892[b,c,f]
Family	Ameiridae
	Nitokra sp. Boeck 1865[g,h,j]
Family	Canuellidae
	Canuella sp. Scott T. and Scott A. 1893[c,g]
Family	Tegastidae
	Parategastes sphaericus Claus 1863[c,i]
Order	Poecilostomatoida
Family	Oncaeidae
	Oncaea conifera Giesbrecht 1891[b,c,f], *Oncaea venusta* Philippi 1843[b,c,l]
	Oncaea sp. Philippi 1843[b,c,f]
Family	Sapphirinidae
	Sapphirina sp. Thompson J. 1829[b,c,f]
Family	Corycaeidae
	Onychocorycaeus agilis Dana 1849[b,c,f,j], *Corycaeus andrewsi* Farran 1911[b,c,f,j], *Onychocorycaeus catus* Dahl F. 1894[b,c,f,j], *Urocorycaeus longistylis* Dana 1849[b,c,f], *Corycaeus speciosus* Dana 1849[b,c,f,j], *Corycaeus danae* Giesbrecht 1891[c,j,l], *Corycaeus* sp. Dana 1845[b,c,f,j], *Farranula concinna* Dana 1849[b,c,f], *Farranula gibbula* Giesbrecht 1891[b,c,f], *Farranula* sp. Wilson C.B. 1932[b,c,f]
Family	Bomolochidae
	Bomolochus sp. Nordmann 1832[b,c,f]
Class	Ostracoda
Subclass	Myodocopa
Order	Halocyprida
Family	Halocyprididae
	Discoconchoecia elegans Sars 1866[b,c,f], *Chonchoecia* sp. Dana 1849[b,c,f]
Order	Myodocopida
Family	Cypridinidae
	Macrocypridina castanea Brady 1897[b,c,f]
Subclass	Podocopa
Order	Podocopida
Family	Cyprididae

(continued)

Table 9.1 (continued)

	Cypris sp. O.F. Müller 1776[b,f,h]
Class	Malacostraca
Subclass	Eumalacostraca
Order	Mysida
Family	Mysidae
	Mesopodopsis orientalis W. Tattersall 1908[b,c,g,i,j]
	Rhopalophthalmus africanus O. Tattersall 1957[c,i]
Order	Amhipoda
Family	Gammaridae
	Gammarus sp. Fabricius 1775[b,c,f,h,j]
Family	Paracalliopiidae
	Paracalliope fluviatilis (Thompson 1879) sensu Chilton 1920[h–j]
Family	Ampeliscidae
	Ampelisca pusilla G.O. Sars 1891[c,i]
Family	Oedicerotidae
	Perioculodes longimanus (Spence Bate and Westwood 1868)[c,i]
Order	Decapoda
Family	Luciferidae
	Belzebub hanseni Nobili 1906[b,c,i,j], *Lucifer* sp. J.V. Thompson 1829[c,g,j,l]
Order	Isopoda
Family	Ligiidae
	Ligia exotica Roux 1828[c,i,j]
Family	Anthuridae
	Apanthura sandalensis Stebbing 1900[c,i]
Family	Leptanthuridae
	Accalathura borradailei Stebbing 1904[c,i]
Order	Cumacea
Family	Diastylidae
	Paradiastylis culicoides Kemp 1916[c,i,j]
Family	Bodotriidae
	Iphinoe sanguinea Kemp 1916[c,i]
Phylum	Chaetognatha
Class	Sagittoidea
Order	Aphragmophora
Family	Sagittidae
	Flaccisagitta enflata Grassi 1881[b,c,f]
	Sagitta sp. Quoy and Gaimard 1827[b,c,i,l]
Phylum	Chordata
Class	Appendicularia
Order	Copelata
Family	Oikopleuridae
	Oikopleura (Vexillaria) dioica Fol 1872[b,c,f,j], *Oikopleura (Vexillaria)* sp. Lohmann 1933[b,c,f,j]

(continued)

Table 9.1 (continued)

Meroplankton	
	Actinula larvae[b,l], alima larvae of *Squilla*[b,l], bivalve veligers[b,d,g,i,l], brachyuran protozoea larvae[l], brachyuran zoea larvae[b,l], brachyuran megalopa larvae[b,f], brachiopod larvae[b,f], caridean larvae[b,f], cirripede cypris[b,f], cirripede nauplii[b,d], copepod nauplii[b,d,i], cyphonautes larvae[b,f], fish egg[b,g,i], fish larvae[b,g,i,l], gastropod veligers[b,d,g,i], isopod larvae[b,f], larvae of mysids[b,f], ophiopluteus larvae[b,f], penaeid prawn larvae[b,g], polychaete larvae[b,f], protozoea of *Lucifer*[b,l], mysis of *Lucifer*[l], tunicate larvae[i]

[a]Mukherjee et al. (2015)
[b]Present study
[c]m, marine
[d]Sahu et al. (2016)
[e]Mukherjee et al. (2018)
[f]New records from the present study
[g]Patnaik (1973)
[h]f, freshwater
[i]Devasundaram and Roy (1954)
[j]b, brackish
[k]Mukherjee et al. (2014)
[l]Rakhesh et al. (2015)

has a major influence on the species diversity recovered from any survey including proportion of developmental stages present in a sample. Therefore, the data generated in the present study was not directly comparable to earlier studies. Our study documented higher zooplankton diversity due to a systematic monitoring at monthly scale over the period of 5 years which was crucial for recovering the maximum species richness from the Chilika.

3.2 *Holoplankton*

3.2.1 Ciliophora

The ecological roles of planktonic ciliates (20–200 μm) in the pelagic food web of the aquatic environment are well-recognized. They often represent an essential component of the microzooplankton population in several coastal lagoons (Godhantaraman and Uye 2003; Sahu et al. 2016). They also act as a trophic intermediate from lower trophic level (e.g., pico- and nanoplankton) to higher trophic level (e.g., meso- and macro-carnivores) (Corliss 2002). Furthermore, ciliates are important phytoplankton grazers, nutrient re-mineralizers, and regenerators in coastal systems. In addition, ciliates have been used as bioindicator in evaluating biotic stress and pollution (Xu et al. 2014). Generally, environmental variables such as salinity, temperature, nutrient, food availability, and grazing activities determine the composition, abundance, and distribution of ciliates (Nche-Fambo et al. 2016; Rakshit et al. 2017; Basuri et al. 2020).

There are few studies which have reported planktonic Ciliophora in Chilika (Patnaik 1973; Mukherjee et al. 2015, 2018; Sahu et al. 2016). Ciliophora investigation started with the study of Patnaik (1973) which documented three marine species (i.e., *Codonella* sp., *Tintinnopsis* sp., *Cyttarocylis* sp.) (Table 9.1). Mukherjee et al. (2015) studied the diversity and distribution of Ciliophora and documented 27 species belonging to 8 genera and 5 families. Subsequently, Sahu et al. (2016) carried out a survey on the microzooplankton and provided a detailed taxonomic account of Ciliophora. They have reported 19 species of Ciliophora of which genus *Tintinnopsis* was the major one and consisted of 14 species. Recently, Mukherjee et al. (2018) carried out an investigation on microplankton dynamics with interactive effect of environmental parameters and recorded 15 species. The present study reported a total of 22 species belonging to 5 families, of which, 8 species (*Tintinnopsis mortensenii*, *Tintinnopsis tenuis*, *Tintinnopsis acuminata*, *Tintinnopsis dadayi*, *Favella brevis*, *Favella* sp., *Dictyocysta seshaiyai*, *Luminella* sp.) serve as first reports from the lagoon. Thus, so far 51 species of Ciliophora have been recorded from the lagoon. The predominance of *Tintinnopsis* in the present study could be attributed to their more flexible adaptive strategies (Reynolds 1997). Other adaptive mechanisms which could contribute to the survival of *Tintinnopsis* in estuarine ecosystems could be the production of resting cysts which usually sink down and rest in the sediments (Krinsic 1987). Once the environmental conditions become conducive, excystment and reproduction occur rapidly leading to the proliferation of *Tintinnopsis*.

3.2.2 Foraminifera

Foraminifera (heterotrophic protists) are unicellular organisms with shells or tests. In general, their shells are composed of organic compounds, sand grains, and crystalline calcites. Foraminifera have been used extensively as an effective proxy for evaluation of environmental perturbations in lagoon ecosystems such as Santa Gilla lagoon (Cagliari, Italy) (Frontalini et al. 2009). The distribution and diversity of foraminifers is usually controlled by environmental parameters, especially salinity, DO, sediment texture, and organic carbon across different marine environments (Murray 2006).

In Chilika, among the two forms (planktonic and benthic) of Foraminifera, benthic foraminifers have been studied extensively (Sen and Bhadury 2016; Gupta et al. 2019). However, the study of Devasundaram and Roy (1954) was the first report of benthic Foraminifera in zooplankton and documented *Elphidium* sp. as a sole member of the community. In the present study, ten benthic (*Ammonia* sp., *Bolivina* sp., *Discorbis* sp., *Nonionella* sp., *Flabellammina* sp., *Textularia* sp., *Spiroloculina* sp., *Quinqueloculina* sp., *Triloculina* sp., *Ammodiscus* sp.) and two planktonic (*Globigerina bulloides*, *Globigerina* sp.) foraminifers have been identified (Table 9.1). The observation of marine planktonic foraminifers in the present study could be due to tidal influx from BoB into the lagoon (Barik et al. 2019).

3.2.3 Tubulinea

Tubulinea (Amoebozoa) commonly termed as testate amoebae are unicellular protists that are partially enclosed in a simple test (shell). They have a wide distribution in estuaries, lakes, rivers, and wetlands as planktonic or benthic forms (Felipe Machado Velho et al. 2000; Qin et al. 2013). Testate amoebae species respond quickly to changes in environmental conditions due to their short generation time.

In context to Indian estuarine ecosystems, there are only few studies which reported Tubulinea in the zooplankton communities (Saraswathi and Sumithra 2016; Kumari et al. 2017). In Chilika, this particular group is understudied, and a single species of Tubulinea represented by *Difflugia* sp. has been reported earlier (Devasundaram and Roy 1954). The present study documented a total of five species of Tubulinea, of which four (*Arcella discoides*, *Arcella* sp., *Centropyxis* sp., *Difflugia corona*) were the first records from Chilika (Table 9.1). Of these, *Difflugia* and *Arcella* are known as indicators of water pollution (Kumari et al. 2017).

3.2.4 Rotifera

Rotifera are the microscopic metazoans (~50–2000 μm) commonly known as "wheel animalcules." Rotifera possess several characteristic features such as an apical field, a muscular pharynx, and a syncytial body wall. Rotifera may be truly planktonic, benthic, or periphytic. Rotifera are found in a broad salinity regime ranging from freshwater to estuarine and marine. However, they are mostly abundant in the freshwater environment with limited occurrences in the marine environment (Sharma and Naik 1996). Rotifera are abundant in aquatic ecosystems due to their rapid reproductive rates among the metazoans (Herzig 1983). Rotifera are herbivores and efficiently feed on algae, bacteria, and flagellates. Rotifera also act as bioindicator in the ecotoxicological studies, eutrophy, and pollution monitoring (Edmondson and Litt 1982; Abdel-Aziz et al. 2011). The distribution and composition of Rotifera depend on the variability of salinity, temperature, turbidity, and chlorophyll (Azemar et al. 2010; Ezz et al. 2014).

Patnaik (1973) initially documented three genera of Rotifera (*Brachionus*, *Filinia*, and *Keratella*) from Chilika (Table 9.1). Their study revealed that rotifers were largely abundant in the NS and CS zones. Later, Mukherjee et al. (2014) investigated Rotifera (distribution, abundance, and diversity) and documented 23 species during 2012–2013. Mukherjee et al. (2014) have also demonstrated that environmental variables such as salinity, transparency, silicate, and total hardness were the important drivers controlling the Rotifera distribution in the lagoon. Sahu et al. (2016) listed 13 species of Rotifera, of which, six species (*Polyarthra* sp., *Trichocerca* sp., *Brachionus rubens*, *Kellicottia longispina*, *Asplanchna* sp., *Lepadella* sp.) were new records. A survey conducted between 2012 and 2015 on the microplankton dynamics reported ten species of Rotifera (Mukherjee et al. 2018). Their study also showed that distribution of *Brachionus bidentata*, *Lecane batilifer*, *Monostyla bulla*, and *Monostyla luna* was controlled by nitrate and transparency, while salinity played a crucial role in regulating the distribution of *Lecane styrax*. The distribution of *Hexarthra* sp., *Lecane inopinata*, *Filinia* sp., and *Brachionus falcatus* was controlled by the variation of free CO_2. Our study reported a total of 17 Rotifera species of which 7 species (*Anuraeopsis fissa*, *Anuraeopsis* sp., *Brachionus dichotomus reductus*, *Keratella tecta*, *Asplanchna brightwellii*, *Dicranophorus* sp., *Polyarthra vulgaris*) were the first records from Chilika (Table 9.1). *Brachionus* and *Keratella* are α-ß mesosaprobic genera and are indicative of moderate to high organic pollution in estuarine ecosystems (Sladecek 1983; Tackx et al. 2004). Further, *Brachionus* sp. has been reported as an indicator of sulfide pollution in the Kadinamkulam estuary, Kerala (India) (Nandan and Azis 1994).

3.2.5 Hydrozoa

Hydrozoa exist as either single or colonial form in different life stages such as polypoid, medusoid, or both. In Chilika, only one species (*Obelia* sp.) has been recorded for the first time by our study which highlighted the need for a comprehensive monitoring to examine the planktonic hydrozoan diversity.

3.2.6 Ctenophora

Ctenophora, commonly known as comb jellies or sea walnut, are composed of soft, fragile, and gelatinous body. Further, bioluminescence is a common feature in most species of ctenophores. They are characterized by rows of cilia arrays, which are utilized for mobility (Pang and Martindale 2008). In general, ctenophores are carnivorous and predate on a diverse zooplankton such as copepods, amphipods, annelids, appendicularians, fish eggs, and larvae.

The qualitative and quantitative study of the ctenophores is challenging mainly because of their fragile body (Mianzan 1999). Specific nondestructive sampling methods are highly recommended. Consequently, ctenophores remain understudied worldwide including Chilika. Our study has reported a single species represented by *Pleurobrachia pileus* from the lagoon (Table 9.1, Plate 9.1). Ctenophores are understudied with respect to their detailed understanding on community composition, physiology, faunal interaction, metabolism, and their environmental drivers and need further investigation from the Chilika.

3.2.7 Nematoda

Nematoda are found either as free-living, or embedded in bottom sediments, or associated as parasites to a variety of biota. In general, they are occasionally observed in plankton samples. Further, zooplankton such as medusae, copepods, amphipods, and chaetognaths predate on immature nematodes. They exhibit elongated, transparent, bilaterally symmetrical body structures and lack cilia or flagella. Our study recorded only a single Nematoda species represented by *Belbolla* sp. in plankton samples.

3.2.8 Annelida

Annelida is a broad phylum of segmented worms that are characterized by a body cavity or coelom. They possess setae or chaetae for locomotion. Annelida is subdivided into Oligochaeta and Polychaeta. Polychaeta are often found in planktonic communities, and only three tychoplanktonic polychaetes, viz., *Nereis chilkaensis*, *Neanthes glandicincta*, and *Perinereis marjorii*, have been reported from Chilika (Devasundaram and Roy 1954) (Table 9.1).

3.2.9 Gastropoda

Gastropoda is the largest class of molluscs that encompasses both planktonic and the benthic forms. Only few studies have reported Gastropoda from Chilika, and so far eight tychoplanktonic species have been documented (Devasundaram and Roy 1954). Our study has reported a total of four truly planktonic Gastropoda, viz., *Creseis acicula*, *Heliconoides inflatus*, *Atlanta* sp., and *Janthina* sp., as new records from the lagoon (Table 9.1).

3.2.10 Bivalvia

Bivalvia, the second largest molluscan class, is commonly known as Lamellibranchia or Pelecypoda. Majority of Bivalvia are benthic, either attached to hard structures or buried in the substratum. Devasundaram and Roy (1954) have reported five species of tychoplanktonic bivalves represented by three families such as Mytilidae (*Modiola undulatus* var. *crassicostata*), Veneridae (*Clementia annandalei*, *Meretrix casta*, *Marcia opima*), and Semelidae (*Theora opalina*) from Chilika (Table 9.1).

3.2.11 Cladocera

Cladocerans (water fleas) are small crustaceans and are recognized by a large compound eye. They belong to the class Branchiopoda and occur exclusively in freshwater, although some taxa can also tolerate higher salinity. The survival of cladocerans in estuarine ecosystems depends on their adaptation to the rapid changes in environmental factors (Haridevan et al. 2015). Most of the cladocerans are herbivorous. Conversely, cladocerans also act as food source for copepods, mysids, small fish, and larval and juvenile stages of larger fishes. Cladocerans exhibit both parthenogenetic and gamogenetic reproduction during favorable and unfavorable environmental conditions, respectively (Egloff et al. 1997; Rivier 1998; Achuthankutty et al. 2000). Spatiotemporal variation in Cladocera is mostly controlled by salinity dynamics of the ecosystems. For example, salinity controlled the distribution, population structure, size, and grazing rates of cladocerans in Cochin backwaters (India) (Achuthankutty et al. 2000; Haridevan et al. 2015).

Devasundaram and Roy (1954) and Patnaik (1973) have reported one species each, namely, *Evadne* sp. and *Moina* sp., from Cladocera group. Our study has reported a total of 12 species (*Pseudevadne tergestina*, *Evadne nordmanni*, *Penilia avirostris*, *Diaphanosoma excisum*, *Diaphanosoma* sp., *Chydorus sphaericus*, *Chydorus* sp., *Alona* sp., *Bosminopsis deitersi*, *Macrothrix* sp., *Moina micrura*, *Moina* sp.) belonging to 3 orders (Onychopoda, Ctenopoda, and Anomopoda) (Table 9.1). Thus, Chilika remains an understudied system with respect to the cladoceran ecology despite their crucial role in fish diets.

3.2.12 Copepoda

Copepods (phylum, Arthropoda; class, Hexanauplia; subclass, Copepoda) are small crustaceans that are highly diverse and biologically important zooplankton group in all aquatic ecosystems. Copepoda is composed of a total ten orders of which Calanoida, Cyclopoida, Harpacticoida, and Poecilostomatoida are dominant ones. To date, ~12,000 copepod species have been identified (Bron et al. 2011). Copepods live either as free-living (pelagic or benthic) or parasitic lifestyle. Copepoda community structure is regulated by both abiotic and biotic environmental variables in estuarine and lagoon ecosystems (Dalal and Goswami 2001; Antony et al. 2020).

To date, 95 Copepoda taxa have been recorded from Chilika that include 55 Calanoida, 12 Cyclopoida, 13 Harpacticoida, and 15 Poecilostomatoida (Table 9.1). Devasundaram and Roy (1954) investigated copepod between 1950 and 1951 at few stations (Balugaon, Kalupadaghat, Rambha, Satpara, and Arkhakuda) and documented 12 species of copepods. Later, a survey during 2004–2005 on mesozooplankton focused on small-sized copepods' dynamics and recorded ten taxa (Rakhesh et al. 2015). In contrast to previous studies, the diversity of species obtained in our survey was relatively higher. Copepoda population in our study was comprised of 80 species representing marine, brackish, and freshwater forms. These assemblages were categorized into four orders: Calanoida (47 species), Cyclopoida (8 species), Harpacticoida (10 species), and Poecilostomatoida (15 species). The dominance of Calanoida could be related to their continuous breeding, rapid larval development, and adaptation to a wide range of environmental conditions (Ramaiah and Nair 1997).

3.2.13 Ostracoda

Ecologically, ostracods can be considered as both zooplankton and benthos. Ostracoda are small crustaceans that are easily distinguished by bivalve carapace. Planktonic ostracods are opportunistic feeders and primarily feed on detritus.

They are also considered as potential indicators of climate change (Lord et al. 2012). Our study has reported a total of four taxa of Ostracoda as new records from the lagoon (Table 9.1). Among the reported species, *Discoconchoecia elegans*, *Chonchoecia* sp., and *Macrocypridina castanea* were representative of marine forms, while *Cypris* sp. was representative of freshwater forms. However, distributional and ecological studies on ostracods have not been conducted so far from Chilika.

3.2.14 Malacostraca

Malacostraca is the largest class within the phylum arthropod that has characteristics of four body regions, i.e., head, pereon, pleon, and urosome. Based on the available literature as well as our study, a total of 13 Malacostraca taxa belonging to 5 orders (Mysida, Amphipoda, Decapoda, Isopoda, Cumacea) have been reported from Chilika. Devasundaram and Roy (1954) have recorded ten tychoplanktonic/benthic Malacostraca and one planktonic Malacostraca. Later, Patnaik (1973) and Rakhesh et al. (2015) have documented two and one Malacostraca species, respectively. Our study has reported three species (*Gammarus* sp., *Belzebub hanseni*, and *Mesopodopsis orientalis*) of Malacostraca (Table 9.1).

3.2.15 Chaetognatha

Chaetognaths (also known as arrow worm) have a tubular elongated transparent body and are commonly present in marine, estuarine, and coastal lagoon habitats. Most of the chaetognaths are pelagic but few benthic species also exist. They are active predators and capture their prey with rigid hooks (Casanova 1999). In Chilika, only two species of Chaetognatha have been reported (Table 9.1). Devasundaram and Roy (1954) and Rakhesh et al. (2015) have reported the occurrence of only one species represented by *Sagitta* sp. The present study has reported two species, namely, *Flaccisagitta enflata* and *Sagitta* sp., from the lagoon.

3.2.16 Chordata

Planktonic chordates are represented mostly by two main classes, namely, Thaliacea and Appendicularia. Thaliacea include three main groups: dolioloids, salps, and pyrosomes. Appendicularia (also known as Larvacea) include three groups: Oikopleuridae (the most studied appendicularians), Fritillariidae, and Kowalewskiidae. In Chilika, earlier studies have not reported planktonic chordates. Our study has reported two species represented by *Oikopleura dioica* and *Oikopleura* sp. as new records from the lagoon (Table 9.1).

3.3 *Meroplankton*

Meroplankton (or temporary plankton) are mainly composed of the larval stages of benthic, littoral, and nektonic organisms and are crucial for the recruitment of new individuals in the benthic community (Mileikovsky 1971). These larvae are classified as long-life planktotrophic (their duration in the plankton phase can vary from 1 week to 3 months), short-life planktotrophic (vary from 1 week or less), and lecithotrophic (a large yolk providing energy until metamorphosis) (Thorson 1946; Grahame and Branch 1985). In addition, these larvae might be either feeding or nonfeeding. They also serve as a necessary feedstuff for larger zooplankton and fishes (Maksimenkov 1982; Pennington et al. 1986). Meroplankton has been observed as a substantial part of the zooplankton community in many coastal lagoons (Miron et al. 2014; Ziadi et al. 2015). The abiotic factors and food availability have been shown to determine the distribution of meroplankton in the lagoon ecosystems (Santangelo et al. 2007; Ziadi et al. 2015). For instance, higher abundances of meroplankton associated with increased salinity have been reported in Imboassica Lagoon (southeastern Brazil) (Santangelo et al. 2007). In another study, peak abundances of barnacle larvae were found associated with higher phytoplankton density in Ghar El Melh Lagoon (northern Tunisia) (Ziadi et al. 2015).

Devasundaram and Roy (1954) documented six types of meroplankton (i.e., copepod nauplii, bivalve veligers, gastropod veligers, tunicate larvae, fish egg, fish larvae). Later Patnaik (1973) documented five types of meroplankton, among which penaeid prawn larvae were included in existing meroplankton list of Chilika. Rakhesh et al. (2015) reported seven types of meroplankton, of which five forms (protozoea of *Lucifer*, mysis of *Lucifer*, brachyuran protozoea, brachyuran zoea, alima larvae) were new reports. Sahu et al. (2016) recorded two molluscan larvae, i.e., bivalve veliger and gastropod veliger. However, our study has reported a total of 20 types of larval plankton, among which 9 forms were new records. To date, 23 types of meroplankton have been recorded in Chilika (Table 9.1).

3.4 *Microzooplankton Abundances and Community Composition*

The abundances of microzooplankton were significantly higher during monsoon (average 755 ind. l^{-1}) compared to post-monsoon (average 250 ind. l^{-1}) and pre-monsoon (average 347 ind. l^{-1}) (Fig. 9.2). At a spatial scale, the highest and lowest abundances were encountered from SS (average 614 ind. l^{-1}) and NS (average 147 ind. l^{-1}), respectively. This was consistent with earlier studies which have reported maximum microzooplankton abundances during monsoon, whereas minimum abundances were noted from freshwater NS region (Sahu et al. 2016). In general, microzooplankton standing stock is determined by salinity and phytoplankton biomass (Godhantaraman 2001; Jyothibabu et al. 2006). In the present study, higher

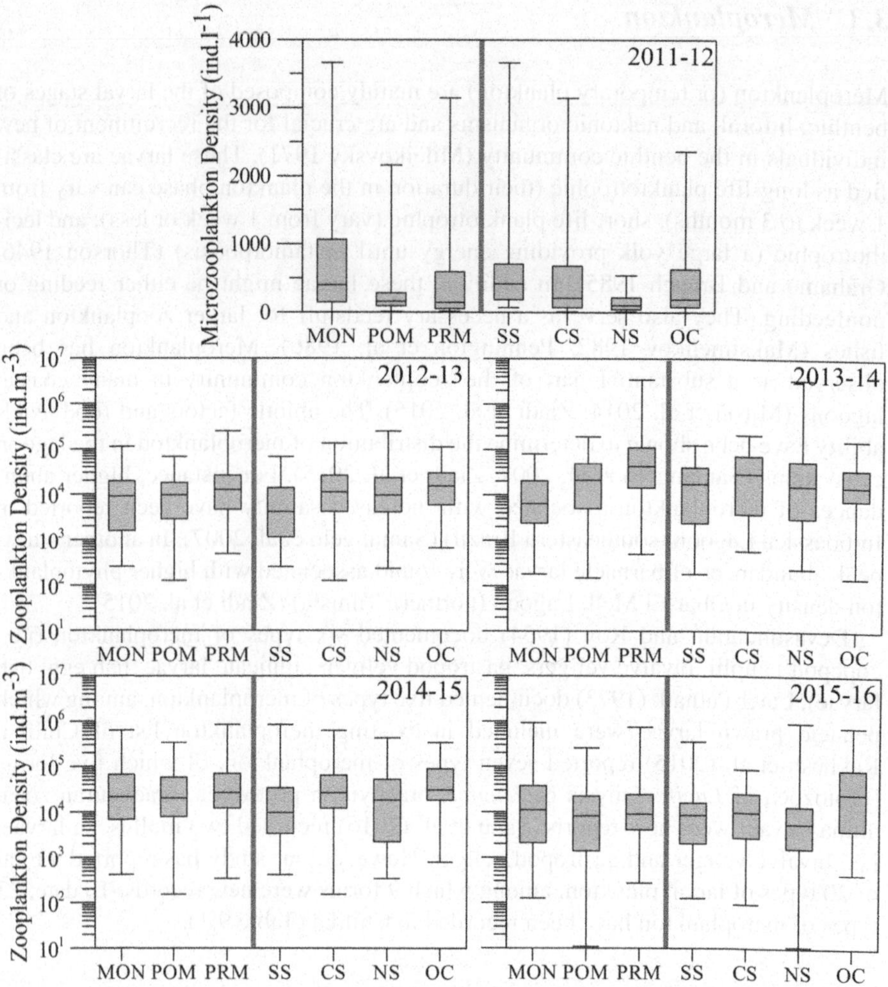

Fig. 9.2 Seasonal (*MON* monsoon, *POM* post-monsoon, *PRM* pre-monsoon) and spatial variability in microzooplankton and zooplankton density during study period. The central bar represents the median. The box represents interval between the 25% and 75% percentiles. The whisker indicates the range

abundances of microzooplankton during monsoon could be due to higher phytoplankton biomass. It has been shown that microzooplankton could consume about 43% of total phytoplankton biomass per day in Cochin backwaters (India) (Jyothibabu et al. 2006). Therefore, one of the explanations for greater microzooplankton abundances during the monsoon period might be the availability of higher phytoplankton biomass (Srichandan et al. 2015a).

The microzooplankton community was composed of Ciliophora, Foraminifera, Rotifera, copepod nauplii, cirripede nauplii, gastropod veliger, and bivalve veliger.

Ciliophora (annual average 63%) were the most abundant microzooplankton, followed by copepod nauplii (30%), Rotifera (4%), and others (3%). Similar dominance of Ciliophora among different groups of microzooplankton has been reported from many Indian estuarine ecosystems (Rakshit et al. 2014; Sooria et al. 2015). A large seasonal variation in Ciliophora (i.e., tintinnid) abundances was observed with higher abundance (average 520 ind. l^{-1}) during monsoon followed by pre-monsoon (average 226 ind. l^{-1}) and post-monsoon (average 123 ind. l^{-1}) seasons (Fig. 9.3). The abundances of Ciliophora observed during the present study were fairly high or low in comparison to the earlier studies from other Indian estuarine ecosystems including Chilika. For instance, earlier studies have reported 48–55 ind. l^{-1} from Chilika (Sahu et al. 2016), 1–17 ind. l^{-1} from Bahuda estuary (Mishra and Panigrahy 1999), 409–3817 ind. l^{-1} from Cochin backwaters (Jyothibabu et al. 2006), 2–420 ind. l^{-1} from Parangipettai estuarine and mangrove waters (Godhantaraman 2002), and 52–1995 ind. l^{-1} from Hooghly estuary (Rakshit et al. 2014, 2017; Rakshit and Sarkar 2016). In general, higher Ciliophora abundance during pre-monsoon season is a common feature in Indian estuarine ecosystems (Godhantaraman 2002; Madhu et al. 2007; Anjusha et al. 2018). In contrast, the maximum abundances of Ciliophora found in Chilika during monsoon could be due to the elevated water temperature and phytoplankton biomass (Srichandan et al. 2015a). Literature suggests that abundance, distribution, and ecology of Ciliophora are primarily governed by food availability (bottom-up control) and predator abundances (top-down control), competitor abundances (e.g., rotifers), temperature, and salinity (Godhantaraman 2002; Biswas et al. 2013; Gauns et al. 2015). Thus, the influence of phytoplankton and

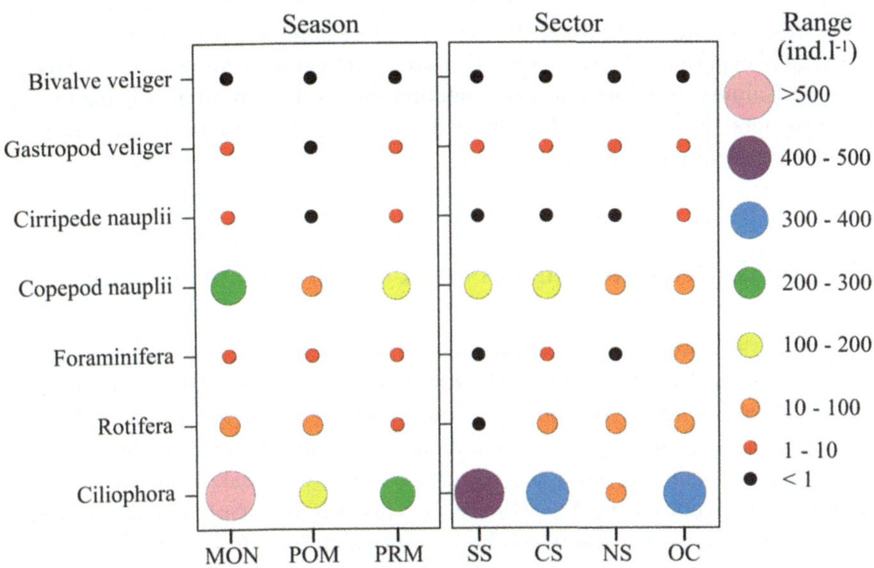

Fig. 9.3 Bubble plot showing seasonal and sectoral variability in microzooplankton communities

temperature in controlling the Ciliophora distribution during the monsoon season seems to be more crucial than other environmental variables.

The distribution of copepod nauplii closely followed the same trend as of Ciliophora with their highest and lowest abundances during monsoon (average 207 ind. l^{-1}) and post-monsoon (average 93 ind. l^{-1}), respectively (Fig. 9.3). Spatially, the highest copepod nauplii abundances were observed in CS (average 198 ind. l^{-1}) followed by SS (average 177 ind. l^{-1}), OC (average 88 ind. l^{-1}), and NS (average 64 ind. l^{-1}). The reason for the large contribution of copepod nauplii to the total microzooplankton might be due to the presence of older stage copepods (copepodites and adults) in higher abundances (maximum up to 571 ind. l^{-1}) in Chilika. Similar large proportion of copepod nauplii in total microzooplankton population has been observed in a brackish water lagoon of Japan (Godhantaraman and Uye 2003).

Rotifera responds quickly to the favorable environmental conditions by parthenogenetic reproduction. In contrast, population size of Rotifera often decline immediately under adverse environmental conditions (Sanders 1987). In Chilika, contribution of Rotifera was lesser in comparison to Ciliophora and copepod nauplii. Rotifera population exhibited a wide range of seasonal fluctuation from 2 (pre-monsoon) to 30 (monsoon) ind. l^{-1} (Fig. 9.3). A clear spatial pattern was also evident in the distribution of Rotifera. The highest abundance of Rotifera was found in the low saline upper reaches (NS) of Chilika, whereas they were completely absent in SS which has higher stable salinity regime. Similar dominance of Rotifera has been recorded in the upper estuarine region (oligohaline to limnetic conditions) of Cochin backwaters (India) (Anjusha et al. 2018). In OC of Chilika, a sharp drop in the salinity occurs during the monsoon months of September and October when there is unidirectional flow of water from lagoon to sea. The drop in salinity of OC could have allowed the appearance of rotifers community in monsoon, although this sector is in close proximity to the BoB. In CS, rotifers appeared particularly at station CS3 which experienced lower salinity during monsoon (salinity 5.8) and post-monsoon (salinity 5). Other microzooplankton such as Foraminifera, cirripede nauplii, gastropod veliger, and bivalve veliger showed a minor contribution at spatiotemporal scales in the lagoon.

3.5 *Zooplankton Abundances and Community Composition*

A significant variability in zooplankton density between different sectors, seasons, and years was evident in this study. The zooplankton abundances were substantially higher during Y–2 (65×10^3 ind. m^{-3}) followed by Y–3 (62×10^3 ind. m^{-3}), Y–4 (38×10^3 ind. m^{-3}), and Y–1 (19×10^3 ind. m^{-3}). The annual variability in zooplankton abundances followed unimodal seasonal pattern with peak abundances during pre-monsoon except during Y–4 (Fig. 9.2). This was in corroboration with other studies from Indian estuaries, which have observed maximum zooplankton density during pre-monsoon (Madhu et al. 2007; Bhattacharya et al. 2015). The reason for the higher zooplankton abundances during pre-monsoon could be attributed to

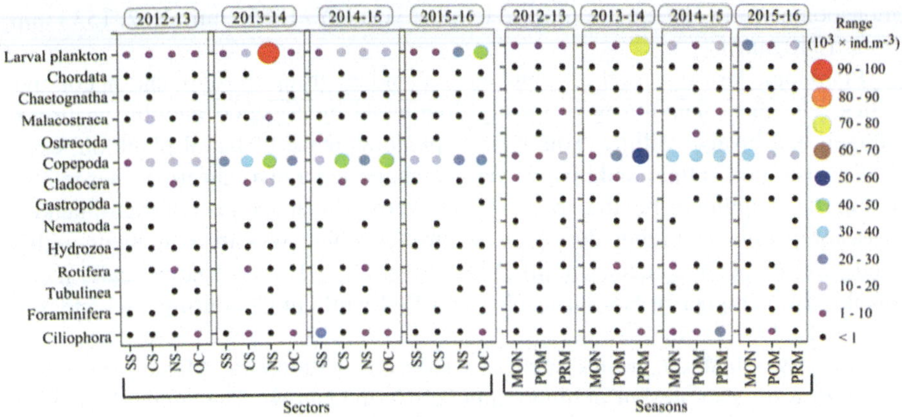

Fig. 9.4 Bubble plot showing seasonal and sectoral variability in zooplankton communities during study years

increased salinity supporting intrusion of marine zooplankton into the lagoon (Madhu et al. 2007). In addition, increased salinity during pre-monsoon could result recruitment of zooplankton population in the lagoon due to rapid multiplication (Venkataramana et al. 2017). The reason for the lower abundances of zooplankton during monsoon might be due to unidirectional flow of water from lagoon to sea resulting concurrent flushing of zooplankton. Similar lower zooplankton abundances during monsoon due to high flushing rate have been observed from Cochin backwaters (India) (Madhupratap 1987; Sooria et al. 2015).

Zooplankton communities in Chilika were distributed into 15 diverse categories, namely, Ciliophora, Foraminifera, Rotifera, Tubulinea, Hydrozoa, Ctenophora, Gastropoda, Cladocera, Copepoda, Ostracoda, Malacostraca, Chaetognatha, Chordata, Nematoda, and planktonic larvae. Copepoda constituted the most dominant zooplankton group irrespective of seasons, sectors, and study year which was in accordance with other coastal lagoons (Naik et al. 2008; Etile et al. 2009; Miron et al. 2014; Rakhesh et al. 2015; Ziadi et al. 2015; Antony et al. 2020). For instance, 81% of copepods' contribution to total zooplankton has been noted in Grand-Lahou lagoon (West Africa) (Etile et al. 2009). In general, increase in salinity is believed to be an important factor for raising the copepod abundances during pre-monsoon season (Vineetha et al. 2015). Copepoda abundances during Y–1 and Y–2 had similar seasonal patterns with higher abundances during pre-monsoon (Fig. 9.4). During Y–3, copepod abundances showed different pattern with much higher abundances during post-monsoon (average 36×10^3 ind. m^{-3}) than pre-monsoon (average 33×10^3 ind. m^{-3}) and monsoon (average 34×10^3 ind. m^{-3}). However, during Y–4, copepod abundances during the monsoon (average 32×10^3 ind. m^{-3}) were prominently higher than post-monsoon (average 12×10^3 ind. m^{-3}) and pre-monsoon (average 13×10^3 ind. m^{-3}). These contrasting response of copepods could be attributed to an increase in salinity (average 13.6) due to relatively lower rainfall during monsoon of Y–4 (710 mm) compared to other years (Y–1, 855 mm; Y–2, 1533 mm; Y–3, 1340 mm).

Planktonic larvae were the second most abundant group in zooplankton communities. The meroplankton were mostly dominated by copepod nauplii, gastropod veliger, and bivalve veliger. This type of preponderance of larval plankton, especially gastropod veliger and bivalve veliger, suggested a pivotal role of meroplankton in the coupling of benthic–pelagic food webs. The abundance of meroplankton was comparatively higher during pre-monsoon which was in agreement with a study from Cochin estuary (India) (Vineetha et al. 2015). At spatial scale, meroplankton was higher in NS during Y–1 and Y–2 while in OC during Y–3 and Y–4 (Fig. 9.4).

Other zooplankton groups such as Cladocera, Ciliophora, Malacostraca, and Rotifera were also present in higher numbers in the lagoon. The annual variability in Cladocera and Rotifera followed an unimodal pattern with peak abundances during monsoon except for Y–2 (Fig. 9.4). In Y–2, maximum abundances of Cladocera and Rotifera were noticed during pre-monsoon and post-monsoon seasons, respectively. The reason for this unusual condition could be attributed to the reduction in salinity in the aftermath of cyclone *Phailin* (October 2013). The low salinity values recorded in CS (salinity 9; station CS4; February 2014) and NS (salinity 1; station NS1; March 2014) favored the development of a large number of oligohaline Rotifera and Cladocera. Furthermore, due to heavy rainfall and land runoff during *Phailin*, a copious amount of freshwater entered into Chilika which reduced the salinity of the lagoon, drastically (Srichandan et al. 2015b). Eventually, Cyanophyta became the most abundant group in CS as well as NS throughout Y–2, which may have favored the growth of Cladocera and Rotifera (Mukherjee et al. 2018). The freshwater brings large organic matter including bacterial load, which may serve as a good source of food for cladocerans (Venkataramana et al. 2017). Spatially, higher abundances of Rotifera and Cladocera were registered in NS and CS, while they were almost absent in SS over the study period (Fig. 9.4). Distribution of Malacostraca showed unimodality with peak abundances during pre-monsoon except for Y–4. Spatially, Malacostraca were comparatively higher in CS and NS as compared to SS and OC over the study period (Fig. 9.4).

3.6 Hydrography and Phytoplankton

Chilika is characterized by a large seasonal and spatial variability in physicochemical factors attributed to the reversing tropical monsoon (southwest monsoon and northeast monsoon). Over the study period, a clear seasonal pattern of rainfall was observed, with the highest during southwest monsoon. Salinity was lowest during monsoon and highest during pre-monsoon over the study period. Annual mean salinity in Y–4 (16) was significantly higher than in Y–1 (13), Y–2 (10), and Y–3 (9). The pH remained mostly alkaline (annual average 7.8–8.4) which could be due to extensive buffering capacity of seawater causing the change of pH within a very narrow limit (Srichandan et al. 2015a). The overall observed DO showed marked variation ranging from 3.87 to 14.0 mg l^{-1}. The overall NO_3^-, PO_4^{3-}, and SiO_4^{4-} concentrations were recorded in the range of 0.0–35.2, 0.01–4.0, and 0.0–258.9 μmol l^{-1}, respectively. A distinct spatiotemporal heterogeneity in distribution of nutrients was observed over the study period. The overall trend in distribution of NO_3^- showed higher values during pre-monsoon, which could be ascribed to the higher residence time during pre-monsoon (325 days) than monsoon (56 days) (Muduli et al. 2013). SiO_4^{4-} was highest during monsoon, which was linked to the increased river influx containing soil and silt particles (Srichandan et al. 2015a). Over the study period, phytoplankton density varied in between 54 and 464,160 cells l^{-1} with significant spatiotemporal variations. In this study, seven phytoplankton classes, Bacillariophyta, Chlorophyta, Euglenophyta, Dinophyta, Cyanophyta, Chrysophyta, and Haptophyta, were identified.

3.7 *Environmental Drivers of Microzooplankton and Zooplankton Communities*

CCA biplots showed that salinity was the key driver controlling the microzooplankton components especially Rotifera. A negative correlation was observed between the freshwater zooplankton group Rotifera and salinity ($r = -0.330$, p-value <0.01) which was consistent with several estuarine ecosystems including Chilika (Park and Marshall 2000; Anjusha et al. 2018; Mukherjee et al. 2018). The abundances of Ciliophora and Dinophyta were positively correlated ($r = 0.322$, p-value <0.01) which was in accordance with a study from Hooghly River estuary (India) (Rakshit et al. 2014) (Fig. 9.5). In addition, Ciliophora exhibited a negative correlation with NO_3^-, PO_4^{3-}, and SiO_4^{4-}. Apart from environmental variables, Ciliophora also showed a negative correlation with Rotifera which corroborated with earlier reports from Rhode River estuary of Chesapeake Bay (Dolan and Gallegos 1992). The negative relationship could be due to competition between Ciliophora and Rotifera for their preferred foods such as bacterioplankton (Buikema et al. 1978).

CCA further showed the influence of environmental variables on the zooplankton community composition. Salinity showed a positive correlation with Copepoda which agreed with other studies from estuarine systems (Miron et al. 2014; Bhattacharya et al. 2015; Vineetha et al. 2015). Generally, any monodiet of Bacillariophyta or Dinophyta is nutritionally inadequate for the growth and reproduction of copepods (Jones and Flynn 2005). CCA showed that Copepoda were mostly associated with both Bacillariophyta and Dinophyta which often are considered the most abundant food for copepods (Liu et al. 2010) (Fig. 9.5). Further, Dinophyta are important food material for copepods due to their higher volume-specific organic content (Kleppel 1993). It has been shown that copepods on a Dinophyta diet increase their egg production and survival rates (Shin et al. 2003; Sushchik et al. 2004).

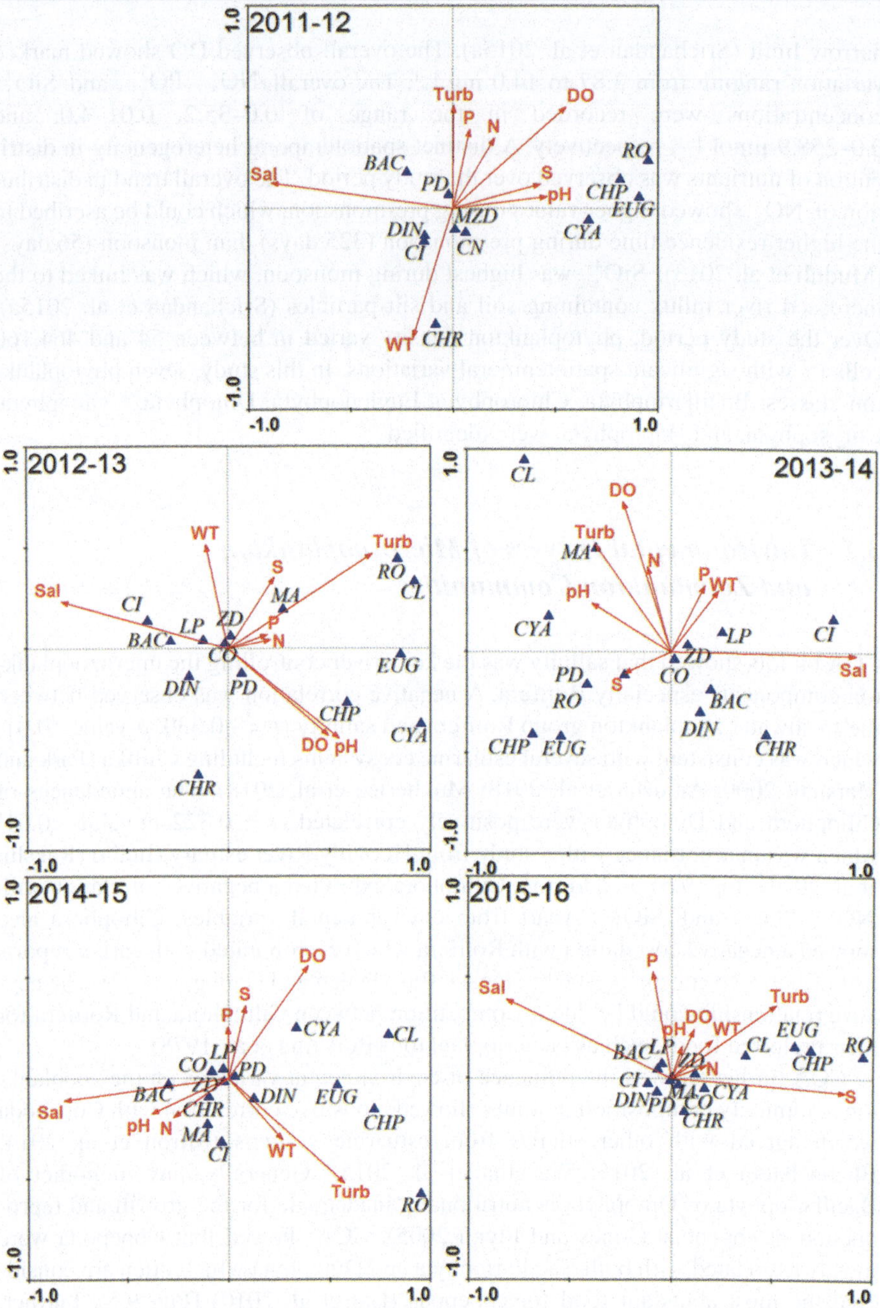

Fig. 9.5 CCA biplots of biological (dominant microzooplankton, zooplankton, and phytoplankton groups) and environmental variables. *WT* water temperature, *DO* dissolved oxygen, *Sal* salinity, *Turb* turbidity, *N* nitrate, *P* phosphate, *S* silicate, *PD* phytoplankton density, *MZD* microzooplankton density, *ZD* zooplankton density, *CI* Ciliophora, *RO* Rotifera, *CN* copepod nauplii, *CL* Cladocera, *CO* Copepoda, *MA* Malacostraca, *LP* larval plankton, *BAC* Bacillariophyta, *DIN* Dinophyta, *CYA* Cyanophyta, *CHP* Chlorophyta, *EUG* Euglenophyta, *CHR* Chrysophyta

In Chilika, multiple environmental variables influenced the distribution and abundances of rotifers. For instance, CCA plot showed a significant positive correlation of rotifers with turbidity, NO_3^-, PO_4^{3-}, and SiO_4^{4-} during Y–1. However, during Y–2 (*Phailin* cyclone year), SiO_4^{4-}, phytoplankton abundances, Chlorophyta, and Euglenophyta showed a positive correlation with rotifers (Fig. 9.5). In addition, salinity was negatively correlated with rotifers during Y–2. During Y–3 (*Hudhud* cyclone year), both correlation matrix and CCA analyses showed that water temperature, turbidity, and Chlorophyta were the key drivers of rotifers distribution. During Y–4, rotifers were positively correlated with several biotic (Chlorophyta, Euglenophyta) and abiotic (water temperature, turbidity, dissolved oxygen, pH, NO_3^-) factors. These abiotic and biotic factors have been shown to control the rotifer community structures in many estuarine ecosystems (Gopakumar and Jayaprakash 2003; Azemar et al. 2010; Varghese and Krishnan 2011; Garcia and Bonel 2014; Wei and Xu 2014; Mukherjee et al. 2018). For example, salinity, SiO_4^{4-}, and phytoplankton biomass were the main controlling factors of the rotifer community in Schelde estuary (Belgium) (Azemar et al. 2010). In another study, turbidity and PO_4^{3-} were the main factors determining the rotifers communities in Cochin backwaters (India) (Varghese and Krishnan 2011). Literature also suggests that rotifers are adapted to thrive under high turbidity as the adverse consequences of competition and predation are partly reduced due to low visibility (Thorp and Mantovani 2005).

In Chilika, cladocerans showed a positive relationship with turbidity in most of the study years which could be attributed to their sensitivity to visual predation (Pangle and Peacor 2009). Both CCA and correlation matrix showed a significant positive correlation of cladocerans with Cyanophyta during Y–2 and Y–3, whereas during Y–4 it was positively correlated with Euglenophyta (Fig. 9.5). The positive relationship between Cladocera and Euglenophyta suggested that the latter could be a good food source for Cladocera (Kawecka and Eloranta 1994). It has been shown that cladocerans graze on colonial or filamentous Cyanophyta (Ka et al. 2012; Tonno et al. 2016). CCA also showed a negative correlation between Cladocera abundance and salinity during Y–2 and Y–3 signifying prevalence of limnophilic forms. Malacostraca were observed in close association with turbidity which was consistent with a study from Gironde estuary (France) (David et al. 2005).

4 Conclusion

The present study is the first compilation on the diversity, composition, and distribution of zooplankton communities from Chilika. To date, 263 species of holoplankton (51 Ciliophora, 13 Foraminifera, 5 Tubulinea, 42 Rotifera, 1 Hydrozoa, 1 Ctenophora, 1 Nematoda, 3 Polychaeta, 12 Gastropoda, 5 Bivalvia, 13 Cladocera, 95 Copepoda, 4 Ostracoda, 13 Malacostraca, 2 Chaetognatha, 2 Chordata) and 23 types of meroplankton have been documented. The present study documented a total of 186 zooplankton taxa, of which 131 were first records from the lagoon. A strong spatial–seasonal variation was evidenced in the zooplankton community which was attributed to the variability in biotic and abiotic variables. A clear seasonal cycle with pre-monsoon maxima was observed in zooplankton abundances over the study period. Copepoda, the most diverse and dominant zooplankton taxon, was represented by calanoids, cyclopoids, harpacticoids, and poecilostomatoids. Other zooplankton groups such as Rotifera, Ciliophora, Cladocera, Malacostraca, and larval plankton also showed higher abundances at spatiotemporal scales. Biotic–abiotic interactions revealed through CCA showed the combined effects of environmental variables and availability of sufficient phytoplankton diet such as Bacillariophyta and Dinophyta as a major factor controlling the composition of Copepoda. CCA also revealed that biotic (Chlorophyta, Euglenophyta) and abiotic variables (water temperature, salinity, turbidity, dissolved oxygen, pH, NO_3^-, SiO_4^{4-}) were the key factors responsible for controlling the distribution of Rotifera. Salinity and availability of food sources played an important role in controlling the abundances, distribution, and diversity of cladocerans. Turbidity played a significant role in controlling the abundance of Malacostraca. This study provided detailed information on the microzooplankton community of Chilika which enhanced our understanding regarding their crucial role in this lagoon. Generally, species diversity and composition is the most recognized facet, but attempts are also essential, specifically with respect to the medusae including jellyfish that are understudied in Chilika. In addition, fine-scale (diurnal and tidal) monitoring is also important to gain deeper insights on the zooplankton ecology. Further, studies on identifying indicator zooplankton taxa may help in discerning the effect of climate change on hydrobiological regimes of the lagoon.

Acknowledgments The study was supported from the funding of World Bank (Credit No. 4765-IN) awarded to Chilika Development Authority (Government of Odisha) under the Integrated Coastal Zone Management Project (ICZMP). Mr. Bibhuti Bhusan Dora, Research Fellow, GIS Cell of CDA, is acknowledged for the preparation of sampling map. The staff of the CDA is acknowledged for their generous help in sampling.

References

Abdel-Aziz NE, Aboul Ezz SM, Abou Zaid MM, Abo-Taleb HA (2011) Temporal and spatial dynamics of rotifers in the Rosetta Estuary, Egypt. Egypt J Aquat Res 37(1):59–70

Achuthankutty CT, Shrivastava Y, Mahambre GG, Goswami SC, Madhupratap M (2000) Parthenogenetic reproduction of Diaphanosoma celebensis (Crustacea: Cladocera): influence of salinity on feeding, survival, growth and neonate production. Mar Biol 137(1):19–22

Altaff K (2004) A manual of zooplankton. University Grants Commission, New Delhi, pp 1–155

Al-Yamani FY, Skryabin V, Gubanova A, Khvorov S, Prusova I (2011) Marine zooplankton practical guide for the Northwestern Arabian Gulf. Kuwait Institute for Scientific Research, Kuwait, p 399. ISBN: 978-99966-95-07-0

Anjusha A, Jyothibabu R, Jagadeesan L, Arunpandi N (2018) Role of rotifers in microzooplankton community in a large monsoonal estuary (Cochin backwaters) along the west coast of India. Environ Monit Assess 190(5):295

Antony S, Ajayan A, Dev VV, Mahadevan H, Kaliraj S, Krishnan KA (2020) Environmental influences on zooplankton diversity in the Kavaratti lagoon and offshore, Lakshadweep Archipelago, India. Reg Stud Mar Sci 37:101330

Arreola-Lizarraga JA, Padilla-Arredondo G, Medina-Galvan J, Méndez-Rodriguez L, Mendoza-Salgado R, Cordoba-Matson MV (2016) Analysis of hydrobiological responses to anthropogenic and natural influences in a lagoon system in the Gulf of California. Oceanol Hydrobiol Stud 45(1):111–120

Azemar F, Maris T, Mialet B, Segers H, Van Damme S, Meire P, Tackx M (2010) Rotifers in the Schelde estuary (Belgium): a test of taxonomic relevance. J Plankton Res 32(7):981–997

Barik SS, Singh RK, Jena PS, Tripathy S, Sharma K, Prusty P (2019) Spatio-temporal variations in ecosystem and CO_2 sequestration in coastal lagoon: a foraminiferal perspective. Mar Micropaleontol 147:43–56

Basuri CK, Pazhaniyappan E, Munnooru K, Chandrasekaran M, Vinjamuri RR, Karri R, Mallavarapu RV (2020) Composition and distribution of planktonic ciliates with indications to water quality in a shallow hypersaline lagoon (Pulicat Lake, India). Environ Sci Pollut Res 27:1–14

Battish SK (1992) Freshwater zooplankton of India. Oxford & IBH Publishing Company, New Delhi

Bhattacharya BD, Hwang JS, Sarkar SK, Rakhsit D, Murugan K, Tseng LC (2015) Community structure of mesozooplankton in coastal waters of Sundarban mangrove wetland, India: a multivariate approach. J Mar Syst 141:112–121

Biswas SN, Godhantaraman N, Rakshit D, Sarkar SK (2013) Community composition, abundance, biomass and production rates of Tintinnids (Ciliata: Protozoa) in the coastal regions of Sundarban Mangrove wetland, India. Indian J Geomarine Sci 42(2):163–173

Bron JE, Frisch D, Goetze E, Johnson SC, Lee CE, Wyngaard GA (2011) Observing copepods through a genomic lens. Front Zool 8:1–15

Buikema AL Jr, Miller JD, Yongue WH Jr (1978) Effects of algae and protozoans on the dynamics of Polyarthra vulgaris. Internationale Vereinigung für theoretische und angewandte Limnologie: Verhandlungen 20(4):2395–2399

Carter JL, Schindler DE, Francis TB (2017) Effects of climate change on zooplankton community interactions in an Alaskan lake. Clim Change Responses 4(1):3

Casanova JP (1999) Chaetognatha. In: Boltovskoy D (ed) South Atlantic zooplankton. Backhuys Publishers, Leiden, pp 1353–1374

Conway DV, White RG, Hugues-Dit-Ciles J, Gallienne CP, Robins DB (2003) Guide to the coastal and surface zooplankton of the south-western Indian Ocean. Occasional Publication of the Marine Biological Association of the United Kingdom, Plymouth, p 15

Corliss JO (2002) Biodiversity and biocomplexity of the protists and an overview of their significant roles in maintenance of our biosphere. Acta Protozool 41(3):199–220

Dalal SG, Goswami SC (2001) Temporal and ephemeral variations in copepod community in the estuaries of Mandovi and Zuari—west coast of India. J Plankton Res 23(1):19–26

Dam HG (2013) Evolutionary adaptation of marine zooplankton to global change. Annu Rev Mar Sci 5:349–370

David V, Sautour B, Chardy P, Leconte M (2005) Long-term changes of the zooplankton variability in a turbid environment: the Gironde estuary (France). Estuar Coast Shelf Sci 64:171–184

Devasundaram MP, Roy JC (1954) A preliminary study of the plankton of the Chilka Lake for the years 1950 & 1951. In: Symposium on marine and fresh-water plankton in the Indo-Pacific, Indo-Pacific Fish. Com., pp 48–54

Dolan JR, Gallegos CC (1992) Trophic role of planktonic rotifers in the Rhode River Estuary, spring-summer 1991. Mar Ecol Prog Ser (Oldendorf) 85(1):187–199

Edmondson WT, Litt AH (1982) Daphnia in lake Washington 1. Limnol Oceanogr 27(2):272–293

Egloff DA, Fofonoff PW, Onbe T (1997) Reproductive biology of marine cladocerans. In: Advances in marine biology, vol 31. Academic Press, London, pp 79–167

Etile RND, Kouassi AM, Aka MNG, Pagano M, N'douba V (2009) Spatio-temporal variations of the zooplankton abundance and composition in a West African tropical coastal lagoon (Grand-Lahou, Cote d'Ivoire). Hydrobiologia 624(1):171–189

Ezz SMA, Aziz NEA, Abou Zaid MM, El Raey M, Abo-Taleb HA (2014) Environmental assessment of El-Mex Bay, Southeastern Mediterranean by using Rotifera as a plankton bio-indicator. Egypt J Aquat Res 40(1):43–57

Felipe Machado Velho L, Amodeo Lansac-Toha F, Costa Bonecker C, Zimmermann-Callegari MC (2000) On the occurrence of testate amoebae (Protozoa, Rhizopoda) in Brazilian inland waters. II. Families Centropyxidae, Trigonopyxidae and Plagiopyxidae. Acta Scientiarum Biol Sci 22:365–374

Frontalini F, Buosi C, Da Pelo S, Coccioni R, Cherchi A, Bucci C (2009) Benthic foraminifera as bio-indicators of trace element pollution in the heavily contaminated Santa Gilla lagoon (Cagliari, Italy). Mar Pollut Bull 58(6):858–877

Ganguly D, Patra S, Muduli PR, Vardhan KV, Abhilash KR, Robin RS, Subramanian BR (2015) Influence of nutrient input on the trophic state of a tropical brackish water lagoon. J Earth Syst Sci 124(5):1005–1017

Gao X, Song J, Li X (2011) Zooplankton spatial and diurnal variations in the Changjiang River estuary before operation of the Three Gorges Dam. Chin J Oceanol Limnol 29(3):591–602

Gao F, Warren A, Zhang Q, Gong J, Miao M, Sun P, Xu D, Huang J, Yi Z, Song W (2016) The all-data-based evolutionary hypothesis of ciliated protists with a revised classification of the phylum Ciliophora (Eukaryota, Alveolata). Sci Rep 6:24874

Garcia MD, Bonel N (2014) Environmental modulation of the plankton community composition and size-structure along the eutrophic intertidal coast of the Rio de la Plata estuary, Argentina. J Limnol 73(3):562–573

Gauns M, Mochemadkar S, Patil S, Pratihary A, Naqvi SWA, Madhupratap M (2015) Seasonal variations in abundance, biomass and grazing rates of microzooplankton in a tropical monsoonal estuary. J Oceanogr 71(4):345–359

Godhantaraman N (2001) Seasonal variations in taxonomic composition, abundance and food web relationship of microzooplankton in estuarine and mangrove waters, Parangipettai region, southeast coast of India. Indian J Mar Sci 30:151–160

Godhantaraman N (2002) Seasonal variations in species composition, abundance, biomass and estimated production rates of tintinnids at tropical estuarine and mangrove waters, Parangipettai, southeast coast of India. J Mar Syst 36:161–171

Godhantaraman N, Uye S (2003) Geographical and seasonal variations in taxonomic composition, abundance and biomass of microzooplankton across a brackish-water lagoonal system of Japan. J Plankton Res 25(5):465–482

Gopakumar G, Jayaprakash V (2003) Community structure and succession of brackish water rotifers in relation to ecological parameters. J Mar Biol Assoc India 45(1):20–30

Gopko M, Telesh IV (2013) Estuarine trophic state assessment: new plankton index based on morphology of Keratella rotifers. Estuar Coast Shelf Sci 130:222–230

Grahame J, Branch GM (1985) Reproductive patterns of marine invertebrates. Oceanogr Mar Biol 23:373–398

Gupta VK, Sen A, Pattnaik AK, Rastogi G, Bhadury P (2019) Long-term monitoring of benthic foraminiferal assemblages from Asia's largest tropical coastal lagoon, Chilika, India. J Mar Biol Assoc U K 99(2):311–330

Gutierrez MF, Tavsanoglu UN, Vidal N, Yu J, Teixeira-de Mello F, Cakiroglu AI, He H, Liu Z, Jeppesen E (2018) Salinity shapes zooplankton communities and functional diversity and has complex effects on size structure in lakes. Hydrobiologia 813(1):237–255

Haridevan G, Jyothibabu R, Arunpandi N, Jagadeesan L, Biju A (2015) Influence of salinity on the life table demography of a rare Cladocera *Latonopsis australis*. Environ Monit Assess 187(10):643

Herzig A (1983) Comparative studies on the relationship between temperature and duration of embryonic development of rotifers. In: Biology of rotifers. Springer, Dordrecht, pp 237–246

Isari S, Psarra S, Pitta P, Mara P, Tomprou MO, Ramfos A, Somarakis S, Tselepides A, Koutsikopoulos C, Fragopoulu N (2007) Differential patterns of mesozooplankters' distribution in relation to physical and biological variables of the northeastern Aegean Sea (eastern Mediterranean). Mar Biol 151(3):1035–1050

Jones RH, Flynn KJ (2005) Nutritional status and diet composition affect the value of diatoms as copepod prey. Science 307(5714):1457–1459

Jyothibabu R, Madhu NV, Jayalakshmi KV, Balachandran KK, Shiyas CA, Martin GD, Nair KKC (2006) Impact of freshwater influx on microzooplankton mediated food web in a tropical estuary (Cochin backwaters–India). Estuar Coast Shelf Sci 69(3–4):505–518

Ka S, Mendoza-Vera JM, Bouvy M, Champalbert G, N'Gom-Ka R, Pagano M (2012) Can tropical freshwater zooplankton graze efficiently on cyanobacteria? Hydrobiologia 679(1):119–138

Kasturirangan LR (1963) A key for the identification of the more common planktonic Copepoda: of Indian coastal waters (no. 2). Council of Scientific & Industrial Research

Kawecka B, Eloranta PV (1994) The outline of algae ecology in freshwater and terrestrial environments. Wyd. Nauk. PWN, Warszawa, p 147

Kleppel GS (1993) On the diets of calanoid copepods. Mar Ecol Prog Ser 99:183–195

Kofoid CA, Campbell AS (1929) A conspectus of the marine and fresh-water Ciliata belonging to the suborder Tintinnoinea, with description of the suborder Tintinnoinea, with description of new species principally from Agassiz expedition to the eastern tropical Pacific, 1904–1905. Univ Calif Publ Zool 34:1–403

Krinsic F (1987) On the ecology of tintinnines (Ciliata—Oligotrichida, Tintinnina) in the open waters of the south Adriatic. Mar Biol 68:83–90

Kumar A, Mishra DR, Equeenuddin SM, Cho HJ, Rastogi G (2016) Differential impact of anniversary-severe cyclones on the water quality of a tropical coastal lagoon. Estuar Coasts 40(2):317–342

Kumari KB, Anbalagan R, Sivakami R (2017) Seasonal protozoan diversity in Agniar Estuary, Tamil Nadu, India. Int J Pure Appl Biosci 5(2):634–637

Liu H, Chen M, Suzuki K, Wong CK, Chen B (2010) Mesozooplankton selective feeding in subtropical coastal waters as revealed by HPLC pigment analysis. Mar Ecol Prog Ser 407:111–123

Lord AR, Boomer I, Brouwers E, Whittaker JE (2012) Ostracod taxa as palaeoclimate indicators in the quaternary. In: Developments in quaternary sciences, vol 17. Elsevier, Amsterdam, pp 37–45

Madhu NV, Jyothibabu R, Balachandran KK, Honey UK, Martin GD, Vijay JG, Shiyas CA, Gupta GVM, Achuthankutty CT (2007) Monsoonal impact on planktonic standing stock and abundance in a tropical estuary (Cochin backwaters–India). Estuar Coast Shelf Sci 73(1–2):54–64

Madhupratap M (1987) Status and strategy zooplankton of tropical Indian estuaries: a review. Bull Plankton Soc Jpn 34:65–81

Maeda M (1986) An illustrated guide to the species of the family Halteridae and Strombilidae (Oligotrichida, Ciliopora), free swimming protozoa common in aquatic environments. Bull Ocean Res Inst Univ Tokyo 19:67

Maksimenkov VV (1982) Correlation of abundance of food (for herring larvae) zooplankton and water temperature in the Korfo-Karaginsk region of Bering Sea [USSR, Clupea pallasi]. Soviet J Mar Biol 8:132–135

Mianzan HW (1999) Ctenophora. In: South Atlantic zooplankton. Backhuys Publishers, Leiden, pp 51–573

Mileikovsky SA (1971) Types of larval development in marine bottom invertebrates, their distribution and ecological significance: a re-evaluation. Mar Biol 10(3):193–213

Miron MIBD, Castellanos-Paez ME, Garza-Mourino G, Ferrara-Guerrero MJ, Pagano M (2014) Spatiotemporal variations of zooplankton community in a shallow tropical brackish lagoon (Sontecomapan, Veracruz, Mexico). Zool Stud 53(1):59

Mishra S, Panigrahy RC (1999) The tintinnids (Protozoa: Ciliata) of the Bahuda estuary, east coast of India. Indian J Mar Sci 28:219–221

Mitsch WJ, Gosselink JG (1993) Wetlands, 2nd edn. Van Nostrand Reinhold Press, New York

Molinero JC, Ibanez F, Nival P, Buecher E, Souissi S (2005) North Atlantic climate and northwestern Mediterranean plankton variability. Limnol Oceanogr 50(4):1213–1220

Muduli PR, Pattnaik AK (2020) Spatio-temporal variation in physicochemical parameters of water in the Chilika lagoon. In: Ecology, conservation, and restoration of Chilika lagoon, India. Springer, Cham, pp 203–229

Muduli PR, Kanuri VV, Robin RS, Kumar BC, Patra S, Raman AV, Rao GN, Subramanian BR (2013) Distribution of dissolved inorganic carbon and net ecosystem production in a tropical brackish water lagoon, India. Cont Shelf Res 64:75–87

Mukherjee M, Banik SK, Suresh VR, Manna RK, Panda D, Sharma AP (2014) Rotifers, their distribution, abundance and seasonal variation in Chilika lagoon. J Inland Fish Soc India 46(1):29–37

Mukherjee M, Banik SK, Pradhan SK, Sharma AP, Suresh VR, Manna RK, Panda D, Roshith CM, Mandal S (2015) Diversity and distribution of Tintinnids in Chilika lagoon with description of new records. Indian J Fish 62(1):25–32

Mukherjee M, Suresh VR, Manna RK (2018) Microplankton dynamics of a coastal lagoon, Chilika: interactive effect of environmental parameters on microplankton groups. Environ Monit Assess 190(11):689

Murray JW (2006) Ecology and applications of benthic foraminifera. Cambridge University Press, Cambridge, 426pp

Naik S, Panigrahy RC, Mohapatra A (2008) Spatio-temporal distribution of zooplankton in Chilka lake—a Ramsar site on the Indian east coast. Indian J Sci Technol 1(3):1–5

Nandan SB, Azis PKA (1994) Zooplankton of the retting zones in the Kadinam-Kulam Estuary, Kerala. Mahasagar 27(1):59–65

Nche-Fambo FA, Tirok K, Scharler UM (2016) Hypersaline conditions cause distinct ciliate community structure in a South African estuarine lake system. J Plankton Res 38(4):878–887

Pang K, Martindale MQ (2008) Comb jellies (Ctenophora): a model for basal metazoan evolution and development. Cold Spring Harbor Protocols 2008(11):pdb-emo106

Pangle KL, Peacor SD (2009) Light-dependent predation by the invertebrate planktivore *Bythotrephes longimanus*. Can J Fish Aquat Sci 66(10):1748–1757

Park GS, Marshall HG (2000) The trophic contributions of rotifers in tidal freshwater and estuarine habitats. Estuar Coast Shelf Sci 51(6):729–742

Patnaik S (1973) Observation on the seasonal fluctuating of plankton in the Chilika lake. Indian J Fish 20:43–45

Pattanaik PK, Sarma ALN (1997) Qualitative and quantitative composition of zooplankton communities of Rambha Bay (Chilika Lake, Bay of Bengal). J Mar Biol Assoc India 39(1–2):49–58

Pennington JT, Rumrill SS, Chia FS (1986) Stage-specific predation upon embryos and larvae of the Pacific sand dollar, Dendraster excentricus, by 11 species of common zooplanktonic predators. Bull Mar Sci 39(2):234–240

Perez-Ruzafa A, Marcos C, Perez-Ruzafa IM, Pérez-Marcos M (2011) Coastal lagoons: "transitional ecosystems" between transitional and coastal waters. J Coast Conserv 15(3):369–392

Qin Y, Fournier B, Lara E, Gu Y, Wang H, Cui Y, Zhang X, Mitchell EAD (2013) Relationships between testate amoeba communities and water quality in Lake Donghu, a large alkaline lake in Wuhan, China. Front Earth Sci 7(2):182–190

Rakhesh M, Madhavirani KSVKS, Kumar BC, Raman AV, Kalavati C, Rao YP, Rosamma S, Rao VR, Gupta GVM, Subramanian BR (2015) Trophic–salinity gradients and environmental redundancy resolve mesozooplankton dynamics in a large tropical coastal lagoon. Reg Stud Mar Sci 1:72–84

Rakshit D, Sarkar SK (2016) Diversity, distribution and polymorphism of loricate ciliate tintinnids along Hooghly estuary, India. J Mar Biol Assoc India 58(2):61–68

Rakshit D, Biswas SN, Sarkar SK, Bhattacharya BD, Godhantaraman N, Satpathy KK (2014) Seasonal variations in species composition, abundance, biomass and production rate of tintinnids (Ciliata: Protozoa) along the Hooghly (Ganges) River Estuary, India: a multivariate approach. Environ Monit Assess 186(5):3063–3078

Rakshit D, Murugan K, Biswas JK, Satpathy KK, Ganesh PS, Sarkar SK (2017) Environmental impact on diversity and distribution of tintinnid (Ciliata: Protozoa) along Hooghly Estuary, India: a multivariate approach. Reg Stud Mar Sci 12:1–10

Ramaiah N, Nair VR (1997) Distribution and abundance of copepods in the pollution gradient zones of Bombay Harbour-Thana creek-Bassein creek, west coast of India. Indian J Mar Sci 26:20–25

Reynolds CS (1997) Vegetation processes in the pelagic: a model for ecosystem theory, vol 9. Ecology Institute, Oldendorf

Rice E, Dam HG, Stewart G (2015) Impact of climate change on estuarine zooplankton: surface water warming in Long Island Sound is associated with changes in copepod size and community structure. Estuar Coasts 38(1):13–23

Rivier IK (1998) The predatory Cladocera (Onychopoda: Podonidae, Polyphemidae, Cercopagidae) and Leptodorida of the world. Guides to the identification of the micro-invertebrates of the continental waters of the world

Sahu BK, Srichandan S, Panigrahy RC (2016) A preliminary study on the microzooplankton of Chilika Lake, a brackish water lagoon on the east coast of India. Environ Monit Assess 188(1):69

Sanders RW (1987) Tintinnids and other microzooplankton—seasonal distributions and relationships to resources and hydrography in a Maine estuary. J Plankton Res 9(1):65–77

Santangelo JM, Rocha ADM, Bozelli RL, Carneiro LS, Esteves FDA (2007) Zooplankton responses to sandbar opening in a tropical eutrophic coastal lagoon. Estuar Coast Shelf Sci 71(3–4):657–668

Saraswathi R, Sumithra P (2016) An assessment of protozoan assemblage in Kottaipattinam Seasonal estuary, Pudukkottai District, Tamil Nadu. Int J Curr Microbiol App Sci 5(2):490–494

Sen A, Bhadury P (2016) Exploring the seasonal dynamics within the benthic foraminiferal biocoenosis in a tropical monsoon-influenced coastal lagoon. Aquat Biol 25:121–138

Sharma BK, Naik LP (1996) Results on planktonic rotifers in the Narmada river (Madhya Pradesh, Central India). In: Perspectives in tropical limnology. SPB Academic Publishing, Amsterdam, pp 189–198

Shin K, Jang MC, Jang PK, Ju SJ, Lee TK, Chang M (2003) Influence of food quality on egg production and viability of the marine planktonic copepod *Acartia omorii*. Prog Oceanogr 57(3–4):265–277

Sladecek V (1983) Rotifers as indicators of water quality. Hydrobiologia 100(1):169–201

Sooria PM, Jyothibabu R, Anjusha A, Vineetha G, Vinita J, Lallu KR, Paul M, Jagadeesan L (2015) Plankton food web and its seasonal dynamics in a large monsoonal estuary (Cochin backwaters, India)-significance of mesohaline region. Environ Monit Assess 187(7):427

Souza LCE, Branco CWC, Domingos P, Bonecker SLC (2011) Zooplankton of an urban coastal lagoon: composition and association with environmental factors and summer fish kill. Zoologia (Curitiba) 28(3):357–364

Srichandan S, Rastogi G (2020) Spatiotemporal assessment of phytoplankton communities in the Chilika lagoon. In: Ecology, conservation, and restoration of Chilika lagoon, India. Springer, Cham, pp 251–294

Srichandan S, Kim JY, Bhadury P, Barik SK, Muduli PR, Samal RN, Pattnaik AK, Rastogi G (2015a) Spatiotemporal distribution and composition of phytoplankton assemblages in a coastal tropical lagoon: Chilika, India. Environ Monit Assess 187:47

Srichandan S, Kim JY, Kumar A, Mishra DR, Bhadury P, Muduli PR, Pattnaik AK, Rastogi G (2015b) Interannual and cyclone-driven variability in phytoplankton communities of a tropical coastal lagoon. Mar Pollut Bull 101:39–52

Steinberg DK, Landry MR (2017) Zooplankton and the ocean carbon cycle. Annu Rev Mar Sci 9:413–444

Sushchik NN, Gladyshev MI, Makhutova ON, Kalachova GS, Kravchuk ES, Ivanova EA (2004) Associating particulate essential fatty acids of the ω3 family with phytoplankton species composition in a Siberian reservoir. Freshw Biol 49(9):1206–1219

Tackx ML, De Pauw N, Van Mieghem R, Azemar F, Hannouti A, Van Damme S, Fiers F, Daro N, Meire P (2004) Zooplankton in the Schelde estuary, Belgium and the Netherlands. Spatial and temporal patterns. J Plankton Res 26(2):133–141
Thorp JH, Mantovani S (2005) Zooplankton of turbid and hydrologically dynamic prairie rivers. Freshw Biol 50(9):1474–1491
Thorson G (1946) Reproduction and larval development of Danish marine bottom invertebrates with special reference to the planktonic larvae in the sound (Oresund). Meddr Kommn Danm Fisk og Havun ders Ser Plankton 4(1):1–523
Tonno I, Agasild H, Koiv T, Freiberg R, Noges P, Noges T (2016) Algal diet of small-bodied crustacean zooplankton in a cyanobacteria-dominated eutrophic lake. PLoS One 11(4):e0154526
Varghese M, Krishnan L (2011) Ecology of rotifers in Cochin backwaters, Kerala, India. Indian J Fish 58(3):109–115
Varghese M, Jasmine S, Sreenath KR, Ranjith L, Thomas VJ (2018) Plankton productivity in lagoons of Agatti and Bangaram atolls of Lakshadweep Archipelago, India. Fish Technol 55:149–151
Venkataramana V, Sarma VVSS, Reddy AM (2017) River discharge as a major driving force on spatial and temporal variations in zooplankton biomass and community structure in the Godavari estuary India. Environ Monit Assess 189(9):474
Vineetha G, Madhu NV, Kusum KK, Sooria PM (2015) Seasonal dynamics of the copepod community in a tropical monsoonal estuary and the role of sex ratio in their abundance pattern. Zool Stud 54(1):54
Wei N, Xu RL (2014) Distinct difference of littoral rotifer community structure in two mangrove wetlands of Qi'ao Island, Pearl River estuary, China. Zool Stud 53(1):30
Xu H, Zhang W, Jiang Y, Yang EJ (2014) Use of biofilm-dwelling ciliate communities to determine environmental quality status of coastal waters. Sci Total Environ 470:511–518
Ziadi B, Dhib A, Turki S, Aleya L (2015) Factors driving the seasonal distribution of zooplankton in a eutrophicated Mediterranean Lagoon. Mar Pollut Bull 97(1–2):224–233

Chapter 10
Metal Transport and Its Impact on Coastal Ecosystem

Piyush Tripathi, Anjali Singhal, and Pawan Kumar Jha

Abstract Heavy metal pollution has risen as an alarming threat in the aquatic systems, including coastal ecosystems comprising of mangroves, salt marshes, wetlands, bays, and estuaries. Heavy metals that pollute the coastal ecosystems mainly consist of metals like arsenic (As), lead (Pb), cadmium (Cd), chromium (Cr), zinc (Zn), copper (Cu), nickel (Ni), and manganese (Mn). The transport of heavy metals in coastal ecosystems occurs through various natural as well as anthropogenic sources. The natural sources comprise natural leaching of bedrocks, transportation from land, and input from freshwater systems, while the anthropogenic sources include mining, smelting, and industrial effluent, followed by agricultural and domestic runoff. Rapid economic growth has further accelerated the transport of heavy metals in coastal ecosystems. The pollution caused by heavy metals not only is restricted to the water but also affects the sediments and biological systems. The heavy metals are not degraded naturally as organic matter and are frequently returned to the system through physicochemical and biological processes, posing risk to the health of humans and the ecosystem. The heavy metals have a strong affinity for particle surfaces; hence the majority is deposited in the sediments because of processes like adsorption and coprecipitation. The mobility, speciation, and bioavailability of these heavy metals are dependent on physical and chemical properties such as pH, redox potential, organic content, and salinity, rendering them as potential pollutants. The slack water conditions in coastal areas encourage heavy metal accumulation; high levels of organic, clay, and sulfide content enhance the adsorption of these metals, while the high rate of sedimentation enhances permanent deposition of locally formed metal sulfides and refractory metal-organic complexes. Such transformations pose a great threat as due to bioavailability they enter the food chain and biological systems causing adverse effects on biological and ecosystem health. Several research works present the health impacts and ecological effects caused due to the contamination of heavy metals by assessment of enrichment factor, ecological risk index, geo-accumulation index, and pollution load index. However, owing to deteriorating water quality, more extensive studies are

P. Tripathi · P. K. Jha (✉)
Centre of Environmental Studies, University of Allahabad, Prayagraj, India

A. Singhal
Department of Botany, University of Allahabad, Prayagraj, India

S. Madhav et al. (eds.), *Coastal Ecosystems*, Coastal Research Library 38,
https://doi.org/10.1007/978-3-030-84255-0_10

required. Therefore, it is extremely important to look into the sources, processes, fate, and consequences of the heavy metals in coastal ecosystems, and design appropriate management policies to save the ecosystem from being further polluted through the contamination of heavy metals.

Keywords Coastal water · Coastal sediments · Heavy metals · Metal transport · Environmental risk assessment · Human health

1 Introduction

Earth comprises 71% of water and 29% of the land, which interact along 1,634,701-km-long coastline with 84% of the countries having open oceans, inland seas, or both (Martínez et al. 2007). Coastal ecosystems aid millions of people's sustenance and well-being by playing a vital role in providing several resources and ecological services. However, the global environmental change has affected the marine and terrestrial ecosystem services and has also impacted the coastal communities that directly depend on them (Lau et al. 2019). The coastal ecosystems comprise the most extensively utilized, and hence among the most threatened, natural systems of the world, with either loss or degradation of 50% of salt marshes, 35% mangroves, 30% coral reefs, and 29% seagrasses (Barbier et al. 2011).

Coastal ecosystems are places where water and land join together to form a unique habitat with a unique composition, variation, and natural flow of moisture. These ecosystems vary significantly in type from being relatively static to highly dynamic and rich in wildlife. They range from relatively dry to being moist (Viles and Spencer 1995). Coastal zones have been widely exploited because of their abundantly accessible resources and cultural and recreational activities. The increasing population and associated infrastructure buildup in coastal areas have resulted in 15 of the 20 global megacities being located in these zones (Mehvar et al. 2018). Such developmental activities support economic growth but also result in the deterioration of natural resources by introducing organic and inorganic pollutants, including metals.

Metals occur as a natural component of water, sediments, soil, and rocks, along with an anthropogenic contribution from activities like agriculture, smelting, mining, printing, municipal discharge, aquaculture, electronic waste, and the petrochemical industry (Wang et al. 2013). Such pollution in the coastal regions causes severe social as well as ecological impacts. Heavy metals refer to the metallic elements with an atomic density greater than 4 g/cm^3 or five times or greater than the density of water (Hawkes 1997; Tchounwou et al. 2012). However, the role of chemical properties outplays the significance of heavy metals' density as their properties show a diversity of effects on various components of the environment, including coastal ecosystems. The most common heavy metal pollutants include Hg, Cd, Cr, Fe, As, Ni, and Pb (Susana Villanueva and Botello 1998; Tchounwou et al. 2012).

The heavy metals released in the environment are bioaccumulated by the coastal organisms and undergo subsequent biomagnification via the food chain, ultimately threatening living organisms' health (Rainbow and Luoma 2011). Heavy metal pollution in the coastal ecosystems has emerged as a chronic environmental concern over the last 50 decades (Batista et al. 2014). The coastal ecosystems contain heavy metals either in the dissolved form with the water or as a deposited matter on the sediment, according to the nature of particular species and physicochemical characteristics such as salinity, conductivity, pH, and organic matter (Chakraborty et al. 2014). The sediments serve as a store for most of the metal pollutants in the aquatic ecosystem and, hence, are used as an indicator to analyze the pollution of heavy metals (Zhao et al. 2018). The sediment-bound heavy metals adsorb and concentrate the fine particles that deposit in certain areas causing toxicity (Zhang et al. 2012). These heavy metals are responsible for denaturing enzymes and proteins, disturbed cellular activities, and production of reactive oxygen species, resulting in impairment of lipids, proteins, and DNA leading to oxidative stress and subsequent cell death, thereby collapse of the living systems. Such problems raise concerns about the economic activity and public health concerns that get impacted by such metal pollution and the environment that bears the same adverse effect. Therefore, a better and more in-depth knowledge of metal transport and its impact is significant for economic, environmental, ecological, and public health aspects to save these precious resources.

2 Coastal Ecosystem: An Overview

The coastal ecosystem holds great importance due to its significant ecological services. Coastal ecosystems include marshes, mangrove forests, coral reefs, mudflats, seagrass beds, and dunes (Granek et al. 2010). The coastal environment maintains the ecological equilibrium and high productivity through the nutrients supplied to it from several sources, including nitrogen fixation, riverine input, precipitation, upwelling from deeper waters, and nutrient regeneration (Warren Flint 1985; Nazneen et al. 2019; Oelsner and Stets 2019). It is a complex ecosystem influenced by several physical, chemical, and biological phenomena, but the extensive utilization of coastal resources by the man in recent years has destabilized its equilibrium (Ngoile and Horrill 1993; Oelsner and Stets 2019). These areas are particularly significant because of the beneficial provisioning, regulating, and recreational services that they provide to the coastal populations aiding in ecological and economic well-being (Nobre 2009). The Millennium Ecosystem Assessment (MEA) has categorized these ecosystem services mainly into four parts, i.e., provisioning services including food from fisheries and aquaculture, fuel from mangrove woods and offshore oil and gas, alternative energy from offshore wave and wind, natural products like pearls and sand, pharmaceutical products, and space for ports; regulating services including weather regulation, carbon sequestration, shoreline stabilization, and protection from natural hazards like floods, storms, and hurricanes; supporting services like the formation of soil, sand, and sediment, photosynthesis, and nutrient cycling; and cultural services that include aesthetics, tourism, education, recreation, and spiritual values (Lau 2013).

The coastal ecosystem also provides several ecological, functional, and social benefits (Milcu et al. 2013). The mangrove forests, salt marshes, and seagrass beds act as a buffer from storm erosion, store carbon, and provide nursery grounds for commercially viable fish species (He et al. 2014; Islam et al. 2018). Out of a global area of about 16.4 million ha, more than one-third of mangrove forests are found in Southeast Asia (Estoque et al. 2018). However, the modern world is losing this treasure rapidly due to developmental activities like urbanization, industrialization, and other associated activities that result in water pollution, ocean acidification, and sea temperature rise (Rao et al. 2015; Sannigrahi et al. 2020). This global deterioration in coastal systems has considerably declined three significant services, viz., the number of important fisheries by 33%; the provision of nursery habitats like wetlands, seagrass beds, and oyster reefs by 69%; and filtering and detoxification function by submerged vegetation, wetlands, and suspension feeders by 63% (Braatz et al. 2007; Cochard et al. 2008; Koch et al. 2009). Such a decline of biodiversity, coastal vegetation, and ecosystem functions has resulted in deteriorated water quality, biological invasions, reduced coastal protection from storm and flood events, coastal pollution by heavy metals, and so on. Table 10.1 provides information about the ranges of heavy metals found in various ecosystems of the world. In addition to these ecological services, the coastal ecosystems are of significant economic values as about 50% of the global population inhabiting the coastal areas are benefitted from access to trade, land development, oil and gas extraction, and food production, boosting the per capita income of coastal inhabitants as compared to those residing in landlocked areas (Gallup et al. 1999; Feldmann 2009). It is also worth mentioning that the coastal zones contribute more than 60% of the biosphere's total economic value (Liquete et al. 2013).

The past centuries have witnessed immense discharge of metals into the coastal waters due to rapid industrialization and related developmental activities, where sediments act as the primary repository and source of metals in the coastal environment (Yan et al. 2010). Several studies aiming at metal pollution have been conducted owing to the significance of the aquatic ecosystem. The types of coastal ecosystem range from coral reefs to mangroves, seagrass meadows, lagoons, salt marshes, and estuaries (Sullivan et al. 2005). These are discussed below in detail:

2.1 Mangroves

The extensive root system of mangroves acts as physical traps for fine substances and the transported metals (Sundaramanickam et al. 2016). The uptake of heavy metals by trees of the mangrove ecosystem is dependent on their biochemical and physiological properties, including their composition, distribution, bioavailability, soil texture, and grain size (Khan et al. 2020). The sediments of mangroves play the

Table 10.1 Ranges of metal contaminants' concentration measured in different coastal ecosystem

Component	Fe	Zn	Cu	Mn	Cd	Pb	Cr	Ni	References
Mangrove ecosystem									
Sediments (μg/g dry weight)	100–33,492	0.28–379	0.3–75	1.23–640	0.1–2.39	1–650	0.71–75.70	0.03–102	Peters et al. (1997), Borrell et al. (2016), Dudani et al. (2017), Alzahrani et al. (2018)
Suspended material (μg/L)	195–2808	18–595	62–76	466–788	2.85–3.2	21–139	–	–	
Biological species (μg/g)	1.6–9.5	9.37–177.5	0.78–60.72	0.16–0.82	0.007–0.22	0.02–0.8	1.26–4.01	0.003–0.07	
Coral reef ecosystems									
Water column (mg/L)	1.0–5.93	0.02–1.5	0.01–1.8	–	0.13–0.43	0.18	–	2.33–5.80	Peters et al. (1997), Abdel-Hamid et al. (2011), Hwang et al. (2018)
Sediment (μg/g)	237–11,445	7.6–40	2.2–17	–	1.97–4.30	18–45	–	74–122.6	
Coral skeleton (μg/g)	0–560	0.08–25	0.24–18	–	–	0.04–39	–	0–126	
Coral tissue (μg/g dry weight)	–	0–126	7.5–18	3.98–13.3	0.44–1.89	–	–	–	
Seagrass ecosystem									
Seagrass tissue (μg/g)	604–7208	8.67–424.1	44.36–86.76	349.84–1180.4	1.04–3.88	0.15–60.9	34.48–138.2	0.82–48.1	Nobi et al. (2010), Mishra et al. (2020)
Sediment		7.5–54.7	0.2–116		0.06–1	1.77–60		0.85–52.4	
Estuarine ecosystem									
Water (μg/L)	–	–	15.8	–	5.6	16.9	12.5	35	Ramesh and Subramanian (1988), Ananthan et al. (1992), Chan and Wang (2019), Karthikeyan et al. (2020)
Sediment (μg/g)	10–1511	14.5–1482	3.5–69	174–6978	7.6	0–4	0–174	0–149	
Biological species (μg/g)	–	0–5669	14–1234	–	0.01–2.9	0–4.1	0.01–16.4	0.3–17.8	

role of buffer between possible sources of pollutants and the marine ecosystem, as they might have a high adsorption capacity for heavy metals (Analuddin et al. 2017). Metals like Cu and Zn are highly mobile in sediments, which results from the presence of organic matter that elevates the mobilization of these metals (Marchand et al. 2016). The anoxic environment of mangroves characterized by negative redox potential and high concentration of sulfide, iron, and organic matter makes sediment be a sink of heavy metals. Factors such as dry periods or changes in salinity may cause metals' mobilization by losing their metal-binding ability. Once entering the biological system, these heavy metals cause severe impacts like inhibition of plant growth, change in photosynthetic pigments, restricted enzymatic activities, reduced carbon assimilation, disturbed reproductive cycle, and abnormal plant growth (Yan et al. 2017). The mangrove fishes are suffering from the effects of metal bioaccumulation in their gills, and other edible parts. The accumulation of zinc was found to be highest in most of the fish species inhabiting mangrove ecosystems (Kulkarni et al. 2018). The elevated endocrine damage and carcinogenicity were also observed in the mangrove gastropods, mollusks, and crabs (Bayen 2012). Such bioaccumulation may get transferred via trophic levels and ultimately affect higher levels of the food chain. Thus, these valuable resources are of significant economic and ecological importance and need to be conserved to avoid further exploitation.

2.2 *Coral Reefs*

Anthropogenic activities have introduced unwanted toxic metals in the coral ecosystem leading to several severe consequences (van Oppen et al. 2017). The heavy metals are absorbed by the corals leading to toxicity, coral bleaching, inhibited growth, or ultimately death (Yang et al. 2020). The metals accumulate in the skeleton of corals' crystal lattice by replacing the ions of calcium with that of other metals, through the matter trapped in cavities, particulate matter in mucus, uptake of organic matter, and feeding. Metals like Cu, Cd, and Fe may cause the corals' bleaching (van der Schyff et al. 2020). These metals are introduced by the use of boat paints, agricultural fertilizers, and aquaculture by-products (Yang et al. 2020). The lab experiment results indicated that the presence of Cu and Cd may lead to loss of symbiotic zooxanthellae from the coral assembly and result in coral bleaching (Sabdono 2009; van Dam et al. 2011). The warming of seas and oceans associated with climate change has led to increasing metal availability and absorption of heavy metals by the coral reefs (Guzmán and Jiménez 1992; Ali et al. 2011). Such contamination subsequently affects other organisms of the food chain, the most vulnerable targets being cnidarians and mollusks (Pitacco et al. 2017).

2.3 *Seagrass Meadows*

They are of immense importance by supporting fisheries, climate change mitigation, and coastal protection worldwide and providing food security (Unsworth et al. 2019). The anthropogenic activities have degraded these meadows' quality, ranking them among the most threatened ecosystems with a global loss rate increasing from 0.9% annually in 1940 to 7% by the twentieth century (Carmen et al. 2019). Heavy metals can be found accumulated in the seagrass tissues from both the water and sediments (Lee et al. 2019). The contamination by heavy metals may lead to the death of phytoplankton and other producers resulted in increase turbidity in the water column and reduce light penetration. The decrease in light penetration reduces the rate of photosynthesis and restricts seagrass distribution in the shallow water region (Papathanasiou et al. 2015). Heavy metals are responsible for several cytotoxic effects that, in turn, hamper the growth of the seagrasses (Lin et al. 2018). The significant impacts of heavy metal contamination in seagrass are observed on the energy metabolism, photosynthetic mechanism, carbon fixation, and defense mechanism, mainly due to disturbed gene expressions and protein abundance (Mohammadi 2019). Such effects interfere with vital pathways and are either lethal or hindrance to the developmental processes by accumulating in the tissues of seagrass (Prange and Dennison 2000).

2.4 Lagoon

It refers to the shallow water bodies and transitional ecosystems between continental aquatic systems, transitional waters, and coastal marine ecosystems and supports rich biodiversity providing critical socio-ecological services including well-being, livelihood, and welfare to humans (Newton et al. 2018; Pérez-Ruzafa et al. 2019). The heavy metals; persistent pollutants such as DDT, hexachlorobenzene, and lindane, emerging from agricultural chemicals; and sewage discharge from human inhabitations comprise the most significant pollutants. Because of their restricted exchange with the ocean, the lagoons also serve as the repository of organic pollutants by trapping organic and inorganic matter from specific sources (Pinto et al. 2016; Leruste et al. 2016). Exposure to metals may result in oxidative stress by promoting the production of harmful and mutagenic reactive oxygen species, damaging the biomolecules, thus hampering the physiological processes (Bejaoui et al. 2020). The absorption of heavy metals by phytoplankton is a serious concern as it subsequently leads to the exposure of a large population of fish to heavy metal contamination (Fernandes et al. 1994; Santhanam 2011; Nikolenko and Fedonenko 2020). Also, the concentration of metals in benthic fauna such as *Mactra lilacea, Nassarius arcularia, Bullia annulata, and Tritia mutabilis* affects the higher trophic levels by successive transfer of heavy metals via the food chain (Abdelhady et al. 2019). This precious ecosystem needs immediate attention to safeguard the economic, ecological, and environmental health for sustained survival and well-being.

3 Major Sources of Heavy Metals

Coastal ecosystems are extremely dynamic systems and respond quickly to the changes occurring within them. It constitutes the essential food, economic, pharmaceutical, aesthetic, and other ecological resources and is the ultimate receptacle of various pollutants, including heavy metals (Maanan 2008). These metals accumulate in water and, at higher concentrations, prove toxic for biota and man. The sources of these metals can either be natural, i.e., from physical or chemical weathering of parent rocks and transportation process, or anthropogenic by activities that are considered as the primary reason for the degrading coastal environment (Callender 2003; Maanan et al. 2015). The main anthropogenic activities that introduce metals in the coastal ecosystem include agricultural runoff, metal-working techniques, mining, smelting of metalliferous ores, industrial and municipal discharge, atmospheric deposition, and leaching from dumps (Rai 2008; Wei et al. 2008; El-Serehy et al. 2012). Heavy metal dumping, which takes place on land, in riverine areas, and near the sea, also contributes to coastal pollution. Coastal mining activities introduce Cu, Zn, Cd, and Hg, while oil spills are responsible for Cu, Zn, Pb, and Cr, and paint manufacturing results in the pollution of Cu, Cr, Zn, As, and Hg (Lu et al. 2018). The release of metals may occur in either dissolved or particulate form; in the latter case, it can be adsorbed and deposited in the sediment (Audry et al. 2004). The sulfide-rich sediment of the coastal systems has a high metal-binding capacity and hence is responsible for the accumulation of heavy metals in sediments and plant tissues as well (Abohassan 2013). The heavy metals get deposited along with the sediment, and it continuously gets exchanged between the water and sediment phase due to constant dynamic interactions (Ali et al. 2019).

However, many scientists and environmental advocates feel that the best way to reduce these problems is to prevent pollution before it even begins in the first place (Spiegel and Maystre 1998; Selvi et al. 2019). The reduction of heavy metal concentration at the source itself can help in managing this menace. Thus, it is of utmost importance to study the sources of heavy metals to understand the kind and intensity of metal pollution, which subsequently have adverse effects on biotic and abiotic factors of the environment.

4 Factors Affecting the Mobility of Metals

Heavy metals are widely distributed in both the aqueous and sediment phase in the coastal ecosystem, depending upon sediment's chemical form and geochemical properties (Zhang et al. 2014). The sediments act as a source and sink of the heavy metals. The metals attached to particles can be mobilized to the aqueous phase, where their fate is controlled by hydrodynamic conditions (Premier et al. 2019). The free ionic species of metals are the most toxic forms, which are highly mobile and bioavailable. In contrast, the metals in the crystal lattice of silicate minerals are usually inert and non-bioavailable. Also, anthropogenic origin metals are more mobile than those arising from lithogenic sources; hence the former quickly enters the food chain. The presence of organic matter and its oxidation by microbial action are a significant factor in the mobilization of heavy metals through the reductive dissolution of Fe-Mn (oxyhydr)oxides (Jokinen et al. 2020). The heavy metals into the underlying sediment are immobilized by several physicochemical processes like sedimentation, coprecipitation, hydrolysis, adsorption, and ligand exchange, where only a small part of ions remain dissolved in the water phase while the sediments retain a significant part of the metal ions, thereby threatening the biological system as well as an ecosystem by bioaccumulation and biomagnification (Lau and Chu 2000; Bastami et al. 2015). The sedimentation contributes to the metal mobilization along with the transport of sediment, while in adsorption the heavy metals are adhered to the soil particles and transported resulting in the mobilization of heavy metals (Violante et al. 2010). The complex soil-sediment-water interactions in the hydrodynamic zones may result in multiple effects, including mobilization, accumulation, and dispersion of heavy metals at short as well as longer time scales (Arakel and Hongjun 1992; Liaghati et al. 2003). The existence of metals in the environment can be varied as they may be transformed from one form to another or may exist in different forms, depending on environmental conditions. These forms or chemical speciation influence the bioavailability, fate, and risk of the metals (Martínez-Sánchez et al. 2008). However, organic matter and ion-exchange materials like clay also significantly affect metals' mobility by the processes of chelation, precipitation, adsorption, and ion exchange (Williams et al. 1994; Yi et al. 2019). It has also been found that the mobility of metals adsorbed on sediments, sedimentary organic matter, carbonate phases, Fe-Mn oxides, and other minerals is dependent upon pH and redox potential as well (Eggleton and Thomas 2004). Most metals are known to mobilize generally at low pH, while some of them undergo complete sorption at a pH of 7 (Wang et al. 2016). The metals that occur naturally in the sediments are chiefly related to the silicates and primary minerals, thereby showing limited mobility. Hence, this fraction does not play a vital role in pollution. However, metals originating from the anthropogenic activities, affects the biogeochemical cycling of elements by bonding with carbonates, organic matter, sulfides, and oxides of Fe-Mn (Marinho et al. 2019). In the deep sediments, sulfide formation and re-oxidation, carbonate decomposition, and reduction and oxidation of Fe and Mn also play an important role in mobility, whereas, in brackish waters, such as those of mangroves, salts are known to enhance the metal mobility in the oxidized sediment layers (Du Laing et al. 2008). The concentration of metals in estuaries can be significantly remobilized, given the physicochemical characteristics of the system. The metal concentration in estuarine sediments is generally higher than in other natural environments (de Souza Machado et al. 2018). The estuaries, with their high salinity, have a considerable effect on the mobility, toxicity, and deposition of metals and metalloids in estuarine wetlands. Increased water salinity could potentially affect the levels of arsenic in water, while high nitrogen values can significantly elevate the concentration of zinc (Bai et al. 2019; Liu et al. 2019). Thus, numerous factors affect the mobility of heavy metals through dynamic physical, chemical, geological, biological, and environmental interactions in the soil, sediment, and water interface.

5 Distribution of Heavy Metals

The distribution of heavy metals is widespread in the world's coastal ecosystems in both the dissolved and suspended phases. Once discharged into water, these heavy metals may get adsorbed from the water phase to fine particles and reach the sediments and ultimately enter the food chain resulting in health risks, depending on the environmental and hydrodynamic conditions (Rahman et al. 2014). The distribution of heavy metals is influenced by the type of coastal areas. Some of them are discussed below:

The port areas of major parts of the world are becoming increasingly polluted by heavy metals. Rio de Janeiro Harbor shows an immensely high concentration of heavy metals, mainly due to the naval activities and pollution load from rivers (Neto et al. 2006). The port areas of Trieste in the northern Adriatic Sea have a high concentration of Zn, Cu, and Hg, with Hg posing a severe threat (Petranich et al. 2018). Recent studies found the port of Santos in Brazil to be contaminated with heavy metals due to the oil extraction activities. The major metals include Cu, Cr, Zn, and Cd, which have been affecting aquatic plants and animals through bioaccumulation (Zampieri et al. 2020).

The bay areas including that of the southwestern coast of Spain have also witnessed serious heavy metal pollution along the coast, which is contributed by Tinto and Odiel rivers, with Zn showing the highest mobility, Mn showing intermediate mobility, and Cd showing least mobility (Morillo et al. 2004). Reports of heavy metal pollution from the Tianjin Bohai Bay of China also showed high levels of Pb and Zn, primarily from river discharge and atmospheric deposition (Wei et al. 2008). The Izmit Bay of Turkey also presents heavy metal pollution due to natural geochemical and anthropogenic inputs (Ergin et al. 1991; Pekey 2006). Saudi Arabia has witnessed some dramatic increase in the anthropogenic developmental activities in the past decade, particularly in the coastal regions of Jeddah, Yanbu, and Rabigh, leading to the water and sediment of these areas being heavily polluted with metals (Badr et al. 2009). The Newcastle region in the northeastern New South Wales of Australia is heavily industrialized and urbanized, resulting in serious pollution of metals like Ag, As, Cd, Cr, Hg, Ni, Pb Sb, and Zn (Lottermoser 1998; Jahan and Strezov 2019). Jeddah, situated on the western coast of Saudi Arabia, is an industrial city threatened by heavy metal pollution. The major metals responsible for pollution in the area include Cr, Mn, Fe, Cu, Zn, and Pb. These metals have caused widespread deterioration in the sediment quality of the adjoining Red Sea and continue to create problems in recent years due to industrial and urban activities in the concerned region (Ding et al. 2018). The Persian Gulf is also not untouched by the growing menace of heavy metal pollution. In a study conducted by Arfaeinia et al. (2019), it was revealed that the sediments along the Persian Gulf coast were "heavily polluted" with metals imparting severe negative impacts on the health of both the environment and humans. They also reported the concentration of metals in decreasing order: Pb > Cu > Zn > Cr > Cd > Ni. In the Southern China region, Hong Kong and Pearl River Estuary are considered the "hot spots" of heavy metal pollution (Wang et al. 2013).

The mangrove sediments along the coastal regions of Pakistan are severely contaminated with heavy metals, particularly in the Port Qasim area, due to pollutant discharge from the point and nonpoint sources (Khattak et al. 2012). The mangrove forest areas of Tamil Nadu, India, are also not untouched by this menace. Heavy metals like Cd, Pb, Fe, Cr, Zn, and Cu are quite widespread due to the vast amount of waste generation and dumping from various electronic, chemical, and agronomical industries, ultimately disturbing ecological health (Agoramoorthy et al. 2008). The Gulf of Kutch in India is also seriously impacted by the heavy metals transported into the mangrove environment (Chakraborty et al. 2014). The mangrove sediments of the Panchagangavali Estuary of Karnataka are also enriched with metals such as Fe, Zn, Ni, Cu, Co, and Cr, indicating recent advances in anthropogenic activities in the catchment area, as earlier only the metals of lithogenic origin existed in the area (Fernandes and Nayak 2020).

The minor estuaries of Goa along Terekhol, Chapora, Sal, and Talpona rivers have the highest particulate-metal concentrations in regions of low salinity of the estuaries in the wet season resulting in metal accumulation and pollution of the coastal region (Fernandes et al. 2019). Kali estuary in the central west coast of India is also contaminated with major and trace elements, the concentration of which is elevated during pre- and post-monsoon periods (Suja et al. 2017). The eastern coast of India is also moderately polluted with heavy metals, the majority of which can be attributed to Fe, Co, Zn, and Cu, that, in turn, accumulate in living systems and cause health and environmental hazards by direct health impacts, disturbed biogeochemical cycles, and contamination of food chain via bioaccumulation and biomagnification (Kumar et al. 2017). Table 10.2 presents the region-specific distribution of major heavy metal pollutants and their sources in the coastal ecosystems around the world.

6 Health Implications

6.1 On Flora

The absorption and accumulation of heavy metals may both be energy-dependent or independent. Some algae may accumulate these metals as intranuclear compounds hampering normal functions, rendering them toxic (Rai and Gaur 2012). The increased concentration of heavy metal impacts the flora in several ways including stunted growth, chlorosis, reduced yield, disturbed nutrient uptake, and metabolic disorders (Guala et al. 2010).

Table 10.2 Region-specific heavy metal contamination in coastal ecosystems of the world

Region	Major metal pollutant	Source
America		
Peninsula La Esperanza, Puerto Rico	Hg, Pb, Zn, Cu, Cr	Anthropogenic
Sinnamary and Kaw, Brazil	Hg, Zn	Lithogenic
Guanabara Bay, Brazil	Cd, Cr, Cu, Pb, Zn, Hg	Anthropogenic
Mazatlán Harbor, Gulf of California	Zn, Pb, Cu, Cd, Hg	Anthropogenic
Africa		
Mtoni, Kunduchi, Mbweni, Chwaka, Makoba, and Rufiji, Tanzania	Cu, Pb, Fe, Al, Sn, Zn, Cr, Ni	Anthropogenic (Cu, Pb), Lithogenic (Fe, Al)
Togo	Cd, Cr, U	Anthropogenic
Fadiouth, Senegal	Cd, Hg, Ni	Anthropogenic
Australia		
Homebush Bay, Sydney	Pb	Anthropogenic
Pumicestone, Queensland	Zn, V, Cr	Lithogenic
Asia		
Sundarbans, India	As, Pb, Co, Cu, Cd	Anthropogenic
Bhitarkanika, India	Fe > Pb > Cr > Cd > Mn > Zn	Anthropogenic
Pichavaram, India	Fe, Mn, Cr, Ni, Pb, Zn	Anthropogenic
Muthupet mangroves, India	Pb, Zn	Anthropogenic, Lithogenic
Eastern Indonesia	Ni, Cr, Cu	Lithogenic
Western Indonesia	Pb, Cd	Anthropogenic
Tanjung Lumpur, Malaysia	Pb	Anthropogenic

Source: Sharifuzzaman et al. (2016), Kulkarni et al. (2018)

6.2 On Nekton

Nekton may be contaminated by heavy metals emerging from various sources like industrial, agricultural, and domestic and contributing rivers. These metals pose major food safety threats as they move up the food chain (Gu et al. 2015). Acute toxicity of Cd, Cu, Zn, and Cr has been observed in nekton species widely from several estuaries of China (Yang et al. 2021).

6.3 On Benthos

The benthic fauna around the world is facing the adverse effect of metal pollution, particularly Cu, Cr, Zn, and Ni posing the major threat (Ryu et al. 2011). The contamination of Cr, Cu, Ni, Zn, and Cd has significantly affected the foraminifera and crustaceans decreasing their population (Bergin et al. 2006). The enrichment of heavy metals in Norway has caused the shift of benthic foraminifera causing a decrease in their population (Alve 1991; Frontalini et al. 2010).

6.4 On Planktons

Phytoplanktons are an immensely significant part of the food chain. Any change in their population is bound to affect the higher trophic levels. The phytoplanktons also show a decrease in population due to heavy metal pollution. Some species, like cyanobacteria, are extremely sensitive to metals like Cu, Zn, and Ni (Chakraborty et al. 2010).

6.5 On Humans

When heavy metals enter the human body through inhalation, absorption through the skin, and ingestion via the food, they are deposited in the organs to harm the body (Pandey and Madhuri 2014). Over time, these metals continue to accumulate in the organs so that they block normal cell growth and production, damage cells and organs, and result in health problems. The emerging problem of heavy metal pollution degrades the quality of the environment and poses a risk to human health by accumulating in vital organs like the liver, kidney, and bones (Bosch et al. 2016). The heavy metals, when exceeding their recommended concentrations, are known to cause toxicity. The individual metals have unique symptoms of toxicity; however, the general signs of heavy metals like Pb, Cd, As, Hg, Cu, Zn, and Al include stomatitis, diarrhea, gastrointestinal disorders, paralysis, ataxia, convulsion, vomiting, and other neurotoxic, mutagenic, carcinogenic, or teratogenic effects (Duruibe et al. 2007). Pb is among the most important toxic heavy metals that can be ingested through water, food, or inhalation. Exposure to Pb, a nonessential element, can cause numerous problems related to endocrine, circulatory, skeletal, nervous, and immune systems (Pascaud et al. 2014). It is also known to cause severe teratogenic effects, inhibition of hemoglobin synthesis, and dysfunctions in the reproductive system, joints, and kidneys, but the most severe impact is observed in the form of acute and chronic damage to the central and peripheral nervous systems causing psychosis and other disorders (Ferner 2001). The major uptake of Pb in the human body occurs into the kidney, followed by the liver, heart, and brain, leading to nervous disorders like loss of memory, headache, dullness, and poor attention (Tchounwou et al. 2012).

As forms complexes with coenzymes by coagulating proteins and consequently inhibiting the synthesis of ATP during respiration. Toxicity of As can result in skin cancer, dermal lesions, vascular diseases, and peripheral neuropathy (Hsiang Tan et al. 2016). It also causes a peculiar anti-immune disorder that results in inflammation of nerves and muscle weakness (Kantor 2006). Large intake of As can result in gastrointestinal problems like severe vomiting, blood circulation disruption, nervous system damage, and ultimately death. If not fatal, such high doses may reduce blood cell production, enlarge the liver, disrupt RBCs, produce tingling and loss of sensation in limbs, and damage the brain. Chronic exposure to inorganic As has caused Blackfoot disease in Taiwan, characterized by severe damage of lower limb blood vessels leading to gangrene (Mahurpawar 2015). Cd can cause toxicity even at deficient concentrations. Chronic exposure to Cd is known to cause prostatic lesions, kidney dysfunction, lung cancer, cadmium pneumonitis, pulmonary adenocarcinomas, and bone defects like osteoporosis, osteomalacia, and frequent fractures (Young 2005; Wu et al. 2018). Cd is transported in bound form by albumin and metallothionein proteins. Upon reaching the gastrointestinal tract, it disrupts the liver and biliary tract's normal functioning, while if stored in the kidney for a long time, it can lead to tubular necrosis (Rehman et al. 2018).

Zn and Pb show almost similar signs of toxicity, making it difficult to distinguish between the two. Zn is considered comparatively nontoxic, but excessive concentrations may impair growth and reproduction, diarrhea, icterus, kidney and liver failure, anemia, and vomiting (Fosmire 1990; Nolan et al. 2003; Plum et al. 2010). Another significant heavy metal is Hg, which does not have any particular role in human physiology and biochemistry. In excess amounts, Hg is toxic and is considered to have severe effects like gastrointestinal disorders, spontaneous abortions, congenital malfunctions, neurological disorders, insomnia, gingivitis, acrodynia, stomatitis, and brain and CNS dysfunctions. The main target organ of Hg toxicity is the brain; however, it is known to affect other organs, nerves, and muscles as well. It interrupts calcium homeostasis, alters cellular functions, and intervenes with transcription and translation phenomena, while its vapors cause bronchitis, asthma, and other respiratory problems (Jaishankar et al. 2014). The organic form of Hg, i.e., methyl mercury, is a potent neurotoxin that has caused thousands of people's death due to Minamata disease in Japan (Chouhan et al. 2016). The health impacts caused by different heavy metals are presented in Table 10.3. Thus, these heavy metals can cause severe health implications through direct-indirect, chronic-acute, or synergistic-antagonistic effects, thereby threatening human beings' survival.

7 Mitigation Strategies

Several measures have been undertaken to mitigate the problem of heavy metal pollution in coastal ecosystems and safeguard these valuable resources. Some of them are discussed below:

Table 10.3 Impact of heavy metals on human health

Metal	Permissible limit (mg/L)	Recommending agency	Health effects
Arsenic	0.01	EPA	"Pins and needles" sensation in hands and feet, "warts" on palms, soles, and torso
Cadmium	0.005	EPA	Kidney, lung, gastrointestinal, and skeletal disorders
Chromium	0.1	EPA	Breathing difficulty, skin allergies, damage to liver, kidney, and nervous tissues
Copper	0.05	EPA	Wilson's disease, reproductive and developmental disorders
Lead	0.015	EPA	Severe damage to brain, kidney, and reproductive system
Manganese	0.05	EPA	Central and peripheral neuropathy
Nickel	0.1	EPA	Cancer, dermatitis

Source: Martin and Griswold (2009), Mahurpawar (2015)

7.1 *Proper Treatment at Source*

The effluents containing heavy metals and toxic substances need to be treated properly at the source itself, including the industrial, domestic, and commercial waste so that it does not further contaminate the aquatic systems, where it is finally dumped. The on-site and in situ precipitation techniques involve the treatment of excavated soil and applying chemicals directly to the soil, respectively, to reduce metal mobility. The four main methods of such precipitation or reduction include the sulfide process, cellulose xanthate process, sodium borohydride process, and lime or carbonate or hydroxide process (Sodango et al. 2018). Also, the proper maintenance of pipelines and associated drainage systems help in the reduction of unwanted heavy metal input into the coastal bodies (Izah et al. 2016).

7.2 *Chemical-Biological Remediation*

The economic feasibility and environment-friendly nature of this method render it one of the most popular mitigation measures. The proper understanding and implementation of this method can help overcome the demerits of individual chemical or biological processes. It generally includes the biological treatment followed by chemical treatment, which proves to be an effective integrated technique for the abatement of heavy metal pollution in coastal ecosystems (Selvi et al. 2019). The remediation includes in situ and ex situ techniques. In situ remediation involves enhancing the stabilization of mobile metals, while ex situ remediation aims at the removal of mobile metals (Peng et al. 2009). The in situ and ex situ techniques under this method include soil flushing, surface capping, electrokinetic extraction, solidification, landfilling, soil washing, vitrification, and other biological remediation techniques. These employ extraction, removal, containment, and immobilization techniques to avoid contamination through various physical, biological, chemical, electrical, and thermal processes (Liu et al. 2018). Solidification uses encapsulation of contaminated soil in a solid matrix of cement, asphalt, bitumen, and thermoplastic binders to reduce contaminant mobility. These techniques are cost-effective and eco-friendly (Li et al. 2019).

7.3 Bioremediation

The presence of microbes play a significant role in biogeochemical cycles of transformations of metals among soluble and insoluble species, resulting in metal-microbe interactions that can be beneficial or harmful. Bioremediation involves the use of dead or living biomass to remove or convert hazardous and toxic heavy metals into less hazardous forms. It is applied for rehabilitating the heavy metal-contaminated soils providing an effective alternative for the restoring of contaminated soils, owing to its socially acceptable, economically viable, and environmentally feasible nature. In this regard, several approaches have been suggested, including the application of both the microbial communities and plant species from diverse sources of origin. These may prove to be efficient in managing the metal-polluted soils (Sobariu et al. 2017). The basic principle involves reducing the solubility of contaminants by changing pH, redox conditions, and adsorption processes. The redox reactions convert the toxic substances to less mobile and inert forms, particularly the heavy metals present in soil and sediments, such as As, Hg, Cr, and Se. The physicochemical properties of the medium, affected by organic and inorganic inputs, have a considerable impact on redox reactions (Ojuederie and Babalola 2017). This process can be applied to both water and sediment phases through various in situ and ex situ techniques. The application of plant species (phytoremediation), in particular, has gained popularity over recent years. These plant species have the potential to accumulate metals in their roots and other parts. The microbial biofilms with high resistance and tolerance capacity for metals can also mediate bioremediation. However, poor selectivity and low efficiencies of reusing biomass prove to be major challenges of this technique (Rai 2008; Kapahi and Sachdeva 2019).

7.4 Public Awareness and Legislations

The most important role in conserving these precious resources can be played by both the public and the authorities. The general public can be made aware of various campaigns, advertisements, and programs to encourage them to contribute toward coastal protection on an individual level. Also, it is the responsibility of the legislative authorities to frame stricter laws and policies followed by proper monitoring and regulation of discharges of industrial and domestic discharge for abating heavy metal pollution and for sustainable management of coastal ecosystems (Wu et al. 2014).

8 Conclusion

The developmental activities are bound to increase with time, demanding more and more resources to satisfy the growing population's needs. This need would ultimately result in the further exploitation of resources by more urbanization, industrialization, agriculture, and other economic activities. However, this need should be fulfilled sustainably so that the precious natural resources, like the coastal ecosystems, face the minimal threat. The pollution of these systems by heavy metals reflects some severe consequences on human and ecological health. The direct impacts are seen on biodiversity, human health, water quality, ecological, and economic functions. The major concern lies in the dynamic nature of interactions between the sediment and aqueous phase, where several factors are involved in the mobilization and transport of these heavy metals. These factors include various physical, chemical, geological, and biological processes, including adsorption, precipitation, hydrolysis, pH, redox potential, chelation, and presence of organic matter. They are mobilized, transported, and deposited to sediment and aqueous phase, where bioaccumulation and subsequent biomagnification occur. These processes, along with other physicochemical, biological, and environmental conditions, affect the ecosystem's different components in both short and longer time scales. It causes various ecological damages by disturbing the biodiversity and ecological balance, economic setbacks by loss of economically important species, and environmental impacts by rendering the water unfit for usage. It also causes several health implications in humans by disrupting the nervous, cardiovascular, gastrointestinal, skeletal, and immunological systems. Thus, heavy metal pollution, transport, and associated health implications need urgent attention from the general public, scientific, and administrative communities to reduce pollution and safeguard our coastal ecosystems for the future.

Acknowledgments Piyush Tripathi thanks UGC for providing fellowship. Pawan Kumar Jha thanks the University Grants Commission (UGC), Govt. of India, for providing financial support (No. F.30-373/2017(BSR)). Authors thank the University of Allahabad for providing the necessary facilities.

References

Abdelhady AA, Khalil MM, Ismail E, Mohamed RS, Ali A, Snousy MG et al (2019) Potential biodiversity threats associated with the metal pollution in the Nile–Delta ecosystem (Manzala lagoon, Egypt). Ecol Indic 98:844–853

Abdel-Hamid AM, Emara HM, Pekhet MA (2011) An attempt for reducing lead content in Tilapia and Mugil during preparing and cooking of fish. J Food Dairy Sci 2(8):513–541

Abohassan RA (2013) Heavy metal pollution in Avicennia marina mangrove systems on the Red Sea Coast of Saudi Arabia. J King Abdulaziz Univ Meteorol Environ Arid Land Agric Sci 24(1):35–53

Agoramoorthy G, Chen FA, Hsu MJ (2008) Threat of heavy metal pollution in halophytic and mangrove plants of Tamil Nadu, India. Environ Pollut 155(2):320–326. https://doi.org/10.1016/j.envpol.2007.11.011

Ali AAM, Hamed MA, Abd El-Azim H (2011) Heavy metals distribution in the coral reef ecosystems of the northern Red Sea. Helgol Mar Res 65:67–80. https://doi.org/10.1007/s10152-010-0202-7

Ali H, Khan E, Ilahi I (2019) Environmental chemistry and ecotoxicology of hazardous heavy metals: environmental persistence, toxicity, and bioaccumulation. J Chem 2019:14

Alve E (1991) Benthic foraminifera in sediment cores reflecting heavy metal pollution in Sorfjord, western Norway. J Foraminiferal Res 21(1):1–19

Alzahrani DA, Selim EMM, El-Sherbiny MM (2018) Ecological assessment of heavy metals in the grey mangrove (Avicennia marina) and associated sediments along the Red Sea coast of Saudi Arabia. Oceanologia 60(4):513–526. https://doi.org/10.1016/j.oceano.2018.04.002

Analuddin K, Sharma S, Jamili SA, Sahidin I, Rianse U, Nadaoka K (2017) Heavy metal bioaccumulation in mangrove ecosystem at the coral triangle ecoregion, Southeast Sulawesi, Indonesia. Mar Pollut Bull 125(1–2):472–480. https://doi.org/10.1016/j.marpolbul.2017.07.065

Ananthan G, Ganesan M, Sampathkumar P, Matheven Pillai M, Kannan L (1992) Distribution of trace metals in water, sediment and plankton of the Vellar estuary. Seaweed Res Util 15(1&2):69–75

Arakel AV, Hongjun T (1992) Heavy metal geochemistry and dispersion pattern in coastal sediments, soil, and water of Kedron Brook floodplain area, Brisbane, Australia. Environ Geol Water Sci 20(3):219–231. https://doi.org/10.1007/BF01706165

Arfaeinia H, Dobaradaran S, Moradi M, Pasalari H, Mehrizi EA, Taghizadeh F, Esmaili A, Ansarizadeh M (2019) The effect of land use configurations on concentration, spatial distribution, and ecological risk of heavy metals in coastal sediments of northern part along the Persian Gulf. Sci Total Environ 653:783–791. https://doi.org/10.1016/j.scitotenv.2018.11.009

Audry S, Schäfer J, Blanc G, Jouanneau JM (2004) Fifty-year sedimentary record of heavy metal pollution (Cd, Zn, Cu, Pb) in the Lot River reservoirs (France). Environ Pollut 132(3):413–426

Badr NBE, El-Fiky AA, Mostafa AR, Al-Mur BA (2009) Metal pollution records in core sediments of some Red Sea coastal areas, Kingdom of Saudi Arabia. Environ Monit Assess 155(1–4):509–526. https://doi.org/10.1007/s10661-008-0452-x

Bai J, Zhao Q, Wang W, Wang X, Jia J, Cui B, Liu X (2019) Arsenic and heavy metals pollution along a salinity gradient in drained coastal wetland soils: depth distributions, sources and toxic risks. Ecol Indic 96:91–98. https://doi.org/10.1016/j.ecolind.2018.08.026

Barbier EB, Hacker SD, Kennedy C, Koch EW, Stier AC, Silliman BR (2011) The value of estuarine and coastal ecosystem services. Ecol Monogr 81:169–193

Bastami KD, Neyestani MR, Shemirani F, Soltani F, Haghparast S, Akbari A (2015) Heavy metal pollution assessment in relation to sediment properties in the coastal sediments of the southern Caspian Sea. Mar Pollut Bull 92(1–2):237–243. https://doi.org/10.1016/j.marpolbul.2014.12.035

Batista D, Muricy G, Rocha RC, Miekeley NF (2014) Marine sponges with contrasting life histories can be complementary biomonitors of heavy metal pollution in coastal ecosystems. Environ Sci Pollut Res 21(9):5785–5794. https://doi.org/10.1007/s11356-014-2530-7

Bayen S (2012) Occurrence, bioavailability and toxic effects of trace metals and organic contaminants in mangrove ecosystems: a review. Environ Int 48:84–101

Bergin F, Kucuksezgin F, Uluturhan E, Barut IF, Meric E, Avsar NİYAZİ, Nazik ATİKE (2006) The response of benthic foraminifera and ostracoda to heavy metal pollution in Gulf of Izmir (Eastern Aegean Sea). Estuar Coast Shelf Sci 66(3–4):368–386

Bejaoui S, Michán C, Telahigue K, Nechi S, El Cafsi M, Soudani N, Alhama J (2020) Metal body burden and tissue oxidative status in the bivalve Venerupis decussata from Tunisian coastal lagoons. Mar Environ Res 159, 105000
Braatz S, Fortuna S, Broadhead J, Leslie R (2007) Coastal protection in the aftermath of the Indian Ocean tsunami: What role for forests and trees?. RAP Publication (FAO).
Borrell A, Tornero V, Bhattacharjee D, Aguilar A (2016) Trace element accumulation and trophic relationships in aquatic organisms of the Sundarbans mangrove ecosystem (Bangladesh). Sci Total Environ 545–546:414–423. https://doi.org/10.1016/j.scitotenv.2015.12.046
Bosch AC, O'Neill B, Sigge GO, Kerwath SE, Hoffman LC (2016) Heavy metals in marine fish meat and consumer health: a review. J Sci Food Agric 96:32–48
Carmen B, Krause-Jensen D, Alcoverro T, Marbà N, Duarte CM, Van Katwijk MM, Santos R (2019) Recent trend reversal for declining European seagrass meadows. Nat Commun 10(1):1–8
Callender E (2003) Heavy Metals in the Environment—Historical Trends. Editor(s): Heinrich D. Holland, Karl K. Turekian. Treatise on Geochemistry. Pergamon (9):67–105. https://doi.org/10.1016/B0-08-043751-6/09161-1
Chakraborty P, Babu PR, Acharyya T, Bandyopadhyay D (2010) Stress and toxicity of biologically important transition metals (Co, Ni, Cu and Zn) on phytoplankton in a tropical freshwater system: an investigation with pigment analysis by HPLC. Chemosphere 80(5):548–553
Chakraborty S, Bhattacharya T, Singh G, Maity JP (2014) Benthic macroalgae as biological indicators of heavy metal pollution in the marine environments: a biomonitoring approach for pollution assessment. Ecotoxicol Environ Saf 100(1):61–68. https://doi.org/10.1016/j.ecoenv.2013.12.003
Chan CY, Wang WX (2019) Biomarker responses in oysters Crassostrea hongkongensis in relation to metal contamination patterns in the Pearl River Estuary, southern China. Environ Pollut 251:264–276
Chouhan B, Meena P, Poonar N (2016) Effect of heavy metal ions in water on human health. Int J Sci Eng Res 4(12):2015–2017
Cochard R, Ranamukhaarachchi SL, Shivakoti GP, Shipin OV, Edwards PJ, Seeland KT (2008) The 2004 tsunami in Aceh and southern Thailand: a review on coastal ecosystems, wave hazards and vulnerability. Perspect Plant Ecol Evol Syst 10(1):3–40. https://doi.org/10.1016/j.ppees.2007.11.001
de Souza Machado AA, Spencer KL, Zarfl C, O'Shea FT (2018) Unravelling metal mobility under complex contaminant signatures. Sci Total Environ 622–623:373–384. https://doi.org/10.1016/j.scitotenv.2017.11.239
Ding X, Ye S, Yuan H, Krauss KW (2018) Spatial distribution and ecological risk assessment of heavy metals in coastal surface sediments in the Hebei Province offshore area, Bohai Sea, China. Mar Pollut Bull 131:655–661. https://doi.org/10.1016/j.marpolbul.2018.04.060
Du Laing G, Van De Moortel A, Lesage E, Tack FMG, Verloo MG (2008) Factors affecting metal accumulation, mobility and availability in intertidal wetlands of the scheldt estuary (Belgium). In: Wastewater treatment, plant dynamics and management in constructed and natural wetlands. Springer, Dordrecht, pp 121–133
Dudani SN, Lakhmapurkar J, Gavali D, Patel T (2017) Heavy metal accumulation in the mangrove ecosystem of South Gujarat Coast, India. Turk J Fish Aquat Sci 17:755–766
Duruibe JO, Ogwuegbu MO, Egwurugwu JN (2007) Heavy metal pollution and human biotoxic effects. Int J Phys Sci 2(5):112–118
Eggleton J, Thomas KV (2004) A review of factors affecting the release and bioavailability of contaminants during sediment disturbance events. Environ Int 30:973–980
Ergin M, Saydam C, Baştürk Ö, Erdem E, Yörük R (1991) Heavy metal concentrations in surface sediments from the two coastal inlets (Golden Horn Estuary and İzmit Bay) of the northeastern Sea of Marmara. Chem Geol 91(3):269–285. https://doi.org/10.1016/0009-2541(91)90004-B
Estoque RC, Myint SW, Wang C, Ishtiaque A, Aung TT, Emerton L, Ooba M, Hijioka Y, Mon MS, Wang Z, Fan C (2018) Assessing environmental impacts and change in Myanmar's mangrove

ecosystem service value due to deforestation (2000–2014). Glob Chang Biol 24(11):5391–5410. https://doi.org/10.1111/gcb.14409
El-Serehy HA, Aboulela H, Al-Misned F, Kaiser M, Al-Rasheid K, El-Din HE (2012) Heavy metals contamination of a Mediterranean coastal ecosystem, Eastern Nile Delta, Egypt. Turk J Fish Aquat Sci 12(4)
Feldmann H (2009) Geography and labor market performance. Eastern Econ J 35:190–208
Fernandes HM, Bidone ED, Veiga LHS, Patchineelam SR (1994) Heavy-metal pollution assessment in the coastal lagoons of Jacarepaguá, Rio de Janeiro, Brazil. Environ Pollut 85(3):259–264
Fernandes LL, Kessarkar PM, Rao VP, Suja S, Parthiban G, Kurian S (2019) Seasonal distribution of trace metals in suspended particulate and bottom sediments of four microtidal river estuaries, west coast of India. Hydrol Sci J 64(12):1519–1534. https://doi.org/10.1080/02626667.2019.1655147
Fernandes MC, Nayak GN (2020) Depositional environment and metal distribution in mangrove sediments within middle region of tropical estuaries, Karnataka, west coast of India. Reg Stud Mar Sci 39: 101473
Ferner DJ (2001) Toxicity of heavy metals. eMed J 2:1
Fosmire GJ (1990) Zinc toxicity. Am J Clin Nutr 51(2):225–227. https://doi.org/10.1093/ajcn/51.2.225
Frontalini F, Coccioni R, Bucci C (2010) Benthic foraminiferal assemblages and trace element contents from the lagoons of Orbetello and Lesina. Environ Monit Assess 170:245–260. https://doi.org/10.1007/s10661-009-1229-6
Gallup JL, Sachs JD, Mellinger AD (1999) Geography and economic development. In: International regional science review. SAGE, London, pp 179–232
Granek EF, Polasky S, Kappel CV, Reed DJ, Stoms DM, Koch EW, Kennedy CJ, Cramer LA, Hacker SD, Barbier EB, Aswani S, Ruckelshaus M, Perillo GME, Silliman BR, Muthiga N, Bael D, Wolanski E (2010) Ecosystem services as a common language for coastal ecosystem-based management. Conserv Biol 24(1):207–216. https://doi.org/10.1111/j.1523-1739.2009.01355.x
Gu YG, Lin Q, Yu ZL, Wang XN, Ke CL, Ning JJ (2015) Speciation and risk of heavy metals in sediments and human health implications of heavy metals in edible nekton in Beibu Gulf, China: a case study of Qinzhou Bay. Mar Pollut Bull 101(2):852–859
Guala SD, Vega FA, Covelo EF (2010) Heavy metal concentrations in plants and different harvestable parts: a soil–plant equilibrium model. Environ Pollut 158(8):2659–2663
Guzmán HM, Jiménez CE (1992) Contamination of coral reefs by heavy metals along the Caribbean coast of Central America (Costa Rica and Panama). Mar Pollut Bull 24(11):554–561
Hawkes SJ (1997) What is a "heavy metal"? J Chem Educ 74(11):1374. https://doi.org/10.1021/ed074p1374
He Q, Bertness MD, Bruno JF, Li B, Chen G, Coverdale TC, Altieri AH, Bai J, Sun T, Pennings SC, Liu J, Ehrlich PR, Cui B (2014) Economic development and coastal ecosystem change in China. Sci Rep 4. https://doi.org/10.1038/srep05995
Hsiang Tan W, Tair R, Aishah S, Ali M, Talibe A, Sualin F, Payus C (2016) Distribution of heavy metals in seawater and surface sediment in coastal area of Tuaran, Sabah
Hwang F, Xu Y, Tan Z, Wu Z, Xu H, Shen L, Hu Z (2018) Assessment of pollutions and identification of sources of heavy metals in sediments from west coast of Shenzhen, China. Environ Sci Pollut Res 25(4):3647–3656
Islam MM, Sunny AR, Hossain MM, Friess DA (2018) Drivers of mangrove ecosystem service change in the Sundarbans of Bangladesh. Singap J Trop Geogr 39(2):244–265. https://doi.org/10.1111/sjtg.12241
Izah SC, Chakrabarty N, Srivastav AL (2016) A review on heavy metal concentration in potable water sources in Nigeria: human health effects and mitigating measures. Expo Health 8:285–304
Jahan S, Strezov V (2019) Assessment of trace elements pollution in the sea ports of New South Wales (NSW), Australia using oysters as bioindicators. Sci Rep 9(1):1416. https://doi.org/10.1038/s41598-018-38196-w

Jaishankar M, Tseten T, Anbalagan N, Mathew BB, Beeregowda KN (2014) Toxicity, mechanism and health effects of some heavy metals. Interdiscip Toxicol 7:60–72

Jokinen SA, Jilbert T, Tiihonen-Filppula R, Koho K (2020) Terrestrial organic matter input drives sedimentary trace metal sequestration in a human-impacted boreal estuary. Sci Tot Environ 717: 137047

Kantor D (2006) Guillain–Barre syndrome: the medical encyclopedia.Bethesda, MD: National Library of Medicine and National Institute of Health. http://www.nlm.nih.gov/medlineplus/

Kapahi M, Sachdeva S (2019) Bioremediation options for heavy metal pollution. J Health Pollut 9:1–20

Khan AF, Srinivasamoorthy K, Rabina C (2020) Hydrochemical characteristics and quality assessment of groundwater along the coastal tracts of Tamil Nadu and Puducherry, India. Appl Water Sci 10(2):1–21

Karthikeyan P, Marigoudar SR, Mohan D, Nagarjuna A, Sharma KV (2020) Ecological risk from heavy metals in Ennore estuary, south east coast of India. Environ Chem Ecotoxicol 2:182–193

Khattak MI, Khattak MI, Mohibullah M (2012) Study of heavy metal pollution in mangrove sediments reference to marine environment along the coastal areas of Pakistan. Pak J Bot 44(1):373–378

Koch EW, Barbier EB, Silliman BR, Reed DJ, Perillo GME, Hacker SD, Granek EF, Primavera JH, Muthiga N, Polasky S, Halpern BS, Kennedy CJ, Kappel CV, Wolanski E (2009) Non-linearity in ecosystem services: temporal and spatial variability in coastal protection. Front Ecol Environ 7:29–37

Kulkarni R, Deobagkar D, Zinjarde S (2018) Metals in mangrove ecosystems and associated biota: a global perspective. Ecotoxicol Environ Saf 153:215–228

Kumar SB, Padhi RK, Mohanty AK, Satpathy KK (2017) Elemental distribution and trace metal contamination in the surface sediment of south east coast of India. Mar Pollut Bull 114(2):1164–1170. https://doi.org/10.1016/j.marpolbul.2016.10.038

Lau WWY (2013) Beyond carbon: conceptualizing payments for ecosystem services in blue forests on carbon and other marine and coastal ecosystem services. Ocean Coast Manag 83:5–14. https://doi.org/10.1016/j.ocecoaman.2012.03.011

Lau SSS, Chu LM (2000) The significance of sediment contamination in a coastal wetland, Hong Kong, China. Water Res 34(2):379–386. https://doi.org/10.1016/S0043-1354(99)00145-1

Lau JD, Hicks CC, Gurney GG, Cinner JE (2019) What matters to whom and why? Understanding the importance of coastal ecosystem services in developing coastal communities. Ecosyst Serv 35:219–230. https://doi.org/10.1016/j.ecoser.2018.12.012

Lee G, Suonan Z, Kim SH, Hwang DW, Lee KS (2019) Heavy metal accumulation and phytoremediation potential by transplants of the seagrass Zostera marina in the polluted bay systems. Mar Pollut Bull 149, 110509

Leruste A, Malet N, Munaron D, Derolez V, Hatey E, Collos Y, De Wit R, Bec B (2016) First steps of ecological restoration in Mediterranean lagoons: shifts in phytoplankton communities. Estuar Coast Shelf Sci 180:190–203. https://doi.org/10.1016/j.ecss.2016.06.029

Li C, Zhou K, Qin W, Tian C, Qi M, Yan X, Han W (2019) A review on heavy metals contamination in soil: effects, sources, and remediation techniques. Soil Sediment Contam Int J 28(4):380–394

Liaghati T, Preda M, Cox M (2003) Heavy metal distribution and controlling factors within coastal plain sediments, Bells Creek catchment, Southeast Queensland, Australia. Environ Int 29:935–948

Liquete C, Piroddi C, Drakou EG, Gurney L, Katsanevakis S, Charef A, Egoh B (2013) Current status and future prospects for the assessment of marine and coastal ecosystem services: a systematic review. PLoS One 8(7). https://doi.org/10.1371/journal.pone.0067737

Liu L, Li W, Song W, Guo M (2018) Remediation techniques for heavy metal-contaminated soils: principles and applicability. Sci Total Environ 633:206–219

Lin H, Sun T, Adams MP, Zhou Y, Zhang X, Xu S, Gu R (2018) Seasonal dynamics of trace elements in sediment and seagrass tissues in the largest Zostera japonica habitat, the Yellow River Estuary, northern China. Mar Pollut Bull 134:5–13

Liu JJ, Diao ZH, Xu XR, Xie Q (2019) Effects of dissolved oxygen, salinity, nitrogen and phosphorus on the release of heavy metals from coastal sediments. Sci Total Environ 666:894–901. https://doi.org/10.1016/j.scitotenv.2019.02.288

Lottermoser BG (1998) Heavy metal pollution of coastal river sediments, north-eastern New South Wales, Australia: lead isotope and chemical evidence. Environ Geol 36(1–2):118–126. https://doi.org/10.1007/s002540050327

Lu Y, Yuan J, Lu X, Su C, Zhang Y, Wang C et al (2018) Major threats of pollution and climate change to global coastal ecosystems and enhanced management for sustainability. Environ Pollut 239:670–680

Maanan M (2008) Heavy metal concentrations in marine molluscs from the Moroccan coastal region. Environ Pollut 153(1):176–183. https://doi.org/10.1016/j.envpol.2007.07.024

Maanan M, Saddik M, Maanan M, Chaibi M, Assobhei O, Zourarah B (2015) Environmental and ecological risk assessment of heavy metals in sediments of Nador lagoon, Morocco. Ecol Indic 48:616–626. https://doi.org/10.1016/j.ecolind.2014.09.034

Mahurpawar M (2015) Effects of heavy metals on human health. Int J Res GRANTHAALAYAH 3(9SE):1–7

Marchand C, Fernandez JM, Moreton B (2016) Trace metal geochemistry in mangrove sediments and their transfer to mangrove plants (New Caledonia). Sci Total Environ 562:216–227. https://doi.org/10.1016/j.scitotenv.2016.03.206

Marinho CH, Giarratano E, Domini CE, Garrido M, Gil MN (2019) Potential mobility assessment of metals in salt marsh sediments from San Antonio Bay. Environ Monit Assess 191(12). https://doi.org/10.1007/s10661-019-7895-0

Martin S, Griswold W (2009) Human health effects of heavy metals. Environ Sci Technol Briefs Citizens 15:1–6

Martínez ML, Intralawan A, Vázquez G, Pérez-Maqueo O, Sutton P, Landgrave R (2007) The coasts of our world: ecological, economic and social importance. Ecol Econ 63(2–3):254–272. https://doi.org/10.1016/j.ecolecon.2006.10.022

Martínez-Sánchez MJ, Navarro MC, Pérez-Sirvent C, Marimón J, Vidal J, García-Lorenzo ML, Bech J (2008) Assessment of the mobility of metals in a mining-impacted coastal area (Spain, Western Mediterranean). J Geochem Explor 96(2–3):171–182. https://doi.org/10.1016/j.gexplo.2007.04.006

Mehvar S, Filatova T, Dastgheib A, de Ruyter van Steveninck E, Ranasinghe R (2018) Quantifying economic value of coastal ecosystem services: a review. J Mar Sci Eng 6:5

Milcu AI, Hanspach J, Abson D, Fischer J (2013) Cultural ecosystem services: a literature review and prospects for future research. Ecol Soc 18(3). https://doi.org/10.5751/ES-05790-180344

Mishra AK, Santos R, Hall-Spencer JM (2020) Elevated trace elements in sediments and seagrasses at CO2 seeps. Mar Environ Res 153:104810

Mohammadi NS (2019) Molecular and physiological investigation of trace metal stress in seagrass, Zostera muelleri

Morillo J, Usero J, Gracia I (2004) Heavy metal distribution in marine sediments from the southwest coast of Spain. Chemosphere 55(3):431–442. https://doi.org/10.1016/j.chemosphere.2003.10.047

Nazneen S, Singh S, Raju NJ (2019) Heavy metal fractionation in core sediments and potential biological risk assessment from Chilika lagoon, Odisha state, India. Quat Int 507:370–388

Neto JAB, Gingele FX, Leipe T, Brehme I (2006) Spatial distribution of heavy metals in surficial sediments from Guanabara Bay: Rio de Janeiro, Brazil. Environ Geol 49(7):1051–1063. https://doi.org/10.1007/s00254-005-0149-1

Newton A, Brito AC, Icely JD, Derolez V, Clara I, Angus S, Schernewski G, Inácio M, Lillebø AI, Sousa AI, Béjaoui B, Solidoro C, Tosic M, Cañedo-Argüelles M, Yamamuro M, Reizopoulou S, Tseng HC, Canu D, Roselli L, Maanan M, Cristina S, Ruiz-Fernández AC, Lima RFD,

Kjerfve B, Rubio-Cisneros N, Pérez-Ruzafa A, Marcos C, Pastres R, Pranovi F, Snoussi M, Turpie J, Tuchkovenko Y, Dyack B, Brookes J, Povilanskas R, Khokhlov V (2018) Assessing, quantifying and valuing the ecosystem services of coastal lagoons. J Nat Conserv 44:50–65

Ngoile MAK, Horrill CJ (1993) Coastal ecosystems, productivity and ecosystem protection: coastal ecosystem management. Ambio 22(7):461–467. https://doi.org/10.2307/4314127

Nikolenko Y, Fedonenko E (2020) Analysis of the content of heavy metals in phytoplankton of the Zaporizhia reservoir. Sci J 3(24):12–17. https://doi.org/10.15587/2519-8025.2020.210095

Nobi EP, Dilipan E, Thangaradjou T, Sivakumar K, Kannan L (2010) Geochemical and geo-statistical assessment of heavy metal concentration in the sediments of different coastal ecosystems of Andaman Islands, India. Estuar Coast Shelf Sci 87(2):253–264. https://doi.org/10.1016/j.ecss.2009.12.019

Nobre AM (2009) An ecological and economic assessment methodology for coastal ecosystem management. Environ Manag 44(1):185–204. https://doi.org/10.1007/s00267-009-9291-y

Nolan AL, Lombi E, McLaughlin MJ (2003) Metal bioaccumulation and toxicity in soils—why bother with speciation? Aust J Chem 56:77–91

Oelsner GP, Stets EG (2019) Recent trends in nutrient and sediment loading to coastal areas of the conterminous U.S.: insights and global context. Sci Total Environ 654:1225–1240. https://doi.org/10.1016/j.scitotenv.2018.10.437

Ojuederie OB, Babalola OO (2017) Microbial and plant-assisted bioremediation of heavy metal polluted environments: a review. Int J Environ Res Public Health 14:1504

Pandey G, Madhuri S (2014) Heavy metals causing toxicity in animals and fishes. Res J Anim Vet Fish Sci 2(2):17–23

Papathanasiou V, Orfanidis S, Brown MT (2015) Intra-specific responses of Cymodocea nodosa to macro-nutrient, irradiance and copper exposure. J Exp Mar Biol Ecol 469:113–122

Pascaud G, Leveque T, Soubrand M, Boussen S, Joussein E, Dumat C (2014) Environmental and health risk assessment of Pb, Zn, As and Sb in soccer field soils and sediments from mine tailings: solid speciation and bioaccessibility. Environ Sci Pollut Res 21(6):4254–4264. https://doi.org/10.1007/s11356-013-2297-2

Pekey H (2006) Heavy metal pollution assessment in sediments of the Izmit Bay, Turkey. Environ Monit Assess 123:219–231

Peng JF, Song YH, Yuan P, Cui XY, Qiu GL (2009) The remediation of heavy metals contaminated sediment. J Hazard Mater 161(2–3):633–640

Pérez-Ruzafa A, Pérez-Ruzafa IM, Newton A, Marcos C (2019) Coastal lagoons: environmental variability, ecosystem complexity, and goods and services uniformity. In: Coasts and estuaries: the future. Elsevier, Amsterdam, pp 253–276

Peters EC, Gassman NJ, Firman JC, Richmond RH, Power EA (1997) Ecotoxicology of tropical marine ecosystems. Environ Toxicol Chem 16:12–40

Petranich E, Croce S, Crosera M, Pavoni E, Faganeli J, Adami G, Covelli S (2018) Mobility of metal (loid) s at the sediment-water interface in two tourist port areas of the Gulf of Trieste (northern Adriatic Sea). Environ Sci Pollut Res 25(27):26887–26902

Pinto MI, Burrows HD, Sontag G, Vale C, Noronha JP (2016) Priority pesticides in sediments of European coastal lagoons: a review. Mar Pollut Bull 112:6–16

Pitacco V, Crocetta F, Orlando-Bonaca M, Mavrič B, Lipej L (2017) The Mediterranean stony coral Cladocora caespitosa (Linnaeus, 1767) as habitat provider for molluscs: colony size effect. J Sea Res 129:1–11

Plum LM, Rink L, Haase H (2010) (2010) the essential toxin: impact of zinc on human health. Int J Environ Res Public Health 7(4):1342–1365. https://doi.org/10.3390/ijerph7041342

Prange JA, Dennison WC (2000) Physiological responses of five seagrass species to trace metals. Mar Pollut Bull 41(7–12):327–336

Premier V, de Souza Machado AA, Mitchell S, Zarfl C, Spencer K, Toffolon M (2019) A model-based analysis of metal fate in the thames estuary. Estuar Coasts 42(4):1185–1201. https://doi.org/10.1007/s12237-019-00544-y

Rahman MS, Saha N, Molla AH, Al-Reza SM (2014) Assessment of anthropogenic influence on heavy metals contamination in the aquatic ecosystem components: water, sediment, and fish. Soil and Sediment Contamination: An Int J 23(4):353–373

Rai PK (2008) Heavy metal pollution in aquatic ecosystems and its phytoremediation using wetland plants: an ecosustainable approach. Int J Phytoremediation 10:133–160

Rai LC, Gaur JP (eds) (2012) Algal adaptation to environmental stresses: physiological, biochemical and molecular mechanisms. Springer Science & Business Media, Berlin

Rainbow PS, Luoma SN (2011) Trace metals in aquatic invertebrates. In: Environmental contaminants in biota. CRC Press, Boca Raton, FL, pp 231–254

Ramesh R, Subramanian V (1988) Temporal, spatial and size variation in the sediment transport in the Krishna River basin, India. J Hydrol 98(1–2):53–65. https://doi.org/10.1016/0022-1694(88)90205-3

Rao NS, Ghermandi A, Portela R, Wang X (2015) Global values of coastal ecosystem services: a spatial economic analysis of shoreline protection values. Ecosyst Serv 11:95–105. https://doi.org/10.1016/j.ecoser.2014.11.011

Rehman K, Fatima F, Waheed I, Akash MSH (2018) Prevalence of exposure of heavy metals and their impact on health consequences. J Cell Biochem 119(1):157–184. https://doi.org/10.1002/jcb.26234

Ryu J, Khim JS, Kang SG, Kang D, Lee CH, Koh CH (2011) The impact of heavy metal pollution gradients in sediments on benthic macrofauna at population and community levels. Environ Pollut 159(10):2622–2629

Sabdono A (2009) Heavy metal levels and their potential toxic effect on coral Galaxea fascicularis from Java Sea, Indonesia. Res J Environ Sci 3:96–102

Sannigrahi S, Zhang Q, Pilla F, Joshi PK, Basu B, Keesstra S, Roy PS, Wang Y, Sutton PC, Chakraborti S, Paul SK, Sen S (2020) Responses of ecosystem services to natural and anthropogenic forcings: a spatial regression based assessment in the world's largest mangrove ecosystem. Sci Total Environ 715. https://doi.org/10.1016/j.scitotenv.2020.137004

Santhanam P (2011) An investigation on heavy metals accumulation in water, sediment and small marine food chain (plankton and fish) from Coromandel Coast, southeast coast of India. Indian J Nat Sci Int Bimonthly. ISSN 976, 0997

Selvi A, Rajasekar A, Theerthagiri J, Ananthaselvam A, Sathishkumar K, Madhavan J, Rahman PKSM (2019) Integrated remediation processes toward heavy metal removal/recovery from various environments—a review. Front Environ Sci 7. https://doi.org/10.3389/fenvs.2019.00066

Sharifuzzaman SM, Rahman H, Ashekuzzaman SM, Islam MM, Chowdhury SR, Hossain MS (2016) Heavy metals accumulation in coastal sediments. In: Environmental remediation technologies for metal-contaminated soils. Springer, Berlin, pp 21–42

Sobariu DL, Fertu DIT, Diaconu M, Pavel LV, Hlihor RM, Drăgoi EN, Curteanu S, Lenz M, Corvini PFX, Gavrilescu M (2017) Rhizobacteria and plant symbiosis in heavy metal uptake and its implications for soil bioremediation. New Biotechnol 39:125–134. https://doi.org/10.1016/j.nbt.2016.09.002

Sodango TH, Li X, Sha J, Bao Z (2018) Review of the spatial distribution, source and extent of heavy metal pollution of soil in China: impacts and mitigation approaches. J Health Pollut 8:53–70

Spiegel J, Maystre LY (1998) Environmental pollution control and prevention. In: Stellman JM, editor. Encyclopedia of Occupational Health and Safety. 4th ed.Vol 2. Geneva: International Labour Office; 1998. Section 55.

Suja S, Kessarkar PM, Fernandes LL, Kurian S, Tomer A (2017) Spatial and temporal distribution of metals in suspended particulate matter of the Kali estuary, India. Estuar Coast Shelf Sci 196:10–21. https://doi.org/10.1016/j.ecss.2017.06.024

Sullivan JJ, Timmins SM, Williams PA (2005) Movement of exotic plants into coastal native forests from gardens in northern New Zealand. N Z J Ecol 29(1):1–10

Sundaramanickam A, Shanmugam N, Cholan S, Kumaresan S, Madeswaran P, Balasubramanian T (2016) Spatial variability of heavy metals in estuarine, mangrove and coastal ecosystems along Parangipettai, southeast coast of India. Environ Pollut 218:186–195

Susana Villanueva F, Botello AV (1998) Metal pollution in coastal areas of Mexico. Rev Environ Contam Toxicol 157:53–94. https://doi.org/10.1007/978-1-4612-0625-5_3

Tchounwou PB, Yedjou CG, Patlolla AK, Sutton DJ (2012) Heavy metal toxicity and the environment. EXS 101:133–164

Unsworth RKF, Nordlund LM, Cullen-Unsworth LC (2019) Seagrass meadows support global fisheries production. Conserv Lett 12(1). https://doi.org/10.1111/conl.12566

van Dam J, Andrew N, Uthicke S, Mueller J (2011). Chemical pollution on coral reefs: exposure and ecological effects. https://doi.org/10.2174/978160805121210187

van der Schyff V, Yive N, Bouwman H (2020) Metal concentrations in corals from South Africa and the Mascarene Basin: a first assessment for the Western Indian Ocean. Chemosphere 239. https://doi.org/10.1016/j.chemosphere.2019.124784

van Oppen MJH, Gates RD, Blackall LL, Cantin N, Chakravarti LJ, Chan WY, Cormick C, Crean A, Damjanovic K, Epstein H, Harrison PL, Jones TA, Miller M, Pears RJ, Peplow LM, Raftos DA, Schaffelke B, Stewart K, Torda G, Wachenfeld D, Weeks AR, Putnam HM (2017) Shifting paradigms in restoration of the world's coral reefs. Glob Chang Biol 23:3437–3448

Viles H, Spencer T (1995) Coastal problems: geomorphology, ecology and society at the coast. Arnold, London. https://doi.org/10.2307/3059890

Violante A, Cozzolino V, Perelomov L, Caporale AG, Pigna M (2010) Mobility and bioavailability of heavy metals and metalloids in soil environments. J Soil Sci Plant Nutr 10(3):268–292

Wang SL, Xu XR, Sun YX, Liu JL, Li HB (2013) Heavy metal pollution in coastal areas of South China: a review. Mar Pollut Bull 76:7–15

Wang Y, Jiao JJ, Zhang K, Zhou Y (2016) Enrichment and mechanisms of heavy metal mobility in a coastal quaternary groundwater system of the Pearl River Delta, China. Sci Total Environ 545–546:493–502. https://doi.org/10.1016/j.scitotenv.2015.12.019

Warren Flint R (1985) Coastal ecosystem dynamics: relevance of benthic processes. Mar Chem 16(4):351–367. https://doi.org/10.1016/0304-4203(85)90056-8

Wei M, Yanwen Q, Zheng B, Zhang L (2008) Heavy metal pollution in Tianjin Bohai bay, China. J Environ Sci 20(7):814–819

Williams TP, Bubb JM, Lester JN (1994) Metal accumulation within salt marsh environments: a review. Mar Pollut Bull 28:277–290

Wu Q, Tam NFY, Leung JYS, Zhou X, Fu J, Yao B, Huang X, Xia L (2014) Ecological risk and pollution history of heavy metals in Nansha mangrove, South China. Ecotoxicol Environ Saf 104(1):143–151. https://doi.org/10.1016/j.ecoenv.2014.02.017

Wu W, Wu P, Yang F, Sun DL, Zhang DX, Zhou YK (2018) Assessment of heavy metal pollution and human health risks in urban soils around an electronics manufacturing facility. Sci Total Environ 630:53–61. https://doi.org/10.1016/j.scitotenv.2018.02.183

Yan C, Li Q, Zhang X, Li G (2010) Mobility and ecological risk assessment of heavy metals in surface sediments of Xiamen Bay and its adjacent areas, China. Environ Earth Sci 60(7):1469–1479. https://doi.org/10.1007/s12665-009-0282-3

Yan Z, Sun X, Xu Y, Zhang Q, Li X (2017) Accumulation and tolerance of mangroves to heavy metals: a review. Curr Pollut Rep 3:302–317

Yang T, Diao X, Cheng H, Wang H, Zhou H, Zhao H, Chen CM (2020) Comparative study of polycyclic aromatic hydrocarbons (PAHs) and heavy metals (HMs) in corals, sediments and seawater from coral reefs of Hainan, China. Environ Pollut 264:114719

Yang CP, Liu Y, Shan BB, Xu J, Yu W, Sun DR, Zhang ZW (2021) Heavy metal concentrations and associated health risks in edible tissues of marine nekton from the outer Pearl River Estuary, South China Sea. Environ Sci Pollut Res 28(2):2108–2118

Yi Y, Xiao M, Mostofa KMG, Xu S, Wang Z (2019) Spatial variations of trace metals and their complexation behavior with DOM in the water of Dianchi Lake, China. Int J Environ Res Public Health 16:4919. https://doi.org/10.3390/ijerph16244919

Young RA (2005) Toxicity profiles: toxicity summary for cadmium, risk assessment information system. RAIS, Univ Tennessee
Zampieri BDB, da Costa Andrade V, Chinellato RM, Garcia CAB, de Oliveira MA, Brucha G, de Oliveira AJFC (2020) Heavy metal concentrations in Brazilian port areas and their relationships with microorganisms: can pollution in these areas change the microbial community? Environ Monit Assess 192(8):1–17
Zhang W, Liu X, Cheng H, Zeng EY, Hu Y (2012) Heavy metal pollution in sediments of a typical mariculture zone in South China. Mar Pollut Bull 64(4):712–720. https://doi.org/10.1016/j.marpolbul.2012.01.042
Zhang C, Yu Z-G, Zeng G-M, Jiang M, Yang Z-Z, Cui F, Zhu M-Y, Shen L-Q, Hu L (2014) Effects of sediment geochemical properties on heavy metal bioavailability. Environ Int 73:270–281
Zhao Y, Xu M, Liu Q, Wang Z, Zhao L, Chen Y (2018) Study of heavy metal pollution, ecological risk and source apportionment in the surface water and sediments of the Jiangsu coastal region, China: a case study of the Sheyang Estuary. Mar Pollut Bull 137:601–609. https://doi.org/10.1016/j.marpolbul.2018.10.044

Chapter 11
A Holistic Study on Impact of Anthropogenic Activities over the Mangrove Ecosystem and Their Conservation Strategies

Monika and Abhinav Yadav

Abstract Despite their importance, the mangrove ecosystem is one of the highly vulnerable ecosystems in the Anthropocene era. Mangrove ecosystems lie in an intertidal zone of subtropics and tropics regions. They provide ecological and economic services to the coastal communities. Mangrove provides multifaceted advantages to the local ecosystem such as it reduces the severity of the hurricane, storm surge, cyclone, and tsunami, prepares a perfect bed for spawning marine fishes, and also plays a major role in carbon sequestration. Deterioration in global estimates of mangrove covers ~150,000 km^2 is the consequence of exponentially increasing urbanization and industrialization. These two major anthropogenic activities induce numerous problems such as an increment in the intensity of natural calamities, local inhabitant losing their livelihood, and many marine species standing on the verge of extinction. An integrated approach is required for the preservation and management of mangrove biotopes with an amalgamation of local inhabitants, researchers, and government. Conservation techniques include afforestation, legislation, policies, application of remote sensing and geoinformation system (GIS), and development of parks and reserves for protection. This chapter is a consolidated approach to study the sources and impact of anthropogenic threats on mangrove forests from a global and Indian perspective with holistic conservation strategies.

Keywords Mangrove ecosystem · Anthropogenic · Impacts · Conservation

Monika · A. Yadav (✉)
Institute of Environmental and Sustainable Development, Varanasi, India

S. Madhav et al. (eds.), *Coastal Ecosystems*, Coastal Research Library 38,
https://doi.org/10.1007/978-3-030-84255-0_11

1 Introduction

1.1 A Brief Outlay of Mangrove Ecosystem (ME)

Mangrove ecosystems are one of the most fragile coastal forests. It is commonly found near the cliff of lagoons, shores, estuaries, and riverside in the subtropics and tropics zones. Maiti and Chowdhury (2013) explained the term "mangrove"; according to them mangrove trees or shrubs have specialized capabilities to adapt in tidal regions. Mangroves are salt-tolerant plant species. Which uptake water through high osmotic potential but after absorption of saline water, its excreted salt through their salt glands which are located in leaves (Parida and Jha 2010). Mangroves are tough, hard, and woody halophytic plants, and their salt-tolerant characteristics angrove ecosystems salt-tolerant characteristics help them to exist in highly saline, extreme tidal, strong windy, high temperature, and muddy-anaerobic soil conditions (Guo et al. 2017). Mangrove is indigenous in halophytic condition with the help of supportive root system, pneumatophores, stilt root, and leathery and evergreen sclerophyllous foliage with salt excretion glands and recessed stomata. Out of the total percentage of mangrove in India, more than half are existing on the east coastline of the Bay of Bengal, only 20% on the west shore of the Arabian Sea, and a major chunk of 13% on the Andaman and Nicobar Islands (Kathiresan et al. 2018). According to the India State of Forest Report (2009), mangrove ecosystem range in India is categorized as highly dense (more than 70% plant cover, 140,500 ha), moderately dense (40–70%, 165,900 ha), and open type (10–40%, 157,500 ha) forest types. It is explained in previous studies; Sundarbans is a very diverse mangrove forest and enriched with more than 62 species. The Indian-Malaysian region is considered to be the center for the progression of the mangrove ecosystem (Maiti and Chowdhury 2013).

Mangroves are considered as a linkage between freshwater and marine ecosystems. Mangroves plays multiple roles in maintaining ecological services such as (1) pollution sink; (2) source of nutrient exchange in coastal environment; (3) increasing sediment accretion and stabilizing shorelines; (4) binding nutrient and heavy metals to improve quality; (5) serving as a reservoir of food, fuel, fodder, and medicine for coastal inhabitants; (6) providing a breeding bed for aquatic species like amphibians, fishes, and crustaceans; (7) providing shelters for reptiles, birds, and mammals; and (8) reducing the risk of some natural calamities, e.g., cyclones, typhoons, or tsunamis (Valiela et al. 1974; Banus et al. 1975; Mitchell 1978; Giblin et al. 1980; Chu et al. 2000). Because of human activity's overextension, the mangrove environment continues to be in grave danger despite these ecological, social, and economic benefits. As a result, in many parts of the world, the species diversity index of mangroves is decreasing steadily (1–2 percent each year), in many regions of the globe. Gradual loss of mangroves in the entire globe will project to 60% by the year 2030 (Chaudhuri et al. 2015).

According to a study by Di Nitto et al.'s (2014) global warming may raise sea levels, causing mangroves to migrate closer to the land report illustrated global warming could increase the sea level and, as a repercussion of that, mangroves can move toward land. However, across the world, mangroves are unlikely to shift landward because of human intervention at the terrestrial boundary, and the major patch of the forest cover is likely to diminish (Ward et al. 2016). A drastic paradigm shift can be observed in mangrove habitat; it alters because of neighboring circumstances such as the condition of wetland and its type, geomorphic setting, as well as most important substantial activities in the wetland. It is predicted that the sea rise may range from ~450 to 650 mm per century (Mimura 2013). However, mangroves can only survive in ~80–90 mm rise in sea level, and hence the predicted threat moves faster toward the mangrove to disintegrate the ecosystems (Sasmito et al. 2016).

This study aims to point out how the mangrove ecology is adversely affected by pollution, population, overexploitation, habitat encroachment, invasion, land-use change, management, and conservation programs through different techniques. Mangrove nurtures the pivotal balance of the integrity of the environment, and it acts vitally in maintaining equilibrium between nutrient cycling in the estuarine and coastal ecosystem. Mangrove restoration activities have been conducted in the Tutuila Island, American Samoa. Countries involved in the rehabilitation of mangrove, and they tried to sustain the coherence of the ecosystem, for instance, substrate conditions, salinity regime, depth of inundation, wave energy, tidal velocity, soil and water pH, and sediment matrix (Maiti and Chowdhury 2013).

2 Benefits of Mangrove Ecosystem

Mangrove ecosystems provide multiple services: provisioning, regulating, cultural, and ecological services. Provisioning ecological services include the use of mangrove-grown fuel, such as Nypa leaves, lumber, and charcoal. Mangrove forests contribute significantly to carbon sequestration/storage and provide ecological benefits (Cornell et al. 2018) and deliver its ecological services. In addition, supporting services of that mangrove provide opportunities for recreation and moderation of extreme events. It reduces the intensity of floods, cyclones, coastal erosion, tsunami, etc.

2.1 *Provisioning Services*

Mangrove forests are the host of thousands of people. It supplies many essential products such as firewood, charcoal, timber, honey, etc. to the local villagers and flourishes the fishery businesses in this mangrove to deliver their provisioning services (Aye et al. 2019). Mangrove woods and their charcoal product have high calorific value than Indian coal. Its product producing more heat and less smoke might be because of the high content of tannin (Sathe et al. 2013). Each part of the mangrove has its importance. Its uprooted pneumatophores are used for manufacturing bottleneck stopper; their leaves are thatched as roofs; shells are used for manufacturing lime. Mangrove facilitates apiculture activities; around 2000 people are engaged in more than a hundred tons of annual honey and wax production in Sundarbans, India (Kathiresan 2018). Some specific mangrove such as *Avicennia* produces cheap and nutritive feed for ruminants (Sathe et al. 2015).

Mangrove ecosystems are very sensitive to hydrological changes; minor variations in tidal regimes trigger remarkable variation. Therefore, it acts as a precursor of rising in sea level (Blasco et al. 1996). Mangrove ecosystem plays a pivotal role in the maintenance of nutrients by restoring them in dead modified roots varying from 36% to 88% of total viable tree biomass (Alongi et al. 2003), unlike terrestrial forests where a significant percentage of nutrient capital can restore in floor litter.

2.2 Ecological Services

Mangroves can protect the coastal communities from harmful solar UV-B radiation. *Avicennia* is a species of mangrove that grows in high sunlight endowed area and well adapted to arid zones (Moorthy and Kathiresan 1997). The mangrove foliage is a good source of flavonoids that screen the UV-B and reduce the detrimental effect of ultraviolet (Moorthy and Kathiresan 1998). IUCN report illustrates mangroves are the sink of carbon dioxide and up to some extent curtail the greenhouse gas. A sufficient amount of carbon dioxide doubles the biomass. *Rhizophora mangle* aboveground biomass (shoot) got doubled. On the contrary, the root system, above and belowground biomass not as responsive as shoot biomass, and the proportion between responsiveness is about 2.5:1. This might be because of the significant amount of organic carbon storage in sediments (Estrada and Soares 2017). Mangrove wood has thick organically rich sediments in its substrata. Except in the deltaic area, the majority of mangrove peat substrate in the tropics is derived by mangrove roots. This entire belowground coalition in forests has great productivity as well as are nutrient-rich. The whole setup of mangrove forest is good for carbon sequestration not only above- but also belowground (Alongi 2014).

Mangrove grows in swampy beds; it is a good source of nutrients for fishes. The swampy bed provides breeding grounds and nurseries for marine Pisces (Mandal et al. 2013). This ecosystem is the ecotone between terrestrial and marine systems so comparatively more productive than the agricultural fields. It is 25 times fertile than the plane paddy cultivated area (Tripathi et al. 2016).

2.3 Supporting Services

Mangrove forest reduces the intensity of cyclone and tidal storms (Krauss and Osland 2020). When the 310 km/h super cyclone struck the Odisha coast in India on October 29, 1999 and killed approximately 10,000 people and devasted the mangrove-depleted region, it was the finest example to illustrate its significance (Das and Vincent 2009). The thick mangrove gallery, on the other hand, sustained just little damage occurred at the dense mangrove gallery. Mangroves, rather than the "cleared" zone, may be used to protect hectares of land in cyclone and tsunami-prone regions. Mangrove plantations are capable of diminishing the fury of the tsunami as well as controlling the shoreline against damage (Das and Vincent 2009) (Table 11.1). Dense mangrove growth avoids the intensity of several disasters in the Bay of Bengal, West Bengal, Bangladesh, Myanmar, Maldives, etc. There is one incidence in Thailand's coastline where mangroves breach the lethality of the tsunami in Surin Island (WHO Report 2004). A hydraulic experiment was done by Harada et al. (2002), to study the tsunami reduction effect on coastlines with five major models of mangroves. According to the result found, the coastal forest is more effective than the seawall to resist the sea wave in the disaster like tsunami, to protect the houses and livestock. To tranquilize the amplitude and energy of the wave, dense vegetation has also been effective via wave forces measurements and fluid dynamics modeling (Massel et al. 1999). For example, according to the analytical model, the agitation caused by tsunami flow pressure may be reduced by as much as 90% when 30 trees from 10sq meter are planted in a 0.1km broad strip. But this method only works when the wave height is between 0m and 5m (Tanaka 2009).

Mangroves not only protect us from tsunami and cyclones but also provide flood resistance in the coastline and reduce seawater intrusion in groundwater. It has a prominent root system and pneumatophores that spread out in the immense area which provide stability and promote sedimentation reducing flood expectancy (Srikanth et al. 2016). Mangroves act as a boon for coastal inhabitants, supporting in maintaining purity of underground water.

Mangrove ecosystem provides commendable services to minimize coastal erosion due to large waves. Its lush growth reduces the severity of large waves and coastal erosion. The restoration of mangroves in Vietnam's Red River Delta has taken many stages and cost millions of dollars (Hai et al. 2020). Around 100-m-wide belt of mangroves can be helpful in protecting the adjacent mangrove area (Albers and Schmitt 2015). However, the diminishing of immense mangrove forest causes huge sediment erosion in the coastal zone. The mangrove induces sedimentation with the help of pneumatophores and thus causes land expansion (Kathiresan 2018). In various cases, there has been evidence of yearly sedimentation rate, lying between 0.1 and 0.8 cm in mangrove zones, which causes expansion of land (Woodroffe et al. 2016) (Table 11.1).

Table 11.1 Services provided by mangrove ecosystem

Benefits of mangrove ecosystem	Provisioning services	• A good source of firewood, charcoal, timber, honey, etc. • Facilitate apiculture and aquaculture
	Ecological services	• Protect from harmful solar radiation • Act as sink for carbon dioxide • Provide swampy bed for fish spawning
	Supporting services	• Reduce intensity of tsunami and cyclone • Act as seawall • Avoid mixing of saline water and fresh water

3 Anthropogenic Threats for Mangrove Ecosystem

Human intervention plays a vital role in mangrove deforestation in past decades. Approximately 73% declined over the 16 years' (2000–2016) period. However, a large proportion of mangrove loss was reported in the first epoch around 1186 km^2, whereas a comparatively slow degradation of 314 km^2 occurred in the last epoch (Goldberg et al. 2020). Anthropogenic activities are responsible for 80% (2065 km^2) loss of the mangrove ecosystem in just six countries: Indonesia, Malaysia, Myanmar, Thailand, Philippines, and Vietnam (Goldberg et al. 2020). Other than these six nations, Southeast Asian countries are also facing the major deterioration of the ecosystem primarily due to widespread mangrove transition to aquaculture ponds and agricultural field (Richards and Friess 2016). According to Synthesis Report on Ten ASEAN Countries Disaster Risks Assessment (2010), coastal mangrove hotspot was widely spread in South Asian countries such as the Rakhine State of Myanmar, Mekong Delta in Vietnam, and the Kalimantan and Sulawesi regions of Indonesia (FAO 2007). A substantial (77%) decrease has been observed in mangrove percentage area from previous year data in last decades (Goldberg et al. 2020). There are many incidences available of nonproductive conversion (NPC), such as petroleum extraction in the Niger Delta causing 20 km^2 area loss and mining activities in Grasberg mine tailing in Papua, Indonesia, inducing 5 km^2 loss (Alonzo et al. 2016). NPC-stimulated losses somehow declined by 46% from 268 km^2 in 2005 to 129 km^2 in 2016 (Toumbourou et al. 2020). Nonproductive conversions (NPC) approximately cause 12% (398 ± 29 km^2) of global loss but reclaimed land for human colonies only representing 3% (96 ± 15 km^2) of global loss extent (Goldberg et al. 2020).

The remarkable amount of mangrove density decreases; many drivers justify this change such as leading aquaculture industries, agriculture practices, forestry, oil palm plantation, rice cultivation, urban sprawling, and irrational industrial establishment (Webb et al. 2014; Lai et al. 2015). These factors are basically responsible for land cover change along with mangrove exhaustion.

3.1 Contribution of Aquaculture in Mangrove Loss

The involvement of aquaculture in the drivers for mangrove deterioration is very high in the last decades. The huge area is converted into fish or shrimp ponds (Valiela et al. 2001; Hamilton 2013). The government of various coastal nations, such as Thailand, Vietnam, Indonesia, and the Philippines, is encouraging the conversion of forest to fish and shrimp ponds in order to increase food security and livelihood. Conversion of mangrove areas into aquaculture is now in trend especially at Kalimantan and Sulawesi, Indonesia. Deforestation accounts for 54% of mangrove for aquaculture since 1980 to the 1990s (Hamilton 2013), and the percentage is going to rise during 2010–2011 (Richards and Friess 2016). Local people got involved in aquacultural development at the site of mangroves in 2010–2011 (Richards and Friess 2016). The current Indonesian government has backed aquaculture it is important to mention that, not only aquaculture is a factor, but production of rice also contributes to mangrove loss.

3.2 *Enhancement of Rice Cultivation*

In recent years, Myanmar has promoted rice cultivation, but at the cost of Southeast Asia's mangrove forests. Numerous studies reported about natives of the Ayeyarwady Delta affecting the mangrove ecosystem for paddy cultivation (Webb 2014), but the present situation deepens in whole Myanmar for expansion of rice agriculture. Such examples act as a driving force for the fastest rate of mangrove deforestation of any country in Southeast Asia. After this incidence Myanmar government realizes their fault and trim all the activities affecting mangroves. They have taken various steps for the conservation of assets of Myanmar in the form of mangrove wood (Aung 2007) (Fig. 11.1). The government of various countries is enhancing rice production through technical assistance and conventional village level expansion targets to improve their food security (Okamoto 2007; Matsuda 2009). The nation benefited financially from rice cultivation, aquaculture, and oil palm culture (Webb 2014; Richards and Friess 2016).

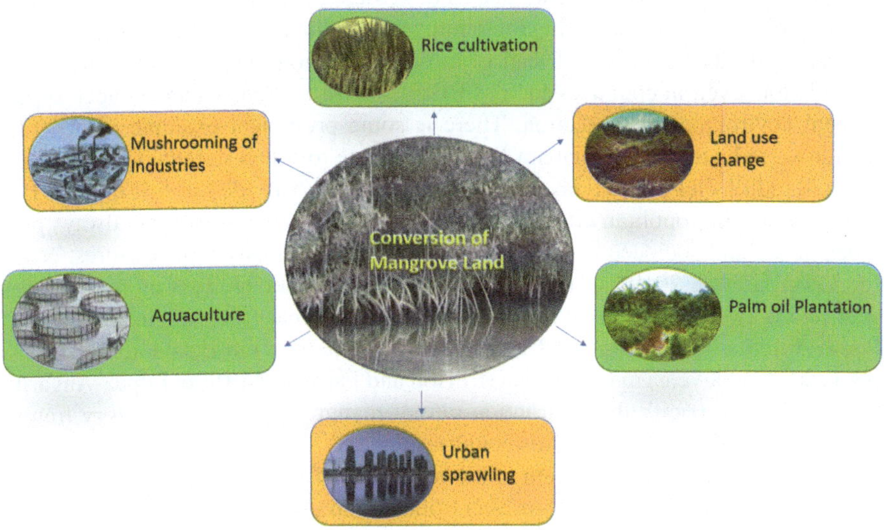

Fig. 11.1 Anthropogenic threat over mangrove land

3.3 *Increase of Oil Palm Plantation*

The increasing trend of oil palm plantations in Indonesia and Malaysia in the coastal region is also responsible for mangrove replacement and their loss (Koh and Wilcove 2008; Koh et al. 2011). According to Lee et al.'s (2014) study, most of the areas of Malaysia, Sumatra, and Indonesia contribute to oil palm production on the replacement of mangrove forest. Numerous governments of Southeast Asian countries support palm oil production companies for energy independence and economic benefits (Wicke et al. 2011) (Fig. 11.1). Now is the time for intertidal inhabitants and government agencies to initiate monitoring and supervising mangrove density and rehabilitation.

3.4 *Elevated Trends of Urban Sprawling and Industrialization*

Urban settlement and mushrooming of industries are migrating towards the coastal regions because mangroves are associated with job creation, land reclamation and recreation activities.Despite these many people dependent on the coastal ecosystem for their livelihood (Thuo 2013). There are few reasons which triggered the change in land-use pattern in the coastal zone such as overexploitation of mangrove for wood, timber (Rahman et al. 2010; Sahu et al. 2015) (Fig. 11.1). These factors are listed in the major drivers of non-replenish mangrove loss. The rate of rising urbanization and industrial development on the sea coast might result in the disposal of wastewater, irrational solid waste disposal, sewage generation, as well as invasion in mangrove zone and converting them into metropolitan space (Ibharim et al. 2015; Hasnat et al. 2018). This will lead to rapid devastation, deforestation, and forest degradation, even in countries such as Brazil, where mangroves are subject to permanent environmental protection. There is some prominent example of 5000 ha mangrove reduction in Guanabara Bay, Brazil, due to urban settlement and expansion of landfill sites (7000 t/day) (Godoy and Lacerda 2015; de Lacerda et al. 2019). Extension of metropolitan cities in seaside always attracts tourists and always provides peace and socioeconomic benefits, but it tremendously reduces forest cover (de Lacerda et al. 2019). Caribbean forest faces considerable loss up to 1.7% per year and 0.2% per year in islands, 2.2% (10,702 ha/year) per year in Central and South America in Panama, 1.48% per year (5358 ha/year) in Ecuador, and 1.23% per year (794 ha/year) in Costa Rica (Ellison and Farnsworth 1996; López-Angarita et al. 2016). To pacify the destruction rate of mangroves, some regulatory frameworks are implemented in various countries.

Oil spill is a common problem in the proximity of unregulated industries which forms a layer over the sea and causes eutrophication. An oil spill incident happened in Panama on April 27, 1986, where approximately 5,962,024 L of medium-weight crude oil percolated into coastal lines and expressed deleterious effects on mangrove forest (Cubit et al. 1987). This event triggered an extensive regulation that involved oil spill preparedness on offshore and onshore support activities (Arbo and Thủy 2016). About 54% of vessel incidents, 21% of pipelines, and 14% of shore tanks have already happened, but now the number of oil spill cases is decreasing, and their ill impact on mangrove ecosystem has also reduced in three decades of the twentieth century (Duke 2016). However, when past and current centenaries are compared, it must be seen that the frequency of occurrences (oil spills) has increased significantly, and the total mangrove area has been significantly reduced by these incidents (de Lacerda et al. 2019).

3.5 *Extensive Agriculture*

Extensive agriculture contributes to nutrient enhancement, but its consequences deprive the water quality and ultimately cause mangrove degradation (Maryantika and Lin 2017). Extensive agriculture includes intensive use of pesticides, damming of rivers, and diversion of waterways shifting the sedimentation erosion equilibrium ratio of coastal land and augmenting the salinity of groundwater (Kusmana 2014) (Fig. 11.1). Agriculture-induced mangrove loss is observed in Colombia, due to sudden alteration in water quality and its chemistry because of agrochemicals. This mangrove disappearance stimulated the most successful restoration experiment of mangroves in the USA (de Lacerda et al. 2019). Regionwide lessening of sediment transport to the coast is the repercussion of damming over the river. Damming causes topsoil erosion in coastal forests in arid and semiarid littoral zone. The salt intrusion has prolonged the saline impact upriver triggering mangrove migration inland (Godoy and de Lacerda 2014). Unbridled use of mangrove resources such as timber, wood, bark, medicinal use, fisheries use, and salt adversely affects the percentage of mangrove. Mangrove wood product has high economic and cultural values all over the world, and this specialty increases its utility as well as deforestation in Northern and Southern America (Venezuela). Approximately 3/4 of the area of Venezuelan mangroves was lost during the twentieth century (Villate Daza et al. 2020).

4 Current Global Status of Mangrove

Currently, a paradigm shift is observed in the density of mangrove forests throughout the world due to anthropogenic actions such as urban sprawling, population burst, the establishment of several industries, sewage discharge, and irregular disposal of municipal solid waste. These mentioned activities cause considerable change in the mangrove population everywhere. Chen et al. (2009) reported a drop of mangrove cover by more than 40% to approximately 22,000 ha in 2001 from more than 50,000 ha in 1950. Numerous threats and possibilities in the environment such as human-induced oil spill, aquaculture, agriculture, surface runoff, excessive use of mangrove products, etc. disturb the mangrove ecosystem (Liu et al. 2008; Chaudhuri et al. 2015). In Myanmar, around 168,500 ha of mangrove were reduced in previous decades (Giri et al. 2011; Webb et al. 2014). Indonesian mangrove forest is one of the largest coastlines belonging to 257 cities. According to the Ministry of Forestry (2007) report estimation, 7.8 million ha (30.7% in good condition, 27.4% moderately disturbed, 41.9% heavily disturbed) potential areas are covered with mangroves. The Agency of Survey Coordination and National Mapping, Republic of Indonesia (2009), estimated at 3.2 million ha the mangrove area. Maximum exploitation of mangroves in Indonesia has occurred due to extensive fishery culture, shrimp farming, salt ponds, mining, and expansion of urban activities (Kusmana 2014). More than 50% of mangrove areas are ruined by human activities in Segara Anakan Lagoon, Indonesia (Ardli and Wolff 2009; Hinrichs et al. 2009). Major destruction of the mangrove ecosystem in the Philippines happened because of the invasion of human activities. Land-use change is responsible for 4500 km^2 in 1920 to 1200 km^2 in 1994 ecosystem cover (Primavera 2000; Chaudhuri et al. 2015). Mangrove forests are spread on the southern and eastern coasts of the Gulf of Thailand mostly concentrated in the Andaman Sea. Coastline mangrove ecosystem forms two-story forest cover: the upper layer has around 20 m height and is dominated by *Xylocarpus mekongensis* (syn. *X. moluccensis*), *Rhizophora apiculata, Heritiera littoralis (ngon kai)*, and *Rhizophora mucronata, and the lower layer consists of Bruguiera parviflora, Bruguiera sexangula, Ceriops decandra*, and *Ceriops tagal*. Mangrove covered around 2 million Rai (1 Rai = 0.16 ha), till 1975, but the land cover changed since 1996, so the mangrove number decreased by 50% (Pumijumnong 2014). Following the strict implementation of conservation and rehabilitation strategies in 2004, the mangrove area increased to 1.5 million Rai, with the remaining mangrove area encroaching for shrimp cultivation. Mangrove land was also used for human settlement expansion, industrial expansion, and road construction in mangrove areas following the increase. In 2007, through remote sensing technology and interpretation of Landsat 5 satellite images, mangrove cover was estimated at 18.55% out of the total area in Thailand, and the largest mangrove cover is in Phang Nga, Thailand (Pumijumnong 2014).

In Brazil, nearly 0.5 thousand km^2 area of shoreside mangrove was deforested in the last 25 years mainly for aquaculture and farming (Giesen et al. 2007). Mangrove cover was dramatically decreased by 18% within 25 years at a mean rate of 0.7% per year due to improper legislation and habitat moderations (Kirui et al. 2013) During the last decade, mangrove forest was diminished by ~15 km^2 in Tanzania due to land reclamation (Wang et al. 2002). In Bangladesh problems have deepened more; Sundarbans' mangroves have lost 45% of their total coverage due to uninterrupted encroachment due to logging, shrimp farming, and natural disasters (Islam and Gnauck 2008; Roy 2014). Despite urbanization and industrialisation, certain nations, such as Australia and New Zealand, continue to invest heavily in the management and restoration of mangroves. These efforts continuously increase the restoration area and lead to the expansion of mangrove boundaries at a constant rate over the last few decades. The pace of extension doubled, from 240 ha in 1943 to 545 ha in 1999, in Tauranga Harbour, New Zealand (Ghosh et al. 2015). Mangrove land cover was improved by approximately 3.8%, 32.8%, and 55% in Gosford (New South Wales), Botany Bay (Sydney), and Phillip Island (Victoria) from 1954 to 1995, 1956 to 1996, and 1939 to 1999, respectively (Harty and Cheng 2003; Harty 2009).

5 Indian Status

The coasts of nine maritime and four Indian union territories are well flourished at 4740 km^2 of mangroves along the bank of estuaries. Indian mangroves are broadly classified into three main categories: (1) deltaic, (2) estuarine and backwater, and (3) insular (Andaman and Nicobar Islands). 58% of total Indian mangrove (which is 4740 km^2) can be found along the eastern coast (Bay of Bengal), 29% along the west coast, and 13% along the Andaman and Nicobar Islands. In mangroves, spatio-temporal changes had started 200 years ago at the degradation rate of 4% per year, which in the recent decades got intensified to an enhanced rate of degradation. Mangrove forests in transboundary Sundarbans reserve forests are drastically degrading for the last 20 years. Unfortunately, before 1870, no reliable data for mangroves is available. So, from the available records, it can be stated that the previous 200 years were more crucial for mangrove degradation. In addition to this, from 1873 to 1933, total forest cover was reduced to an extent of 1500 km^2. The last 2 years were more crucial for using remote sensing techniques in mangroves for mapping and understanding the spatiotemporal extent of mangrove forest especially in the perspective of natural disasters and anthropogenic forces. Such, site-specific or short-duration (2 years) investigation is extremely uncommon for the Indian Sundarbans world heritage site, particularly in the southwestern region. Among numerous studies, the majority were related to change in coastal geomorphic patterns, ocean level elevation (Jayappa et al. 2006). Change in vegetation dynamics is one of the least discussed issues in micro-/meso-level studies. Thus, using conventional NDVI methods, an attempt has been made to monitor and assess net changes in the vegetation of the entire integrated Sundarbans.

5.1 Mangrove Status in Southern Parts of India

Kerala has a 590-km-long coastal line. In this state, the mangrove vegetations are spread in the form of patches or continuous form along the banks of estuaries, in nearby areas of backwater channels, and near the water bodies. In this state, the mangrove is supported by 41 perennial rivers which create a favoring ecological environment for the development of mangroves on the fringes of estuaries, bank water, and creeks. Mangroves in the Andaman and Nicobar Islands (ANI) are abundant and very contrasted with other mangrove wood in India (Dagar et al. 1991; Mandal and Naskar 2008; Goutham-Bharathi et al. 2014). However, mangrove zones in the ANI encountered an extreme decrease in the most recent decade, potentially prompting changes in floristic organization and region termination of certain species. A region of around 54 km^2 has been degraded somewhere in the years 2003 and 2013, and especially somewhere in the range of 2011–2013, an overall zone of 13,000 m^2 of huge mangrove chunk has been disturbed (FSI 2015). It was determined that the geological and morphological changes caused by the massive seismic earthquake and subsequent tidal wave in December 2004 were the causal components of the ANI's subsequent loss of mangroves. As per the most recent measure by the Forest Survey of India (FSI 2015), the absolute mangrove zone in India is around 4740 km^2, of which 617 km^2 happens in the ANI. Of that area, 616 km^2 area in the Andaman Islands and the Nicobar Islands represents 1 km^2. There has been a net increment of 13 km^2 in the mangrove front of the ANI when contrasted with the 2013 evaluation. Even though mangroves of the ANI have seen an expanding pattern recently, the fast abatement of mangrove regions of the ANI during the most recent decade has become a significant concern regarding protection ratios of such a resource-rich island coastal environment. The mangroves of the ANI have been concentrated by numerous specialists; however, there is no agreement on the mangrove floristics of the ANI. Uncertainty exists concerning the qualification of major or genuine mangroves from minor and mangrove-related species, especially at their conventional levels. A few genera/species, viz., *Acrostichum*, *Acanthus*, *Pemphis acidula*, *Phoenix paludosa*, *Cynometra*, and *Dolichandrone spathacea*, are around the world considered as evident mangrove species (Duke 1992; Polidoro et al. 2010), though the previously mentioned species were dynamically grouped by Dagar et al. (1991), Singh (2003), Debnath (2004), and Dam Roy et al. (2009) in the ANI. Furthermore, the taxonomical character and incident of certain mangrove species in the ANI stay uncertain. Because of this, mangrove floristics of the ANI is regularly confounded and the right picture on mangrove plant variety of the ANI is under question.

6 Conservation and Management Strategies

Successful examples of mangrove conservation can be found throughout the world. Several innovative management strategies concerning the people's needs alongside the mangrove forest are reported in studies (Romañach et al. 2018). Concern over the lack of mangrove biological systems frequently centers around the disturbance or disruption in the arrangement of natural administrations by mangrove forests (Datta et al. 2012), for example, the security of coastal improvement against storms and floods that harm property and cause passing and injury, just as buffering environmental change impacts induced via ocean level rise, saltwater interruption, and coastal disintegration (Whiteley 2011). Valuation of biological services has progressively been accelerated and used in preservation and management strategy (Watson et al. 2018).

6.1 Global Approaches to Mangrove Conservation

The advancement of mangrove protection and the improvement of human affluence and occupations simultaneously have been the two essential needs of the United Nations Sustainable Development Goals plan (United Nations 2015). These two clashing desires can be met by developments in eco-cultivating inside mangrove forests to accomplish agreement between people's resources and mangrove health. Coinciding with the Millennium Ecosystem Assessment concept (MA 2005), natural system services might be characterized comprehensively as the resources given to people by specific natural environments. Natural system services should consequently relate to human well-being and social-financial values, a cycle known as Ecosystem Service Economic Valuation (ESEV). The development of ESEV methods has been driven by the growing requirement to control natural system deterioration globally, and valuation perspectives have been pushed to assist dynamic and environmental managers. ESEV gives helpful data about the social-monetary advantages and costs related to elective coastal arrangements, encouraging the evaluation of compromises and collaborations inherent in ecosystem-based management. Significant difficulties that appeared by ESEV in developing nations originate from the lack of information, absence of funding, and absence of institutional responsibility (Torres and Hanley 2017). ESEV can be utilized to recognize who gets the advantages, and who faces the expenses, especially over the social-financial gap. Assessment of nonuse values should be straightforward and performed with coordinated effort among social, characteristic, and political specialists to assemble trust and to lessen incompatible circumstances (Torres and Hanley 2017). The adjacent clients' perplexing view of the scene, their all-encompassing feeling of prosperity, and their setting explicit socio-economical valuation of mangrove ESEV past the money-related worth are essential standards to be joined into protection approaches. A decent comprehension of the overwhelming interrelationships among social and normal frameworks and of the numerous measurements and diverse time sizes of biological system administrations is crucial. Such a methodology is steady with the United Nations Sustainable Development Goals of improving human prosperity and of advancing the protection of marine biological systems (United Nations 2015).

6.2 Inclusion of Human Needs

The United Nations Sustainable Development Goals (UN SDG) recognize the agreeable conjunction of normal environments and people and needed the simultaneous improvement of human well-being and occupations while advancing the preservation of marine biological systems (United Nations 2015). Numerous efforts have been made around the world to achieve this harmonious relationship through ways of improvement in sustainable aquaculture practices in mangrove forests. We quickly present three contextual investigations in Guangxi of China, in Ca Mau of the Mekong Delta, and in the Volta estuary of Ghana, to show the capability of mangrove protection and reclamation confronting expanding requests on mangrove resources because of population development in the Guangxi province of China – farmers have acquired the aquaculture system, which does not require mangrove deforestation and industrial nutrient input. This agriculture practice has succeeded in achieving the conflict between the economic return of aquaculture and mangrove conservation. This environment-friendly agriculture practice is beneficial in several other ways, i.e., this innovative practice is facilitating ecotourism, increasing the farmer's income and promoting the UN SDG program. Such agriculture practices are low in management cost and are easy to operate while providing high-quality products. Several natural events including increasing sea level, extreme weather, climatic changes, and coastal erosion had impacted mangrove habitats seriously (Sippo et al. 2018). Additionally, land-use changes caused by anthropogenic activities have arisen the extreme level of challenges before policymakers (Díaz et al. 2019). Some improper anthropogenic activities in the mangrove region are induced by population demands. For example, the modification of the mangrove ecosystem into agriculture and aquaculture for food production and industrial growth has grown in recent decades at the cost of the environmental health of mangroves.

7 The Role of Traditional Knowledge and GIS Is of Great Use in the Management of the Mangrove Ecosystem

Mangrove preservation will remain the top priority for the restoration scientist in the coming decade, and its implementation success will depend on the microscale management of the restoration sites (Doody 2008). The intensity of implementation may depend on the total valuation of the mangroves including cost-benefit analysis of restoration practices (Turner et al. 2003). Since mangroves were related to several cultural and ethnic aspects of societies, harvesting the knowledge of natural resources, i.e., specific use of plants, may be a critical challenge (Datta et al. 2012). In several nations, mangrove propagules are consistently bought from neighboring locals and village inhabitants for afforestation and protection purposes (Islam and Wahab 2005). Geographic Information System (GIS)-based complete data set methodology will be significant for fruitful micro size estimations (Shinde et al. 2010). Mangrove environments are consistently situated in a blocked-off zone because of their zone of occurrences. Mangrove natural system is frequently immersed with flowing water. The use of remote sensing will give valuable and viable constant data for identification, portrayal, planning, and observing of mangrove conditions (Moffett et al. 2015). This will give an understanding of information over a distant region. A combination of remote sensing and GIS is discovered to be profoundly advantageous in distinguishing evidence and planning of particular mangrove environments (Kuenzer et al. 2011). For example, small changes in land use and land cover in an inaccessible district can without much of a stretch be identified by remote sensing methods. A specific redirection of the flow of water because of the development of a dam may accelerate the rate of disintegration in a specific region and the pace of sedimentation in some different territories. These data are extremely critical to plan a micro-level management for mangrove preservation.

8 Conclusion

Mangrove forests have vital importance in a coastal ecosystem. It provides multifaceted benefits and ecological services to the coastal inhabitants and improves their livelihood. It enters in every respect in millions of people's lives and plays a significant role in protecting from solar UV-B radiation, tsunami, hurricane, cyclone, floods, and coastal erosion, and its product has medicinal use. Instead of their importance, many anthropogenic-induced threats are hovering over its density. Unrestricted growth of population, pollution, encroachment, land-use change, and overexploitation reduce the number of mangrove forests at a fast rate. Land-use changes are profoundly affected by different demographic groups of people in various circumstances. Observations in a disturbed mangrove stressed by human interferences can result in massive sedimentation and coastline receding, which further shown that the pace of flow velocities and sediment deposition rates is accompanied by a fragmented and scattered vegetation pattern. Human intervention causes sediment dryness and coastline shrinking and has evidently diminished the mangrove's resilience. Degradation of mangrove increases the intensity of disasters, creates the non-conducive condition for marine fishes and their existence, promotes coastal erosion, and negatively affects carbon cycling and groundwater purity. Excessive loss of mangroves is non-concomitant of environmental integrity. To reduce the rate of loss, ESEV studies introduce conservation and restoration. Researchers and decision-makers suggest the use of remote sensing and GIS in the management of the mangrove forest cover. This technique represents the exact situation of land-use change. These insights help the government and decision-makers in developing successful schemes and management schemes. Such global activities help to ensure the sustainability of mangrove forests.

References

Agency of Survey Coordination and National Mapping, Republic of Indonesia (2009) Peta mangroves Indonesia. Pusat Survey Sumber Daya Alam Laut Badan Koordinasi Survey dan Pemetaan Nasional, Cibinong

Albers T, Schmitt K (2015) Dyke design, floodplain restoration and mangrove co-management as parts of an area coastal protection strategy for the mud coasts of the Mekong Delta, Vietnam. Wetl Ecol Manag 23(6):991–1004. https://doi.org/10.1007/s11273-015-9441-3

Alongi DM, Clough BF, Dixon P, Tirendi F (2003) Nutrient partitioning and storage in arid-zone forests of the mangroves Rhizophora stylosa and Avicennia marina. Trees 17(1):51–60. https://doi.org/10.1007/s00468-002-0206-2

Alonzo M, Van Den Hoek J, Ahmed N (2016) Capturing coupled riparian and coastal disturbance from industrial mining using cloud-resilient satellite time series analysis. Sci Rep 6:35129. https://doi.org/10.1007/s00468-002-0206-2

Arbo P, Thủy PTT (2016) Use conflicts in marine ecosystem-based management—the case of oil versus fisheries. Ocean Coast Manag 122:77–86. https://doi.org/10.1016/j.ocecoaman.2016.01.008

Ardli ER, Wolff M (2009) Land use and land cover change affecting habitat distribution in the Segara Anakan lagoon, Java, Indonesia. Reg Environ Change 9(4):235. https://doi.org/10.1007/s10113-008-0072-6

Aung UM (2007) Policy and practice in Myanmar's protected area system. J Environ Manag 84(2):188–203. https://doi.org/10.1016/j.jenvman.2006.05.016

Aye WN, Wen Y, Marin K, Thapa S, Tun AW (2019) Contribution of mangrove forest to the livelihood of local communities in Ayeyarwaddy Region, Myanmar. Forests 10(5):414. https://doi.org/10.3390/f10050414

Banus MD, Valiela I, Teal JM (1975) Lead, zinc and cadmium budgets in experimentally enriched salt marsh ecosystems. Estuar Coast Mar Sci 3(4):421–430. https://doi.org/10.1016/0302-3524(75)90042-0

Blasco F, Saenger P, Janodet E (1996) Mangroves as indicators of coastal change. Catena 27(3–4):167–178. https://doi.org/10.1016/0302-3524(75)90042-0

Chaudhuri P, Ghosh S, Bakshi M, Bhattacharyya S, Nath B (2015) A review of threats and vulnerabilities to mangrove habitats: with special emphasis on east coast of India. J Earth Sci Clim Change 6(4). https://doi.org/10.4172/2157-7617.1000270

Chen L, Wang W, Zhang Y, Lin G (2009) Recent progresses in mangrove conservation, restoration and research in China. J Plant Ecol 2(2):45–54. https://doi.org/10.1093/jpe/rtp009

Chu HY, Tam NFY, Lam SKS, Wong YS (2000) Retention of pollutants by mangrove soil and the effects of pollutants on Kandelia candel. Environ Technol 21(7):755–764. https://doi.org/10.1080/09593330.2000.9618961

Cubit JD, Getter CD, Jackson JB, Garrity SD, Caffey HM, Thompson RC, Weil E, Marshall MJ (1987) An oil spill affecting coral reefs and mangroves on the Caribbean coast of Panama. In: International oil spill conference, vol 1987(1). American Petroleum Institute, Washington, DC, pp 401–406. https://doi.org/10.7901/2169-3358-1987-1-401

Das S, Vincent JR (2009) Mangroves protected villages and reduced death toll during Indian super cyclone. Proc Natl Acad Sci 106(18):7357–7360. https://doi.org/10.1073/pnas.0810440106

Datta D, Chattopadhyay RN, Guha P (2012) Community based mangrove management: a review on status and sustainability. J Environ Manag 107:84–95. https://doi.org/10.1016/j.jenvman.2012.04.013

De Lacerda LD, Borges R, Ferreira AC (2019) Neotropical mangroves: conservation and sustainable use in a scenario of global climate change. Aquat Conserv Mar Freshwat Ecosyst 29(8):1347–1364. https://doi.org/10.1002/aqc.3119

Di Nitto D, Neukermans G, Koedam N, Defever H, Pattyn F, Kairo JG, Dahdouh-Guebas F (2014) Mangroves facing climate change: landward migration potential in response to projected scenarios of sea level rise. Biogeosciences 11(3):857–871. https://doi.org/10.5194/bg-11-857-2014

Díaz SM, Settele J, Brondízio E, Ngo H, Guèze M, Agard J et al (2019) The global assessment report on biodiversity and ecosystem services: summary for policy makers

Doody JP (2008) Saltmarsh conservation, management and restoration, vol 12. Springer Science & Business Media, London. ISBN: 978-1-4020-5748-9

Duke NC (2016) Oil spill impacts on mangroves: recommendations for operational planning and action based on a global review. Mar Pollut Bull 109(2):700–715. https://doi.org/10.1016/j.marpolbul.2016.06.082

Ellison AM, Farnsworth EJ (1996) Anthropogenic disturbance of Caribbean mangrove ecosystems: past impacts, present trends, and future predictions. Biotropica:549–565. https://doi.org/10.2307/2389096

Estrada GC, Soares ML (2017) Global patterns of aboveground carbon stock and sequestration in mangroves. An Acad Bras Cienc 89(2):973–989. https://doi.org/10.1590/0001-3765201720160357

FAO (2007) Mangroves of Asia 1980-2005: country reports. Forest Resources Assessment Programme. Forestry Department, Food and Agriculture Organization of the United Nations. http://www.fao.org

Giblin AE, Bourg A, Valiela I, Teal JM (1980) Uptake and losses of heavy metals in sewage sludge by a New England salt marsh. Am J Bot 67(7):1059–1068. https://doi.org/10.1002/j.1537-2197.1980.tb07738.x

Giesen W, Wulffraat S, Zieren M, Scholten L (2007) Mangrove guidebook for Southeast Asia. Regional Office for Asia and the Pacific, Wetlands International, Bangkok

Giri C, Ochieng E, Tieszen LL, Zhu Z, Singh A, Loveland T, Masek J, Duke N (2011) Status and distribution of mangrove forests of the world using earth observation satellite data. Glob Ecol Biogeogr 20(1):154–159. https://doi.org/10.1111/j.1466-8238.2010.00584.x

Godoy MDP, de Lacerda LD (2014) River-island morphological response to basin land-use change within the Jaguaribe River estuary, NE Brazil. J Coast Res 30(2):399–410. https://doi.org/10.2112/JCOASTRES-D-13-00059.1

Godoy MD, Lacerda LDD (2015) Mangroves response to climate change: a review of recent findings on mangrove extension and distribution. An Acad Bras Cienc 87(2):651–667. https://doi.org/10.1590/0001-3765201520150055

Goldberg L, Lagomasino D, Thomas N, Fatoyinbo T (2020) Global declines in human-driven mangrove loss. Glob Chang Biol 26(10):5844–5855. https://doi.org/10.1111/gcb.15275

Goutham-Bharathi MP, Roy SD, Krishnan P, Kaliyamoorthy M, Immanuel T (2014) Species diversity and distribution of mangroves in Andaman and Nicobar Islands, India. Bot Mar 57(6):421–432. https://doi.org/10.1515/bot-2014-0033

Guo W, Wu H, Zhang Z, Yang C, Hu L, Shi X, Jian S, Shi S, Huang Y (2017) Comparative analysis of transcriptomes in Rhizophoraceae provides insights into the origin and adaptive evolution of mangrove plants in intertidal environments. Front Plant Sci 8:795. https://doi.org/10.3389/fpls.2017.00795

Hai NT, Dell B, Phuong VT, Harper RJ (2020) Towards a more robust approach for the restoration of mangroves in Vietnam. Ann Forest Sci 77(1):18. https://doi.org/10.1007/s13595-020-0921-0

Hamilton S (2013) Assessing the role of commercial aquaculture in displacing mangrove forest. Bull Mar Sci 89(2):585–601. https://doi.org/10.5343/bms.2012.1069

Harada K, Imamura F, Hiraishi TL (2002) Experimental study on the effect in reducing tsunami by the coastal permeable structures. In: The twelfth international offshore and polar engineering conference. International Society of Offshore and Polar Engineers, Cupertino, CA

Harty C (2009) Mangrove planning and management in New Zealand and South East Australia–A reflection on approaches. Ocean Coast Manag 52(5):278–286. https://doi.org/10.1016/j.ocecoaman.2009.03.001

Harty C, Cheng D (2003) Ecological assessment and strategies for the management of mangroves in Brisbane Water—Gosford, New South Wales, Australia. Landsc Urban Plan 62(4):219–240. https://doi.org/10.1016/S0169-2046(02)00151-2

Hasnat GT, Kabir MA, Hossain MA (2018) Major environmental issues and problems of South Asia, particularly Bangladesh. In: Handbook of environmental materials management. Springer, Cham, pp 1–40. https://doi.org/10.1007/978-3-319-58538-3_7-1

Hinrichs S, Nordhaus I, Geist SJ (2009) Status, diversity and distribution patterns of mangrove vegetation in the Segara Anakan lagoon, Java, Indonesia. Reg Environ Change 9(4):275. https://doi.org/10.1007/s10113-008-0074-4

Ibharim NA, Mustapha MA, Lihan T, Mazlan AG (2015) Mapping mangrove changes in the Matang Mangrove Forest using multi temporal satellite imageries. Ocean Coast Manag 114:64–76. https://doi.org/10.1016/j.ocecoaman.2015.06.005

India State of Forest Report (2009) Forest survey of India

Islam SN, Gnauck A (2008) Mangrove wetland ecosystems in Ganges-Brahmaputra delta in Bangladesh. Front Earth Sci China 2(4):439–448. https://doi.org/10.1007/s11707-008-0049-2

Islam MS, Wahab MA (2005) A review on the present status and management of mangrove wetland habitat resources in Bangladesh with emphasis on mangrove fisheries and aquaculture. In: Aquatic biodiversity II. Springer, Dordrecht, pp 165–190

Jingchun L, Chongling Y, Macnair MR, Jun H, Yuhong L (2006) Distribution and speciation of some metals in mangrove sediments from Jiulong River Estuary, People's Republic of China. Bull Environ Contam Toxicol 76(5):815–822. https://doi.org/10.1007/s00128-006-0992-0

Kathiresan K (2018) Mangrove forests of India. Curr Sci 114(5):976–981. https://doi.org/10.18520/cs/v114/i05/976-981

Kirui KB, Kairo JG, Bosire J, Viergever KM, Rudra S, Huxham M, Briers RA (2013) Mapping of mangrove forest land cover change along the Kenya coastline using Landsat imagery. Ocean Coast Manag 83:19–24. https://doi.org/10.1016/j.ocecoaman.2011.12.004

Koh LP, Wilcove DS (2008) Is oil palm agriculture really destroying tropical biodiversity? Conserv Lett 1:60–64. https://doi.org/10.1111/j.1755-263X.2008.00011.x

Koh LP, Miettinen J, Liew SC, Ghazoul J (2011) Remotely sensed evidence of tropical peatland conversion to oil palm. Proc Natl Acad Sci 108(12):5127–5132. https://doi.org/10.1073/pnas.1018776108

Krauss KW, Osland MJ (2020) Tropical cyclones and the organization of mangrove forests: a review. Ann Bot 125(2):213–234. https://doi.org/10.1093/aob/mcz161

Kuenzer C, Bluemel A, Gebhardt S, Quoc TV, Dech S (2011) Remote sensing of mangrove ecosystems: a review. Remote Sens 3(5):878–928. https://doi.org/10.3390/rs3050878

Kusmana C (2014) Distribution and current status of mangrove forests in Indonesia. In: Mangrove ecosystems of Asia. Springer, New York, pp 37–60

Lai S, Loke LH, Hilton MJ, Bouma TJ, Todd PA (2015) The effects of urbanisation on coastal habitats and the potential for ecological engineering: a Singapore case study. Ocean Coast Manag 103:78–85. https://doi.org/10.1016/j.ocecoaman.2014.11.006

Lee JSH, Abood S, Ghazoul J, Barus B, Obidzinski K, Koh LP (2014) Environmental impacts of large-scale oil palm enterprises exceed that of smallholdings in Indonesia. Conserv Lett 7(1):25–33. https://doi.org/10.1111/conl.12039

López-Angarita J, Roberts CM, Tilley A, Hawkins JP, Cooke RG (2016) Mangroves and people: lessons from a history of use and abuse in four Latin American countries. For Ecol Manag 368:151–162. https://doi.org/10.1016/j.foreco.2016.03.020

Maiti SK, Chowdhury A (2013) Effects of anthropogenic pollution on mangrove biodiversity: a review. J Environ Prot. https://doi.org/10.4236/jep.2013.412163

Mandal B, Mukherjee A, Banerjee S (2013) A review on the ichthyofaunal diversity in mangrove based estuary of Sundarbans. Rev Fish Biol Fish 23(3):365–374. https://doi.org/10.1007/s11160-012-9300-8

Maryantika N, Lin C (2017) Exploring changes of land use and mangrove distribution in the economic area of Sidoarjo District, East Java using multi-temporal Landsat images. Inform Proc Agric 4(4):321–332. https://doi.org/10.1016/j.inpa.2017.06.003

Massel SR, Furukawa K, Brinkman RM (1999) Surface wave propagation in mangrove forests. Fluid Dyn Res 24(4):219. https://doi.org/10.9753/icce.v33.waves.40

Matsuda M (2009) Dynamics of rice production development in Myanmar: growth centers, technological changes, and driving forces. Trop Agric Dev 53(1):14–27. https://doi.org/10.11248/jsta.53.14

Millennium Ecosystem Assessment (2005) Ecosystems and human well-being: wetlands and water. World Resources Institute, Washington, DC

Mimura N (2013) Sea-level rise caused by climate change and its implications for society. Proc Jpn Acad Ser B 89(7):281–301. https://doi.org/10.2183/pjab.89.281

Mitchell DS (1978) The potential for wastewater treatment by aquatic plants in Australia. Water (Australia) 5(3):15–17

Moffett KB, Nardin W, Silvestri S, Wang C, Temmerman S (2015) Multiple stable states and catastrophic shifts in coastal wetlands: progress, challenges, and opportunities in validating theory using remote sensing and other methods. Remote Sens 7(8):10184–10226. https://doi.org/10.3390/rs70810184

Moorthy P, Kathiresan K (1997) Photosynthetic pigments in tropical mangroves: impacts of seasonal flux of UV-B radiation and other environmental attributes. Bot Mar 40(1–6):341–350. https://doi.org/10.1515/botm.1997.40.1-6.341

Moorthy P, Kathiresan K (1998) Influence of ultraviolet-B radiation on photosynthetic and biochemical characteristics of a mangrove Rhizophora apiculata. Photosynthetica 34(3):465–471. https://doi.org/10.1023/A:1006828505050

Okamoto I (2007) Transforming Myanmar's rice marketing, Myanmar

Parida AK, Jha B (2010) Salt tolerance mechanisms in mangroves: a review. Trees 24(2):199–217. https://doi.org/10.1007/s00468-010-0417-x

Primavera JH (2000) Development and conservation of Philippine mangroves: institutional issues. Ecol Econ 35(1):91–106. https://doi.org/10.1016/S0921-8009(00)00170-1

Pumijumnong N (2014) Mangrove forests in Thailand. In: Mangrove ecosystems of Asia. Springer, New York, pp 61–79

Rahman MM, Rahman MM, Islam KS (2010) The causes of deterioration of Sundarban mangrove forest ecosystem of Bangladesh: conservation and sustainable management issues. Aquacult Aquar Conserv Legisl 3(2):77–90

Richards DR, Friess DA (2016) Rates and drivers of mangrove deforestation in Southeast Asia, 2000–2012. Proc Natl Acad Sci 113(2):344–349. https://doi.org/10.1073/pnas.1510272113

Romañach SS, DeAngelis DL, Koh HL, Li Y, Teh SY, Barizan RSR, Zhai L (2018) Conservation and restoration of mangroves: global status, perspectives, and prognosis. Ocean Coast Manag 154:72–82. https://doi.org/10.1016/j.ocecoaman.2018.01.009

Roy AKD (2014) Determinants of participation of mangrove-dependent communities in mangrove conservation practices. Ocean Coast Manag 98:70–78. https://doi.org/10.1016/j.ocecoaman.2014.06.001

Sahu SC, Suresh HS, Murthy IK, Ravindranath NH (2015) Mangrove area assessment in India: implications of loss of mangroves. J Earth Sci Clim Change 6(5):1. https://doi.org/10.4172/2157-7617.1000280

Sasmito SD, Murdiyarso D, Friess DA, Kurnianto S (2016) Can mangroves keep pace with contemporary sea level rise? A global data reviews. Wetl Ecol Manag 24(2):263–278. https://doi.org/10.1007/s11273-015-9466-7

Sathe SS, Lavate RA, Bhosale LJ (2013) Mangrove as source of energy for Rural development with special reference to Ratnagiri and Sindhudarg district (MS) India. Biosci Discov 4(2):198–201

Sathe SS, Lavate RA, Sajjan MB (2015) The role of organic constituents of Avicennia in animal nutrition. Biosci Discov 6(2):145–151

Shinde V, Tiwari KN, Singh M (2010) Prioritization of micro watersheds on the basis of soil erosion hazard using remote sensing and geographic information system. Int J Water Resour Environ Eng 5(2):130–136

Sippo JZ, Lovelock CE, Santos IR, Sanders CJ, Maher DT (2018) Mangrove mortality in a changing climate: an overview. Estuar Coast Shelf Sci 215:241–249. https://doi.org/10.1016/j.ecss.2018.10.011

Srikanth S, Lum SKY, Chen Z (2016) Mangrove root: adaptations and ecological importance. Trees 30(2):451–465. https://doi.org/10.1007/s00468-015-1233-0

Stead SM (2019) Using systems thinking and open innovation to strengthen aquaculture policy for the United Nations Sustainable Development Goals. J Fish Biol 94(6):837–844. https://doi.org/10.1111/jfb.13970

Synthesis Report on Ten ASEAN Countries Disaster Risks Assessment (2010) ASEAN disaster risk management initiative. https://www.preventionweb.net
Tanaka N (2009) Vegetation bioshields for tsunami mitigation: review of effectiveness, limitations, construction, and sustainable management. Landsc Ecol Eng 5(1):71–79. https://doi.org/10.1007/s11355-008-0058-z
The Tsunami (2004) World Health Organization. www.who.org
Thuo ADM (2013) Impacts of urbanization on land use planning, livelihood and environment in the Nairobi rural-urban fringe, Kenya
Torres C, Hanley N (2017) Communicating research on the economic valuation of coastal and marine ecosystem services. Mar Policy 75:99–107. https://doi.org/10.1016/j.marpol.2016.10.017
Toumbourou T, Muhdar M, Werner T, Bebbington A (2020) Political ecologies of the post-mining landscape: activism, resistance, and legal struggles over Kalimantan's coal mines. Energy Res Soc Sci 65:101476. https://doi.org/10.1016/j.erss.2020.101476
Tripathi R, Shukla AK, Shahid M, Nayak D, Puree C, Mohanty S, Raja R, Lal B, Gautam P, Bhattacharyya P, Panda BB (2016) Soil quality in mangrove ecosystem deteriorates due to rice cultivation. Ecol Eng 90:163–169. https://doi.org/10.1016/j.ecoleng.2016.01.062
Turner RK, Paavola J, Cooper P, Farber S, Jessamy V, Georgiou S (2003) Valuing nature: lessons learned and future research directions. Ecol Econ 46(3):493–510. https://doi.org/10.1016/S0921-8009(03)00189-7
United Nations. International Strategy for Disaster Reduction. Secretariat (2015) Global assessment report on disaster risk reduction 2015: making development sustainable: the future of disaster risk management. UN
Valiela I, Banus MD, Teal JM (1974) Response of salt marsh bivalves to enrichment with metal-containing sewage sludge and retention of lead, zinc and cadmium by marsh sediments. Environ Pollut (1970) 7(2):149–157. https://doi.org/10.1016/0013-9327(74)90082-2
Valiela I, Bowen JL, York JK (2001) Mangrove forests: one of the world's threatened major tropical environments: at least 35% of the area of mangrove forests has been lost in the past two decades, losses that exceed those for tropical rain forests and coral reefs, two other well-known threatened environments. Bioscience 51(10):807–815. https://doi.org/10.1641/0006-3568(2001)051
Villate Daza DA, Sánchez Moreno H, Portz L, Portantiolo Manzolli R, Bolívar-Anillo HJ, Anfuso G (2020) Mangrove forests evolution and threats in the Caribbean sea of Colombia. Water 12(4):1113. https://doi.org/10.3390/w12041113
Wang BS, Liao BW, Wang YJ, Zan QJ (2002) Mangrove forest ecosystem and its sustainable development in Shenzhen Bay. Science Press, Beijing
Ward RD, Friess DA, Day RH, MacKenzie RA (2016) Impacts of climate change on mangrove ecosystems: a region-by-region overview. Ecosyst Health Sustain 2(4):e01211. https://doi.org/10.1002/ehs2.1211
Watson JE, Evans T, Venter O, Williams B, Tulloch A, Stewart C et al (2018) The exceptional value of intact forest ecosystems. Nat Ecol Evol 2(4):599–610. https://doi.org/10.1038/s41559-018-0490-x
Webb EL, Jachowski NR, Phelps J, Friess DA, Than MM, Ziegler AD (2014) Deforestation in the Ayeyarwady Delta and the conservation implications of an internationally-engaged Myanmar. Glob Environ Chang 24:321–333. https://doi.org/10.1016/j.gloenvcha.2013.10.007
Whiteley NM (2011) Physiological and ecological responses of crustaceans to ocean acidification. Mar Ecol Prog Ser 430:257–271. https://doi.org/10.3354/meps09185
Wicke B, Sikkema R, Dornburg V, Faaij A (2011) Exploring land use changes and the role of palm oil production in Indonesia and Malaysia. Land Use Policy 28(1):193–206. https://doi.org/10.1016/j.landusepol.2010.06.001
Woodroffe CD, Rogers K, McKee KL, Lovelock CE, Mendelssohn IA, Saintilan N (2016) Mangrove sedimentation and response to relative sea-level rise. Annu Rev Mar Sci 8:243–266. https://doi.org/10.1146/annurev-marine-122414-034025

Chapter 12
Assessment of Total Petroleum Hydrocarbon Accumulation in Crabs of Chilika Lagoon, India

Prasannajit Acharya, Pradipta R. Muduli, Mira Das, and Amrit Kumar Mishra

Abstract Globally, the assessment of total petroleum hydrocarbons (TPHCs) and their accumulation in biotic components has received considerable scientific attention due to their carcinogenic nature and health risks. Long-term accumulation of TPHCs in sediment-associated biota such as fish or crabs could be hazardous to consumers, once the threshold levels are breached. Chilika lagoon in India is one of the largest lagoon ecosystems in Asia and supports the livelihood of more than 0.2 million coastal communities through fishery and tourism. The use of motorized boats operating for both fishing and tourism activities is a major source of TPHCs in the lagoon. The proposed study quantified the concentration of these TPHCs in the tissues of three edible crabs *Portunus pelagicus*, *Scylla serrata*, and *Scylla tranquebarica* from Chilika lagoon that are the major food source of the coastal communities. Along with crab tissues, the concentration of TPHCs was also quantified in the sediment and surface water samples to assess the bioaccumulation rates. The estimated dietary intake (EDI) of TPHCs for humans through crab consumption and the associated health risk index (HRI) were also quantified.

P. Acharya (✉)
Chilika Development Authority, Wetland Research and Training Centre, Balugaon, Odisha, India

Department of Chemistry, Institute of Technical Education and Research, Siksha 'O' Anusandhan (Deemed to be University), Bhubaneswar, Odisha, India

P. R. Muduli
Chilika Development Authority, Wetland Research and Training Centre, Balugaon, Odisha, India

M. Das
Department of Chemistry, Institute of Technical Education and Research, Siksha 'O' Anusandhan (Deemed to be University), Bhubaneswar, Odisha, India

A. K. Mishra
School of Earth Ocean and Climate Sciences, Indian Institute of Technology Bhubaneswar, Jatni, Odisha, India

S. Madhav et al. (eds.), *Coastal Ecosystems*, Coastal Research Library 38,
https://doi.org/10.1007/978-3-030-84255-0_12

The mean TPHC concentration ranged from 0.246 to 9.409 μg/L with avg. of 1.672 ± 1.518 μg/L in the water, 0.036 to 2.520 μg/g with avg. of 0.767 ± 0.558 μg/L in the sediment, and 0.05–7.03 ng/g DW with avg. 2.576 ± 1.655 ng/g in crab tissue. In wet season TPHC in the water varied from 0.280 to 5.124 with an avg. of 1.119 ± 0.884 μg/L and in sediment 0.151 to 1.768 with avg. of 0.758 ± 0.442 μg/L, whereas in dry season TPHC concentration of water ranged from 0.246 to 9.409 with avg. of 2.224 ± 1.809 μg/L and in sediment it varied from 0.036 to 2.525 with avg. of 0.776 ± 0.661 μg/L. The trend of TPHC concentration followed the sequence of *Portunus pelagicus* > *Scylla serrata* > *Scylla tranquebarica*, and *Portunus pelagicus* accumulated the highest concentration of TPHC in the dry season. Also, sediment and water TPHC concentration showed higher in the dry season. The concentration of TPHCs in crab tissues was significant ($p < 0.01$) between seasons. It was estimated that the general population through crab consumption was exposed to 0.03 and 0.07 ng/kg of body weight/day of TPHC as indicated by EDI values. The human heath-associated risks due to consumption of TPHC as indicated by HRI ranged from 0.0002 to 0.0351. The bioaccumulation factors (BAFs) values were lower than the standard limit of 1 μg/g prescribed by the US Environmental Protection Agency (USEPA). The various factors or indices such as BAF, EDI, and HRI indicated that the TPHC concentration in the three crab species studied was within the safe limit (<1) as per USEPA. The BAFs were relatively low in water than in sediment, with water having BAFs of 0.0017 μg/g and sediment having BAFs of 0.0033 μg/g. Because the crab is a bottom feeder, it displayed a minor increase in BAFs linked with sediment. This suggested the three crab species of the Chilika lagoon currently have no health risks and are safe for human consumption. However, constant monitoring of TPHCs in the surface water, sediment, and biota of the Chilika lagoon is essential to avoid any human health hazards.

Keywords Anthropogenic pollution · Bioaccumulation · Human health risk · Crab · Lagoon ecosystem

1 Introduction

Petroleum hydrocarbons (PHCs) are considered as priority contaminants due to their high usage in various industries across the world, and these contaminants are also the source of various wide-scale environmental threats. Total petroleum hydrocarbons (TPHCs) are considered as the mixture of hydrocarbons contained in crude oil. The major sources of these TPHCs are transport, production, shipping activities, coastal oil refining, off-shore oil production, and accidental spillages (Varjani and Upasani 2016). Annually, 60–600 million gallons of TPHCs are subjected to accidental spillages around the world that end up in the marine environment (NRC 2003; Kvenvolden et al. 2003). All forms of petroleum hydrocarbons, i.e. TPHC, reach the aquatic ecosystem through the way of accidents or boat spills or as by-products of industrial effluents that use TPHCs. These TPHC extracts have the propensity to float and generate thin films on the water's surface. As the shipping industry is ever-expanding, the chances of more oil spills and leakage are also increasing, as the majority of the shipping industry uses diesel as a propulsion fuel (Al-Shwafi et al. 2008). TPHC can affect aquatic systems by unenhanced petroleum products such as lubricants, fuel, and gasoline, as well as through exhaust fumes from the by-product of incomplete combustion (Ashiru et al. 2019). The presence of TPHC and its toxic properties increases the risk of loss of biodiversity and fishery resources of coastal ecosystems through huge oil spills (Hardy and Higgins 1992; Chase et al. 2013). Secondly, these TPHCs can bioaccumulate in organisms like fish or crabs and can result in trophic transfer of these contaminants and their toxicity to higher organisms (Adeniji et al. 2017; Porte et al. 2000). This accumulation and transfer of TPHC through marine organisms are well understood in the case of oil spills in the short term. It is predictable that the long-term effects will be more dangerous and could cause permanent damage (Saadoun 2015) to the fin and shellfish population. In general, following an oil spill, the lipid-soluble TPHCs are broken down first and then consumed by various finfish and shellfish populations (Gobas et al. 1999). Long-term exposure to oil spill-generated PHCs can result in both physiological and morphological changes in finfish and shellfish (Mazhar et al. 1987; Anderson et al. 1974). However, in extreme cases, both finfish and shellfish populations can die or develop genetic mutations. Several reports have shown that TPHCs are also harmful to human health (Rose et al. 2012; Asuquo et al. 2004). Because marine finfish and shellfish are a major source of protein-rich food for millions of coastal communities, TPHCs in these organisms can eventually reach humans and cause serious health-related issues, including cancer (Das et al. 2011; Connell et al. 1980; Ghauch et al. 2000; Oluwatobi et al. 2019). Various properties of these TPHCs include lipophilicity and ingenuity and occur in the build-up of these compounds inside the tissues of nontarget organisms, where the high toxicity of TPHCs causes immediate consequences (Rao et al. 2016; Ogunfowokan et al. 2003).

A significant number of scientific studies have been conducted in recent decades on environmental pollution and contamination caused by oil spills and oil-related industries. When oil is spilled or is discharged from industry or boats, it forms a thin layer. The thin surface layer separates into droplets, which are dispersed by wave action (Veerasingam et al. 2011). These TPHCs become easier to gather throughout the tissues of finfish, shellfish, and marine animals as they disseminate (Zhou et al. 2015). Furthermore, the maritime industry and anthropogenic pollution also play a major role in TPHCs entering the aquatic ecosystem through small-scale oil spills or leakages and industrial disposal (Vandermeulen et al. 1985). In the marine ecosystem, significant portion of TPHC is detached through evaporation, whereas a small fraction gets dispersed in water that results in accumulation of TPHC in the sediment, which can be transferred to sediment-associated biota (Chouksey et al. 2004). Other heavy fractions settle in the benthic compartment of aquatic organisms, where contaminants may have an impact on bottom-feeding fish and organisms (Ololade et al. 2008). When TPHC enters the marine ecosystem, it is consumed by the system's various food networks and can be bioaccumulated in different trophic levels and organism tissues (Varanasi et al. 1989; Khan et al. 2005).

Due to the fact that TPHC contaminants have a tendency to accumulate in the organism than in the environment (Lee et al. 1976), it has become essential to monitor their concentration in the marine ecosystem using biological organisms such as fish and crabs (Copat et al. 2012). The study showed that as the fish and crabs seem to be at top of the benthic food web in the marine environment, these are suitable for TPHC accumulation (Alkindi et al. 1996). Invertebrates, such as crabs, do not metabolize TPHC efficiently, and as a result, their accumulation capacity in tissues is higher than in fish (Mironov et al. 1980). These TPHCs are accumulated via gills, skin, and food sources. However, this accumulation in fish and crab tissues depends on their feeding preference, general behaviour, and trophic level (Ansari et al. 2012).

Crabs are decapod crustaceans of the suborder Brachyura, and they are known for having a very short projecting "tail" that is usually hidden completely under the thorax. These are the most dominant vertebrate elements in coastal ecosystems throughout the world. The faeces of the crabs provide a valuable source of nutrition for other eaters since it is high in nitrogen, carbon, phosphorus, and trace metals (Rice et al. 1985). Mangrove seedlings benefit from their burrowing behaviour because it improves aeration and the free circulation of water, which supports the growth of young trees. The crawling sections are extremely beneficial in the recycling of nutrients through ploughing (Mohapatra et al. 2005). The exposure of particulate organic matter to microbes also aids in the breakdown of the organic matter in the particles. The burrowing nature of mangrove wetlands aids in the oxidation of sulphide, which accumulates as a result of the high rate of organic decomposition in the swamps (Barrento et al. 2010). They provide food for a variety of birds, snakes, and predatory fish, and their larvae are devoured by a variety of carnivores, including humans. Crabs are important predators of molluscs, small crustaceans, and other invertebrates in some ecosystems, but they also serve as a source of food for fish, decapods, and some terrestrial vertebrates in others (Burns et al. 1976). Consequently, crabs serve a critical part in the food chain of coastal habitats, including lagoons, by consuming a variety of different foods.

Sugars, starches, and fibre are found in abundance in crabs, and these carbohydrates serve as a key source of energy for animals. Carbohydrates in fish are devoid of dietary fibre, instead consisting primarily of glucosides, the majority of which are made up of glucose. It generally contains small amounts of glucose, fructose, sucrose, and various mono- and disaccharides in addition to the other ingredients (Okuzumi et al. 2000). The important amino acid and vital fatty acid levels of seafood products, particularly marine crabs, are strongly connected with their taste, nutritional quality, and health advantages (Chen et al. 2007). As well as being needed for physical functions, including physiology, biochemistry, and immunology, amino acids are also necessary for human growth and development (Maria et al. 2007). Extremely high amounts of amino acids have been linked to the development of several disorders, including Crohn's disease (Shoda et al. 1996) and inflammation (Shoda et al. 1996; Gil et al. 2002). The nutritional quality of crab proteins compares favourably to that of muscular meat from animals such as mutton, chicken, duck, and fish (Newcombe et al. 1994). The term "carbohydrate" refers to a category of organic substances that includes sugars, starches, and fibre. Carbohydrates are the key source of energy for both humans and animals. Carbohydrates in fish are devoid of dietary fibre, instead consisting primarily of glucides, the majority of which are made up of glucose. Crab is a significant source of protein and solid lipids for humans, and the long-chain omega-3 unsaturated fats found in crab have been shown to have a variety of beneficial effects on human health. As a result of the fish's response to natural change, it is feasible to utilize it as a biological marker for natural contamination. Shellfish, such as crayfish, crabs, oysters, and mussels, are high in iron, zinc, magnesium, vitamin B12, omega-3 fatty acids, lean protein, healthy fats, and minerals, and they are also a good source of protein. Shellfish are popular owing to their high protein and nutrition content, and they may help with weight loss, immunity, and brain and heart health, among other things.

Petroleum pollutants, as we know, prefer to accumulate in organisms rather than in the surrounding environment especially shellfish which is a major component of global seafood production and makes it reasonable to use as a marker for natural contamination (Batvari et al. 2007). In addition to the numerous nutritional benefits of fish and crab as a diet, the potential health risks associated with regular consumption of fish and crab are a major source of concern (FAO 2010). While fish and crabs have numerous nutritional benefits, all fish and crabs consume oil hydrocarbons legitimately or inadvertently as food and dregs, resulting in massive pulverized concrete of aquatic biota (Asuquo et al. 2004). However, fish and crabs account for 40% of animal protein consumption on the east coast (Barik 2017). Fisheries and crab products are of considerable interest in coastal regions such as estuaries and lagoons, owing to increased awareness of their value in the local diet and also the fact that they are more affordable than substitutes (FAO 2010). On India's east coast, Chilika, which is Asia's largest brackish water lagoon, is no exception to this phenomenon. More than 150,000 fishing communities and surrounding people depend on the Chilika lagoon for their livelihoods and nutritional necessities. The Chilika lagoon is one of the world's major reservoirs of aquatic biodiversity, as well as a reliable source of fishing (Ghosh et al. 2006). As India's first Ramsar site, the lagoon is rich in diversity and dynamic diversity of invertebrates and crustaceans from coastal, brackish, and freshwater ecosystems (CDA 2005). According to Suresh et al. (2018), Chilika has 336 finfish, 29 prawn and shrimp, and 35 crab species. According to Sahoo et al. (2013), a total of 21 species of crabs belonging to 16 genera and 8 families are observed in the lagoon, out of which 9 species of crabs belonging to 8 genera under 9 families of Portunidae are caught during all seasons, including *Portunus pelagicus*, *Scylla serrata*, and *Scylla tranquebarica*. The annual landing of these crabs during 2014–2015 was 209.18 tonnes (Mohanty et al. 2008). Average unit price for *Portunus pelagicus* is Indian rupees (INR) 90–100 per kg and also for *Scylla serrata*; *Scylla tranquebarica* is INR 350–1000 per kg (Suresh et al. 2018). These crab species also get exported to all the commercial hubs of India, such as Chennai, Kolkata, Hyderabad, etc. (Suresh et al. 2018). As a result, it is critical from a livelihood standpoint to keep the Chilika crabs toxic-free, which is why the TPHC level assessment is necessary. On the contamination front, studies on Chilika fishes and shellfish are limited and thus are essential (Parida et al. 2017).

Therefore, the purpose of this study was to provide baseline data on TPHC levels in various crab species. The study was expanded to analyse the potential of harmful health consequences from crab consumption based on current ingestion rates, as well as to give useful information for environmental assessment activities for the marine environment. This study is the first-ever integrated approach to assess the TPHC accumulation level in some economically and commercially available crabs of Chilika Lake. We aim to establish the relationship with TPHC level in water and sediment of the Chilika lagoon with the crab species.

2 Materials and Methods

2.1 Study Area

Chilika is India's largest coastal lagoon and, after the New Caledonian barrier reef, the world's second largest brackish water lagoon (Panda et al. 2008). It is India's largest coastal lagoon and has been proposed as a UNESCO World Heritage site. Chilika lagoon has been recognized as Ramsar Convention site no. 229, rendering it the first Indian wetland to be certified as having international significance by the Ramsar Convention (1981). With an average depth of 1.5 m, a linear axis of 64.3 km, and a mean width of 20.1 km, it is an extremely dynamic shallow coastal habitat (Panda et al. 2008). During summer, the lagoon covers 704 km^2 and expands to 1020 km^2 during the monsoon (Muduli et al. 2012). The Chilika lagoon is interconnected with the Bay of Bengal over the outer channel and Palur canal, from which salt water reaches the lagoon while 12 main rivers discharge freshwater into the lagoon (Muduli et al. 2013, Fig. 12.1). Based on hydrological difference, a total of four ecological sectors exists within the lagoon: the southern sector (SS, $n = 3$ stations), central sector (CS, $n = 5$ stations), northern sector (NS, $n = 3$ stations), and the outer channel (OC, $n = 3$ stations) (Barik et al. 2017; Muduli et al. 2017). Muduli and Pattnaik (2020) provide additional information on the Chilika lagoon's hydrological parameters and fluctuations in water quality. With an estimated annual production of 10,000 metric tonnes, the lagoon provides significant fishery resources (Mohapatra et al. 2007). As a result, the bulk of the local population is completely reliant on the fishing industry, which leads to an increased number of boats within the lagoon. According to inquiries (during March 2021) from boat associations (11, 5, and 1 from Puri, Khurda, and Ganjam districts in Odisha, respectively) near the Chilika lagoon, there are around 8000 boats (6400 for fishing and 1600 for tourism) operating every day in the Chilika lagoon. Tourism, after fishing, is the local community's second most important source of income. Given these various ecological

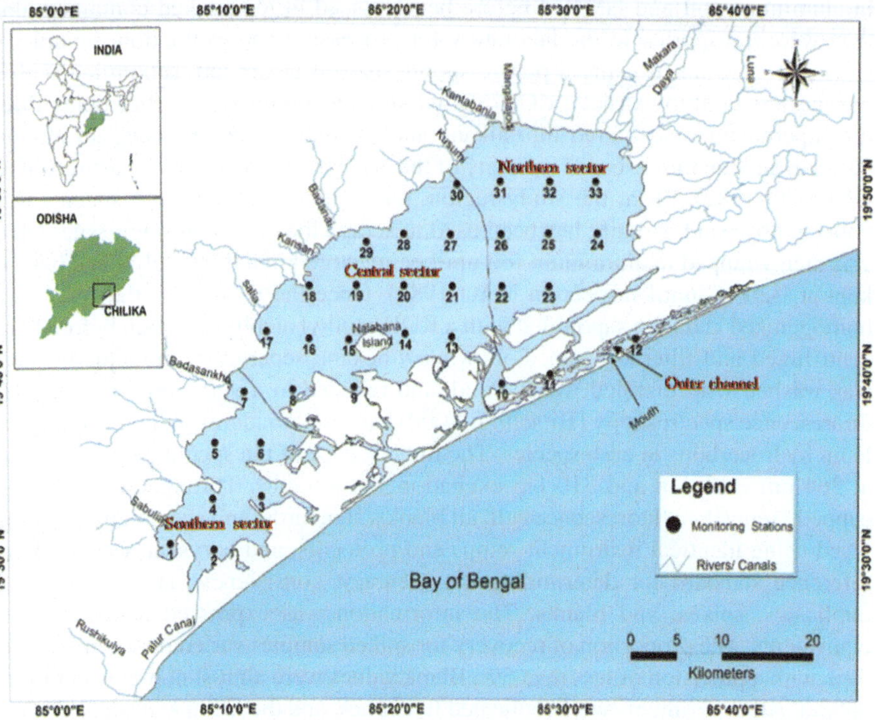

Fig. 12.1 Map showing the 33 sampling locations in the four sectors: southern, northern, central, and outer channel sector of the Chilika lagoon

and socioeconomic circumstances, the current study is essential. The study aims to investigate the TPHC level in this ecosystem's fishing products.

2.2 *Sampling and Analysis*

To evaluate the levels of TPHC in commercially available crabs from Chilika lagoon, fresh biospecimen samples of *Portunus pelagicus* (n = 3/station), *Scylla serrata* (n = 3/station), and *Scylla tranquebarica* (n = 3/stations) were collected from local fishermen in July 2019 (wet season) and November 2018 (dry season). These biological samples were chosen for their commercial value, ease of availability, and widespread use by locals in the Chilika region and beyond. A subsurface water sampler was used to collect surface water samples. Two-litre amber glass bottles were used to collect surface water samples. After that, the aqueous phase was collected twice with *n*-hexane (50 mL each time) to remove the TPHC and then allowed to dry using anhydrous Na_2SO_4. To collect sediment samples from each station, a Van Veen grab sampler was used to collect the samples. The samples were wrapped in aluminium foil and labelled before being placed in ice-packed containers until they were transported to the laboratory for processing and evaluation. Samples of sediment were taken using a plastic spoon, thawed in the lab, saponified with *n*-hexane was used to extract a KOH/methyl alcohol combination. After being dried, the supernatant was filtered into alkane and aromatic fractions using an alumina column, and the fluorescence intensity of the aromatic fraction was measured (IOC-UNESCO 1982). From the study region, ten samples of each crab species were collected by hand. A knife has been used to scrape the crab's muscle tissue, which was then wrapped in aluminium foil until being stored in a thick polythene bag and kept at −20 °C until inspection (FAO 1982). Once the tissue was thawed, it was homogenized and then saponified with a KOH/methyl alcohol solution before being centrifuged and filtered again. Following *n*-hexane separation, the aqueous layer was washed with distilled water, dried, and reduced to a very small amount. UV fluorescence spectroscopy (Hitachi, F-7000) was used to identify the level of petroleum hydrocarbons in crab species. The fluorescence of the specimen was analysed at 364 nm emission and 310 nm excitation wavelengths. Throughout the Teflon-capped 1-cm silica fluorescence cell, all blanks, standards, and specimens were analysed using identical instrument setups and protocols, and chrysene was used as a reference standard for determining the accuracy, consistency, and readability of duplicates, spikes, and blanks. The information was expressed using chrysene equivalents. The proportion of recovery for spiked samples varied between 96% and 99% with a precision of less than 5%. Blank values were almost non-existent in this dataset. All experiments were replicated five times, and the mean and standard deviations (SD) were determined for each of them. The significance of the findings was determined statistically using the student's *t*-test ($p < 0.05$). A two-way analysis of variance (ANOVA) was conducted to see whether there were any differences between the environmental matrices (sediment, crab tissues, and water). SPSS 18 and Origin 8Pro were used for the statistical analysis.

2.3 Potential Human Health Risk (Olayinka et al. 2019)

The human health risk index related to TPHC consumption through crab dietary intake was measured by determining crab dietary intake of TPHC on a regular basis (Olayinka et al. 2019). The total daily intake of TPHC from crab tissues was estimated by measuring the overall TPHC content in each sample by the rate of consumption of an average-weight adult human (70 kg) for a period of 1 week (Kumar et al. 2005). Based on the assumption that an average Asian country consumes 25 kg of fish per year and 0.0016–0.0219 kg of crabs and shrimp per day, respectively (Laili et al. 2013)

$$\begin{aligned}&\text{The dietary intake concentration of crab} = \\&\text{Concentration of TPHC} \times 0.0016 \text{ kg/day}\end{aligned} \tag{12.1}$$

2.4 *Bioaccumulation Factors*

The BAF in crab tissues was calculated by using the equation below (Akinola et al. 2019).

$$BAF = \text{Conc. of TPHC in crab tissue } (\text{mg}/\text{kg}) / \text{Conc. of TPHC in water } (\text{mg}/\text{L}) \text{ or Sediment} \tag{12.2}$$

3 Results and Discussion

Table 12.1 shows a summary of the TPHC levels detected in crabs collected throughout the wet and dry seasons. For each biological sample, Table 12.2 displays the 95% confidence interval, statistics, and two-tailed probability findings. Figure 12.2b indicates the concentrations of TPHC in three crab species in both wet and dry seasons. The findings indicated that the TPHC level in *Portunus pelagicus* during the

Table 12.1 Statistics of TPHC levels in crab tissues, estimated daily intake (EDI), and health risk index (HRI). PP (*Portunus pelagicus*), SS (*Scylla serrata*), ST (*Scylla tranquebarica*)

	Wet season			Dry season		
Parameter	PP	SS	ST	PP	SS	ST
TPHC (ng/g)	3.257 ± 1.48	2.331 ± 1.93	2.273 ± 0.71	3.307 ± 2.32	2.241 ± 1.31	2.049 ± 1.62
Range (ng/g)	1.521–6.740	0.263–6.510	1.230–3.897	0.410–7.030	0.77–4.72	0.050–4.78
EDI	0.00521	0.00373	0.00364	0.00529	0.00358	0.00327
HRI	0.0165	0.0119	0.0102	0.0162	0.0121	0.0114
Weight (g)	50.8–212.6	50–120	39.7–150.0	55.8–112.6	65–130	29–170
95% CI of mean	1.23–4.20	7.20–8.73	1.92–2.51	5.43–7.94	4.59–8.32	5.71–9.21
Variance	0.73	0.65	0.71	1.24	2.03	1.93
SE	0.19	0.45	0.12	0.79	0.82	0.92
CV%	16	13	11	32	29	41
Range	2.19	4.02	1.11	6.92	5.34	7.02
IQR	1.61	1.23	1.97	3.91	2.95	5.6
25th Percentile	1.17	1.04	0.84	2.77	4.9	3.61
50th Percentile	2.04	3.21	2.94	5.67	10.93	9.51
75th Percentile	3.82	2.19	1.78	0.812	0.721	0.922
Shapiro	0.921	0.41	0.262	0.856	0.932	0.043
Kurtosis	0.872	0.593	−1.231	0.121	−0.412	0.41

Table 12.2 Mean concentration ± SD of TPHC levels in wet and dry seasons. Statistics are presented for paired sample *t*-test

TPHC concentration	PP	SS	ST
Wet season (ng/g)	3.257 ± 1.487	2.331 ± 1.931	2.273 ± 0.714
Dry season (ng/g)	3.307 ± 2.328	2.241 ± 1.318	2.049 ± 1.625
95% confidence interval	2.451–4.561	1.032–4.12	0.21–5.813
t-Statistics	4.91	5.61	7.55
2-tailed *p* level	0.0004	0.0007	0.0001

wet season ranged from 1.52 to 6.74 ng/g DW with a mean concentration of 3.25 ± 1.48 ng/g DW. However, in the dry season, the TPHC concentration ranged from 0.41 to 7.03 ng/g DW with an average concentration of 3.30 ± 1.48 ng/g DW (Table 12.2). When comparing the wet and dry seasons, the concentration of TPHC in *Scylla serrata* was twofold greater. Concentrations varied between 0.26 and 6.51 ng/g dry weight (DW), with an average concentration of 2.33 ng/g dry weight (DW) in *S. serrata* (95% confidence limit of 7.20–8.73). As a result, during the wet season, the TPHC concentration in *S. serrata* tissues ranged from 0.77 to 4.72 ng/g DW, with a mean of 2.24 ± 1.31 ng/g DW (95% confidence interval = 4.59–8.32) during the dry season (Table 12.1). The TPHC concentrations in *Scylla tranquebarica* ranged from 1.23 to 3.89 ng/g DW with a mean value of 2.27 ± 0.71 ng/g DW during the wet season, whereas, in the dry season, the TPHC concentration ranged from 0.05 to 4.78 ng/g DW with an average of 2.04 ± 1.62 ng/g DW. When the TPHC concentrations of our results were compared to the Central Pollution Control Board (CPCB) of Odisha-derived thresholds, the TPHC concentrations in three crab species studied were remarkably low despite the fact that the TPHC concentration was considerably higher during the wet season than the dry season. Mean TPHC concentrations in crab tissues of the Chilika lagoon varied from 0.05 to 7.03 ng/g during the dry season, with an average of 2.532 ± 1.832 ng/g DW, and ranged from 0.26 to 6.74 ng/g during the wet season, with an average of 2.620 ± 1.487 ng/g DW. The value obtained in this study is less than the recently quantified range (0.52–2.05 g/g) from the coast of Tamil Nadu (Veerasingam et al. 2011) and crab tissues (0.47–3.77 g/g) from the north and central Arabian Sea (Veerasingam et al. 2011). Table 12.3 shows the mean concentrations of TPHC found in surface water and sediment samples collected over a period.

In the dry season, the level of TPHCs in the surface water was 1.119 ± 0.884 μg/L, whereas in the wet season, the concentration of TPHC was 2.224 ± 1.808 μg/L. In the sediment, the concentration of TPHCs in the dry season was 0.758 ± 0.442 μg/g, whereas in the wet season, the TPHC concentration was 0.776 ± 0.661 μg/g. Our data represent baseline data on the most prominent TPHC pollutants responsible for environmental contamination and degradation of coastal lagoon ecosystems. As a high concentration of TPHCs in the crab tissues can lead to health hazards for the coastal communities and the general public, our results provide valuable information that can be used to monitor future coastal living resource management of the Chilika lagoon with other crab and fish species. The three commercially important

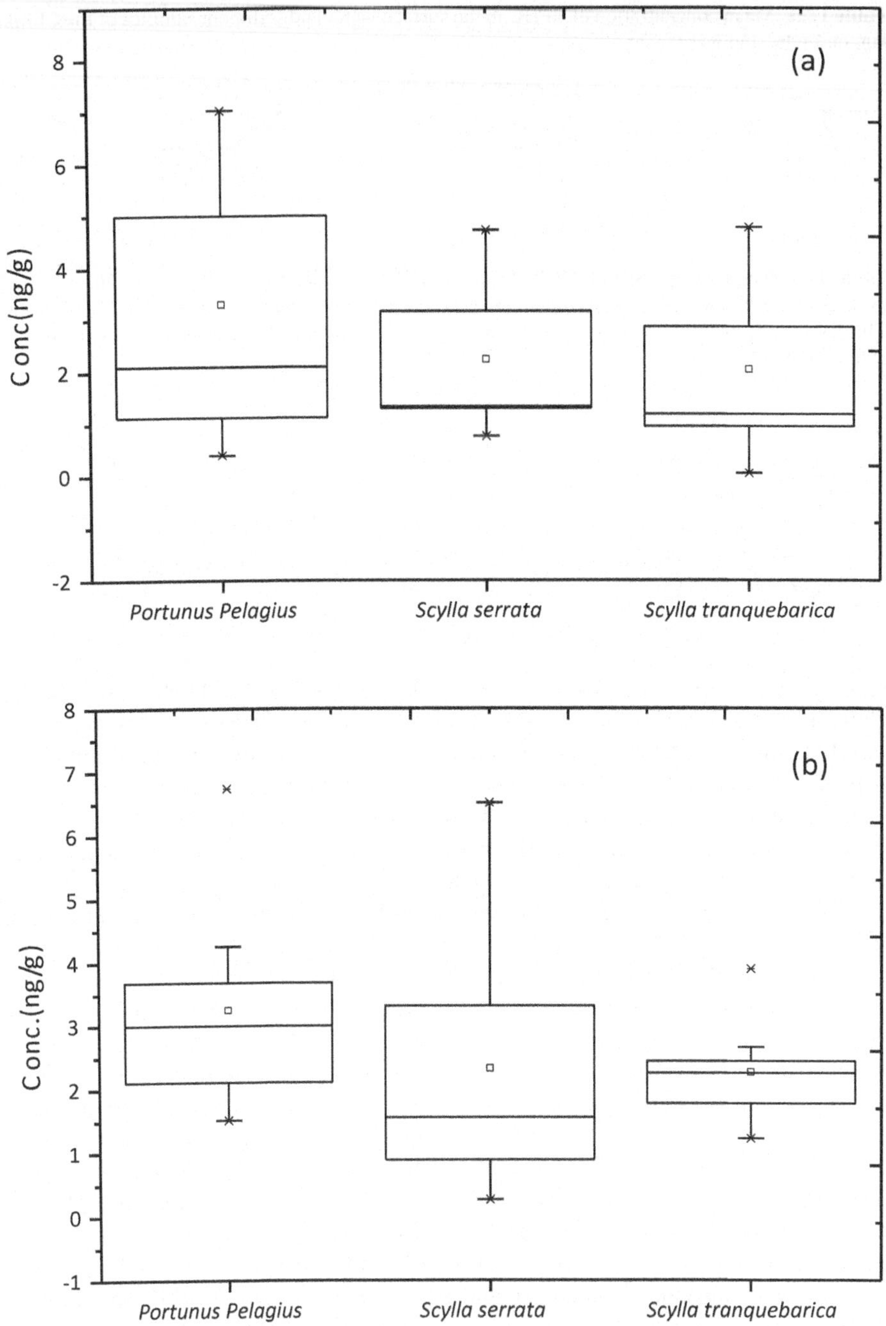

Fig. 12.2 TPHC levels in biospecimen samples from Chilika Lake during the (**a**) dry season, (**b**) wet season

Table 12.3 Mean concentration of TPHC in the surface water and sediment samples of the Chilika lagoon in dry and wet season

	TPHC concentration	
Season	Water (μg/L)	Sediment (μg/g)
Wet season	2.224 ± 1.808	0.776 ± 0.661
Dry season	1.119 ± 0.884	0.758 ± 0.442

Table 12.4 The relationships of TPHC levels in three crab species *PP* (*Portunus pelagicus*), *SS* (*Scylla serrata*), and *ST* (*Scylla tranquebarica*) with surface water and sediment TPHC in the Chilika lagoon in surface water. *SED* sediment, *r* correlation coefficient, *R** coefficient of determination

	Slope (*Y*)	*r*	*R** (%)
P. pelagicus vs water	$0.93x + 3.57$	0.81	54.23
P. pelagicus vs sediment	$0.03x + 2.97$	0.93	67.91
S. serrata vs water	$0.07x + 6.21$	0.54	72.01
S. serrata vs sediment	$0.27x + 1.89$	0.23	83.67
S. tranquebarica vs water	$0.56x + 6.21$	0.82	77.41
S. tranquebarica vs sediment	$0.07x + 7.91$	0.71	56.23

crab species were studied in order to establish a baseline for TPHC levels in the Chilika lagoon which are the important food source for Odisha's coastal communities. In comparison, there was a statistically significant difference in concentrations between the dry and wet seasons, according to the findings of the sample "*t*"-test analysis of seasonal mean concentrations. Table 12.4 shows that the correlation between the concentrations of TPHC in the three crab species and those in the water and sediment was statistically significant.

It was found that all species of crabs had significantly different bioaccumulation factors for TPHC, with *Portunus pelagicus* having a higher BAF than *Scylla serrata* having a lower BAF than *Scylla tranquebarica* having BAF values less than 1 μg/g over the duration of the study period (USEPA 2000). The BAFs values in water varied from 0.0011 μg/g in the dry season and 0.0023 μg/g in the wet season, whereas the BAFs values in sediment varied from 0.0032 μg/g in the dry season to 0.0034 μg/g in the wet season, with the lowest value in the dry season and the highest value in the wet season. It is possible that the high lipid content and solubility of *Portunus pelagicus*, as well as its hydrophilic nature, contribute to the absorption and bioaccumulation of these chemical compounds in the fish (Gobas et al. 1999). When the species under investigation were compared to the standard threshold, there was no evidence of bioaccumulation found in it. However, due to the fact that the chemical substance was absorbable by the species, it is possible that the low TPHC concentration in the water and sediment played a crucial role.

Because of the long-term exposure of the aquatic biota to this TPHC, it is possible that bioaccumulation and biomagnification will occur, which will be harmful to both the aquatic species and the entire ecosystem. Food consumption has been determined to be a major route in human exposure to petroleum hydrocarbons (PHCs), and various other reports have highlighted that petroleum hydrocarbons are hazardous to human health (Raoux et al. 1999). Recently, studies have found that the majority of human cancers, such as prostate and liver cancer, may be directly connected to dietary factors, such as dietary intake of TPHCs, which are a type of organic pollutant (Kucuksezgin et al. 2006). Because of their exceptional hydrophobic characteristics, they are particularly resistant to biodegradation (Jack et al. 2005). These properties allow hydrocarbons in the pelagic column to be biotransferred into the tissues of marine animals by consuming hydrocarbon-contaminated water (Soclo et al. 2000). Specifically, various aquatic indigenous species, such as crabs and oysters that live in nearshore sediments, as well as a few bottom feeders, seem to be highly vulnerable to the epistatic effects of hydrocarbons on their habitats (Moles et al. 1998). As a result, biomonitoring of TPHCs through surface water, sediment, and biota might appear to become an accurate and convenient approach for determining the long-term viability of marine habitats affected by hydrocarbon contamination (Kucuksezgin et al. 2006; Chindah et al. 2004). Significantly more hydrocarbon absorption in crabs may have far-reaching environmental consequences due to bioconcentration in their tissues (Benson et al. 2014). The significantly higher amounts of TPHC in biota tissues than in the pelagic column showed that the studied biological entities can bioaccumulate hydrocarbons. This might have occurred through direct absorption and ingestion (Micheel and Zengel 1998). Several aquatic species have been discovered as effective bioaccumulators of organic and inorganic pollutants, including spine fauna, crabs, and finfish (King et al. 2003). Because of their tendency to accumulate and bioconcentrate organic contaminants, TPHCs, as well as heavy metals in their tissues to variable concentrations greater than detection limits, biota such as molluscs, crabs, and prawns have also been identified as significant indicators of TPHCs (Osibanjo et al. 1994; Etuk et al. 2000). Crabs have high lipid levels, which increases their ability to absorb more hydrocarbon chemicals, particularly ones that are difficult to break down or eliminate (Olayinka et al. 2019).

The TPHC accumulation reported in the current study could also be related to the TPHC present in the lagoon's water and sediment, where TPHC is sourced by the movement of boats for fishing and tourism (Mohanty et al. 2016, 2017). The TPHC reported for Chilika crabs is much lower than the values found for crabs (*Callinectes sapidus*; 101.10–151.49 μg/g) gathered from other coastal habitats such as the Ondo State coast in Nigeria (Olayinka et al. 2019). The high amounts of TPHC in crabs are most likely the result of bioaccumulation in the food chain and can be harmful to human health if consumed. Along with the water column and sediment, food absorption is thought to be a major route of exposure to these pollutants in the crab. It may be that crabs are bottom feeders and also because crabs are excellent bioaccumulators of organic and inorganic pollutants (Eisler 1987). In the case of much greater TPHC concentrations in crabs, this may have broad environmental ramifications. Incapacity to metabolize TPHC properly has a negative impact on bioconcentration through their tissues. According to the current study, however, Chilika crabs are harmless in this aspect. According to the current study, the dietary consumption of TPHC seems to be as follows: *Portunus pelagicus* (dry, 0.00529 ± 0.0037 ng/g; wet, 0.00521 ± 0.0028 ng/g); *Scylla serrata* (dry, 0.00311 ± 0.0021 ng/g; wet, 0.00321 ± 0.0030 ng/g). This study found that *Portunus pelagicus* (crab), which feeds on contaminated sediments, accumulates TPHC inside its tissue throughout the year, including during the dry and wet seasons (Fig. 12.2a, b). The hydrocarbon bioaccumulation potential of the other two crab species was also significant. However, in all cases, the levels of TPHC acquired during the wet season are significantly higher than during the dry season. The reported variations in concentration over time could be attributed to factors such as feeding, tidal impacts, and the type of sediments scavenged (Chindah et al. 2004). In addition, TPHC concentrations were higher in the wet season than in the dry season. Additionally, these species (*Portunus pelagicus, Scylla serrata, and Scylla tranquebarica*) appear to have a higher tendency to bioaccumulate hydrocarbons from their environment and may serve as effective bioindicators of hydrocarbon pollution in the Chilika lagoon. In this research, the amount of TPHC in Chilika crabs was substantially lower than the World Health Organization (WHO) suggested threshold of 0.01 μg/g for the biota (Nozar et al. 2015). This indicates that the ecosystem of Chilika, as well as the crabs, fish, and lagoon shellfish, is in good condition. Chilika fish and shellfish, according to the study, are toxic-free, at least in terms of TPHC, and can be consumed as part of a healthy diet.

The data in Table 12.1 indicates that, between the two seasons, the levels of hydrocarbons of the various species are usually <1.0 μg/g. According to the findings of the study, the specific hydrocarbon succession observed may be due to hydrocarbon biotransformation in crabs. Among fish, the presence of the mixed function oxygenase (MFO) pathway is relatively common (Rice et al. 1985) and has also been observed in crabs (George et al. 1995). These MFOs perform the role of metabolically altering imported complex substances. Heavier hydrocarbons, alkyl substituted hydrocarbons, and TPHC indicating interaction with a significantly weathered petroleum combination dominated the hydrocarbon profile, particularly during the dry season. We hypothesized that part of the increased quantity was likely caused by biogenetically produced hydrocarbons by bacteria, algae, and macrophytes (Eisler 1987; Sauer and Uhler 1994).

As crabs are considered a delicacy by humans, understanding the TPHC concentrations in crabs is remarkably necessary for the development of human health. Because the metabolic processes within crabs are incredibly low and effective at adsorption within crab tissue, this is also confirmed through this study. The higher TPHC concentrations observed through the wet season indicated that, in accumulation to the biogenic source, the discharge of TPHC holding residue into the ecosystem is occurring. Petroleum oil spills from the boat could be a significant cause of TPHC in the ecosystem of Chilika lagoon. Away from the water column and sediment, food absorption is found to be a major way for these contaminants to enter the crab. Crabs are bottom feeders, so this could be the case. Reflecting variations in concentrations between the seasons, the TPHC levels obtained are comparable. The detection and higher amounts of various TPHC require additional investigation to assess the level of toxicity (Table 12.5).

Table 12.5 Range of TPHC concentration (μg/g DW) in the tissues of various crab species derived from literature review

Area	Crab species	TPHC	Reference
Chilika lagoon, India	*Portunus pelagicus, Scylla serrata, Scylla tranquebarica*	0.0005–0.0073	Current study
UAE	*Callinectes amnicola*	0.73–3.5	Tolosa et al. (2005)
Nigeria	*Callinectes amnicola*	101.10–151.49	Olayinka et al. (2019)
Gulf	*Portunus segnis*	0.20–2.9	Nozar et al. (2015)
Iran	*Portunus segnis*	2.7–3	Nozar et al. (2015)
Bahrain	*Callinectes amnicola*	0.49–4.4	De Mora et al. (2010)

4 Conclusions

Assessment of TPHC was carried out in Chilika Lake water, sediments, and three commercially important crab species. The study concludes TPHC level in Chilika is at a safe level; however, there is a need of monitoring at least on a seasonal basis to ensure it's the safe limit of accumulation in Chilika biota. Among the three crab species considered in this study, the TPHC accumulation in *Portunus pelagicus* was highest followed by *Scylla serrata* and *Scylla tranquebarica*. The accumulation level is also significantly influenced by the season as evidenced from the present study. The TPHC in crabs was attributed to the benthic feeding habit and anthropogenic TPHC maintained in the Chilika sediment. TPHC residue levels in all the studied Chilika crabs were considerably lower than the hazardous levels prescribed by regulating authorities of different countries around the globe. Scientific evidence on the safe consumption of Chilika crab would improve the market value chain and livelihood of the fishermen of Chilika lagoon. From the public health point of view, this level must be maintained in a long run. Keeping this in view, the study recommends further investigations focusing on the TPHC level in the rest of the crabs reported for this Asia's brackish water lagoon. Since the source and quantity of the anthropogenic input to the lagoon change with time, monitoring of TPHC is needed from time to time. The current findings on the TPHC as compared to previous monitoring records would be helpful for the policy formulation and sustainable management of the lagoon.

Acknowledgements The authors would like to express their gratitude to the Chief Executive of the Chilika Development Authority for the encouragement to conduct the research. Sample analysis was carried out at the Chilika Development Authority's WRTC laboratory facilities, which were developed with funding assistance from the World Bank (ICZMP, Odisha: Credit no. 4765-IN) and the Chilika Development Authority. Thanks are also due to the CDA research and support staff, who assisted with the sampling and analysis in the laboratory throughout the course of the research.

References

Adeniji AO et al (2017) Petroleum hydrocarbon profiles of water and sediment of Algoa Bay, Eastern Cape, South Africa. Int J Environ Res Public Health 14:1263. https://doi.org/10.3390/ijerph14101263

Akinola JO, Olawusi-Peters OO, Akpambang VOE (2019) Ecological hazards of total petroleum hydrocarbon in brackish water white shrimp Nematopalaemon hastatus (Aurivillus 1898). Egypt J Aquat Res 45(3):205–210

Alkindi AYA et al (1996) Endocrine, osmoregulatory, respiratory and haematological parameters in flounder exposed to the water soluble fraction of crude oil. J Fish Biol 49:1291–1305. https://doi.org/10.1006/jfbi.1996.0254

Al-Shwafi NAA et al (2008) Total petroleum hydrocarbon carcinogens in commercial fish in the Red Sea and Gulf of Aden, Yemen. J King Abdulaziz Univ Mar Sci 19:15–28

Anderson JW et al (1974) Characteristics of dispersion and water-soluble fractions of crude and refined oils and their toxicity to estuarine crustaceans and fish. Mar Biol 27:75–88

Ansari ZA et al (2012) Total petroleum hydrocarbon in the tissues of some commercially important fishes of the Bay of Bengal. Mar Pollut Bull 64(11):2564–2568. https://doi.org/10.1016/j.marpolbul.2012.07.029

Ashiru OR et al (2019) An assessment of total petroleum hydrocarbon and trace metal concentration in the sediments of Ugbo water way, south western Nigeria. Afr J Environ Sci Technol 13:13–21. https://doi.org/10.5897/AJEST2018.2578

Asuquo FE et al (2004) Fish species used as biomarker for heavy metal and hydrocarbon contamination for Cross River, Nigeria. Environmentalist 24:29–37

Barik SK et al (2017) Spatio-temporal variability and the impact of Phailin on water quality of Chilika lagoon. Cont Shelf Res 136:39–56. https://doi.org/10.1016/j.csr.2017.01.019

Barik N (2017) Freshwater fish for nutrition security in India: Evidence from FAO data. Aquac Rep 7. 1–6. https://doi.org/10.1016/j.aqrep.2017.04.001

Barrento S et al (2010) Composition of Atlantic spider crab Maja brachydactyla: human health implications. J Food Comp Anal Chem 23:230–237. https://doi.org/10.1016/j.jfca.2009.10.007

Batvari BP, Kamalakannan S, Viswanathan S, Lee K, Jayaprakash M (2007) Heavy metals in two fish species (Carangoidel malabaricus and Belone stronglurus) from Pulicat Lake, North of Chennai, Southeast Coast of India. Environ Monit Assess 145:167–175. https://doi.org/10.1007/s10661-007-0026-3

Benson NU, Essien JP, Asuquo FE et al. (2014) Occurrence and distribution of polycyclic aromatic hydrocarbons in surface microlayer and subsurface seawater of Lagos Lagoon, Nigeria. Environ Monit Assess 186, 5519–5529. https://doi.org/10.1007/s10661-014-3800-z

Burns KA et al (1976) Hydrocarbon metabolism in the intertidal fiddler crab Uca pugnax. J Fish Environ 11:5–11

CDA (2005) Collection of fish, prawn and crab landing statistics in the Chilika lagoon (annual report-2002–03 and 2003–04). Chilika Development Authority, Orissa, Bull No. 3, 146 pp

Chase DA et al (2013) Bioaccumulation of petroleum hydrocarbons in fiddler crabs (Uca minax) exposed to weathered MC-252 crude oil alone and in mixture with an oil dispersant. Sci Total Environ 444:121–127. https://doi.org/10.1016/j.scitotenv.2012.11.078

Chen D et al (2007) Compositional characteristics and nutritional quality of Chinese mitten crab Eriocheir sinensis. Food Chem 103(4):1343–1349

Chindah AC et al (2004) Distribution of hydrocarbons and heavy metals in sediment and a crustacean (Shrimps: Penaeus Notialis) from the Bonny/New Calabar River Estuary, Niger Delta. AJEAM-RAGE E 9:1–17

Chouksey MK et al (2004) Petroleum hydrocarbon residues in the marine environment of Bassein-Mumbai. Mar Pollut Bull 49:637–647. https://doi.org/10.1016/j.marpolbul.2004.04.007

Connell DW et al (1980) Petroleum hydrocarbons in aquatic ecosystem—behavior and effect of sub lethal concentrations. Part 2. CRC Crit Rev Environ Control 11:37–104

Copat C et al (2012) Heavy metals concentrations in fish from Sicily (Mediterranean Sea) and evaluation of possible health risks to consumers. Bull Environ Contam Toxicol 88:78–83. https://doi.org/10.1007/s00128-011-0433-6
Das N et al (2011) Microbial degradation of petroleum hydrocarbon contaminants: an overview. Biotechnol Res Int. https://doi.org/10.4061/2011/941810
De Mora S, Tolosa I, Fowler SW, Villeneuve JP, Cassi R, Cattini C (2010) Distribution of petroleum hydrocarbons and organochlorinated contaminants in marine biota and coastal sediments from the ROPME Sea area during 2005. Mar Pollut Bull 60(12):2323–2349
Eisler R (1987) Polycyclic aromatic hydrocarbon hazards to fish, wildlife and invertebrates. A synoptic review. Biological Report, 85(1.11). Patuxent Wildlife Research Centre, U.S. Fish and Wildlife Service, Laurel, MD, p 81
Etuk EUI et al (2000) Tissues elemental status of the freshwater clam, Galatea paradoxa (Donacidae) from the Cross River, Nigeria. Ecol Food Nutr 39:1–11
Food and Agriculture Organisation (FAO) (1982) The review of the health of the oceans. FAO/ IMCO/UNESCO/WMO/WHO/ IAEA/UNEP Joint Group of Experts on Scientific Aspects of Marine Pollution (GESAMP)
Food and Agriculture Organization (FAO) (2010) World review of fisheries and aquaculture; Fisheries resources: trends in production, utilization and trade
George SG, Christiansen JS, Killies B, Wright J (1995) Dietary crude oil exposure during sexual maturation induces hepatic missed function oxygenase (CYPIA) activity at very low environmental temperatures in polar cod Boreogadus saida. Mar Ecol Prog Ser 122:307–312
Ghauch A et al (2000) Temperature phosphorescence analysis of PAHs using an imaging sensing system combined with a bifurcated optical filter and a cooled charge coupled device detector. Talanta 51:807–816
Gil MI, Tomás-Barberán FA, Hess-Pierce B, Kader AA (2002) Antioxidant capacities, phenolic compounds, carotenoids, and vitamin C contents of nectarine, peach, and plum cultivars from California. J Agric Food Chem 50(17):4976–4582. https://doi.org/10.1021/jf020136b. PMID: 12166993.
Ghosh AK, Pattnaik AK, Ballatore TJ (2006) Chilika lagoon: restoring ecological balance and livelihoods through re-salinization. Lakes Reserv 11:239–255. https://doi.org/10.1111/j.1440-1770.2006.00306.x
Gobas FAPC et al (1999) Mechanism of biomagnification in fish under laboratory and field conditions. Environ Sci Technol 33:133–141. https://doi.org/10.1021/es980681m
Hardy JA, Higgins GA (1992) Alzheimer's disease: the amyloid cascade hypothesis. Science. Apr 10;256(5054):184–5. https://doi.org/10.1126/science.1566067. PMID: 1566067.
IOC-UNESCO (1982) The determination of petroleum hydrocarbons in sediments. Manuals guides no. 11, pp 13–23
Jack IR et al (2005) Determination of total hydrocarbons levels in some marine organisms from some towns within the Rivers State of Nigeria. J Appl Sci Environ Manag 9:59–61
Khan MAQ et al (2005) Effects of petroleum hydrocarbons on aquatic animals. Environ Sci Technol
King RP et al (2003) Aquatic environmental perturbations and monitoring. African Experience, USA, p 166
Kucuksezgin F et al (2006) Assessment of marine pollution in Izmir Bay: nutrient, heavy metal and total hydrocarbon concentrations. Environ Int 32:41–51
Kvenvolden KA, Cooper CK et al (2003) Natural seepage of crude oil into the marine environment. Geo-Marine Lett 23:140–146. https://doi.org/10.1007/s00367-003-0135-0
Kumar P, Dey M, Paraguas F (2005) Demand for Fish by Species in India: Threestage Budgeting Framework. Agric Econ Res Rev 18
Laili S, Nozar M, Ismail WR, Zakaria MP, Sampling A (2013) Residual concentration of PAHs in seafood from Hormozgan Province, Iran, human health risk assessment for urban population. IJESD 4:393. https://doi.org/10.7763/IJESD.2013.V4.379
Lee RF et al (1976) Fate of petroleum hydrocarbons taken up from food and water by the blue crab Callinectes sapidus. Mar Biol 37:363–370

Maria ALY et al (2007) Protein and amino acid contents in the crab, Chionoecetes opilio. Food Chem 103:1330–1336

Mazhar FM et al (1987) Effect of environmental pollution by crude oil on the Nile fish Clarias lazera. 11—Histopathological features. Proc Zool Soc A R Egypt 14:381–390

Micheel J, Zengel S (1998) Monitoring of oysters and sediments in Acajutla, El Salvador, Mar

Mironov OG et al (1980) Aspects of petroleum hydrocarbon metabolism in marine animals. Helgoland Mar Res 33:292–296

Mohanty SK, Bhatta KS, Mohanty RK, Mishra S, Mohapatra A, Pattnaik AK (2008) Eco-restoration impact on fishery biodiversity and population structure in Chilika Lake. Springer, Dordrecht, pp 1–2. https://doi.org/10.1007/978-1-4020-6646-7_1

Mohanty B et al (2016) Assessment of petroleum hydrocarbon in a tropical brackish water lagoon: Chilika, India. Chem Ecol 32:653–668

Mohanty B et al (2017) Sources and variability of petroleum hydrocarbon residues in sediments of Chilika lagoon, East Coast of India. Bull Environ Contam Toxicol 99:100–107

Mohapatra A et al (2005) Food and feeding habits of mud crab Scylla serrata (Forssakal) from Chilika lagoon

Mohapatra A, Mohanty RK, Mohanty SK, Bhatta KS, Das NR (2007) Fisheries enhancement and biodiversity assessment of fish, prawn and mud crab in Chilika lagoon after ecorestoration through hydrological intervention. Wetl Ecol Manag 15:229–251. https://doi.org/10.1007/s11273-006-9025-3

Moles A et al (1998) Effects of oil-laden sediments on growth and health of juvenile flatfishes. Can J Fish Aquat Sci 55:605–610

Muduli PR, Pattnaik A (2020) Spatio-temporal variation in physicochemical parameters of water in the Chilika Lagoon. In: Ecology conservation and restoration of Chilika lagoon, India. Springer, New York, pp 203–229. https://doi.org/10.1007/978-3-030-33424-6_9

Muduli PR et al (2012) Spatio-temporal variation of CO2 emission from Chilika Lake, a tropical coastal lagoon, on the east coast of India. Estuar Coast Shelf Sci 113:305–313. https://doi.org/10.1016/j.ecss.2012.08.020

Muduli PR et al (2013) Distribution of dissolved inorganic carbon and net ecosystem production in a tropical brackish water lagoon, India. Cont Shelf Res 64:75–87. https://doi.org/10.1016/j.csr.2013.05.014

Muduli PR et al (2017) The impact of tropical cyclone 'Phailin' on the hydrology of Chilika lagoon, India. Int J Environ Sci Nat Resour. https://doi.org/10.19080/IJESNR.2017.04.555632

National Research Council (US) (2003) Committee on Oil in the Sea: Inputs, Fates, and Effects. Oil in the Sea III: Inputs, Fates, and Effects. Washington (DC): National Academies Press (US). PMID: 25057607.

Newcombe CL et al (1994) The nutritional value of sea foods. Virginia Fisheries Laboratory of the College of William and Mary and Commission of Fisheries

Nozar SLM, Pauzi MZ, Salarpouri A, Daghooghi B, Salimizadeh M (2015) Total petroleum hydrocarbons in edible marine biota from Northern Persian Gulf. Environ Monit Assess 187:1–6. https://doi.org/10.1007/s10661-015-4443-4

Ogunfowokan AO et al (2003) Isolation and estimation of PAHs in surface run off and sediments. Water Air Soil Pollut 147:245–261

Okuzumi M et al (2000) Nutritional and functional properties of squid and cuttlefish. Japan National Cooperative Association of Squid Processors, Tokyo, p 223

Olayinka OO, Adewusi AA, Olujimi OO (2019) Polycyclic aromatic hydrocarbons in sediment and health risk of fish, crab and shrimp around Atlas Cove, Nigeria. J Health Pollut 9:191204

Ololade IA et al (2008) Occurrence and toxicity of hydrocarbon residues in crab (Callinectes sapidus) from contaminated site. J Appl Sci Environ Manag 12:19–23

Oluwatobi J et al (2019) Ecological hazards of total petroleum hydrocarbon in brackish water white Shrimp Nematopalaemon hastatus (Aurivillus 1898). Egypt J Aquat Res 45:205–210

Osibanjo O et al (1994) Review of chlorinated hydrocarbon substances in the African aquatic environment. Committee for inland fisheries of Africa. Report of the fourth session of the working party on pollution and fisheries, Accra, Ghana, 18–22 Oct 1993. FAO Fisheries Report No. 502. FAO, Rome, p 45

Panda S, Bhatta KS, Rath KC, Mishra C, Samal RN (2008) The atlas of Chilika. Chilika Development Authority, Bhubaneswar, p 133

Parida S et al (2017) Trace metal concentrations in euryhaline fish species from Chilika lagoon: human health risk assessment. Int J Environ Sci Technol 14:2649–2660. https://doi.org/10.1007/s13762-017-1334-y

Porte C, Biosca X et al (2000) The Aegean Sea oil spill. 2. Temporal study of the hydrocarbons accumulation in bivalves. Environ Sci Technol 34(24):5067–5075

Ramsar Convention (1981). Ramsar Regional Center – East Asia. The Designation and Management of Ramsar Sites – A practitioner's guide. www.ramsar.org and www.rrcea.org

Rao MN et al (2016) A preliminary estimate of total petroleum hydrocarbons in water and some commercially important fish species in the Amba estuary, West Coast of India. Bull Environ Contam Toxicol. https://doi.org/10.1007/s00128-016-1819-2

Raoux C et al (1999) Particle fluxes of aliphatic and aromatic hydrocarbons in near-shore waters to the north-western Mediterranean Sea and the effects of continental runoff. Estuar Coast Shelf Sci 48:605–616

Rice SD et al (1985) Effects of oil on fish. In: Engelhard FR (ed) Petroleum effects in the Arctic environment. Elsevier Applied Science Publishers, Essex, pp 157–182

Rose A et al (2012) Bioaccumulation of polycyclic aromatic hydrocarbons in fish and invertebrates of Lagos lagoon, Nigeria. J Emerg Trends Eng Appl Sci 3(2):287–296

Saadoun IMK (2015) Impact of Oil Spills on Marine Life, Emerging Pollutants in the Environment - Current and Further Implications, Marcelo L. Larramendy and Sonia Soloneski, IntechOpen, https://doi.org/10.5772/60455. Available from: https://www.intechopen.com/chapters/48738

Sahoo DK, Palita SK, Panda S (2013) New records and distribution pattern of brachyuran crabs in Chilika lagoon, Odisha, India. Ecol Environ Conserv 19:409–415

Sauer TC, Uhler AD (1994) Pollutant source identification: advances in hydrocarbon fingerprinting. Remediation 5:25–46

Shoda RK et al (1996) Epidemiologic analysis of chronic disease in Japan, increased dietary intake of n-6 polyunsaturated fatty acids and animal protein relates to the increased incidence of chronic disease in Japan. Am J Clin Nutr 63:741–745

Soclo HH et al (2000) Origin of polycyclic aromatic hydrocarbons (PAHs) in coastal marine sediments: case studies in Cotonou (Benin) and Aquitaine (France) areas. Mar Pollut Bull 40:387–396

Suresh VR et al (2018) Fish and shellfish diversity and its sustainable management in Chilika Lake. CAR—Central Inland Fisheries Research Institute, Barrackpore, Kolkata and Chilika Development Authority, Bhubaneswar, 376p

Tolosa I, De Mora SJ, Fowler SW, Villeneuve J-P, Bartocci J, Cattini C (2005) Aliphatic and aromatic hydrocarbons in marine biota and coastal sediments from the Gulf and the Gulf of Oman. Mar Pollut Bull 50(12):1619–1633

USEPA (United States Environmental Protection Agency) (2000) Guidance for assessing chemical contaminant data for use in fish advisories. Volume 2: risk assessment and fish consumption limits (3rd ed.). EPA 823-B-00-008, Office of Science and Technology, Office of Water, Washington, DC, 383p

Vandermeulen J et al (1985) Toxicity vs mutagenicity of some crude oils, distillates and their water-soluble fractions. Water Res 19:1283–1289. https://doi.org/10.1016/0043-1354(85)90183-6

Varanasi U et al (1989) Biotransformation and disposition of PAHs in fish. In: Varanasi U (ed) Metabolism of PAHs in the aquatic environment. CRC Press Inc., Boca Raton, FL, p 93

Varjani SJ, Upasani VN (2016) Bioresource technology biodegradation of petroleum hydrocarbons by oleophilic strain of Pseudomonas aeruginosa NCIM 5514. Bioresour Technol 222:195–201. https://doi.org/10.1016/j.biortech.2016.10.006

Veerasingam S et al (2011) Petroleum hydrocarbon concentrations in eight mollusc species along Tamilnadu coast, Bay of Bengal, India. J Environ Sci 23:1129–1134

Zhou Z, Guo L, Osburn CL (2015) Fluorescence EEMs and PARAFAC techniques in the analysis of petroleum components in the water column, pp 179–200. https://doi.org/10.1007/8623_2015_137

Chapter 13
Coastal Ecosystem Services of Gujarat, India: Current Challenges and Conservation Needs

Jayendra Lakhmapurkar, Deepa Gavali, and Nilesh Bhatt

Abstract Gujarat is the only state in India with the longest coastline of 1663 km (20% of the country) and the widest shelf zone covering about 184,000 km^2. The main feature includes two gulfs, the Gulf of Khambhat and the Gulf of Kachchh, and the open coast of Saurashtra facing the Arabian Sea. Further, Gujarat is the only state on the west coast of India with coral reefs. The other ecosystems present in the coastline are the seagrass, seaweeds, mangroves, beaches, and coastal dunes.

Ecosystem services are defined as the many and varied benefits to humans provided by the natural environment and from any ecosystems. All ecosystem services can be grouped into four broad categories: provisioning (production of food), regulating (control of climate and diseases), supporting (nutrient cycling), and cultural (recreational benefits). The services provided by different ecosystems are immense and described in length in the paper (Millennium Assessment 2000). The paper describes the ecosystem services provided by corals, mangroves, seagrass, seaweeds, coastal dunes, and others. For example, the corals not only protect the shoreline from erosion but also act as carbon sinks. Likewise, the seagrass meadows provide habitat for threatened *Dugong dugon*. Seaweeds have an important provisional role as raw materials in the pharmaceutical and cosmetics industry. Mangroves are considered as one of the most valuable coastal vegetation providing economic, social, and environmental benefits to the local communities. These mangroves also act as a major carbon sink and provide stability to coastal erosion. The mangroves of the Kori Creek support unique breed of Kharai camels. Large intertidal zone of the Gulfs provide regulatory services, as they act as major sinks of pollutants released into the coastal waters. The sandy beaches not only provide habitat for the green sea and olive ridley turtles but also sustain coastal tourism. The service provided by these coastal ecosystems is enormous, and this article discusses the components of each ecosystem with important services it provides.

J. Lakhmapurkar (✉) · D. Gavali
Gujarat Ecology Society, Vadodara, India

N. Bhatt
Department of Geology, M.S. University of Baroda, Vadodara, India

S. Madhav et al. (eds.), *Coastal Ecosystems*, Coastal Research Library 38,
https://doi.org/10.1007/978-3-030-84255-0_13

Industrial development combined with coastal developmental activities has increased stress on the various ecosystems along the coast, and the paper brings out the facts. The effluent discharges into the coastal region and its impact on the sensitive mangrove system and coastal fishery need to be discussed as the country is to implement the Sustainable Development Goals (SDGs). There is need to mitigate the growing pressures on the ecosystem through policy interventions and need to conserve or enhance the ecosystem services in a way that reduces the negative trade-off with other ecosystem services. There is need to create more conservation areas in the coastal line for the various threatened ecosystems similar to that of the terrestrial system where protected area for threatened species is emphasized. Examples could include seagrass conservation centers or areas with high diversity of seaweeds.

Keywords Gujarat · Salt marshes · Mangroves · Ecosystem services

1 Introduction

The oceans and coastal areas provide both human (fisheries, energy, tourism, and transport/shipping) and environmental (climate regulation, carbon sequestration, habitat for biodiversity) benefits. This is the reason for the presence of human settlements within 100 km of the coastline, and it is estimated that more than 40% of the world's population (>2.8 billion people) live in the coastal region. In Asia, the coastal megacities like Chennai (2011: population 8.65 million), Dhaka (21.00 million), Karachi (11.62 million), Kolkata (14.05 million), and Mumbai (18.4 million) are located only a few meters above sea level (United Nations 2019). India has a long coastline of about 5422.6 km starting from the Gulf of Kachchh in the west and extending up to the Indo-Gangetic delta in the east. Among the various states, Gujarat has the longest coastline of 1663 km and the largest continental shelf of about 184,000 km^2 in the country. Also, the state owns 214,000 km^2 of exclusive economic zone (EEZ) which extends up to 20 nautical miles from the coast as defined by the United Nations Conference on the Law of Sea (Balan et al. 1987).

Geologically and geomorphologically, the coastline of Gujarat is differentiated into four distinct coastal zones, viz. (1) Gulf of Kachchh, (2) Saurashtra coast, (3) Gulf of Khambhat, and (4) South Gujarat coast (GES 1998). The presence of two gulfs, a large continental shelf, very high tidal amplitude (highest in India) (Mitra et al. 2020), a large area under mangroves (Forest Survey of India 2019), and corals are some of the unique features of the state (Dixit et al. 2010). Because of this geomorphological character, the intertidal region of the state is highly diverse and sustaining rich biodiversity. All of these components add together to provide ecosystem services, and the present research article describes the ecosystem services with the conservation needs.

2 Coral Reefs

The coral reef is a unique bio-geological structure formed by a group of coral polyps held together by calcium carbonate. The coral reefs in the Gulf of Kachchh are found between 22°20′ N and 22°40′ N latitudes and 69° and 70° E longitudes.

(Source: Deshmuke et al 2000)

Source: Deshmukhe et al. (2000)

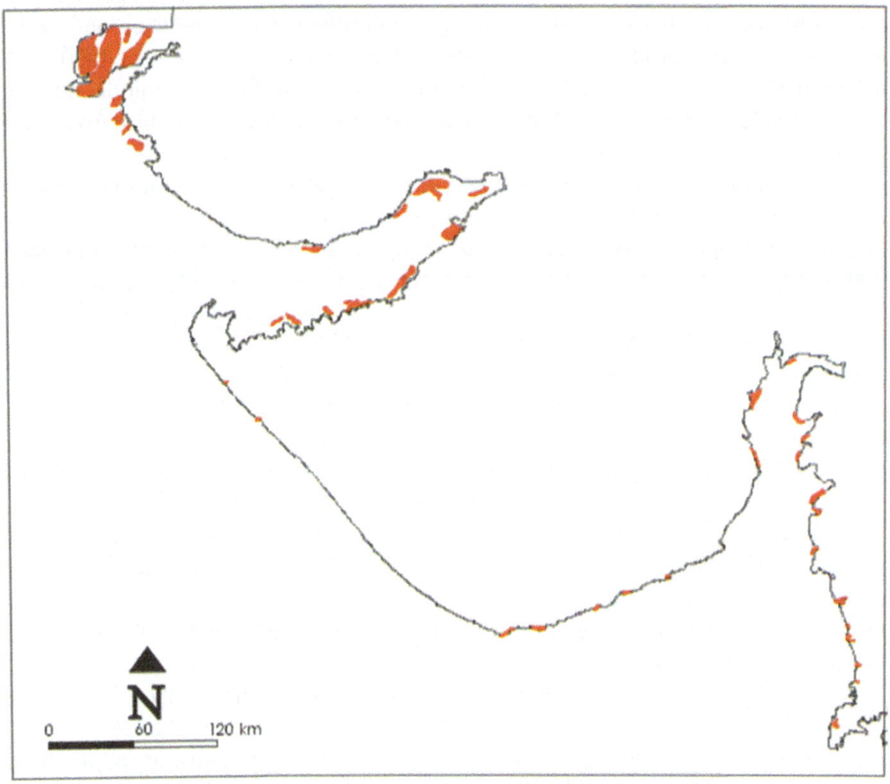

Source: Based on GEC (2010)

The Gulf of Kachchh region has 42 islands on its southern side, 34 of which have corals on one or the other shores (Satyanarayana and Ramakrishna 2009; Lakhmapurkar and Gavali 2018).

2.1 Corals

Various authors have worked on the coral diversity in the Gulf at different time scales. Patel (1978) reported 44 species of scleractinian corals and 12 species of soft corals from the Gulf. The monograph on Biological Diversity of Gujarat (Gujarat Ecology Commission 1996) lists 40 species of stony corals and 3 species of soft corals from the gulf. Gujarat Ecology Society (GES) in a pioneering effort carried out surveys of the subtidal reefs in 13 coral reef islands in the Gulf of Kachchh by scuba divers in 1999 and 2002. The study reported 75–80% subtidal live corals represented by 21 species of stony corals and 12 species of soft corals (Deshmukhe et al. 2000; Sen Gupta et al. 2003). Corals are reported outside the Marine National Park from Shivrajpur located between Okha and Dwarka (Lakhmapurkar and Gavali 2018), and it was declared as a Blue Flag beach in 2020.

The ecosystem services provided by corals reefs include coral fish diversity, coastal protection, fisheries, pharmaceutical, and others. Coral reefs are one of the highly biologically productive ecosystems with estimated productivity at 4200 (g C/m^2/year) (https://apescoralreefs.weebly.com). The corals have symbiotic relation with microalgae wherein the corals provide shelter to the algae and algae give color to the reefs with exchange of nutrients between both the organisms. The various life forms present to enhance the bio-resource of the region and assist to sustain the fishery or pharmaceutical industry; concerning economic significance globally, around 500 million people depend directly or indirectly on the reef ecosystem (Miththapala 2008). About 30 million worldwide depend exclusively on reefs for their food (Wilkinson 2008). Many coral species and species associated with coral reefs possess significant medicinal value (Hunt and Vincent 2006; Demers et al. 2002; Chivian 2006). Some hard corals are also utilized in bone grafts (Demers et al. 2002).

The latest the coral reefs play a major role as net sinks for C, principally as $CaCO_3$ accretion. Oceanic algae absorb carbon dioxide from the atmosphere for photosynthesis. The polyps build exoskeletons of calcium carbonate ($CaCO_3$) consuming these algae rich in carbon. This serves as a long-lasting store of carbon, making coral reefs as carbon sinks. However, several recent studies believe corals release carbon dioxide as a result of alteration of pH of seawater. However, on a global scale, the magnitude of the reef-generated CO_2 is small compared to current human-induced perturbations. Further it is found that deep coral reef lagoons act better sink of carbon compared to intertidal corals (Philben 2016).

Apart from these, coral reef plays a crucial role in protecting the shoreline. It acts as a physical barrier against the tides and helps in protecting coastal erosion, flooding, and loss of infrastructure. The best example of corals assisting in shoreline protection was witnessed in the 2004 tsunami in Tamil Nadu. In absence of coral reefs, extensive damages to the shoreline and human life and property were seen along the Kanyakumari and Nagapattinam coast, whereas the coastal region of Tuticorin and Tirunelveli in the Gulf of Mannar was least affected due to presence of coral reefs in 21 islands of Gulf of Mannar which dissipated the energy of strong tsunami waves, thereby reducing the damages (Kumaraguru et al. 2005).

It has been estimated that coral reefs of the Gulf of Kachchh provide an annual benefit of approximately Rs. 2200 million (Dixit et al. 2010). The coastline of the Gulf of Kachchh is well protected from shoreline erosion due to the presence of corals against the coastline of the South Gujarat coast which is muddy in nature. However, coral reefs are fragile ecosystems and very sensitive to water quality and temperature variation. Water quality parameters like pH, total suspended solids, and nutrient load have very strong impacts on coral survival (Kelmo et al. 2014; Lapointe et al. 2008).

Coral reefs worldwide are under threat from a rise in sea surface water temperature as a result of global warming. In the process of the temperature rise, the sensitive photosynthetic algae die leading to large-scale bleaching events (Hoegh-Guldberg 1999; Done et al. 2003; Great Barrier Reef Marine Park Authority 2017). There are incidences of coral bleaching from the region in 1999 and 2010 (Adhavan et al. 2014). Another important threat to the survival of corals in the region is the presence of an industrial cluster adjoining the coastal areas that release treated effluents into the Gulf of Kachchh (Gujarat Ecology Commission 2017; Panseriya et al. 2020).

The presence of ports and their increased activity in the Gulf of Kachchh have to lead to sedimentation along the southern coast largely due to high dredging activity. Gujarat Ecology Society (2018) has reported the occurrence of petroleum hydrocarbons (PHCs) in the sediments along the coast that bears a direct relationship to the transport activity of crude oil by various ports. This could be detrimental to the growth of benthic diversity. High sediment load reduces light penetration, and this results in reduced photosynthetic activity of zooxanthellae (Anthony and Fabricius 2000). A thin layer of sediments over the coral reef can be cleaned by the coral itself, but such a cleaning process by the corals cannot overcome the high rate of sedimentation (Stafford-Smith 1993; Erftemeijer et al. 2012). This can lead to deposition of thick sediment cover on the coral reefs. The Gulf of Kachchh is highly turbid, and the suspended sediment concentration during the post-monsoon season varies from 0.5 to 674 mg/l (Vethamony and Babu 2010). Studies have reported (Bahuguna and Nayak 1998) a reduction in total coral cover because of the high sediment influx. Sharma et al. (2008) have reported loss in coral area and coral bleaching in Pirotan islands blaming on the high sedimentation rate, loss of mangrove cover, and sand mining.

With the increase in urbanization and expansion of the existing urban zones, disposal of solid waste has become an important issue. Urban centers like Jamnagar, Lalpur, Dwarka, Khambhaliya, Gandhidham, Anjar, and Mandvi located close to the Gulf of Kachchh discharge sewage of 369 tons/day (estimation for 2012) directly or indirectly into the Gulf. Studies carried out by GES have reported the discharge of high nutrient loads from various urban and industrial centers into the Gulf. These discharges lead to algal bloom which hinders coral growth (Sen Gupta and Deshmukhe 2000).

The unregulated tourism in the coral supporting reef area especially the Pirotan islands is posing a threat to the corals. A detailed study (Sharma et al. 2008) on the Pirotan reef indicated that increase in sediment (due to dreding activity) and release of untreated sewage have resluted in increase of algal growth and sandy/muddy deterimental for the occurence of corals and sea grass.

2.2 Restoration of Corals

Coral restoration is a very slow process and requires a lot of scientific interventions. The critical part of coral transplantation is the adaptation by the coral polyps introduced from a donor site to the Gulf of Kachchh. Reported presence of ample quantity of Acropora coral fragment on beach and intertidal in several studies indicated presence of live Acropora corals in the subtidal region (Satyanarayana and Ramakrishna 2009; Dixit et al. 2010). Pillai and Patel (1988) have reported local extinction of *Acropora* from the region because of temperature fluctuations and high sedimentation rate. Considering the loss of *Acropora*, Zoological Survey of India undertook the initiative of improving the population of *Acropora* in the Gulf of Kachchh region, during December 2013 to January 2015 in three different reefs, i.e., Pirotan Island, Narara Reef, and Mithapur reef (Kumar et al. 2017). Three hundred and twelve nubbins of *Acropora* sp. and *Montipora* sp. were transplanted from the Gulf of Mannar. The preliminary results (Kumar et al. 2017) indicated successful rate of survival of transplanted nubbins as the transplanted coral could adopt the high sedimentation load, temperature variation, and strong tidal currents of the restoration site.

Looking at the cost-effectiveness of coral transplants, it would be advisable to conserve the existing coral reefs through conservation measures. Some initiative like pollution control, sustainable tourism (Diedrich 2007), and declaration of certain areas as no human intervention zone can help to protect the corals for a longer run (Ali et al. 2011, Anu et al. 2007, Allers et al. 2013; Larsson et al. 2013). Pollutants like heavy metals reduce abundance of live hard corals as hard coral colonies are susceptible to contaminants dissolved in seawater (Ali et al. 2011). Similarly, many times it has been reported that unplanned tourism leads to activities like human trampling which leads to destruction in coral reef ecosystems (Sarmento and Santos 2012).

2.3 Seagrass

Seagrass belongs to angiosperms group capable of surviving under submerged conditions. Seagrass prefers shallow, sheltered coastal water and in Gujarat is found in the intertidal region of the Gulf of Kachchh. *Halophila beccarii*, *Halodule uninervis*, *Halophila ovalis*, and *Thalassia hemprichii* are the commonly observed species. Maximum seagrass extent is observed in Kalubhar Island, Bhural reef, and Pirotan Island (GeeVarghese et al. 2017).

The presence of seagrass can be seen as an indicator of the overall environmental quality of the coastal zone (Martínez-Crego et al. 2008; Syukur et al. 2017). The ecological services provided by seagrasses are immense. Seagrass serves as food for marine herbivores (Valentine and Heck 1999) and carnivore species like dugongs, manatees, sharks, turtles, tiny seahorses, shrimps, and octopus. Seagrass is known to filter and clean the water. The deep root structures associated with seagrass help in stabilizing the sediment (Ondiviela et al. 2014).

Seagrass meadows produce a variety of goods (finfish and shellfish, sediment) and provide ecological services (maintenance of biodiversity, water quality control, shoreline protection) that are directly used or beneficial to humans. The presence and abundance of seagrass can be considered an indicator of the overall environmental quality of the coastal zone (Mishra and Apte 2020). Hence, their long-term maintenance could be a surrogate target of coastal management strategies aiming at preserving or improving the environmental quality of the coastal zone. There is a need to undertake studies on the ecological significance of seagrass in Gujarat and evaluate the economic value.

Worldwide, the areas of seagrass have been disappearing at a rate of 110 km^2/year since 1980, about 30% since initial records in 1879 (Waycott et al. 2009). Like coral reefs, the seagrasses are threatened by sewage effluent and coastal development projects. Seagrass is sensitive to pollution, and therefore it becomes vital to protect the habitat from sewage and industrial waste pollution. The restoration of seagrass is very tedious and not easy as it prefers serene water conditions and long-term conservation of the areas with seagrass is the best solution.

3 Seaweeds

Seaweeds are macro algae belonging to the groups green algae, red algae, and brown algae. Based on the substratum, the seaweed distribution in Gujarat differs from the Gulf of Kachchh to Saurashtra coast. The Gulf of Kachchh region has records of 89 algal species (Nair 2002), whereas the open Saurashtra coast has higher diversity of seaweeds (198 species; Jha et al. 2009).

Seaweeds are rich in minerals and essential trace elements and used as raw materials in the pharmaceutical and cosmetics industry (Ahmed et al. 2014; Pereira 2018). Important commercially available agar is extracted from red seaweeds and algin from brown seaweeds, and green seaweeds are mostly directly consumed as salads. The economic importance of seaweeds is shown below:

Use	Reference
Production of agar, alginates, and carrageenan	Abraham et al. (2018)
Bio-fertilizer in agriculture	Zodape (2001)
As feed supplement in animal and fish feeds	Ismail (2019)
Rich sources of macro- and micronutrients, trace minerals, alginic acid, vitamins, and amino acids	McHugh (2003)

Apart from the economic value, seaweeds have an ecological role as they provide habitat for invertebrates, fish, mammals, and birds. Seaweeds serve as a source of food for many grazing vertebrates and invertebrates. However, the seaweed distribution is under threat largely from pollution, strong waves, habitat, and overexploitation, viz., collection for commercial purpose. The coastal development projects like ports and jetties are damaging the coastal habitat and thereby affecting the area of seaweed distribution.

One of the effective conservations includes their farming in natural habitat to reduce pressure in the wild (Anon 2019). Seaweed farming of red seaweed (*Gracilaria dura*) is being initiated in two villages of the coastal region of Saurashtra coast by the Central Salt and Marine Chemicals Research Institute (CSMCRI). There is a need to escalate these efforts in other coastal villages as well not only for economic gains but also for ecological purposes. Gujarat Livelihood Promotion Company (GLPC) is engaged in the promotion of seaweed farming for small farmers living near the coast. Other benefits of open ocean seaweed farms include improving the water quality and reducing ocean acidification as the seaweeds are known to absorb five times more carbon than terrestrial plants.

4 Salt Marsh Ecosystem

4.1 Mangroves

The mangrove ecosystem is one of the productive ecosystems and sustains diverse marine forms. The mangroves provide the nursery grounds for fish, crab, shrimps, and molluscs. Mangroves are considered as one of the most valuable coastal vegetation providing economic, social, and environmental benefits to the local communities. They provide a valuable biological resource like fodder and firewood to the coastal community.

In Gujarat mangrove distribution is the second largest in the country covering 1177 km^2, published by Forest Survey of India (2019). There are 15 mangroves and associated species present in Gujarat.

1	*Acanthus ilicifolius* L.
2	*Aegiceras corniculatum* (L.) Blanco
3	*Avicennia alba* Bl.
4	*Avicennia marina* (Forsk.) Vierh
5	*Avicennia officinalis* L.
6	*Bruguiera cylindrica* (L.) Bl.
7	*Bruguiera gymnorhiza* (L.) Savigny
8	*C. decandra* (Griff.) Ding Hou.
9	*Ceriops tagal* (Perr.) Robinson
10	*Excoecaria agallocha* L.

11	*Kandelia candel* (L.) Druce
12	*Lumnitzera racemosa* Willd
13	*Rhizophora mucronata* Lamk.
14	*Rhizophora mucronata* Lam.
15	*Sonneratia apetala* Buch.-Ham.

The Purna estuary has the highest representation, while the Gulf of Kachchh has seven mangrove species. The important ones include *Avicennia* spp., *Rhizophora mucronata*, and saline grasses in the intertidal region (*Aeluropus lagopoides*, *Sporobolus* sp.) (Singh 2000). The intertidal region of the Gulf of Khambhat is muddy in nature and supports small- to medium-size *Avicenna marina*.

The ecological service provided by mangroves is enormous. The root systems of mangroves help form a natural barrier against heavy storms and floods. River sediment trapped by the roots protects coastline areas and slows erosion. In the case of the Gulf of Kachchh, filtering process by mangroves prevents a large amount of sediment from reaching coral reefs and seagrass beds. Mangroves are considered as nature's best solution for carbon sequestration. It has been estimated that mangroves of Gujarat sequestrated 8.116 million tons of carbon, with an average of 88.95 tons sequestration per ha (Pandey and Pandey 2013). The carbon sequestration rate is high in South Gujarat (180.24 tons/ha) because of good mangrove cover and density (Pandey and Pandey 2013). The rate of carbon sequestration is high in the dense mangrove patches (95.3 tons/ha), followed by moderate dense mangroves (39.1 tons/ha) and least in the sparse mangrove patches (19.3 tons/ha).

The foliage of many mangrove species is used as fodder for cattle, camels, and goats and sustains them during incidences of drought. Kharai breed of camel, an indigenous breed of the Kachchh region, feeds on mangroves of Kori Creek. The mangroves are a reliable source for construction and fuelwood; it is hardy and resistant to both rot and insects.

Mangroves are facing threats from various natural and man-made activities like pollution, coastal land diversion, and geomorphic changes, and overexploitation. The diversion of mangrove area for port activities has led to a loss of mangrove cover in the Mundra region between 2006 and 2010. The photographic evidence shows how the removal of mangroves has resulted in the sand deposition in the same area by 2010 (Fig. 13.1). Similarly, there is a loss of mangrove area in the Gulf of Khambhat toward diversion of coastal area for ports and jetties.

Another threat to the mangroves is pollution released from the industrial setup along the coastline. Gujarat Ecology Society studied the impact of industries on the mangroves along the Jamnagar coast in seven selected stations. The study revealed high stress condition of mangroves at Rozi beyt, Salaya, and Sikka, intermediate condition at Narara beyt, and low stress condition from Dhani, Okha, and Pindara. Damages to the mangroves with respect to physiology and damaging symptoms like chlorosis and necrosis were reported from Rozi beyt and Narara beyt (Sankhwal and Gavali 2017). At Rozi beyt, dust accumulation on the leaves was high due to the unloading of coal at the nearby jetty (Fig. 13.2).

Fig. 13.1 Synoptic view of the area in 2005 (with mangroves) and 2010 (with sand layers)

Fig. 13.2 Dust accumulation at Rozi beyt on *A. marina* clicked on 12 June 2016

4.1.1 Restoration

Restoration efforts are being made by government to improve the mangrove cover through compulsory afforestation. There is success of mangrove restoration in the South Gujarat coast; however the ecological services provided by the mature mangrove stand cannot be replaced by the young patch. Mangroves have the ability to absorb heavy metals like cadmium and chromium, and this potentiality can be utilized for bioremediation in the coastal areas with heavy pollution load.

The South Gujarat coast is lined by a large number of perennial and seasonal rivers. Some of them have expansive mudflats, while some have a large estuarine area. Purna River is an important west flowing river of South Gujarat. Extensive mudflats at the mouth and fringing mangroves along the river and islands are important features of the river. It is one of the most diverse mangrove areas in the entire South Gujarat coastal stretch. Six species of mangroves *Avicennia marina Sonneratia apetalaAcanthus ilicifolius Rhizophora mucronata Ceriops tagalBruguiera cylindrica* eleven mangrove associates and *Clerodendrum inerme Salvadora persica Aeluropus lagopoides Derris trifoliate Suaeda fruticosa Porteresia coarctata Suaeda nudiflora Sesuvium portulacastrum Salicornia brachiata Arthrocnemum indicum Cressa cretica* have been recorded in Purna.

4.2 Beaches

Beaches are associated with rocky intertidal at many places, where the rocks form upper or supra tidal zones. The beach sands are dominantly calcareous and biogenic in nature. This indicates high biological productivity along the coasts. Sandy beaches provide ecosystem services like sediment storage and transport; wave dissipation; dynamic response to sea level rise; breakdown of organic materials and pollutants; water filtration; nutrient mineralization and recycling; storage of water in dune aquifers; nursery for juvenile fishes; nesting sites for turtles, shorebirds, and pinnipeds; prey for birds and other terrestrial species; and tourism (Defeo et al. 2009). Beaches act as a buffer against the strong wind, rough seas, and powerful storms.

The coastline of Gujarat is dotted with beaches, and some of the important tourist places include Tithal, Diu, Chorwad Madhavpur, and Mandvi. The sea beach along the Chorwad Madhavpur section is known for nesting grounds of green sea turtle and olive ridley turtle.

Beach ecosystem is one of the most vulnerable coastal ecosystems of human impact. Tourism Corporation of Gujarat Limited has identified 22 beaches and 5 other coastal sites to promote tourism in the state. Absence of regulatory beach tourism is destroying the important ecosystem, and beaches of Ubhrat, Dumas, Hazira, and Mandvi (Kachchh) are already facing issues of solid waste disposal and discharges from industrial effluents.

Beach raking is the mechanized removal of seaweed and other natural materials from the beach and is proving to be a major threat. Removal of nutrient-rich organic layer deposited along the beach can seriously affect the health of the beach and dune. Excessive beach raking promotes sand being blown and deposited in the nearby coastal areas. Of late there are reports of removal of seaweeds deposited along the shore for use as fertilizers. This activity has resulted in sand being blown in into the farmlands along the Saurashtra coast, thereby affecting the farm productivity.

Seawalls created at several places result in narrower intertidal zones and reduce abundances of invertebrates and shorebirds (Dugan and Hubbard 2006). Beach grooming also results in decreased species abundance and biomass (Hubbard et al. 2013; Dugan et al. 2010). For example beach modifications has resulted in large man made structures at Porbandar coast adversely affecting biological productivity of the beach (Illustration 13.1).

Beach management and conservation plan needs to be incorporated and made compulsory for the coastal development projects. The ecological role of the beach cannot be ignored and has to be integrated in the development projects. There is need to delineate the entire coastal zone of Gujarat and mark out no development zone in the context of conservation of beach ecosystem. Likewise, beach sand mining needs to be monitored and strict rules and regulation have to be considered. Special emphasis on the sensitive and valuable natural resources and aquifer conservation needs to be inducted during the beach management projects.

Illustration 13.1 Beach modifications for tourism at Porbandar coast

5 Coastal Dunes

Coastal dunes are geological formations, aeolian in nature, and located behind the beach. Coastal dunes are present on Saurashtra and Kachchh coast.

Sand dune formation is facilitated by strong winds blowing from SW direction during late summer. The wind lifts the beach sand during low tide and transports it landward where this sand gets trapped within the coastal vegetation, leading to sand dune formation. The process of sand dune formation is strong during the dry months of summer and winter. The ecological succession stage gets initiated during the monsoon months and the sand dune gets covered with vegetation. This vegetation over the years stabilizes the dune and such stable sand dunes are observed in Jamnagar and Kachchh district.

Ecological significance is the presence of endemic species like *Cyperus dwarkensis* from Dwarka (Nayar and Sastry 1988). Dune plays critical role protecting land fertility by preventing windblown salinity from the sea. Sand dunes are important freshwater aquifers, providing waters to coastal habitations, and act as a barrier for salinity ingress. Coastal dunes play a vital role in protecting our beaches and coastline from coastal hazards such as erosion, coastal flooding, and storm damage. Sand dunes also provide a future supply of sand to maintain the beach. The wider the band of dunes, the larger is the reservoir of sand.

Overexploitation of groundwater and dune sand mining damages/disturbs this delicate balance between saline sea and terrestrial fresh water, leading to salinization of coastal aquifers and soils. At present communities along Saurashtra coast of Gujarat are facing this situation of salinity ingress in coastal aquifers.

The coastal dunes are under threat as the ecological services provided are ignored. In Gujarat coast most of the dunes are stabilized and infested with *Prosopis juliflora*, which forms thick thickets damaging the native vegetation (GES MSU and GUIDE 2002). Such areas form habitat for carnivorous species like jackal and wolf, threatening the breeding of sea turtle by destroying their eggs. Therefore, control of proliferation of *Prosopis juliflora* in the sand dunes through strategic planning is required.

Conservation and restoration of sand dunes through meticulously planned efforts is required, wherein *Prosopis juliflora* is replaced with native vegetation. Prior taking up of restoration, identification of native flora through survey and historical records should be done. Restoration includes removal of exotic species like *Prosopis* sp. that has to be done manually as using heavy machinery would disturb the due morphology. *Prosopis* removal is not a onetime job, as relict tubers may again lead to their coppicing. These saplings need to be removed every year. Establishment of seral vegetation would start with grass and herbs species. Grass seed spreading during monsoon would help to restore grass species. Grass establishment would help to stabilize and prevent erosion. The grass would attract several herbivores species; though this would put pressure on the newly formed grassland, in return it will benefit the grassland through the organic manure and help in natural regeneration of other palatable species. Next step is to regenerate native shrub and tree species through plantation of saplings keeping appropriate spacing. The whole process of restoration is dependent on water and moisture availability, and it may run in to 5–10 years depending on monsoon.

The ecological role of the sand dune has to be understood and integrated in the coastal development projects. Coastal dune acts as a freshwater reservoir and barrier preventing sea salinity ingress and therefore is important system that needs conservation.

6 Salt Marsh Ecosystems

Tidal flats, along with intertidal salt marshes and mangrove forests, are important ecosystems. They usually support a large population of wildlife and are a key habitat that allows tens of millions of migratory shorebirds to migrate. Nine species of mudskippers are available in India of which seven species were reported from the Gujarat coast (Gujarat Ecology Commission 1996; Barman et al. 2000; Shukla et al. 2014).

6.1 *Mudflats Fishery*

Artisanal fishers doing intertidal fishery are known as Pagadiya fishers. Depending on nature of intertidal zone, they are engaged in fishery of prawn, crab, and mudskippers. The Pagadiyas of Gulf of Kachchh are doing prawn fishery from creeks, while crab fishery is common in Saurashtra. But the most dominant among this is mudskippers fishery in the gulf of Khambhat region (Fig. 13.3). Estuarine region of Narmada and Bhavnagar coast is among the most favored region for fishery of Mudskipper fishery.

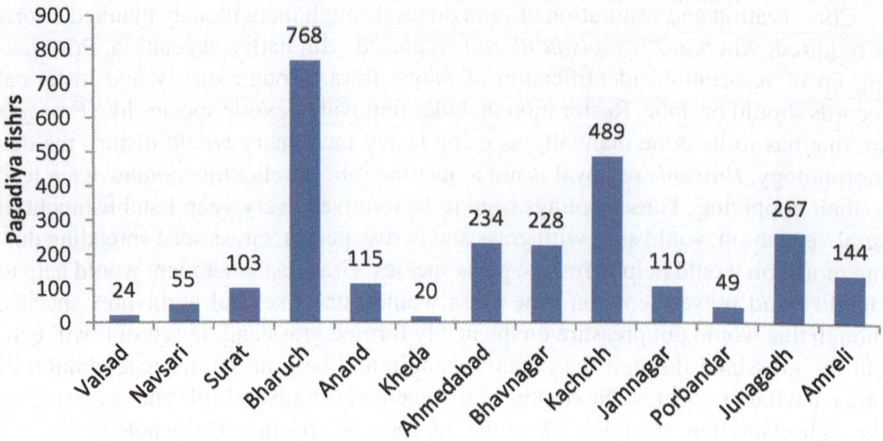

Fig. 13.3 Pagadiya fishers in Gujarat (GOG 2013)

6.2 Paleo-Mudflats

Paleo-mudfalts are represented inform of various landforms like palaeo-flats, salt flat and old deltaic plains. Their presence is indicative of high sea levels in the region in geological past. These paleo-mudflats form larger buffer zone in the state just above the high tide line. These mudflats are formed as a result of Khambhat and Kachchh fault line and tectonic activity (Nayak and Sahai 1983, 1985). Such landforms occur extensively in the entire Great Rann of Kachchh (including Banni plans), Little Rann of Kachchh, and northern as well western coast of the Gulf of Khambhat, also referred as *Bhal* and *Nalkantha* region. There is a small tract on east coast of Gulf of Khambhat in Bharuch region, known as *Bara* tract.

These regions play critical role in protection of human habitations from natural hazards. The raised mudflats provide habitat for unique ecosystem of salt marshes comprising of halophytic plants. The Gulf of Kachchh is represented by 27 halophytes belonging to 23 genera. Some of the important species reported from the mudflats are *Aeluropus lagopoides*, *Cressa cretica*, *Halopyrum* spp., *Ipomoea pes-caprae*, *Sesuvium portulacastrum*, and *Suaeda* spp. This vegetation provides food for a variety of herbivore species. The mudflats provide ample habitat for micro-invertebrates, gastropods, and other vertebrates. These halophytes also form food for the coastal communities and have religious value as well. For example, *Suaeda nudiflora* is consumed during fasting in the month of *Shravan*. These mudflats sustain *Pagadiya* fishery of mudskippers and crabs that is source of income for the coastal communities.

6.3 Estuarine Fishery

The coastal region of South Gujarat and Gulf of Khambhat is having estuaries of major rivers of Gujarat that mainly originated from eastern hilly region of Gujarat and neighboring states. Other parts of the state have seasonal rivers and creeks in which tidal water dominates. Creeks are small tidal channel of the sea into the coast through tidal swarms. Sometimes small seasonal streams drain into the creek in the upstream region.

Estuaries include the region where there is mixing of seawater and fresh waste and therefore have the highest primary productivity in the world. This area supports mangroves and related floral and faunal species. Estuaries also provide free passage of catadromous and anadromous fishes during their breeding cycle. In Gujarat, Narmada estuary is the largest one and sustains good fishery, and important breeding species includes hilsa (*Tenualosa ilisha*) and freshwater giant prawn (*Macrobrachium rosenbergii*). The economic value of the fishery in the estuarine area is high, and it provide livelihood to thousands of fishermen in the area. Another important estuary of Gujarat is of Purna River in South Gujarat with rich mangrove diversity.

The major challenges faced by the estuaries of Gujarat are water pollution as these are the receiver points of land-based pollution activity. The addition of treated and untreated pollution destroys the aquatic life and the Sabarmati and Mahi estuaries are the examples.

Aliya Bet Case of MaldharisAliya Bet in the delta of the Narmada River had good-quality grass for camels. Fishing along the creeks and mudskipper collection do not pose any disturbance to the habitat because these activities are restricted to a certain portion of the section. Extensive mudflats with salt marsh vegetation are important habitats that support wetland birds. The utilization of mangroves by Maldharis for traditional camel rearing has additional conservation significance. The site has been designated a community reserve involving the local fishing community.

7 Ranns the Saline Deserts

The Great Rann of Kachchh forms almost half of the area of the Kachchh, covering almost 45,000 km^2 area, and comprises a flat barren landscape that occurs about 2–6 m above mean sea level (Merh 2005). In local dialect, the term *rann* means "saline wasteland." In general, the Great and Little Ranns are considered to be uplifted floors of the former gulfs (Merh 2005; Maurya et al. 2008).

The Great Rann represents a filled up Holocene basin, marking the sites of the ancient shallow gulf with river mouths. Historical studies suggest that a navigable sea existed at least up to ~2000 year BP (Oldham 1926). At present, the Ranns get submerged annually, i.e., during monsoon, under a thin sheet of water (Roy and Merh 1977). The Ranns are approachable only during the summer months when it dries out with temperatures reaching to 45 °C.

The major geomorphic component of the Great Rann is the flat areas without any gradient and isolated islands (locally called *bet*). The Arabian Sea in the west enters and submerges about two-thirds of the Rann resulting in a thick salt crust (white desert). Inland saline flat is a zone on the east that gets inundated during monsoon by terrestrial inflow from the east and north. The bet zone comprises the flat rann surface in the northwestern part of the Great Rann and shows several bets.

The unique feature of the Great Rann is the flamingo habitat known as Kutch Desert Wildlife Sanctuary, declared as a sanctuary in February 1986. It is the largest wildlife sanctuary in India. Every year thousands of greater flamingo (*Phoenicopterus roseus*) nest in the world-famous "Flamingo City." It is the only area where flamingoes congregate to breed regularly. The Rann also sustains endemic and threatened wild ass, *Equus hemionus khur*, and some part of the area is declared as Indian Wild Ass Sanctuary. Apart from wild ass (*Equus hemionus khur*), the Wild Ass Sanctuary is also a habitat of many migratory birds, like sarus crane, ducks, the Dalmatian pelican, and flamingoes, sand grouse, the francolin, and the Indian bustard.

The climatic condition of the Rann supports the salt formation and is one of the largest salt-producing centers of the State of Gujarat. Marine chemicals like potassium sulfate and liquid bromide are also produced in the Great Rann of Kachchh. Thus, the economic services provided by the Rann are enormous in terms of foreign exchange earned from the export of raw salts to the western countries. Apart from salt works, Little Rann supports a unique seasonal fishery of ginger prawn (*Metapenaeus kutchensis*) done by Pagadiya fisherman. During the monsoon season, the Little Rann gets connected to the Gulf through flood water which allows passage of ginger prawn juveniles into the Rann.

Recently activities like tourism and White Rann and the solar farm have picked up bringing alteration in the socioeconomic conditions of the locals. However, there is need to assess the long-term consequences of these activities on ecosystems of the Rann.

Acknowledgments The authors are grateful to Gujarat Ecology Society for providing the facilities to work on the paper.

References

Abraham A, Afewerki B, Tsegay B, Ghebremedhin H, Teklehaimano B, Reddy KS (2018) Extraction of agar and alginate from marine seaweeds in Red Sea region. Int J Mar Biol Res 3(2):1–8

Adhavan D, Kamboj RD, Marimuthu N, Bhalodi MM (2014) Seasonal variation and climate change influence coral bleaching in Pirotan Island, Gulf of Kachchh Marine National Park, Gujarat. Curr Sci 107(2014):1780–1781

Ahmed ABA, Adel M, Karimi P, Peidayesh M (2014) Pharmaceutical, cosmeceutical, and traditional applications of marine carbohydrates. Adv Food Nutr Res 73:197–220. https://doi.org/10.1016/B978-0-12-800268-1.00010-X

Ali AAM, Hamed MA, El-Azim HA (2011) Heavy metals distribution in the coral reef ecosystems of the northern Red Sea. Helgol Mar Res 65:67–80. https://doi.org/10.1007/s10152-010-0202-7

Allers E, Abed RMM, Wehrmann LM, Wang T, Larsson AI, Purser A, de Beer D (2013) Resistance of Lophelia pertusa to coverage by sediment and petroleum drill cuttings. Mar Pollut Bull 74:132–140
Anon (2019) Seaweed aquaculture and wild harvesting in Jersey, an assessment of potential and viability with recommendations for future management of the resource, Seaweed Resource Assessment 2019, Growth, housing and environment marine resources section Howard Davis Farm
Anthony KRN, Fabricius KE (2000) Shifting roles of heterotrophy and autotrophy in coral energetics under varying turbidity. J Exp Mar Biol Ecol 252:221–253
Anu G, Kumar NC, Jayalakshmi KV, Nair SM (2007) Monitoring of heavy metal partitioning in reef corals of Lakshadweep Archipelago, Indian Ocean. Environ Monit Assess 128:195–208
Bahuguna A, Nayak S (1998) Coral reefs of the Indian Ocean SAC/RSA/RSAG/DOD-COS/SN/16/97. Space Application Centre (ISRO), Ahmedabad
Balan K and Sivaraman P and George KP and Ramachandran M 1987. Appraisal of the Marine Fisheries of Gujarat. *CMFRI Special Publication (38)*. pp. 1–51.
Barman RP, Mukherjee P and Kar S 2000. Marine and estuarine fishes. Fauna of Gujarat State Fauna Series, Zoological Survey of India 8:311–411.
Chivian E (2006) Medicines from natural sources: how human health depends on nature. In: Miththapala S (ed) Conserving medicinal species: securing a healthy future. Ecosystems and Livelihoods Group Asia, IUCN, Colombo, pp 1–7, 184 pp
Defeo O, McLachlan A, Schoeman DS, Schlacher TA, Dugan J, Jones A, Lastra M, Scapini F (2009) Threats to sandy beach ecosystems: a review. Estuar Coast Shelf Sci 81:1–12
Demers C, Hamdy RC, Coris K, Challoat F, Tabrizian M, Yahia L, H. (2002) Natural coral exoskeleton as a bone graft substitute. Biomed Mater Eng 12:15–35
Deshmukhe G, Ramamoorthy K, Sen Gupta R (2000) On the coral reefs of the Gulf of Kachchh. Curr Sci 79:160–162
Diedrich A (2007) The impacts of tourism on coral reef conservation awareness and support in coastal communities in Belize. Coral Reefs 26:985
Dixit AM, Kumar P, Kumar L, Pathak KD, Patel MI (2010) Economic valuation of coral reef systems in Gulf of Kachchh. Final report. World Bank Aided integrated Coastal Zone Management (ICZM) Project, Submitted to Gujarat Ecology Commission, 158 pp
Done T, Whetton P, Jones R, Berkelmans R, Lough J, Skirving W, Wooldridge S (2003) Global climate change and coral bleaching on the great barrier reef, Final report. Queensland Government Department of Natural Resources and Mines, p 49. ISBN: 0 642 32220 1
Dugan JE, Hubbard DM (2006) Ecological responses to coastal armoring on exposed sandy beaches. Shore Beach 74(1):10–16
Dugan JE, Defeo O, Jaramillo E, Jones AR, Lastra M, Nel R, Peterson CH, Scapini F, Schlacher T, Schoeman DS (2010) Give beach ecosystems their day in the sun. Science 329:1146. Retrieved from www.sciencemag.org
Erftemeijer PLA, Riegl B, Hoeksema BW, Todd PA (2012) Environmental impacts of dredging and other sediment disturbances on corals: a review. Mar Pollut Bull 64(9):1737–1765
FSI (2019) State of forest report 2017. Forest Survey of India, Ministry of Environment & Forests
GeeVarghese GA, Akhil B, Magesh G, Krishnan P, Purvaja R, Ramesh R (2017) A comprehensive geospatial assessment of seagrass distribution in India. Ocean Coast Manag. https://doi.org/10.1016/j.ocecoaman.2017.10.032
GEC (2010). Mangrove's atlas of Gujarat state. Prepared by Gujarat Ecology Commission and Bhaskaracharya Institute for Space Applications and Geo-Informatics. Gujarat. 188 pp
GES (1998). Coastal marine environment: a benchmark survey. Final report published by Gujarat Ecological Commission. 245 pp
GES, MSU and GUIDE (2002) Conservation of rare and endangered biodiversity of Gujarat. Final report submitted by Gujarat Ecology Society, MS University of Baroda and Gujarat Institute of Desert Ecology to Gujarat Ecology Commission, sponsored by Gujarat State Forest Department, Jun 2002, 428 pp

GOG (2013) Fisheries statistics of Gujarat 2012–13. Commissioner of Fisheries, Government of Gujarat, Gandhinagar

Great Barrier Reef Marine Park Authority (2017) Final report: 2016 coral bleaching event on the Great Barrier Reef. GBRMPA, Townsville

Gujarat Ecology Commission (1996) Biological diversity of Gujarat. Printed at the M.S. University of Baroda, Press, Baroda, 330 pp

Gujarat Ecology Commission (2017) State of environment report 2017, p 217

Gujarat Ecology Society (2018) Impact of industrial pollution on coastal and marine ecosystem of Gulf of Kachchh between Dwarka and Rozi Beyt. Ministry of Earth Science Project

Hoegh-Guldberg O (1999) Climate change, coral bleaching and the future of the world's coral reefs. Mar Freshw Res 50:839–866

Hubbard DM, Dugan JE, Schooler NK, Viola SM (2013) Local extirpations and regional declines of endemic upper beach invertebrates in southern California. Estuar Coast Shelf Sci. https://doi.org/10.1016/j.ecss.2013.06.017

Hunt B, Vincent A (2006) The use of marine organisms in traditional and allopathic medicine. In: Miththapala S (ed) Conserving medicinal species: securing a healthy future. Ecosystems and Livelihoods Group Asia, IUCN, Colombo, pp 64–75, 184 pp

Ismail MM (2019) Review on seaweed as supplement fish feed. Oceanogr Fish Open Access J 11(2):5557808

Jha B, Reddy CRK, Thakur MC, Rao MU (2009) Seaweeds of India: the diversity and distribution of seaweeds of the Gujarat coast. Springer, Dordrecht, p 198

Kelmo F, Bell JJ, Moraes SS, Gomes RDT, Mariano-Neto E, Attrill MJ (2014) Differential responses of emergent intertidal coral reef fauna to a large-scale El-Niño southern oscillation event: sponge and coral resilience. PLoS One 9(3):e93209, 1–10. www.plosone.org

Kumar JSY, Satyanarayana C, Venkataraman K et al (2017) Studies on survival and growth rate of transplanted Acroporidae in Gulf of Kachchh Marine National Park, India. J Coast Conserv 21:23–34. https://doi.org/10.1007/s11852-016-0465-5

Kumaraguru AK, Jayakumar K, Jerald Wilson J, Ramakritinan CM (2005) Impact of the tsunami of 26 December 2004 on the coral reef environment of Gulf of Mannar and Palk Bay in the southeast coast of India. Curr Sci 89(10):1729–1741

Lakhmapurkar J, Gavali D (2018) Coral reef structures at Kachhigarh, Gujarat. Indian J Geomarine Sci 47(02):503–508

Lapointe BL, Baumberger RE, Hurley SW, Bedford BJ (2008) Effects of land-based nutrient pollution on coral reefs: lessons from the Florida keys. In: Proceedings of the 61st Gulf and Caribbean Fisheries Institute, 10–14 Nov, Gosier, Guadeloupe, French West Indies, pp 250–255

Larsson AI, D van Oevelen, A Purser & L Thomsen 2013. Tolerance to long-term exposure of suspended benthic sediments and drill cuttings in the cold-water coral Lophelia pertusa. Marine Pollution Bulletin 30:176–188.

Mitra A, V Sanil Kumar BK Jena 2020. Khambhat, northern Arabian Sea – based on observation and global tidal model data, Oceanologia , Volume 62, Issue 4, Part A, Pages 443–459, https://doi.org/10.1016/j.oceano.2020.05.002

Martínez-Crego B, Vergés A, Alcoverro T, Romero J (2008) Selection of multiple seagrass indicators for environmental biomonitoring. Mar Ecol Prog Ser 361:93–109

Maurya DM, Thakkar MG, Patidar AK, Bhandari S, Goyal B, Chamyal LS (2008) Late quaternary geomorphic evolution of the coastal zone of Kachchh, western India. J Coast Res 24:746–758

McHugh DJ (2003) A guide to the seaweed industry. FAO fisheries technical paper. No. 441. FAO, Rome, 105p

Merh SS (2005) The great Rann of Kachchh: perceptions of a field geologists. J Geol Soc India 65:9–25

Mishra A, Apte D (2020) The current status of Halophila beccarii: an ecologically significant, yet vulnerable seagrass of India. https://doi.org/10.20944/preprints202008.0126.v1

Miththapala S (2008) Coral reefs, Coastal ecosystems series, vol 1. Ecosystems and Livelihoods Group Asia, IUCN, Colombo, pp 1–36 + iii

Nair VR (2002) Status of Flora and Fauna of Gulf of Kachchh. National Institute of Oceanography, Goa, 157 pp

Nayak SR, Sahai B (1983) Morphological changes in the Mahi estuary. In: Proceedings of the national conference on applications of remote sensing to natural resources. CSRE, IIT Bombay, Mumbai, pp 87–96

Nayak SR, Sahai B (1985) Coastal morphology a case study of the Gulf of Khambhat. Int J Remote Sensing 6:559–567

Nayar MP, Sastry ARK (1988) Red data book of Indian plants, vol 2. Botanical Survey of India, Calcutta, 150 pp

Oldham RD (1926) The Cutch (Kachh) earthquake of 16th June 1819 with a revision of the great earthquake of 12th June 1897. Mem Geol Soc India 46:1–77

Ondiviela B, Losada IJ, Lara JL, Maza M, Galvan C, Bouma TJ, Belzen J (2014) The role of seagrasses in coastal protection in a changing climate. Coast Eng 87:158–168

Pandey CN, Pandey R (2013) The status of mangroves in Gujarat, towards conservation and management of Indian mangroves. Int J Bot Res 3:57–70

Panseriya HZ, Gosai HB, Sankhwal AO, Sachaniya BK, Gavali DJ, Dave BP (2020) Distribution, speciation and risk assessment of heavy metals: geochemical exploration of Gulf of Kachchh, Gujarat, India. Environ Earth Sci 79:213

Patel MI (1978) Generic diversity of Scleractinians around Poshitra point, Gulf of Kutch. Indian J Mar Sci 7:30–32

Pereira L (2018) Seaweeds as source of bioactive substances and skin care therapy—cosmeceuticals, algotheraphy, and thalassotherapy. Cosmetics 5(4):68. https://doi.org/10.3390/cosmetics5040068

Philben M (2016) Do coral reefs help fight climate change? 11 Jul 2016. https://oceanbites.org/

Pillai CSG, Patel PI (1988) Scleractinian corals from the Gulf of Kutch. J Mar Biol Assoc India 30(1&2):54–74

Roy B and Merh S (1977). Geomorphology of the Rann of Kachchh and climatic changes. In: Ecology and Archeology of western India. (Eds. Agarwal, D. P. and B. M. Pande) Ocept. Publ. Co., Delhi. 195-200 pp

Sankhwal AO, Gavali DJ (2017) Spatial variation in physiology of *Avicennia marina* along the Jamnagar coast, Gujarat. Int J Plant Anim Environ Sci 7(1):17–22

Sarmento VC, Santos PJP (2012) Trampling on coral reefs: tourism effects on harpacticoid copepods. Coral Reefs 31:135–146. https://doi.org/10.1007/s00338-011-0827-2

Satyanarayana C, Ramakrishna (2009) Handbook on hard corals of GoK. Zoological Survey of India, Kolkata, 114 p

Sen Gupta R, Deshmukhe G (2000) Coastal and maritime environments of Gujarat: ecology and economics. Gujarat Ecological Society, Vadodara, 148 pp

Sen Gupta R, Patel MI, Ramamoorthy K, Deshmukhe G (2003) Coral reefs of the Gulf of Kachchh, a sub-tidal videography. Gujarat Ecological Society, Vadodara, p 82

Sharma S, Bahuguna A, Chaudhary NR, Nayak S, Chavan S, Pandey CN (2008) Status and monitoring the health of coral reef using Multi-temporal remote sensing—a case study of Pirotan Coral Reef Island, Marine National Park, Gulf of Kachchh, Gujarat, India. In: Proceedings of the 11th international coral reef symposium, session number 17, Ft. Lauderdale, Florida, 7–11 Jul 2008

Shukla M L, Trivedi J N, Soni G M, Patel B K and Vachhrajani K D 2014. Mudskipper (Gobiidae: Oxudercinae) fauna of Northern Gulf of Khambhat with two new record of the species from Gujarat, India, European Journal of Zoological Research, 3(3):67–74 (http://scholarsresearchlibrary.com/archive.html

Singh HS (2000) Growth patterns of mangroves in the Gulf of Kutch. J Bombay Nat Hist Soc 97(2):202–207

Stafford-Smith MG (1993) Sediment-rejection efficiency of 22 species of Australian scleractinian corals. Mar Biol 115:229–243

Syukur A, Wardiatno Y, Muchsin I, Kamal MM (2017) Threats to seagrass ecology and indicators of the importance of seagrass ecological services in the coastal waters of East Lombok, Indonesia. Am J Environ Sci 13(3):251–265

United Nations (2019) World urbanization prospects: the 2018 revision (ST/ESA/SER.A/420). Department of Economic and Social Affairs, Population Division United Nations, New York

Valentine JF, Heck K (1999) Seagrass herbivory: evidence for the continued grazing of marine grasses. Mar Ecol Prog Ser 176:291–302. https://doi.org/10.3354/meps176291

Vethamony P, Babu MT (2010) Physical processes in the Gulf of Kachchh: a review. IJMS 39(4):497–503

Waycott M, Duarte CM, Carruthers TJB, Orth RJ, Dennison WC, Olyarnik S, Calladine A, Fourqurean JW, Heck KL Jr, Hughes AR, Kendrick GA, Kenworthy WJ, Short FT, Williams SL (2009) Accelerating loss of seagrasses across the globe threatens coastal ecosystems. PNAS 106(30):12377–12381. https://doi.org/10.1073/pnas.0905620106

Wilkinson C (2008) Status of coral reefs of the world. Global Coral Reef Monitoring Network and Reef and Rainforest Research Centre, Townsville

Zodape ST (2001) Seaweeds as a biofertilizer. J Sci Ind Res 60(5):378–382

Chapter 14
Macrophyte Diversity and Distribution in Brackish Coastal Lagoons: A Field Survey from Chilika, Odisha

Pramod Kumar Tripathy, Madhusmita Mohapatra, Roma Pattnaik, Lipika Tarafdar, Sudhakar Panda, and Gurdeep Rastogi

Abstract Macrophytes are one of the major components that contribute to the primary production in shallow coastal lagoons. These macroscopic plants play a vital role in maintaining the ecological health of water bodies and also in structuring the biotic communities by providing physical structure for colonization and expanding the habitat complexity and heterogeneity. Chilika, the largest brackish water coastal lagoon of India, is situated on the east coast in the state of Odisha. The lagoon is a biodiversity hotspot with high primary productivity due to a rich and diverse community of aquatic macrophytes and phytoplankton. The present chapter summarized the macrophyte studies from coastal lagoons of India and examined the spatiotemporal distribution, composition, and biomass of macrophytes from Chilika Lagoon based on the data derived from 2 years (2018–2019) of systematic field survey. A total of 22 macrophytes belonging to 14 families were identified from 33 sites which included 4 emergent species (*Alternanthera philoxeroides*, *Ipomoea aquatica*, *Phragmites karka*, and *Schoenoplectus litoralis*), 11 submerged (*Ceratophyllum demersum*, *Najas indica*, *Stuckenia pectinata*, *Potamogeton crispus*, *Potamogeton nodosus*, *Hydrilla verticillata*, *Vallisneria natans*, *Halophila beccarii*, *Halophila ovalis*, *Halodule pinifolia*, and *Ruppia maritima*), 2 rooted with the floating leaves (*Nymphaea pubescens* and *Nymphoides cristata*), and 5 free-floating (*Azolla pinnata*, *Eichhornia crassipes*, *Pistia stratiotes*, *Salvinia cucullata*, and *Spirodela polyrhiza*). *Stuckenia pectinata* and *Najas indica* were found in all sectors except the outer channel. A total of 11 macroalgal taxa comprising 6 species of Chlorophyta and 5 species of Rhodophyta were identified. *Gracilaria verrucosa* was often associated with seagrasses, whereas *Chaetomorpha* sp. and *Ulva* sp. were found growing with hard substratum such as rocks, etc., in the southern and central sectors. The seasonality in total macrophyte biomass revealed that it was the highest in winter (4322.38 g m^{-2}) followed by summer (3056.18 g m^{-2}) and monsoon

P. K. Tripathy · M. Mohapatra · R. Pattnaik · L. Tarafdar · S. Panda · G. Rastogi (✉)
Wetland Research and Training Centre, Chilika Development Authority, Balugaon, Odisha, India

S. Madhav et al. (eds.), *Coastal Ecosystems*, Coastal Research Library 38,
https://doi.org/10.1007/978-3-030-84255-0_14

(1957.14 g m^{-2}). *Salicornia brachiata, Sesuvium portulacastrum,* and *Paspalum distichum* were abundant salt marshes in the Nalabana Bird Sanctuary of the lagoon. The occurrence of six seagrasses, namely, *Halophila ovata, Halophila ovalis, Halophila beccarii, Halodule pinifolia, Halodule uninervis,* and *Ruppia maritima,* was recorded from southern, central, and outer channel sector. A total of 169.2 km^2 area covered by seagrass meadows was mapped through ground survey, which signified the good ecological health of this coastal ecosystem.

Keywords Aquatic macrophytes · Macroalgae · Seagrasses · Salt marsh · Biomass · Chilika Lagoon

1 Introduction

Coastal lagoons are shallow habitats and generally support the extensive growth of aquatic macrophytes due to high photic depth and nutrients (dos Santos Fonseca et al. 2015). These are one of the most productive ecosystems and are renowned for their ecosystem services such as fisheries, food, and habitat for resident and migratory birds and fishes, nutrient cycling, pollutant sequestration, and coastal protection (Prado et al. 2013). They also provide different habitats such as mudflats, marshes, open waters, and fringing wetlands which have outstanding recreational, commercial, as well as ecological importance. However, coastal lagoons are often gravely threatened by human activities (such as industrialization, land reclamation for urbanization and agriculture, disposal of sewage), sea-level rise associated with climate change, and overexploitation of natural resources (Gedan et al. 2009; Kumar et al. 2017; He and Silliman 2019).

Macrophytes are the key constituent of wetlands and other aquatic ecosystems. They are the macroscopic, photosynthetic plants (including macroalgae, bryophytes, pteridophytes, and spermatophytes) that can grow temporarily or permanently in water as floating-leaves, free-floating, submerged, and emergent forms (Rosqvist 2010; Özbay et al. 2019; Pattnaik et al. 2020). The major macrophytes of coastal ecosystems are seagrasses, salt marshes, and mangroves (Cragg et al. 2020). Seagrasses are the submerged marine flowering plants found in coastal and estuarine habitats worldwide and can colonize soft substrates particularly in an area which has high salinity and water clarity (Short et al. 2007; Nobi et al. 2011; Ramesh et al. 2019). Salt marshes include salt-loving (halophytes) grasses, shrubs, and herbs that are inundated with marine or brackish water for at least part of the time and are typically found in the intertidal zone and mudflats. They are the important components of the coastal systems as they are vital in protecting the coast from erosion, providing the habitat for faunal communities, and mitigating pollution through sequestration of heavy metals and carbon (Patro et al. 2017). Seaweeds are commonly known as macroalgae, primitive type of plants devoid of true roots, stems, and leaves, and are an integral component of the coastal vegetation that contribute in sustaining ecosystem services of coastal systems (Kim et al. 2017; Ganesan et al. 2019).

Macrophytes improve water quality by enhancing transparency, providing dissolved oxygen, absorbing nutrients and pollutants, regenerating nutrients, and preventing sediment resuspension and coastal erosion (Dhote 2007; Aubry et al. 2020). Moreover, they can increase the biodiversity of an aquatic ecosystem by providing food and complex habitat for various juveniles, adults, and larval organisms (Solimini et al. 2006; Abubakr 2010). Macrophytes provide nursery ground for a variety of marine organisms such as fishes and birds and form the base of the food webs (Banerjee et al. 2017; Macreadie et al. 2017; Mishra and Apte 2021). The spatiotemporal distribution of macrophytes depends on biotic (competitive interactions with other macrophytes and phytoplankton) and abiotic (physicochemical factors of water and sediments) factors (Dar et al. 2014; Aubry et al. 2020). In estuarine coastal habitat, salinity gradient and nutrients play a key role in determining the macrophyte composition and distribution. Therefore, changes in the quantity and quality of freshwater flow can lead to major changes in the macrophyte diversity and distribution with an effect on the overall productivity and biodiversity of coastal systems (Prado et al. 2013). The eutrophication status of a water body can be assessed by macrophytes as their biomass and species composition are directly linked to the concentration of nutrients (Solimini et al. 2006). Pereira et al. (2012) showed that freshwater macrophytes were prominent indicators of the trophic status of the aquatic environment. Growth of emergent plants such as *Eichhornia crassipes*, *Phragmites karka*, and *Typha angustata* was indicative of high nutrients, whereas free-floating species such as *Salvinia*, *Pistia*, and *Spirodela* indicated medium nutrient concentration. *Hydrilla verticillata*, *Potamogeton crispus*, and *Vallisneria natans* were indicators of aquatic environment with low nutrients. Another study from the coastal lagoons located on the mid-Atlantic coast of the USA observed that the growth of *Gracilaria* was a good indicator of increased nitrogen concentration in the system (Fertig et al. 2009). Thus, macrophyte diversity and distribution could be an efficient bioindicator for assessing the changes in water quality and for conserving biodiversity (Christia et al. 2018).

1.1 Diversity and Distribution of Macrophytes

The studies on diversity, distribution, and community composition of aquatic macrophytes are important for ecological health assessment of water bodies. The diversity of vascular macrophytes is much higher in freshwater ecosystems compared to coastal ecosystems as salinity constrains a selective pressure on the growth and survival of macrophytes (Chappuis et al. 2011). Considering the ecological significance of macrophytes, worldwide, numerous studies have targeted the diversity and distribution of macrophytes in coastal lagoons and estuaries. Obrador et al. (2007) examined the spatial and temporal distribution of rooted macrophytes in S'Albufera des Grau, a coastal lagoon situated in the Mediterranean Sea, and recorded *Ruppia cirrhosa* as the dominant submerged macrophyte with 79% coverage. Christia and Papastergiadou (2007) monitored six coastal lagoons in western Greece and showed that salinity and dissolved oxygen were the major factors that determined the aquatic macrophyte distribution. Christia et al. (2018) showed that salinity as well as nitrate had a significant influence on the distribution and composition of macrophyte assemblages in western Greece coastal lagoons (Rodia, Kleisova, and Araxos). Phan et al. (2018) studied Cau Hai coastal lagoon situated in Vietnam and showed that *Najas indica* and *Halophila beccarii* were the major macrophytes. The study reported that salinity and sand/silt particles were the major drivers for determining the distribution of submerged aquatic vegetation communities that act as indicators and integrators of environmental variations in coastal lagoons.

Studies have also been conducted on the diversity and distribution of macrophytes from Indian lagoonal ecosystems (Table 14.1). Jagtap and Inamdar (1991) conducted aerial mapping of seagrasses in Lakshadweep islands and reported *Thalassia hemprichii* as an abundant species. Nobi et al. (2011) studied the distribution and biomass of seagrass in the Lakshadweep islands and noted that *Halophila decipiens* was restricted to Kalpeni Island. A total of 7 genera and 16 species of seagrass belonging to 3 families (Hydrocharitaceae, Cymodoceaceae, and Ruppiaceae) have been reported from the Indian coast (Thangaradjou and Bhatt 2018; Mishra and Apte 2021). Patro et al. (2017) investigated species diversity, distribution, and threats to seagrasses and salt marsh ecosystems in South Asia and reported the occurrence of 15 species of seagrasses and 14 species of salt marshes from India. Malathi et al. (2018) examined the diversity and distribution of seaweeds in coastal regions of Gulf of Mannar and reported a total of 22 seaweed species distributed under Chlorophyta (4 species), Phaeophyta (9 species), and Rhodophyta (9 species).

Numerous studies have been conducted on macrophytes from Indian lagoons of the east and west coast. Umamaheswari et al. (2009) used IRS-1D 1998 satellite data and GIS technology for the mapping of seagrass meadows from the Gulf of Mannar (an east coast lagoon) and recorded a total of 12 species of seagrasses covering an area of 85.5 km^2. Mathews et al. (2010) investigated the diversity, distribution, and density of seagrasses in Palk Bay and Gulf of Mannar Marine National Park and showed that *Cymodocea serrulata* and *Thalassia hemprichii* were the dominant species covering an area of 76 km^2. Geevarghese et al. (2018) employed Landsat 8 OLI satellite imagery combined with digital classification and contextual editing. Their study reported that the seagrass area in India was 516.59 km^2, of which 85.47 km^2 existed within the Chilika Lagoon. Josephine et al. (2013) studied the diversity and distribution of seaweeds from the Gulf of Mannar and recorded a total of 90 seaweed species belonging to Chlorophyceae (30 species), Phaeophyceae (28 species), and Rhodophyceae (32 species). Bhasha et al. (2015) investigated the phytodiversity of Pulicat Lagoon and documented a total of 180 species comprising 11 submerged macrophytes (2 species of seagrasses and 9 species of freshwater vascular plant), halophytic marshy plants (14 species), and inland plants. A study on aquatic and semiaquatic macrophytic flora of Vembanad Lake (Kerala) recorded

120 species of aquatic and semiaquatic vascular macrophytes along with mangroves and their associates (Vaheeda and Thara 2014). In another study from Kole wetlands in Kerala, 75 vascular plant species belonging to 32 families and 53 genera were identified (Jyothi and Sureshkumar 2014). Based on the available literature, macrophyte studies from other lagoons, namely, Pennar, Bendi, Nizampatnam, Muttukadu, and Muthupet from the east coast and Ashtamudi, Paravur, Ettikulam, Murukumpuzha, Veli, and Talapady from the west coast, have not been conducted yet (Table 14.1).

1.2 Ecological Roles of Macrophytes

The habitat complexity provided by macrophytes increases the richness and abundances of faunal communities and facilitates interspecies interactions for habitat and food (Thomaz and Cunha 2010). Macrophytes in Chilika Lagoon are crucial for sustaining the high biodiversity and fishery resources by providing a complex habitat for the luxurious growth of macroalgae, insects, benthic invertebrates, and fish juveniles (Bhatta and Patra 2018; Pattnaik et al. 2020). Bivalves and polychaetes were abundantly found associated with *Hydrilla verticillata*, *Nymphaea pubescens*, *Ceratophyllum demersum*, *Vallisneria spiralis*, and *Ipomoea aquatica*-dominated areas in the central sector (Mahapatro et al. 2012). Polychaeta and Chironomidae were found associated with a perennial emergent grass *Phragmites karka*, free-floating *Eichhornia crassipes* (water hyacinths), and submerged *Hydrilla verticillata* in the northern sector (Mahapatro et al. 2012). A high abundance of food materials such as macrophytes, macrobenthos, and other invertebrates has been shown to attract birds to congregate in such areas for foraging and resting. For example, gadwall and Eurasian wigeon forage on the fresh seeds and shoots of *Potamogeton* that are found in shallow open water zones of the lagoon (Balachandran et al. 2020). Other species, e.g., black-tailed godwit, northern pintail, jacanas, moorhens, and whistling ducks, prefer freshwater marshes located in the Mangalajodi and Bhusandpur area of the lagoon (Balachandran et al. 2020). Several species of birds forage on the *Salicornia brachiata* that is prevalent in the mudflats of Nalabana Bird Sanctuary (Balachandran et al. 2020). Seagrass beds in the Chilika Lagoon serve as an excellent breeding ground for a variety of invertebrate species, including fish and shellfish (Pattnaik et al. 2020). Pati et al. (2014a) have reported gastropods (*Telescopium telescopium*, *Cerithidea cingulata*, *Oliva oliva*, and *Conus virgo*) and bivalves (*Modiolus striatulus* and *Donax*) associated with seagrasses and seaweeds in the southern sector. Extensive thickets of *Potamogeton pectinatus* located in the central sector of the lagoon offer great shelter for the sea bass. Furthermore, *Enteromorpha compressa*, a macroalgae attached with the rocks in the shoreline of islands, provides diets and nesting habitat to mullets (Pattnaik et al. 2020).

The bioaccumulation of nutrients from water and sediments has been shown by macrophytes and periphyton communities in the aquatic ecosystem (Srivastava et al. 2008). Macrophytes also sequester and eliminate pollutants (e.g., heavy

Table 14.1 Summary of major aquatic macrophyte studies conducted from Indian coastal lagoons

Sl. No	Lagoon	Stud period	Details of the work	Major findings
East coast lagoon				
1	Chilika (Odisha)	1996–1998	Macrophytes were quantitatively assessed by quadrat method from northern and central sector of the lagoon (Shaw et al. 2000)	Inter-seasonal variation of salinity was found to be the highest in the central sector. Western region of central sector remained more saline than the southern sector during summer. 12 macrophyte species, namely, *Eichhornia crassipes*, *Hydrilla verticillata*, *Najas indica*, *Najas graminea*, *Vallisneria spiralis*, *Potamogeton pectinatus*, *Potamogeton natans*, *Potamogeton amplifolius*, *Potamogeton crispus*, *Nymphaea tetragona*, and *Nymphaea odorata,* were identified. *Scirpus littoralis* a prominent weed was confined to western bank of the northern sector. The distribution of the macrophytes and the massiveness of their colonization were dependent on the salinity
		2002	Survey of angiospermic plants of Chilika lake, its island and shorelines (Panda et al. 2002)	A total of 711 macrophytes were recorded from water, island, and shoreline and also reported that the occurrence of *Potamogeton crispus* was a new distributional record for the state of Orissa
		2006–2007	Distribution and succession of macrophytes (Jaikumar et al. 2011).	Seagrass species, namely, *Halodule uninervis*, and *Halophila* sp. and macroalgae, namely, *Chaetomorpha*, *Gracilaria*, and *Enteromorpha* sp. were distributed in the southern and central sectors. The freshwater weeds like *Najas*, *Hydrilla*, *Ceratophyllum* sp., and other aquatic weeds were confined to the northern sector of the lagoon. During monsoon season, growth of the aquatic macrophytes was lower compared to summer
		2008–2010	Monthly inventory of macrophytes and macroalgae were carried out from four sectors (Bhatta and Patra 2018)	The study reported a total of 37 species of macrophytes including macroalgae from Chilika Lagoon. All macrophytes were distributed in northern sector and macroalgae species such as *Chaetomorpha*, *Enteromorpha*, *Gracilaria*, and macrophytes, namely, *Cynodon dactylon* and *Schoenoplectus litoralis* were distributed in all sectors
		2012–2016	Multi-seasonal inventory was carried out from both aquatic and shoreline zone of the lagoon (Kar et al. 2017).	The study added 79 angiospermic macrophytes to the earlier list of 711 flora of the Chilika Lagoon (Panda et al. 2002)
		2013	Studies on biomass of seagrass and seaweed from southern sector (Palur canal, Sidha Gumpha, Nala Muha, and Binchanapali) of the Chilika Lagoon (Pati et al. 2014a).	The average biomasses (dry weight) of *Halophila ovalis*, *Halodule uninervis*, and *Gracilaria verrucosa* were 113, 47.4, and 89 g m^{-2}, respectively, in southern sector.

Sl. No	Lagoon	Stud period	Details of the work	Major findings
		2013	Studies on seagrass in relation to environmental parameters (Pati et al. 2014b)	Seagrasses, namely, *Halodule uninervis*, *Halodule pinifolia*, *Halophila ovalis*, *Halophila ovata*, and *Halophila beccarii,* were found in Rambha, Palur canal, Somolo, Nalabana. *Halodule uninervis*, *Halodule pinifolia*, and *Halophila ovata* were new record from the lagoon
		1975–2014	Monitoring the aquatic vegetation distribution in Chilika Lagoon using GIS and Remote Sensing (Rout et al. 2014)	The emergent and submerged vegetation increased from 18 km^2 to 90.9 km^2 and the 215.26 km^2 to 301.43 km^2 area, respectively, during 1975–2014
2	Gulf of Mannar, Rameswaram Island in the north to Kanyakumari, (Tamil Nadu)	–	Mapping of seagrass meadows by using satellite imagery from 21 islands (Shingle, Krusadai, Pullivasal and Poomarichan, Manoli and Manoliputti, Musai, Mulli, Valai and Talairi, Appa, Poovarasanpatti and Vallimunai, Anaipar, Nallathani, Pulivinichalli, Upputhanni, Karaichalli, Vilanguchalli, Kasuwar, Van) of Gulf of Mannar (Umamaheswari et al. 2009)	The occurrence of 12 seagrass species from 21 islands was verified and mapped using satellite imagery. The study reported a total of 85.5 sq km area covered by seagrass beds in Gulf of Mannar based on IRS-1D 1998 satellite data
		2007–2008	Study on the status of seagrass diversity, distribution, and abundance in Gulf of Mannar marine national park—including the same 21 islands (Mathews et al. 2010)	The total seagrass cover in Gulf of Mannar Marine National Park was 76 km^2. *Thalassia hemprichii* and *Cymodocea serrulata* were the dominant seagrass species
		2012	Macroflora and fauna (Gopalakrishnan et al. 2012)	12 species of seagrasses and 124 species of seaweeds were recorded
		2011–2012	Study on the status of seaweed diversity and their seasonal availability at Hare Island, Gulf of Mannar (Josephine et al. 2013)	90 species of seaweeds were identified of which 30, 28, and 32 belonged to Chlorophyceae, Phaeophyceae, and Rhodophyceae, respectively. Seasonal distribution of seaweeds revealed that the members of Rhodophyceae and Phaeophyceae were the most abundant in all seasons

(continued)

Table 14.1 (continued)

Sl. No	Lagoon	Stud period	Details of the work	Major findings
3	Pulicate Lake (Andhra Pradesh)	2011–2012	Study on spatial and temporal variation in the environmental parameters and its impact on the seagrass and associated macrofauna (Keren and Inbaraj 2013)	3 different seagrass species (*Halophila ovalis*, *Halophila minor*, and *Halodule pinifolia*) were reported. *Halodule pinifolia* was abundant in the northern and southern part of the lagoon. Seagrass biomass and density were high during the pre-monsoon season. Spatial variation of seagrass parameters (density, percent frequency, and biomass) was significant
		2012–2014	Study on wetland flora (Rajyalakshmi and Basha 2016)	A total of 180 species were reported from the wetlands. Of these, 117 species were dicotyledonous, 51 species were monocotyledonous plants, and 12 species were mangroves
		2013–2014	Study the on the eco-degradation and phytodiversity of the lake (Bhasha et al. 2015)	A total of 180 wetland plant species were recorded from Pulicate Lake and its surrounding areas. There were 128 dicots, 48 monocots, and 3 Pteridophytes. Among the 156 inland wetland plants species, 71.6% were annuals and 38.4% were perennials as during pre-monsoon 19.85% of freshwater wetlands disappeared
4	Pennar (Andhra Pradesh)	–	–	–
5	Bendi (Andhra Pradesh)	–	–	–
6	Nizampatnam (Andhra Pradesh)	–	–	–
7	Muttukadu (Tamil Nadu)	–	–	–
8	Muthupet (Tamil Nadu)	1988	Study on the composition, distribution, and standing crop of algae (Balakrishnan et al. 1992)	A total of 19 species from 14 genera of algae were recorded. These were distributed into Chlorophyta (12 species), Phaeophyta (2 species), Rhodophyta (4 species), and Cyanophyta (1 species). Biomass estimated for *Gracilaria verrucosa*, *Hypnea valentiae*, and *Enteromorpha* sp. varied from 905 to 1220 g m^{-2}, 740 to 980 g m^{-2}, and 53 to 72 g m^{-2} respectively

Sl. No	Lagoon	Stud period	Details of the work	Major findings
West coast lagoon				
9	Vembanad Lake (Kerala)	–	Study on the aquatic and semi-aquatic macrophytic flora of brackish waters of Kodungallur, Kerala (Vaheeda and Thara 2014)	A total of 120 species of aquatic and semi-aquatic vascular macrophytes were enumerated. The flora assessment indicated that the aquatic ecosystem has undergone eutrophication and the estuary is experiencing stress
	Kole Wetland, a part of Vembanad-Kole Ramsar site (Kerala)	2011	Documentation of aquatic macrophytes of Kole wetlands of Northern Kerala, India (Jyothi and Sureshkumar 2014)	A total of 75 species of vascular plants under 53 genera and 32 families including vascular cryptogams have been identified
10	Ashtamudi (Kerala)	1980–1981	Ecology of seagrass bed of *Halophila ovalis* (Nair et al. 1983)	Average standing crop of *Halophila ovalis* varied from 3.6 to 48 g m^{-2}
11	Paravur (Kerala)	–	–	–
12	Ettikulam (Kerala)	–	–	–
13	Murukum-puzha (Kerala)	–	–	–
14	Veli (Kerala)	–	–	–
15	Talapady (Karnataka)	–	–	–
16	Lagoons of Lakshadweep (Kerala)	1986–1987	Mapping of seagrass meadows from the Lakshadweep Islands from 6 islands Kadmat, Minicoy, Amini, Kavaratti, Kalpeni, and Agatti (Jagtap and Inamdar 1991)	The dominant seagrass species were *Thalassia hemprichii* while *Cymodocea rotundata*, *Halophila ovata*, *Syringodium isoetifolium*, and *Halodule uninervis* were common to these islands. The total seagrass cover from 6 major islands of Lakshadweep was estimated to be 112 ha.
		2009	Diversity, distribution, biomass, and productivity of seagrasses in the Lakshadweep group of Islands (Nobi et al. 2011)	Out of 7 seagrass species, *Cymodocea serrulata* was found to be dominant in the Lakshadweep group of islands whereas *Halophila decipiens* was found only in Kalpeni Island. Biomass varied widely from 72 ± 09 to 944 ± 99 g m^{-2} across different species of seagrasses

metals, trace elements, pesticides, and phenols) as well as nutrients (e.g., organic carbon, phosphorous, and nitrogen) from the wetlands (Dhote 2007; Olette et al. 2008; Javed et al. 2019; Ali et al. 2020). Macrophytes also limit the growth of phytoplankton by competing for available nutrients and thus increase the water clarity and euphotic depth (Takamura et al. 2003; Aubry et al. 2020). Submerged macrophytes such as *Hydrilla verticillata, Ceratophyllum demersum*, and *Vallisneria natans* also reduce turbidity by absorbing suspended solids on their leaves and increase the light availability in the pelagic column (Takamura et al. 2003). Aquatic macrophytes have been used widely for the cleanup of polluted water bodies and wastewater through the phytoremediation technique. The extensive root system developed by aquatic macrophytes allows them to efficiently accumulate the contaminants in their roots which later get transported to stems and leaves through xylem tissue (Ali et al. 2020). Farias et al. (2018) reported that the seagrasses (*Ruppia megacarpa*) and seaweed (*Ulva australis*) were good accumulators of heavy metals (Zn) and *Zostera muelleri* (Pb) in the Derwent estuary (Tasmania, Australia). *Hydrilla verticillata*, *Eichhornia crassipes*, and *Phragmites karka* have been used widely for the phytoremediation of wastewater to reduce heavy metals, total suspended solids, phosphate, nitrate, and chemical oxygen demand (Singh 2016; Ting et al. 2018; Ali et al. 2020). The release of oxygen and root exudates through the roots of *Phragmites australis* triggers the recruitment of several N_2-fixing bacteria (*Bradyrhizobium*, *Mesorhizobium*, *Rhizobium*), nitrifying bacteria (*Nitrosococcus*, *Nitrosomonas*, *Nitrosospira*, *Nitrospira*), sulfate reducers (*Desulfovibrio*, *Desulfobulbus*, *Desulfotomaculum*, *Desulfoluna*), and methanotrophs (*Methylohalomonas*, *Methylobacterium*, *Methylarcula*, *Methylibium*) in rhizosphere sediments that perform nutrient cycling and degradation of hydrocarbon and sustain plant growth (Vladár et al. 2008; Trias et al. 2012; Faußer et al. 2012; Zhang et al. 2013; Toyama et al. 2015). Many important biogeochemical processes such as nitrification, denitrification, and organic matter mineralization are more active in aquatic plant roots. Several microbes such as *Thiobacillus*, *Methylotenera*, *Bacillus*, *Steroidobacter*, *Escherichia/Shigella*, and *Methanomassiliicoccus* have been reported in high abundances from the rhizosphere sediments of *Phragmites karka* collected from the northern sector of the Chilika Lagoon (Behera et al. 2018). These root-associated microbes play a key role in maintaining the ecological health of coastal wetlands through beneficial macrophyte-microbe interactions, biogeochemical cycling, and biodegradation of pollutants (Behera et al. 2018).

1.3 Macrophyte Assessment from the Chilika Lagoon

Chilika, Asia's largest brackish water lagoon, is located on the east coast of India in the state of Odisha. The lagoon is a designated Ramsar site (no. 229) due to its rich biodiversity and socioeconomic importance (Tarafdar et al. 2021). The lagoon receives the freshwater discharge from 12 major rivers, and distributaries of the Mahanadi River that drain almost 80% of freshwater flow into the lagoon. The shallow depth (~2 m) of the lagoon combined with rich nutrient inputs from the riverine sources leads to profuse growth of freshwater macrophytes (Pattnaik et al. 2019, 2020). These macrophytes constitute the foundation of coastal food webs and store a significant amount of carbon, carry nutrient cycling through their microbes, and promote biodiversity (Pattnaik et al. 2020). The aquatic vegetation of Chilika is highly diverse consisting of freshwater, brackish, and marine plant species (Table 14.1). Shaw et al. (2000) identified a total of 12 macrophyte species and studied their distribution in relationship with salinity. The authors have reported occurrences of freshwater species in the northern sector, while salt-tolerant species were more prevalent in the southern sector. *Potamogeton pectinatus* (facultative halophyte) and *Najas indica* (salt-tolerant) species were abundantly present in the northern sector (Kalupadaghat). Panda et al. (2002) have reported 119 families, 492 genera, and 711 species of angiospermic plants and recorded *Potamogeton crispus* as a new record from the lagoon. The northern sector of the lagoon receives high nutrient inputs from Mahanadi River distributaries leading to the prolific growth of freshwater macrophytes (e.g., *Potamogeton*, *Najas*, *Hydrilla*, and *Ceratophyllum*) and other aquatic weeds indicative of the eutrophic conditions (Jaikumar et al. 2011). Rout et al. (2014) have used GIS and remote sensing tools to monitor aquatic vegetation from Chilika Lagoon and showed that the emergent and submerged vegetation increased in area from 18 to 90.9 km^2 and 215.26 to 301.43 km^2, respectively, between 1975 and 2014. Kar et al. (2017) further added 79 macrophytes to the inventory list of Panda et al. (2002) leading to a total of 790 species from the Chilika Lagoon. Bhatta and Patra (2018) provided the inventory of aquatic macrophytes from four ecological sectors of the lagoon based on the monthly survey. These authors have recorded 37 species of macrophytes (freshwater weeds, seagrasses, and salt marsh grasses) including macroalgae (*Gracilaria verrucosa*, *Gracilaria lichenoides*, *Chaetomorpha linum*, *Enteromorpha intestinalis*, *Polysiphonia subtilissima*, and *Chara braunii*).

Several studies have specifically targeted seagrasses, their biomass, and distribution in relation to environmental parameters from Chilika Lagoon. For instance, Pati et al. (2014a) studied the distribution and biomass of seagrasses in the southern sector of the lagoon and recorded *Halophila ovalis*, *Halodule uninervis*, and *Gracilaria verrucosa*. The average biomass (dry weight) of *Halophila ovalis*, *Halodule uninervis*, and *Gracilaria verrucosa* were 113, 47.4, and 89 g m^2, respectively. Pati et al. (2014b) recorded *Halophila ovalis*, *Halophila ovata*, *Halodule pinifolia*, *Halodule uninervis*, and *Halophila beccarii* from Palur canal, Rambha, Nalabana, and Somolo Island. Although several studies have been carried out on macrophyte distribution and succession, no systematic information is available on seasonal changes in macrophyte distribution and their biomass from the Chilika Lagoon.

The objectives of this present study were to (1) inventorize the spatiotemporal diversity and community composition of macrophytes including macroalgae, (2) estimate biomass of selected dominant macrophytes, and (3) monitor and map the distribution and diversity of seagrass meadows. This study has provided information on the latest status of macrophyte distribution for their conservation and management in the Chilika Lagoon.

2 Methodology

2.1 Site Description

Chilika (19°28′–19°54′ N and 85°06′–85°35′ E) spreads over an area of 1020 km^2 during monsoon and 704 km^2 during summer (Srichandan et al. 2015a; Tarafdar et al. 2021). The lagoon connects with the Bay of Bengal (BoB) in the eastern part through tidal inlets located in the outer channel. The southern sector of the lagoon is also connected with BoB through a 16-km-long "Palur canal." Thus, the outer channel and southern sectors are much more saline than the central and northern sectors (Behera et al. 2017, 2020). The outer channel represents a marine system due to the large and direct flow of seawater from the BoB. The northern sector is mostly oligohaline in character due to large freshwater discharges from Daya, Bhargavi, and Luna Rivers (Srichandan et al. 2015a; Behera et al. 2018). The central sector is mostly brackish due to intermixing of freshwater and seawater and has high variability in salinity due to large open area. During the monsoon, the entire lagoon turns into a freshwater system due to massive riverine discharge into the lagoon. The lagoon falls under the tropical climate with an annual mean temperature between 14.0 and 39.9 °C. The annual average rainfall during the monsoon season in the catchment is 1533 mm (Srichandan et al. 2015b). The northern sector of the lagoon is almost entirely covered with macrophytes that grow luxuriantly during post-monsoon months and decompose during summer when the salinity of the lagoon rises to its peak (Pattnaik et al. 2020). The northern shoreline of the lagoon has thick monoculture stands of *Phragmites karka* covering an area of about 50 km^2 (Behera et al. 2018).

2.2 Inventorization of Macrophytes

The species diversity, abundance, and community composition of macrophytes were determined by collecting plant specimens monthly between 2 years (January 2018 to December 2019) using a handheld rake from 33 stations spanning northern, southern, central, and outer channel sectors of the Chilika Lagoon (Fig. 14.1). At each sampling station, visual inspection and field photographs were used for accurate identification of genus and species of macrophytes. Taxonomic identification of macrophytes was carried out with the help of standard taxonomic keys described by Haines (1921–1925). The plant list (http://www.theplantlist.org/) and International Plant Names Index (http://www.ipni.org/ipni/plantnamesearchpage.do) were also referred for taxonomic identification.

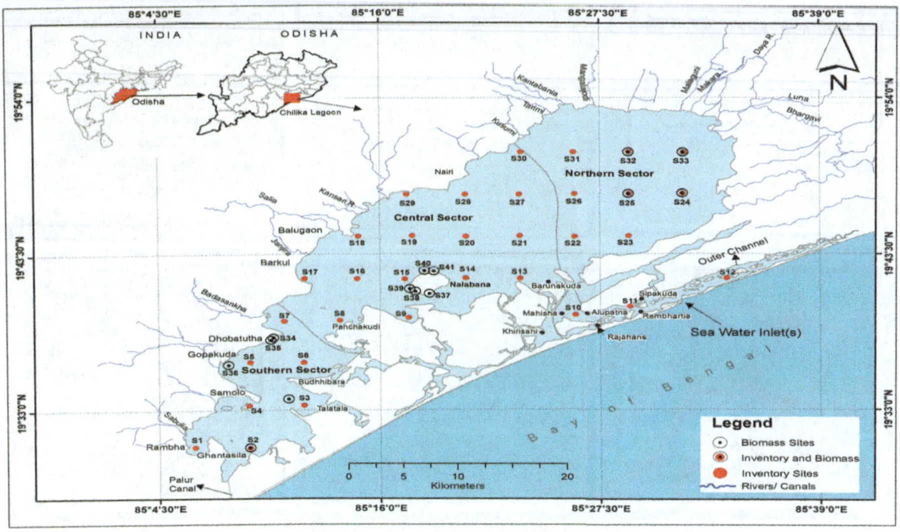

Fig. 14.1 Map showing macrophyte inventory (n = 33) and biomass (n = 14) monitoring sites across four sectors of Chilika Lagoon

2.3 *Macroalgae Associated with Macrophytes*

Macroalgae were collected from rocks submerged with water, jetties, and also those growing on the surface of macrophytes such as *Potamogeton*. The basic data necessary for species identification such as color and the shape of thallus or filaments were documented on-site. Samples were preserved with 4% formalin *on-board* and transported to the laboratory. The detailed identification of algal species based on morphology and anatomy was carried out using Olympus upright light microscope (BX53) using standard keys and research publications (Rath and Adhikary 2008).

2.4 *Biomass and Chlorophyll Estimation*

Primary production in terms of the biomass of aquatic macrophytes was assessed from 14 sampling stations (Fig. 14.1) on a seasonal basis (i.e., once during winter, summer, and monsoon) over 1 year. The key species chosen for biomass assessment were based on their productivity and ecological significance in a particular region. The biomass of key species, namely, *Ceratophyllum demersum*, *Hydrilla verticillata*, *Najas indica*, *Potamogeton crispus*, *Potamogeton nodosus*, *Stuckenia pectinata*, and *Vallisneria natans*, was measured in the northern sector. The biomass of *Halophila ovalis*, *Halophila beccarii*, *Halodule pinifolia*, and *Ruppia maritima* was collected from the southern (shoreline region) and central sector (around Nalabana

Fig. 14.2 Quadrats depicting the percentage cover of *Halophila ovalis*. 10–30% (**a**), 60–80% (**b**), 100% (**c**). *Salicornia brachiata*, 100% (**d**); *Paspalum distichum*, 100% (**e**); and *Sesuvium portulacastrum*, 100% (**f**)

Island). On each seagrass sampling site, six quadrats (50 cm × 50 cm) were positioned randomly (English et al. 1994), and percentage coverage was estimated using visual inspection of the quadrat area (Fig. 14.2).

Plant samples were washed thoroughly with tap water to remove the epiphytes, bivalves, and soil debris and kept on blotting paper to remove the excess water. Samples were then dried for 72 h in a hot air oven at 80 °C, and dry weight was measured for biomass estimation and expressed in g m^{-2}. The matured fresh green leaves of the macrophytes were collected to examine their photosynthetic productivity through total chlorophyll measurements. Plant leaves were rinsed using distilled water, dried in blotted papers, and cut into small pieces. Thereafter, 0.1 g of leaf pieces were soaked in 25 ml of 90% aqueous acetone and stored at 4 °C in the refrigerator for 48 h in a sealed tube. The amount of total chlorophyll (Chl a and Chl b) was measured from leaf extract using Epoch™ Microplate Spectrophotometer (BioTek, India) at 663 to 645 nm wavelength as described by Porra (2002).

2.5 *Physicochemical Parameters Analysis*

Biomass monitoring sites (n = 14), as well as inventory sites (n = 33), were examined for transparency, depth, and salinity. Transparency and depth were measured on-site using a Secchi disk (KC Denmark). The in situ salinity of water samples was recorded using a Thermo Scientific™ Orion™ Star A212 Conductivity Benchtop Meter.

2.6 *Mapping of Seagrasses*

To prepare a spatial distribution map, seagrass meadows were surveyed in various locations from the southern sector (Talatala, Kumarpur, Budhibaranasi, Dhobatutha, Gopakuda, Ghantasila, and Somolo), central sector (Panchakudi and around Nalabana Island), and outer channel (Barunikuda, Alupatna, Sipakuda, Rambhartia, Mahisha, Khirisahi, and Rajhans) during the winter season of 2019. Samples were collected using Van Veen grab from the deeper areas, and species composition was determined by their morphological characteristics. The seagrass distribution map was prepared using ArcMap (v10.5), a GIS (Geographic Information System) tool, by integrating the GPS (Global Positioning System) location of ground-truthing data.

2.7 *Assessment of Major Salt Marsh Grasses from Nalabana*

The growth pattern of *Salicornia brachiata*, *Paspalum distichum*, and *Sesuvium portulacastrum* was studied from three sites located adjacent to the Watch Tower no. 3 of Nalabana Island from January 2019 to June 2019. On each site, three quadrats of 50 cm × 50 cm were laid down in a line transect of 60 m with an interval of ~20 m between them. The species composition and percentage coverage were evaluated by visual inspection (Fig. 14.2). Chlorophyll content and biomass were estimated as described in Sect. 2.4.

2.8 *Statistical Analysis*

The seasonal changes in macrophyte biomass, transparency, depth, temperature, and salinity were analyzed using one-way ANOVA followed by Games-Howell nonparametric post hoc test (SPSS 20.0, IBM software).

3 Results and Discussion

3.1 *Diversity and Distribution of Aquatic Macrophytes*

The monthly inventory revealed that submerged macrophytes were most abundant in all seasons and sectors (except outer channel) in the lagoon. A total of 22 species of aquatic macrophytes were recorded during 2 years of field survey from the 33 stations located within the Chilika Lagoon. Of these, 21 species were from angiosperm and 1 from pteridophytes (Table 14.2). The family Hydrocharitaceae was represented by four genera (two species of *Halophila*, one species each of *Hydrilla*,

Table 14.2 Inventory and distribution of aquatic macrophytes in Chilika Lagoon

Plant type	Species	Sector	Station	2018												2019											
				Jan	Feb	Mar	Apr	May	Jun	Jul	Aug	Sep	Oct	Nov	Dec	Jan	Feb	Mar	Apr	May	Jun	Jul	Aug	Sep	Oct	Nov	Dec
Emergent	*Alternanthera philoxeroides*	NS	S31	■	■	■					■	■	■	■	■	■	■						■	■	■	■	■
			S32	■	■	■	■				■	■	■	■	■	■	■						■	■	■	■	■
			S33	■	■						■	■	■	■	■	■	■						■	■	■	■	■
	Ipomoea aquatica	NS	S33	■								■	■	■	■								■	■	■	■	■
	Phragmites karka	NS	S30	■	■	■	■	■	■	■	■	■	■	■	■	■	■	■	■	■	■	■	■	■	■	■	■
			S31	■	■	■	■	■	■	■	■	■	■	■	■	■	■	■	■	■	■	■	■	■	■	■	■
			S32	■	■	■	■	■	■	■	■	■	■	■	■	■	■	■	■	■	■	■	■	■	■	■	■
	Schoenoplectus litoralis	CS	S9	■	■	■	■	■	■	■	■	■	■	■	■	■	■	■	■	■	■	■	■	■	■	■	■
			S14	■	■	■	■	■	■	■	■	■	■	■	■	■	■	■	■	■	■	■	■	■	■	■	■
Rooted with Floating leaves	*Nymphaea pubescens*	NS	S24	■	■	■	■	■	■	■	■	■	■	■	■	■	■	■	■	■	■	■	■	■	■	■	■
			S33	■	■	■	■	■	■	■	■	■	■	■	■	■	■	■	■	■	■	■	■	■	■	■	■
	Nymphoides cristata	NS	S24	■	■				■	■	■	■	■	■	■	■	■					■	■	■	■	■	■
			S33	■	■					■	■				■	■	■	■				■	■	■	■	■	■
Free-floating	*Azolla pinnata*	NS	S30	■	■									■	■	■	■							■	■	■	■
			S32	■	■							■	■	■	■	■	■					■	■	■	■	■	■
			S33	■	■							■	■	■	■	■	■					■	■	■	■	■	■
	Eichhornia crassipes	NS	S24	■	■						■	■	■	■	■	■	■						■	■	■	■	■
			S25												■										■	■	
			S30	■	■	■					■	■	■	■	■	■	■	■	■	■		■	■				■
			S31	■	■	■				■	■			■	■								■	■	■	■	
			S32	■	■	■						■	■	■	■	■	■	■	■			■	■	■	■	■	■
			S33	■	■	■			■	■	■	■	■	■	■	■	■	■	■			■	■	■	■	■	■
	Pistia stratiotes	NS	S32	■	■		■							■	■	■	■					■	■	■	■	■	■
			S33	■	■	■	■				■	■	■	■	■	■	■	■	■	■		■	■	■	■	■	■
	Salvinia cucullata	NS	S30	■	■	■				■	■	■	■	■	■	■	■	■					■	■	■		
			S32		■	■								■	■		■	■	■			■	■	■	■	■	
			S33	■	■	■				■	■			■	■	■	■	■	■			■	■	■	■	■	■
	Spirodela polyrrhiza	NS	S30	■	■	■					■	■	■	■	■	■	■							■	■	■	■
			S33	■	■	■					■	■	■	■	■	■	■	■				■	■	■	■	■	■
Submerged	*Ceratophyllum demersum*	NS	S25	■	■	■								■	■	■	■	■	■	■							
			S32	■	■	■	■	■	■	■	■	■	■	■	■	■	■	■	■	■	■	■	■	■	■	■	■
			S33	■	■	■	■	■	■	■	■	■	■	■	■	■	■	■	■	■	■	■	■	■	■	■	■
	Halodule pinifolia	SS	S2	■	■	■	■	■	■	■	■	■	■	■	■	■	■	■	■	■	■	■	■	■	■	■	■
		CS	S13																				■	■			
			S14																					■	■	■	■
	Halophila beccarii	CS	S16																								■
	Halophila ovalis	CS	S9			■	■	■	■	■	■	■	■	■	■	■	■	■	■	■							
			S13							■	■	■	■														
			S14	■	■			■	■	■	■	■	■	■	■	■	■	■	■	■	■	■	■	■	■	■	■
			S15	■	■	■	■	■	■	■	■	■	■	■	■	■	■	■	■	■	■	■	■	■	■	■	■
			S16	■			■	■	■	■	■	■	■	■	■	■	■	■	■	■	■	■	■	■	■	■	■
		SS	S1					■	■	■	■	■	■	■	■	■	■	■	■	■	■	■	■	■	■	■	■
			S2	■	■	■	■	■	■	■	■	■	■	■	■	■	■	■	■	■	■	■	■	■	■	■	■
			S3	■	■	■	■	■	■	■	■	■	■	■	■	■	■	■	■	■	■	■	■	■	■	■	■
			S4		■	■	■						■	■	■	■	■	■	■					■	■	■	■
	Hydrilla verticillata	NS	S24	■	■	■	■	■	■	■	■	■	■	■	■	■	■	■	■	■	■	■	■	■	■	■	■
			S25				■	■	■	■	■	■	■	■	■	■	■	■	■	■	■	■	■	■	■	■	■
			S33	■	■	■	■	■	■	■	■	■	■	■	■	■	■	■	■	■	■	■	■	■	■	■	■
	Najas indica	CS	S9						■	■	■	■											■	■			
			S14	■					■	■	■	■											■	■			
			S18		■	■	■	■					■	■	■	■	■	■	■	■							■
		NS	S24	■	■	■	■		■	■	■	■	■	■	■	■	■	■	■								
			S25			■	■		■	■	■							■	■								
			S30			■	■						■	■	■	■	■	■	■								
			S31			■	■	■	■	■	■	■	■	■	■	■	■	■	■	■	■	■	■	■	■	■	■
			S32	■	■	■	■	■	■	■	■	■	■	■	■	■	■	■	■	■	■	■	■	■	■	■	■
			S33	■	■	■	■	■	■	■	■	■	■	■	■	■	■	■	■	■	■	■	■	■	■	■	■
	Potamogeton crispus	NS	S24	■	■	■	■	■	■	■	■	■	■	■	■	■	■	■	■	■	■	■	■	■	■	■	■
			S32	■	■	■	■	■	■	■	■	■	■	■	■	■	■	■	■	■	■	■	■	■	■	■	■
			S33	■	■	■	■	■	■	■	■	■	■	■	■	■	■	■	■	■	■	■	■	■	■	■	■
	Potamogeton nodosus	NS	S24	■	■	■	■	■	■	■	■	■	■	■	■	■	■	■	■	■	■	■	■	■	■	■	■
			S25	■	■	■	■	■	■	■	■	■	■	■	■	■	■	■	■	■	■	■	■	■	■	■	■
			S32	■	■	■	■	■	■	■	■	■	■	■	■	■	■	■	■	■	■	■	■	■	■	■	■
			S33	■	■	■	■	■		■	■	■	■	■	■	■	■	■	■	■	■	■	■	■	■	■	■
	Ruppia maritima	CS	S15										■	■	■	■	■	■	■	■					■	■	■
		SS	S2																			■	■	■	■	■	■

Table. 14.2 (continued)

Plant type	Species	Sector	Station	2018												2019											
				Jan	Feb	Mar	Apr	May	Jun	Jul	Aug	Sep	Oct	Nov	Dec	Jan	Feb	Mar	Apr	May	Jun	Jul	Aug	Sep	Oct	Nov	Dec
Submerged	*Stuckenia pectinata*	CS	S9	■	■	■	■	■	■	■	■	■	■	■	■	■	■	■	■	■	■	■	■	■	■	■	■
			S13	■	■	■	■	■	■	■	■	■	■	■	■	■	■	■	■	■	■	■	■	■	■	■	■
			S14	■	■	■	■	■	■	■	■	■	■	■	■	■	■	■	■	■	■	■	■	■	■	■	■
			S15	■	■	■	■	■	■	■	■	■	■	■	■	■	■	■	■	■	□	□	□	□	■	■	■
			S16	■	■	■	■	□	■	■	■	■	■	■	■	■	■	■	■	□	□	□	□	□	□	□	□
			S17	■	■	■	■	■	■	■	■	■	■	■	■	■	■	■	■	■	■	■	■	■	■	■	■
			S18	■	■	■	■	■	■	■	■	■	■	■	■	■	■	■	■	■	■	■	■	■	■	■	■
			S20	■	■	■	■	■	■	■	■	■	■	■	■	■	■	■	■	■	□	□	□	□	■	■	■
			S21	■	■	■	■	■	■	■	■	■	■	■	■	■	■	■	■	■	□	□	□	□	■	■	■
		NS	S22	■	■	□	□	□	■	■	■	■	■	■	■	■	■	■	■	□	□	□	□	□	□	□	□
			S23	■	■	■	■	■	■	■	■	■	■	■	■	■	■	■	■	■	□	□	□	□	□	□	□
			S25	■	■	■	■	■	□	□	□	□	□	□	□	■	■	■	■	■	□	□	□	□	□	□	□
			S26	■	■	□	□	□	■	■	■	■	■	■	■	■	■	■	■	■	□	□	□	□	□	□	□
			S27	□	□	■	■	■	■	■	■	■	■	■	■	■	■	■	■	■	□	□	□	□	□	□	□
			S28	□	□	■	■	■	■	■	■	■	■	■	■	■	■	■	■	■	□	□	■	■	■	■	□
			S29	■	■	■	■	■	■	■	■	■	■	■	■	■	■	■	■	■	■	■	■	■	■	■	■
			S30	■	■	■	■	■	■	■	■	■	■	■	■	■	■	■	■	■	■	□	□	□	□	■	■
			S31	■	■	■	■	■	■	■	■	■	■	■	■	■	■	■	■	■	□	□	□	□	□	□	□
			S32	■	■	■	■	□	■	■	■	■	■	■	■	■	■	■	■	□	□	□	□	□	□	□	□
		SS	S2	□	□	□	■	■	■	■	■	■	■	■	■	■	■	■	■	■	■	■	■	■	■	■	■
			S3	■	□	□	□	■	■	■	■	■	■	■	■	■	■	■	■	■	■	■	■	■	■	■	■
			S8	■	■	■	■	■	■	■	■	■	■	■	■	■	■	■	■	■	■	■	■	■	■	■	■
	Vallisneria natans	NS	S24	■	■	■	■	■	■	■	■	■	■	■	■	■	■	■	■	■	■	■	■	■	■	■	■
			S32	■	■	■	■	■	■	■	■	■	■	■	■	■	■	■	■	■	■	■	■	■	■	■	■
			S33	■	■	■	■	■	■	■	■	■	■	■	■	■	■	■	■	■	■	■	■	■	■	■	■

NS: northern sector, *SS*: southern sector, *CS*: central sector. Grey: presence; white: absence. S1: Rambha jetty; S2: Palur canal; S3: Malud-talatala; S4: Badakuda; S5: Gopakuda; S6: Budhibara; S7: Malatikuda; S8: Panchakudi; S9: Veteswara; S10: Mahisha; S11: Kianasi; S12: Arakhakuda; S13: Maggarmukh; S14: Nuapada; S15: Nalabana; S16: WRTC-Nalabana; S17: WRTC-Naval Hill; S18: Kalijugeswar; S19, S20, S21: Kalijugeswar-Tuagambhari; S22: Tuagambhari; S23: Tatabandha; S24: Haridaspur; S25, S26, S27: Tinimuhani-Baulabandha; S28: Baulabandha; S29: Bhasaramundia-Nairi; S30: Sorana; S31: Kalupadaghat; S32: Bhusandapur; S33: Tinimuhani. No vegetation was found in S5, S6, S7, S10, S11, S12, and S19 stations.

Najas, and *Vallisneria*) followed by Potamogetonaceae with two genera (two species of *Potamogeton* and one species of *Stuckenia*), Araceae with two genera (one species each of *Pistia* and *Spirodela*), and Salviniaceae with two genera (one species each of *Azolla* and *Salvinia*). Other ten families, i.e., Amaranthaceae, Ceratophyllaceae, Convolvulaceae, Cymodoceaceae, Cyperaceae, Menyanthaceae, Nymphaeaceae, Poaceae, Pontederiaceae, and Ruppiaceae were represented by a single genus and species.

3.1.1 Spatial Distribution of Macrophytes

Macrophyte distribution showed a marked spatial variation across different sectors (Table 14.2). *Halophila ovalis*, *Halophila beccarii*, *Halodule pinifolia*, and *Ruppia maritima* were documented from the southern sector. *Halophila ovalis* was the most abundant seagrass in the lagoon due to their higher tolerance to low salinity and low light than other species (Pattnaik et al. 2020). *Halodule pinifolia* was observed only in the shallow zones of station S2 along with *Ruppia maritima* from July to December 2019 (Table 14.2). *Stuckenia pectinata*, a submerged species, was found in stations S2, S3, and S8 over the entire study period. *Schoenoplectus litoralis* (an emergent macrophyte), *Halophila ovalis*, *Halodule pinifolia*, and freshwater submerged species (*Stuckenia pectinata* and *Najas indica*) were distributed in the central sector. The distribution of *Stuckenia pectinata* was recorded from nine stations in the central sector during the entire study period (Table 14.2). *Halophila beccarii* was only found in S16 during December 2019. These findings were in accordance with the previous study from the lagoon which showed that *Halophila ovalis* and *Halodule pinifolia* were the dominant seagrass species in the central sector (Pattnaik et al. 2020).

In the central sector, *Halophila ovalis* and *Halodule pinifolia* were found in and around the Nalabana Island and Panchakudi. In the outer channel, seagrasses were abundantly found in Rajhans area (Table 14.4). *Halophila ovalis* was distributed in the Khirisahi and Mahisha areas. *Halophila beccarii* was abundantly distributed in the Alupatna, Sipakuda, and Rambhartia areas with fewer occurrences in Mahisha area. A mixed bed of *Halodule pinifolia* and *Halophila beccarii* was abundantly found in the shallow area of Rambhartia and Mahisha. A previous study has also shown that the Barunikuda Island, including Khirisahi and Mahisha, sustains good seagrass meadows in the outer channel (Pattnaik et al. 2020).

Freshwater aquatic macrophytes were enriched in the northern sector due to low salinity and high nutrients caused by the river discharge. The rapid growth of floating species in the northern sector is usually associated with an increase in water nutrients through freshwater riverine discharges (Jaikumar et al. 2011; Behera et al. 2018). A total of 17 major species were encountered from northern sector. Of these, five free-floating (*Azolla pinnata*, *Eichhornia crassipes*, *Pistia stratiotes*, *Salvinia cucullata*, and *Spirodela polyrhiza*), three emergent (*Phragmites karka*, *Ipomoea aquatica*, and *Alternanthera philoxeroides*), two floating-leaves (*Nymphaea pubescens* and *Nymphoides cristata*), and seven submerged (*Ceratophyllum demersum*, *Hydrilla verticillata*, *Najas indica*, *Potamogeton crispus*, *Potamogeton nodosus*, *Stuckenia pectinata*, and *Vallisneria natans*) plants were recorded (Table 14.2, Fig. 14.3). Previous studies have shown that *Phragmites karka* was abundant in the shoreline zone of the northern sector extending from Kalupadaghat to Mangalajodi (Bhatta and Patra 2018; Pattnaik et al. 2020). *Nymphaea pubescens* and *Nymphoides cristata* were restricted to the northern sector and were abundantly growing in Sorana, Mangalajodi, and Kalupadaghat. The major submerged species, namely, *Hydrilla verticillata*, *Potamogeton nodosus*, *Najas indica*, *Vallisneria natans*, and *Potamogeton crispus* have been reported earlier from the northern sector of the lagoon (Pattnaik et al. 2020).

3.1.2 Temporal Distribution of Macrophytes

Seasonal variability in macrophyte distribution was also observed. *Halophila ovalis* was recorded from the southern (S2, S3) and central sector (S15) in all seasons (Table 14.2). *Halophila beccarii* was recorded only from S16 of the central sector during the winter season (December 2019). *Halodule pinifolia* was encountered in central sector (S13, S14). *Stuckenia pectinata* was distributed in both southern (S8, S9) and central sectors (S13, S14, S17, S18), and only in S29 of the northern sector

Fig. 14.3 Dominant species of macrophytes identified from Chilika Lagoon. (**a**) *Nymphaea pubescens*, (**b**) *Nymphoides cristata*, (**c**) *Potamogeton nodosus*, (**d**) *Potamogeton crispus*, (**e**) *Stuckenia pectinata*, (**f**) *Ceratophyllum demersum*, (**g**) *Vallisneria natans*, (**h**) *Hydrilla verticillata*, (**i**) *Najas indica*, (**j**) *Spirodela polyrhiza*, (**k**) *Azolla pinnata*, (**l**) *Eichhornia crassipes*, (**m**) *Phragmites karka*, (**n**) *Ipomoea aquatica*, (**o**) *Alternanthera philoxeroides*, (**p**) *Ruppia maritima*, (**q**) *Halophila ovalis*, (**r**) *Halophila beccarii*, (**s**) *Halodule uninervis*, and (**t**) *Halodule pinifolia*

in all seasons. *Potamogeton crispus*, *Vallisneria natans*, *Potamogeton nodosus*, *Hydrilla verticillata*, *Ceratophyllum demersum*, and *Najas indica* were recorded from the northern sector in all seasons (Table 14.2). All submerged macrophytes except *Halophila beccarii* and *Ruppia maritima* were present in all seasons and could be considered as key aquatic macrophytes.

The free-floating species were not restricted to a particular site in the northern sector, but the profuse growth of *Eichhornia crassipes* trapped them for a long time. The free-floating species, namely, *Azolla pinnata*, *Pistia stratiotes*, *Salvinia cucullata*, and *Spirodela polyrhiza* were only recorded during the winter season from the northern sector (Table 14.2). The dense mats of *Azolla pinnata* and profuse growth of *Eichhornia crassipes* gradually disappeared during summer due to increase in the salinity (Pattnaik et al. 2020).

Phragmites karka was present in the northern sector throughout the year. *Alternanthera philoxeroides* and *Ipomoea aquatica* were recorded from the northern sector during monsoon and winter seasons (Table 14.2). Two rooted with floating-leaves macrophytes, viz., *Nymphaea pubescens* and *Nymphoides cristata* were encountered in all seasons from the northern sector (S24 and S33) of the lagoon. A total of 15 macrophyte species were encountered from most diversified station S33 (Table 14.2).

An extremely severe cyclonic storm *Fani* made its landfall at Satapada in Puri district of Odisha on 3 May 2019. *Fani* was accompanied by heavy precipitation, land runoff, and huge river discharge and severe high-velocity winds which resulted in the complete elimination of free-floating hydrophytes like *Eichhornia*, *Spirodela*, *Salvinia*, and *Pistia* from all stations of the northern sector. After the cyclone, *Stuckenia pectinata* flushed out from most of the areas of the northern and central sectors and started reappearing after September 2019 (Table 14.2).

3.2 Distribution of Macroalgae

A total of 11 species of macroalgae belonging to 7 genera and 2 phyla were identified from the lagoon. Of these, six species belonged to phylum Chlorophyta (e.g., *Chaetomorpha linum*, *Chaetomorpha* sp., *Ulva compressa*, *Ulva flexuosa*, *Ulva intestinalis*, and *Chara braunii*), and other five species belonged to phylum Rhodophyta (e.g., *Ceramium* sp., *Ceramium diaphanum*, *Gracilaria verrucosa*, *Polysiphonia subtilissima*, and *Polysiphonia sertularioides*) (Fig. 14.4). These

Fig. 14.4 Macroscopic and microscopic identification of macroalgae from Chilika Lagoon. Thallus view of (**a**) *Ulva flexuosa*, (**b**) *Ulva compressa*, (**c**) *Ulva intestinalis*, (**d**) *Ceramium* sp., (**e**) *Chaetomorpha linum*, (**f**) *Chaetomorpha* sp., (**g**) *Ceramium diaphanum*, (**h**) *Polysiphonia subtilissima*, (**i**) *Polysiphonia sertularioides*, (**j**) *Chara braunii*, and (**k**) *Gracilaria verrucosa*

findings were in accordance with the previous studies conducted by Sahoo et al. (2003) and Mohanty and Adhikary (2013). *Ulva* sp. was found in almost all stations of the southern sector and central sector (S9, S14, S15, S17, and S18) either attached to the rock substratum or floating freely (Table 14.2) (Sahoo et al. 2003). *Chaetomorpha* sp. was found in association with *Ulva* sp. in stations S1, S2, S6, and S8 of southern; S9, S14, and S15 of central; and S29 of northern sector. *Chara braunii* was found in S18, S34, and S35 of the southern sector during the winter season. *Gracilaria verrucosa* was often associated with seagrass meadows and was confined to the southern and central sectors, especially around the Nalabana Island. *Ceramium* sp., *Polysiphonia subtilissima*, and *Polysiphonia sertularioides* were also reported from S8, S15, and S17 during the study period.

3.3 Biomass Production

Biomass of aquatic macrophytes showed a marked seasonal difference (Fig. 14.5). The average total macrophyte biomass was 169.00 g m^{-2} in the lagoon. The highest biomass was recorded in winter (4322.38 g m^{-2}) followed by summer (3056.18 g m^{-2}) and monsoon (1957.14 g m^{-2}). Of these, the maximum biomass of seagrasses (1730.85 g m^{-2}) and freshwater weed (2591.53 g m^{-2}) was recorded during winter which was ~1.4- and 2.3-fold higher than summer and monsoon, respectively. Jaikumar et al. (2011) have reported lower growth of submerged macrophytes during monsoon compared to summer in the Chilika Lagoon. The seasonal and sectoral variation in salinity, transparency, and depth recorded from biomass monitoring sites are given in Table 14.3. Transparency was the highest (73.00 ± 11.59 cm) during monsoon which could be due to greater water depth of the lagoon during this season and reduced wind speed resulting in less resuspension of bottom sediments due to wind derived churning action and concurrent evaporation of water (Srichandan et al. 2015b; Pattnaik et al. 2020). The lowest salinity (4.23 ± 0.86) was recorded during monsoon which was due to the inflow of a large amount of freshwater into the lagoon from the Mahanadi River catchment. The highest salinity (7.36 ± 1.10) was observed during the summer season which was due to cessation of riverine discharge and higher residence period of water during the low-flow period (Srichandan et al. 2015b). Spatially, water transparency significantly varied between the northern (36.42 ± 4.09 cm) and central sectors (75.73 ± 9.62 cm). The northern sector of the lagoon showed the lowest transparency due to the large sediment load discharged into the lagoon from the Mahanadi River distributaries (Srichandan et al. 2015b). Salinity exhibited a significant spatial variation (p value <0.05) across three different sectors of the Chilika Lagoon (Table 14.3).

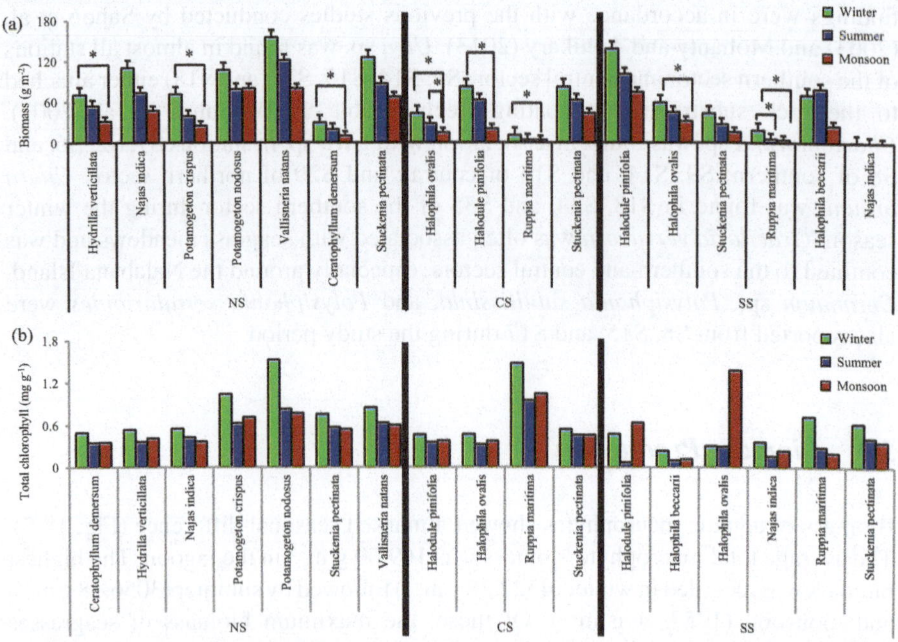

Fig. 14.5 Seasonal variation in the average biomass (**a**) and total chlorophyll contents (**b**) of aquatic macrophytes. The error bars indicate the standard error. Means with significant differences are denoted by *asterisks*

Table 14.3 Physicochemical measurement from biomass monitoring and inventory sites

Site	Sector/ season	Parameters			Temperature (°C)
		Transparency (cm)	Depth (cm)	Salinity	
Biomass monitoring sites ($n = 14$)	NS ($n = 4$)	36.42 ± 4.09^{b}	107.75 ± 13.25^{a}	1.06 ± 0.51^{a}	ND
	CS ($n = 5$)	75.73 ± 9.62^{a}	92.73 ± 8.54^{ab}	6.44 ± 0.52^{b}	ND
	SS ($n = 5$)	51.20 ± 8.94^{ab}	67.40 ± 9.65^{b}	9.04 ± 0.62^{c}	ND
	Winter	48.57 ± 8.98^{ab}	69.86 ± 7.39^{a}	5.91 ± 0.99^{ab}	ND
	Summer	45.64 ± 4.53^{a}	71.71 ± 8.12^{a}	7.36 ± 1.10^{a}	ND
	Monsoon	73.00 ± 11.59^{b}	122.36 ± 11.15^{b}	4.23 ± 0.86^{b}	ND
Inventory sites ($n = 33$)	NS ($n = 9$)	37.11 ± 2.23^{a}	111.22 ± 2.84^{a}	2.21 ± 0.24^{a}	28.69 ± 0.26^{a}
	CS ($n = 14$)	70.48 ± 2.15^{b}	144.90 ± 2.72^{b}	8.05 ± 0.33^{b}	28.28 ± 0.19^{a}
	SS ($n = 7$)	85.43 ± 2.79^{c}	189.42 ± 6.36^{c}	9.31 ± 0.27^{c}	28.72 ± 0.27^{a}
	OC ($n = 3$)	48.10 ± 4.01^{d}	217.01 ± 6.32^{d}	17.80 ± 1.52^{d}	28.59 ± 0.42^{a}
	Winter	59.68 ± 2.13^{a}	136.49 ± 3.80^{a}	5.67 ± 0.32^{b}	24.91 ± 0.21^{a}
	Summer	55.11 ± 1.89^{a}	131.92 ± 3.55^{a}	12.34 ± 0.52^{a}	30.05 ± 0.13^{b}
	Monsoon	76.06 ± 3.18^{b}	190.54 ± 3.83^{b}	5.51 ± 0.38^{b}	30.57 ± 0.11^{c}

The value represents mean with standard error. Mean differences were analyzed by one-way ANOVA followed by Games-Howell test. Means with same alphabets are not significantly different (p value > 0.05)

ND not determined

3.3.1 Biomass Production in the Southern Sector

Halophila ovalis, *Halodule pinifolia*, *Ruppia maritima*, and *Stuckenia pectinata* were the most abundant submerged macrophytes found in shallow areas of the southern sector. In this sector, the maximum biomass of *Halodule pinifolia* was recorded during winter (138.52 g m^{-2}) which was 1.4- and 1.9-fold higher than summer and monsoon season, respectively (Fig. 14.5a). The minimum biomass (0.01 g m^{-2}) was recorded for *Najas indica* during the monsoon which was 20- and 66-fold lower than the summer and winter season, respectively (Fig. 14.5a). In winter, the biomass production in *Halophila ovalis* was 1.3-fold higher than summer and 1.8-fold higher than the monsoon. Biomass production (18.66 g m^{-2}) in *Ruppia maritima* revealed a 3.2- and 4.4-fold increase in winter as compared to summer and monsoon, respectively. The biomass of *Halophila ovalis* and *Ruppia maritima* reached as high as 78.32 and 21.37 g m^{-2}, respectively, in S2 (transparency, 63 cm; depth, 63 cm; and salinity, 9.52) during winter.

3.3.2 Biomass Production in the Central Sector

Stuckenia pectinata, *Halodule pinifolia*, and *Halophila ovalis* including *Ruppia maritima* were recorded from Nalabana. Biomass production in *Halodule pinifolia* and *Stuckenia pectinata* was 1.3-fold higher in winter than summer (Fig. 14.5a). The lowest biomass production was recorded for *Ruppia maritima* during the monsoon season. *Halophila ovalis* showed a 2.0- and 3.0-fold decline in biomass during monsoon compared to the summer and winter seasons, respectively (Fig. 14.5a).

3.3.3 Biomass Production in the Northern Sector

Vallisneria natans displayed 1.3-fold higher biomass in the winter compared to the summer season (Fig. 14.5a). Biomass production in *Ceratophyllum demersum* decreased by 1.9-fold during the monsoon compared to winter. The biomasses of *Potamogeton crispus*, *Hydrilla verticillata*, and *Ceratophyllum demersum* were significantly higher in winter compared to monsoon. The biomass of *Hydrilla verticillata*, *Potamogeton crispus*, and *Ceratophyllum demersum* reached as high as 105.21, 95.40, and 37.08 g m^{-2}, respectively, in S33 (transparency, 33 cm; depth, 55 cm; salinity, 0.34) during winter. Thus, station S33 was most productive both in terms of species diversity and biomass. The higher biomass production in the winter season could be due to high amounts of nutrients, NO_3 and PO_4 in the lagoon (Behera et al. 2017) which promote macrophyte growth.

3.3.4 Biomass Production in Major Salt Marshes of Nalabana

The Nalabana Island covers an area of 15.53 km^2 in the middle of the lagoon. The island constitutes a bird sanctuary and provides a suitable wintering ground for millions of migratory and local birds. The island supports the luxurious growth of

Fig. 14.6 Salt marsh grasses recorded from Nalabana Island. (**a**) *Salicornia brachiata*, (**b**) *Paspalum distichum*, (**c**) *Heliotropium curassavicum*, (**d**) *Cyperus rotundus*, (**e**) *Suaeda maritima*, (**f**) *Alternanthera philoxeroides*, (**g**) *Sesuvium portulacastrum*, (**h**) *Boerhavia diffusa*, and (**i**) *Cynodon dactylon. Salicornia brachiata* with different developmental stages: (**j**) regenerative (mature green shoots), (**k**) early senescence (greenish-yellow shoots), and (**l**) complete senescence (dark-brown plants)

halotolerant grasses, namely, *Salicornia brachiata*, *Paspalum distichum*, *Heliotropium curassavicum*, *Cyperus rotundus*, *Suaeda maritima*, *Alternanthera philoxeroides*, *Sesuvium portulacastrum*, *Boerhavia diffusa*, and *Cynodon dactylon* (Fig. 14.6). *Paspalum distichum* is fast-growing rhizomatous grasses that grow even under partially submerged condition during monsoon and winter (Pattnaik et al. 2020). *Sesuvium portulacastrum*, *Heliotropium curassavicum*, and *Suaeda maritima* have a patchy distribution in the sanctuary. *Salicornia brachiata* usually germinate and sprout after monsoon as plants require low salinity for germination and growth during the early stage. As the water level recedes during summer and salinity rises, *Salicornia brachiata* spread in a huge area of the sanctuary within a short life span of 4–5 months (January to June). The rise in water level during monsoon leads to the inundation of sanctuary and decomposition of *Salicornia brachiata* in the mudflats.

During the field survey, different growth stages of *Salicornia brachiata* were encountered. At first, a regenerative stage (growing green shoot) was observed in January, and then the adult stage appeared by March which has greenish-yellow shoot coloration. The plants reached their full maturity by May and finally turned into dark brown in color and senescence by June (Fig. 14.6j–l). This was consistent with a previous study which has shown that species can attain maturity and succulence and became senescent with increasing salinity and temperature in the summer season (March–June) (Jagtap et al. 2002).

All halophyte grasses showed an increasing trend in their biomass production. The maximum aboveground biomass was recorded during May 2019 for *Salicornia brachiata* (693.80 g m^{-2}), *Paspalum distichum* (930.77 g m^{-2}), and *Sesuvium portulacastrum* (639.03 g m^{-2}). The gradual increase in the biomass of *Salicornia brachiata* from its early growth stage (just after monsoon) to the senescent stage (end of summer) was in accordance with the study of Jagtap et al. (2002).

3.4 Chlorophyll Content in Aquatic Macrophytes and Salt Marsh

The total chlorophyll content is a crucial growth parameter as well as an important ecological index that has a relationship with primary productivity (Liu et al. 2019). The seasonal fluctuations in total chlorophyll contents of macrophytes are shown in Fig. 14.5b. Within seagrasses, maximum total chlorophyll content was found in *Halophila ovalis* followed by *Halodule pinifolia* during monsoon season in the southern sector (Fig. 14.5b). The variation in the chlorophyll content between these seagrass species could be due to the large leaf area of *Halophila ovalis*. However, *Ruppia maritima* showed maximum chlorophyll content during winter season in central sector (Fig. 14.5b). Within freshwater weeds, the higher total chlorophyll content was found in *Potamogeton nodosus* (1.52 mg g^{-1}) followed by *Potamogeton crispus* (1.04 mg g^{-1}) and *Vallisneria natans* (0.84 mg g^{-1}) during the winter season in northern sector (Fig. 14.5b). The maximum chlorophyll content in aquatic macrophytes during winter season was consistent with a previous study which examined the diversity and distribution pattern of aquatic angiosperms from Chilika Lagoon (Regional Plant Resource Centre 2016).

A sharp decline in chlorophyll content with the onset of senescence in *Salicornia brachiata* was noted due to defoliation of leaves. *Paspalum distichum* varied greatly in their total chlorophyll content during the growing season, being highest in February, declining by late March, and increasing again in May. Total leaf chlorophyll concentrations ranged from 0.19 to 0.83 mg g^{-1} in *Paspalum distichum*, 0.09 to 0.11 mg g^{-1} in *Sesuvium portulacastrum*, and 0.02 to 0.21 mg g^{-1} in *Salicornia brachiata*. The total chlorophyll content of *Sesuvium portulacastrum* did not vary much over the growing season.

3.5 Spatial Mapping of Seagrasses

Based on the field survey, the area with seagrasses was estimated to be 169.2 km^2 in the lagoon (Fig. 14.7). Out of 16 seagrass species reported from the Indian coast (Mishra and Apte 2021), 6 species, viz., *Halophila ovalis*, *Halophila beccarii*, *Halophila ovata*, *Halodule pinifolia*, *Halodule uninervis*, and *Ruppia maritima* were recorded from Chilika Lagoon during the study period (Fig. 14.8). Sipakuda, Alupatna, and Rambhartia in the outer channel showed the maximum

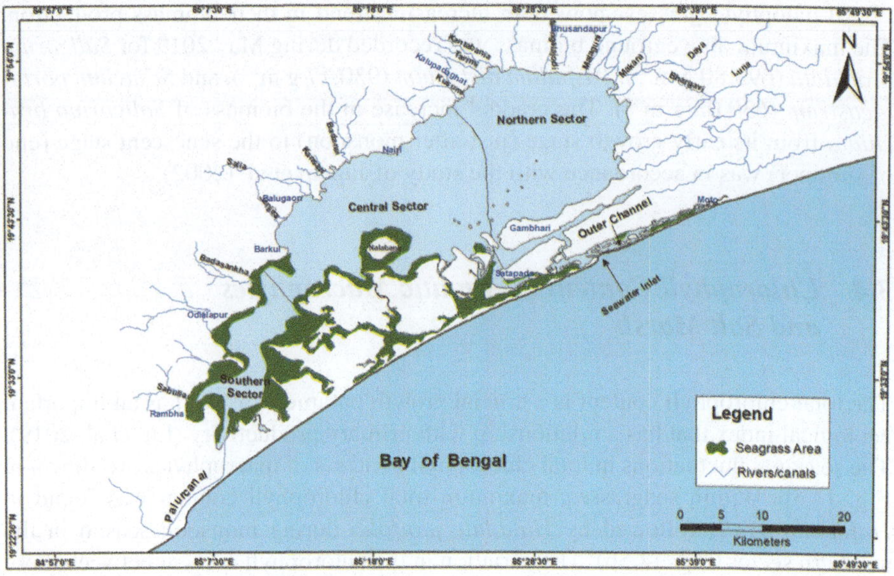

Fig. 14.7 Seagrass distribution map based on field survey carried out in the winter season of year 2019

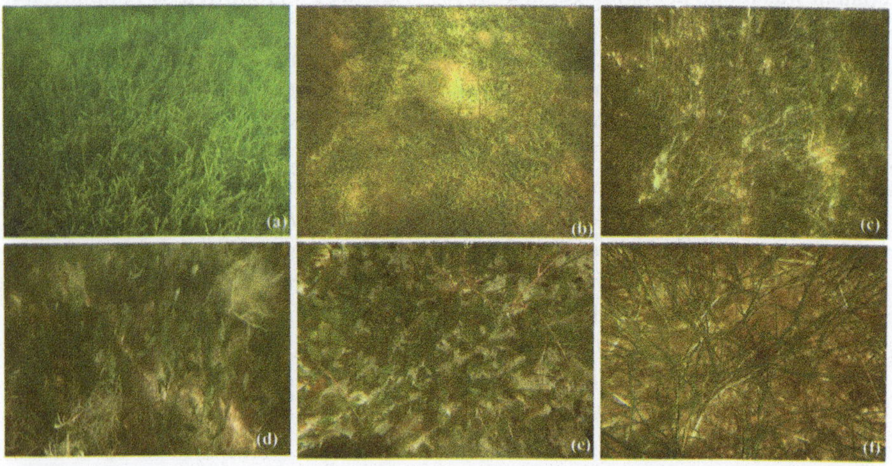

Fig. 14.8 Underwater photographs of seagrass meadows from southern sector of the lagoon. (**a**) *Halodule pinifolia*, (**b**) *Halophila beccarii*, (**c**) mixed bed of *Halodule* and *Halophila*, (**d**) *Halophila ovalis*, (**e**) *Halophila ovata*, and (**f**) *Halodule uninervis*

seagrass cover dominated by *Halophila beccarii*. However, *Halophila ovalis* and *Halodule pinifolia* were abundant in the shallow shoreline region of the southern sector (Table 14.4). The maximum biomass of 127.48 g m^{-2} was recorded for *Halophila beccarii* on exposed sand in the Sipakuda area (Table 14.4). *Halodule pinifolia* was also found extensively in the Mahisha area with the highest biomass of 49.84 g m^{-2}.

Table 14.4 Percentage coverage and biomass of the different seagrass species measured during seagrass mapping in winter 2019. The number in the bracket indicates % coverage

Sector/station	No. of locations (*n*)	No. of locations with % coverage						Biomass (g m^{-2})						Salinity	Trans (cm)	Depth (cm)
		H. ovalis	*H. ovata*	*H. beccarii*	*H. pinifolia*	*H. uninervis*	*R. maritima*	*H. ovalis*	*H. ovata*	*H. beccarii*	*H. pinifolia*	*H. uninervis*	*R. maritima*			
Southern sector																
Talatala	$n = 10$	8 (20–100)	–	–	2 (70–80)	–		30.61	–	–	4.35	–	–	6.35	106.25	106.25
Kumarpur	$n = 15$	5 (30–100)	–	–	8 (50–100)	1 (25)	1 (50)	73.1	–	–	90.4	13	16.3	5.57	68.10	80.80
Budhibaranasi	$n = 4$	2 (100)	1 (30)	–	–	–	1 (70)	–	18.2	–	–	–	29.5	6.42	51.67	51.67
Dhobatutha	$n = 8$	1 (100)	–	5 (60–100)	1 (30)	–	1 (10)	37.62	–	27.09	60.56	–	1.1	7.32	45.71	46.86
Gopakuda	$n = 6$	3 (20–100)	–	–	3 (30–80)	–	–	19.26	–	–	23.68	–	–	5.61	112.50	115.75
Ghantasila	$n = 8$	5 (20–100)	–	–	3 (10–80)	–	–	33.14	–	–	43.25	–	–	6.18	35.20	35.20
Somolo	$n = 6$	4 (30–100)	–	–	1 (100)	–	1 (5)	38.12	–	–	80.42	–	0.4	6.55	59.80	59.80
Central sector																
Panchakudi	$n = 4$	3 (60–100)	–	–	1 (100)	–	–	–	–	–	–	–	–	6.55	22.50	22.50
Nalabana	$n = 11$	5 (50–100)	1 (40)	–	3 (10–30)	–	3 (5–60)	31.34	22.17	–	12.19	–	23	3.74	55.20	67.00
Outer channel																
Barunikuda	$n = 1$	–	–	–	1 (60)	–	–	–	–	–	34.6	–	–	3.22	32.00	32.00
Alupatna	$n = 2$	–	–	2 (100)	–	–	–	–	–	47.04	–	–	–	3.80	31.50	31.50
Sipakuda	$n = 1$	–	–	1 (100)	–	–	–	–	–	127.48	–	–	–	2.98	7.00	7.00
Rambhartia	$n = 3$	–	–	1 (100)	2 (80–100)	–	–	–	–	31.86	64.45	–	–	5.07	14.33	18.33
Mahisha	$n = 8$	1 (10)	–	4 (10–30)	3 (70–100)	–	1 (50)	7.16	–	–	49.84	–	14.2	4.93	22.00	22.00
Khirisahi	$n = 5$	2 (10–60)	–	–	3 (80–100)	–	–	–	–	–	–	–	–	ND	ND	ND
Rajhans	$n = 7$	–	–	–	4 (30–90)	1 (100)	2 (20–60)	–	–	–	–	53.6	31.6	ND	ND	ND

Trans: transparency, – absent, *ND:* Not determined

Analysis of percentage cover of seagrasses showed an increasing trend over the years. Before the opening of a new seawater inlet in 2001, the seagrass area was 24.8 km^2 in 1998 which expanded to 86.84 km^2 after restoration in 2004, 102 km^2 in 2012, and 152 km^2 in 2018 (Pattnaik et al. 2020). The improvement of the salinity regime and water clarity due to the flushing of sediment (Mohanty et al. 2009; Kim et al. 2015) were the main factors responsible for the increase in seagrass area. Among the six seagrass species, *Halophila ovalis* displayed maximum spread in the lagoon followed by *Halodule pinifolia*, *Halophila beccarii*, *Ruppia maritima*, *Halodule uninervis*, and *Halophila ovata* (Table 14.4).

The growth, propagation, and reproduction of seagrasses depend on environmental conditions (Pati et al. 2014b). The extent of seagrasses not only depends on the physiochemical parameters of water but also on sediment characteristics (De Boer 2007). The species composition of seagrasses in Talatala area showed that *Halophila ovalis* and *Halodule pinifolia* were widely distributed in all sites irrespective of depth and transparency (Table 14.4 and Fig. 14.8). Mixed beds of *Halophila ovalis* and *Halodule pinifolia* and monotypic meadow of *Halodule* were observed in most sites in the Kumarpur area (Table 14.4 and Fig. 14.8). *Halophila beccarii* was confined to Dhobatutha area, and *Halodule pinifolia* was distributed in Gopakuda, Ghantasila, Somolo, and Dhobatutha (Table 14.4, Fig. 14.8). The southern sector has been shown to possess a stable salinity regime and the highest water transparency throughout the year as there is minimal effect of freshwater discharge on this sector due to large spatial separation from Mahanadi River distributaries (Srichandan et al. 2015a; Tarafdar et al. 2021). The saltwater intrusion from the Rushikulya Estuary through the "Palur canal" promotes the development of healthy seagrass beds in the southern sector (Pattnaik et al. 2020).

4 Management of Macrophytes

Although macrophytes are the critical component of biodiversity, the vigorous growth of macrophytes such as *Phragmites karka*, *Eichhornia crassipes*, and *Salvinia cucullata* adversely affects the biodiversity and growth of other macrophytes (Pattnaik et al. 2020). Therefore, management measures are needed for controlling their spread or eradicating invasive weeds from the natural environment to re-establish the native community. A study on habitat management of avifauna in Nalabana has also demonstrated that mudflats are getting invaded by *Paspalum distichum* and *Salicornia brachiata* making them unsuitable for the congregation of birds such as waders (Balachandran et al. 2020). The high amount of silt load and increased nutrient loading has led to the extensive growth of macrophytes in the northern sector of the lagoon which supports high growth of *Eichhornia crassipes*, *Salvinia cucullata*, *Azolla pinnata*, and *Phragmites karka* (Kumar et al. 2011; Pattnaik et al. 2020).

Phragmites karka is a major threat to the Chilika Lagoon because of its invasive monocultural growth leading to habitat degradation for both flora and fauna (Kumar et al. 2011; Pattnaik et al. 2020). These macrophytes severely impact the sediment flushing, impediment to navigation and fishing, and movement of water birds for foraging. *Phragmites karka* has been shown to grow in a wide salinity range and can sequester nutrients, pesticides, pharmaceuticals, and heavy metals from the wastewater, soil, and sediments (Badejo et al. 2015). Several studies have shown that reeds can be used in the phytoremediation of polluted water bodies and industrial effluents in constructed wetlands (Zhang et al. 2013; Badejo et al. 2015; Toyama et al. 2015; Almuktar et al. 2018; Rai 2018). Common reeds can mediate several ecosystem services such as shore stabilization, biogeochemical cycling, and phytoremediation in natural wetlands. The rhizosphere microbial communities of *Phragmites karka* mediate many biochemical processes, pollutant biodegradation, and supporting reed growth (Almuktar et al. 2018; Behera et al. 2018). Therefore, ecosystem services offered by reeds should be taken into consideration in the weed management plan of wetlands.

5 Conclusion

The study assessed the diversity, distribution, and biomass of macrophytes which would be useful for macrophyte management and conservation in the lagoon. Seagrass meadows were observed in most of the stations from southern and outer channel sectors and in and around Nalabana Island in the central sector. A total of 22 species of macrophytes were recorded from the lagoon. Of these, *Potamogeton nodosus*, *Potamogeton crispus*, *Vallisneria natans*, *Hydrilla verticillata*, *Ceratophyllum demersum*, *Nymphaea pubescens*, *Nymphoides cristata*, *Alternanthera philoxeroides*, *Ipomoea aquatica*, *Spirodela polyrhiza*, *Azolla pinnata*, *Eichhornia crassipes*, *Salvinia cucullata*, *Phragmites karka*, *Halophila ovalis*, *Halodule pinifolia*, *Stuckenia pectinata*, and *Najas indica* were the most abundant macrophytes. The macroalgal species of Chlorophyta (*Chaetomorpha linum*, *Chaetomorpha* sp., *Ulva compressa*, *Ulva flexuosa*, *Ulva intestinalis*, and *Chara braunii*) and Rhodophyta (*Ceramium* sp., *Ceramium diaphanum*, *Gracilaria verrucosa*, *Polysiphonia subtilissima*, and *Polysiphonia sertularioides*) were identified. A total of six species, viz., *Halophila ovalis*, *Halophila beccarii*, *Halophila ovata*, *Halodule pinifolia*, *Halodule uninervis*, and *Ruppia maritima*, were recorded from Chilika Lagoon. Ground survey estimated an area of 169.2 km^2 covered with seagrasses. Periodic seagrass mapping should be used to measure the spread of seagrass meadows for conservation and wetland management. Future studies should focus on the influence of anthropogenic pressures, eutrophication, and climate change on the macrophytes.

Acknowledgments This work was financially supported by the World Bank (Credit No. 4765-IN) and National Plan for Conservation of Aquatic Ecosystems (NPCA) grants to Chilika Development Authority (CDA). We would like to thank Subhasish Pradhan and Bibhuti Bhusan Dora, Research Fellows at GIS Cell, CDA, for making the sampling maps.

References

Abubakr A (2010) Sanative role of macrophytes in aquatic ecosystems. Nat Environ Pollut Technol 9(4):657–662

Ali S, Abbas Z, Rizwan M, Zaheer IE, Yavaş İ, Ünay A, Abdel-Daim MM, Bin-Jumah M, Hasanuzzaman M, Kalderis D (2020) Application of floating aquatic plants in phytoremediation of heavy metals polluted water: a review. Sustainability 12(5):1927

Almuktar SA, Abed SN, Scholz M (2018) Wetlands for wastewater treatment and subsequent recycling of treated effluent: a review. Environ Sci Pollut Res 25(24):23595–23623

Aubry BF, Acri F, Scarpa GM, Braga F (2020) Phytoplankton–macrophyte interaction in the Lagoon of Venice (Northern Adriatic Sea, Italy). Water 12(10):2810

Badejo AA, Sridhar MK, Coker AO, Ndambuki JM, Kupolati WK (2015) Phytoremediation of water using *Phragmites karka* and *Veteveria nigritana* in constructed wetland. Int J Phytoremediation 17(9):847–852

Balachandran S, Pattnaik AK, Gangaiamaran P, Katti T (2020) Avifauna of Chilika, Odisha: assessment of spatial and temporal changes. In: Ecology, conservation, and restoration of Chilika lagoon, India. Springer, Cham, pp 335–363

Balakrishnan S, Ravichandran M, Kaliaperumal N (1992) Studies on the distribution and standing crop of algae at Muthupet estuary, Tamilnadu. Seaweed Res Util 15(1 & 2):63–68

Banerjee K, Sappal SM, Ramachandran P, Ramesh R (2017) Salt marsh: ecologically important, yet least studied blue carbon ecosystems in India. J Clim Change 3(2):59–72

Behera P, Mahapatra S, Mohapatra M, Kim JY, Adyta TK, Raina V, Suar M, Pattnaik AK, Rastogi G (2017) Salinity and macrophyte drive the biogeography of the sedimentary bacterial communities in a brackish water tropical coastal lagoon. Sci Total Environ 595:472–485

Behera P, Mohapatra M, Adhya TK, Suar M, Pattnaik AK, Rastogi G (2018) Structural and metabolic diversity of rhizosphere microbial communities of *Phragmites karka* in a tropical coastal lagoon. Appl Soil Ecol 125:202–212

Behera P, Mohapatra M, Kim JY, Rastogi G (2020) Benthic archaeal community structure and carbon metabolic profiling of heterotrophic microbial communities in brackish sediments. Sci Total Environ 706:135709

Bhasha SKM, Paul JM, Reddy SP (2015) Eco-degradation and phytodiversity of Pulicat lagoon, east coast of southern India. J Ecol Photon 110:429–438

Bhatta K, Patra HK (2018) Distribution of macrophytes in Chilika. Int J Innov Sci Res Technol 3:95–99

Chappuis E, Ballesteros E, Gacia E (2011) Aquatic macrophytes and vegetation in the Mediterranean area of Catalonia: patterns across an altitudinal gradient. Phytocoenologia 41(1):35–44

Christia C, Papastergiadou ES (2007) Spatial and temporal variations of aquatic macrophytes and water quality in six coastal lagoons of western Greece. Belgian J Bot 140:39–50

Christia C, Giordani G, Papastergiadou E (2018) Environmental variability and macrophyte assemblages in coastal lagoon types of western Greece (Mediterranean Sea). Water 10(2):151

Cragg SM, Friess DA, Gillis LG, Trevathan-Tackett SM, Terrett OM, Watts JE, Distel DL, Dupree P (2020) Vascular plants are globally significant contributors to marine carbon fluxes and sinks. Annu Rev Mar Sci 12:469–497

Dar NA, Pandit AK, Ganai BA (2014) Factors affecting the distribution patterns of aquatic macrophytes. Limnol Rev 14(2):75–81

De Boer WF (2007) Seagrass–sediment interactions, positive feedbacks and critical thresholds for occurrence: a review. Hydrobiologia 591(1):5–24

Dhote S (2007) Role of macrophytes in improving water quality of an aquatic eco-system. J Appl Sci Environ Manag 11(4):133–135

dos Santos Fonseca AL, Marinho CC, de Assis Esteves F (2015) Aquatic macrophytes detritus quality and sulfate availability shape the methane production pattern in a dystrophic coastal lagoon. Am J Plant Sci 6(10):1675

English S, Wilkinson C, Baker V (1994) Survey manual for tropical marine resources. Australian Institute of Marine Science, Townsville

Farias DR, Hurd CL, Eriksen RS, Macleod CK (2018) Macrophytes as bioindicators of heavy metal pollution in estuarine and coastal environments. Mar Pollut Bull 128:175–184

Faußer AC, Hoppert M, Walther P, Kazda M (2012) Roots of the wetland plants *Typha latifolia* and *Phragmites australis* are inhabited by methanotrophic bacteria in biofilms. Flora Morphol Distrib Funct Ecol Plants 207(11):775–782

Fertig B, Carruthers TJB, Dennison WC, Jones AB, Pantus F, Longstaff B (2009) Oyster and macroalgae bioindicators detect elevated δ 15 N in Maryland's coastal bays. Estuar Coasts 32(4):773–786

Ganesan M, Trivedi N, Gupta V, Madhav SV, Reddy CR, Levine IA (2019) Seaweed resources in India–current status of diversity and cultivation: prospects and challenges. Bot Mar 62(5):463–482

Gedan KB, Silliman BR, Bertness MD (2009) Centuries of human-driven change in salt marsh ecosystems. Annu Rev Mar Sci 1:117–141

Geevarghese GA, Akhil B, Magesh G, Krishnan P, Purvaja R, Ramesh R (2018) A comprehensive geospatial assessment of seagrass distribution in India. Ocean Coast Manag 159:16–25

Gopalakrishnan A, Divya PR, Basheer VS, Rajaswaminathan T, Kathirvelpandian A, Bineesh KK, Kumar RG, Jena JK (2012) Macro flora & fauna of the Gulf of Mannar a checklist. National Bureau of Fish Genetic Resources, Lucknow

Haines HH (1921–1925) The botany of Bihar and Orisha, 6 parts. Botanical Survey of India, Calcutta, London

He Q, Silliman BR (2019) Climate change, human impacts, and coastal ecosystems in the anthropocene. Curr Biol 29(19):1021–1035

Jagtap TG, Inamdar SN (1991) Mapping of seagrass meadows from the Lakshadweep islands (India), using aerial photographs. J Indian Soc Remote Sens 19(2):77–82

Jagtap TG, Bhosale SH, Nagle VL (2002) Ecological observations on major *Salicornia* beds from highly saline coastal wetlands of India. Wetlands 22(3):443–450

Jaikumar M, Chellaiyan D, Kanagu L, Kumar PS, Stella C (2011) Distribution and succession of aquatic macrophytes in Chilika Lake-India. J Ecol Nat Environ 3(16):499–508

Javed MT, Tanwir K, Akram MS, Shahid M, Niazi NK, Lindberg S (2019) Phytoremediation of cadmium-polluted water/sediment by aquatic macrophytes: role of plant-induced pH changes. In: Cadmium toxicity and tolerance in plants. Academic Press, London, pp 495–529

Josephine M, Usha R, Maria Victorial Rani S (2013) Current status of seaweed diversity and their seasonal availability at Hare island, Gulf of Mannar. Sci Res Rep 3:146–151

Jyothi PV, Sureshkumar S (2014) Preliminary documentation of aquatic macrophytes of kole wetlands of northern Kerala, India. Int J Environ Sci 5(1):117–122

Kar SK, Tripathy PK, Mohanty SK, Acharya PK, Panda PC (2017) Additions to the flora of Chilika lake and its immediate neighbourhood. J Econ Taxon Bot 40(3):134–150

Keren TL, Inbaraj RM (2013) Spatial and temporal variation in the environmental parameters and its impact on the seagrass and associated macro fauna of Pulicat. The Ecoscan 7(3):115–121

Kim JY, Rastogi G, Do Y, Kim DK, Muduli PR, Samal RN, Pattnaik AK, Joo GJ (2015) Trends in a satellite-derived vegetation index and environmental variables in a restored brackish lagoon. Glob Ecol Conserv 4:614–624

Kim JK, Yarish C, Hwang EK, Park M, Kim Y (2017) Seaweed aquaculture: cultivation technologies, challenges and its ecosystem services. Algae 32(1):1–13

Kumar R, Pattnaik A, Pattnaik P (2011) Management of *Phragmites karka* invasion in Chilika lake, Orissa. In: Consultation workshop on management of *Phragmites karka* invasion in Chilika, 17–18 Jan

Kumar A, Mishra DR, Equeenuddin SM, Cho HJ, Rastogi G (2017) Differential impact of anniversary-severe cyclones on the water quality of a tropical coastal lagoon. Estuar Coasts 40(2):317–342

Liu C, Liu Y, Lu Y, Liao Y, Nie J, Yuan X, Chen F (2019) Use of a leaf chlorophyll content index to improve the prediction of above-ground biomass and productivity. PeerJ 6:e6240
Macreadie PI, Jarvis J, Trevathan-Tackett SM, Bellgrove A (2017) Seagrasses and macroalgae: importance, vulnerability and impacts. In: Climate change impacts on fisheries and aquaculture: a global analysis. Wiley-Blackwell, Chichester, pp 729–770
Mahapatro D, Mishra RK, Samal RN, Patanaik AK (2012) Study of macrobenthos in relation to eutrophication at Chilika lagoon, east coast of India. Mar Sci 2:139–148
Malathi G, Arunprakash S, Kumar MA, Boopathy AB, Rama P (2018) Diversity, and distribution of macroalgae in the Gulf of Mannar. Int J Pharma Res Health Sci 6(1):2264–2268
Mathews G, Diraviyaraj K, Thinesh T, Edward JP, Wilhemsson D (2010) Status of seagrass diversity, distribution and abundance in Gulf of Mannar marine national park and Palk bay (Pamban to Thondi), southeastern India. South Indian Coast Mar Bull 2(2):1–21
Mishra AK, Apte D (2021) The current status of *Halophila beccarii*: an ecologically significant, yet vulnerable seagrass of India. Ocean Coast Manag 200:105484
Mohanty D, Adhikary SP (2013) Assessment of changes in the algal diversity of Chilika lagoon after opening of new mouth to Bay of Bengal. J Water Resour Prot 5:611–623
Mohanty RK, Mohapatra A, Mohanty SK (2009) Assessment of the impacts of a new artificial lake mouth on the hydrobiology and fisheries of Chilika lake, India. Lakes Reserv Res Manag 14(3):231–245
Nair NB, Arunachalam M, Azis PK, Dharmaraj K, Kumar KK (1983) Ecology of Indian estuaries: part 2—Ecology of seagrass bed of *Halophila ovalis* (Hook) in the Ashtamudi estuary, SW coast of India
Nobi EP, Dilipan E, Sivakumar K, Thangaradjou T (2011) Distribution and biology of seagrass resources of Lakshadweep group of islands, India. Indian J Geomarine Sci 40(5):624–634
Obrador B, Pretus JL, Menéndez M (2007) Spatial distribution and biomass of aquatic rooted macrophytes and their relevance in the metabolism of a mediterranean coastal lagoon. Sci Mar 71(1):57–64
Olette R, Couderchet M, Biagianti S, Eullaffroy P (2008) Toxicity and removal of pesticides by selected aquatic plants. Chemosphere 70(8):1414–1421
Özbay H, Yaprak AE, Turan N (2019) Assessing water quality in the Ceyhan river basin (Turkey) with the use of aquatic macrophytes. Chem Ecol 35(10):891–902
Panda PC, Pattnaik AK, Rath J, Patnaik SN (2002) Flora of Chilika lake and its immediate neighbourhood: a checklist. J Econ Taxon Bot 26(1):1–20
Pati MP, Nayak L, Sharma SD (2014a) Studies on biomass of seagrass, seaweed and its associated fauna from Chilika lagoon. Int J Environ Sci 5(2):423–431
Pati MP, Nayak L, Sharma SD, Sahoo J, Behera DP (2014b) Studies on seagrasses in relation to some environmental variables from Chilika lagoon, Odisha, India. Int J Environ Sci 3(11):92–101
Patro S, Krishnan P, Samuel VD, Purvaja R, Ramesh R (2017) Seagrass and salt marsh ecosystems in South Asia: an overview of diversity, distribution, threats and conservation status. In: Wetland Science, pp 87–104
Pattnaik R, Tarafdar L, Mohapatra M, Behera P, Mohanty P, Rastogi G (2019) Biodiversity of Chilika through the lens of foldscope. J Emerg Technol Innov Res 6:865–874
Pattnaik AK, Panda PC, Rastogi G (2020) Survey, characterization, ecology, and management of macrophytes in Chilika lagoon. In: Ecology, conservation, and restoration of Chilika Lagoon, India. Springer, Cham, pp 415–438
Pereira SA, Trindade CRT, Albertoni EF, Palma-Silva C (2012) Aquatic macrophytes as indicators of water quality in subtropical shallow lakes, southern Brazil. Acta Limnol Bras 24(1):52–63
Phan TH, Stiers I, Nguyen TH, Pham TT, Ton TP, Luong QD, Triest L (2018) Spatial and temporal distribution of submerged aquatic vegetation in a tropical coastal lagoon habitat in Viet Nam. Bot Mar 61(3):213–224
Porra RJ (2002) The chequered history of the development and use of simultaneous equations for the accurate determination of chlorophylls a and b. Photosynth Res 73(1–3):149–156

Prado P, Caiola N, Ibáñez C (2013) Spatio-temporal patterns of submerged macrophytes in three hydrologically altered Mediterranean coastal lagoons. Estuar Coasts 36(2):414–429

Rai PK (2018) Phytoremediation of emerging contaminants in wetlands. CRC Press, Boca Raton

Rajyalakshmi E, Basha SKM (2016) Wetland flora of Pulicat lake southern coastal wetland of SPSR Nellore district of Andhra Pradesh, India. Int J Sci Eng Res 7(2):1333–1346

Ramesh R, Banerjee K, Paneerselvam A, Raghuraman R, Purvaja R, Lakshmi A (2019) Importance of seagrass management for effective mitigation of climate change. In: Coastal management. Academic Press, London, pp 283–299

Rath J, Adhikary SP (2008) Biodiversity assessment of algae in Chilika lake, east coast of India. In: Monitoring and modelling lakes and coastal environments. Springer, Dordrecht, pp 22–33

Regional Plant Resource Centre (2016) Final technical report on "Qualitative and quantitative assessment of biological diversity and distribution pattern of macrophytes (angiospermic plants) of Chilika lagoon and its adjoining regions" under Integrated Coastal Zone Management Project

Rosqvist K (2010) Distribution and role of macrophytes in coastal lagoons: implications of critical shifts. Doctoral thesis, University of Åbo, Finland

Rout J, Ojha A, Srichandan S, Tripathy M, Bhatt KS, Samal RN, Pattnaik AK, Rajesh G (2014) Monitoring aquatic vegetation distribution in Chilika lagoon with G-tech. J Space Sci Technol 3(1):17–23

Sahoo D, Sahu N, Sahoo D (2003) A critical survey of seaweed diversity of Chilika lake, India. ALGAE-INCHON 18(1):1–12

Shaw BP, Rout NP, Barman BC, Choudhury SB, Rao KH (2000) Distribution of macrophytic vegetation in relation to salinity in the Chilika lake, a lagoon along east coast of India. Indian J Mar Sci 29:144–148

Short F, Carruthers T, Dennison W, Waycott M (2007) Global seagrass distribution and diversity: a bioregional model. J Exp Mar Biol Ecol 350(1–2):3–20

Singh MM (2016) Phytoremediation measures for heavy metal contamination of Loktak lake, Manipur, India. Doctoral dissertation, Mizoram University

Solimini AG, Cardoso AC, Heiskanen AS (2006) Indicators and methods for the ecological status assessment under the Water Framework Directive. Linkages between chemical and biological quality of surface waters. Joint Research Centre, European Commission, pp 1–262

Srichandan S, Kim JY, Bhadury P, Barik SK, Muduli PR, Samal RN, Pattnaik AK, Rastogi G (2015a) Spatiotemporal distribution and composition of phytoplankton assemblages in a coastal tropical lagoon: Chilika, India. Environ Monit Assess 187(2):47

Srichandan S, Kim JY, Kumar A, Mishra DR, Bhadury P, Muduli PR, Pattnaik AK, Rastogi G (2015b) Interannual and cyclone-driven variability in phytoplankton communities of a tropical coastal lagoon. Mar Pollut Bull 101(1):39–52

Srivastava J, Gupta A, Chandra H (2008) Managing water quality with aquatic macrophytes. Rev Environ Sci Biotechnol 7(3):255–266

Takamura N, Kadono Y, Fukushima M, Nakagawa M, Kim BH (2003) Effects of aquatic macrophytes on water quality and phytoplankton communities in shallow lakes. Ecol Res 18(4):381–395

Tarafdar L, Kim JY, Srichandan S, Mohapatra M, Muduli PR, Kumar A, Mishra DR, Rastogi G (2021) Responses of phytoplankton community structure and association to variability in environmental drivers in a tropical coastal lagoon. Sci Total Environ 783:146873

Thangaradjou T, Bhatt JR (2018) Status of seagrass ecosystems in India. Ocean Coast Manag 159:7–15

Thomaz SM, Cunha ERD (2010) The role of macrophytes in habitat structuring in aquatic ecosystems: methods of measurement, causes and consequences on animal assemblages' composition and biodiversity. Acta Limnol Bras 22(2):218–236

Ting WHT, Tan IAW, Salleh SF, Wahab NA (2018) Application of water hyacinth (Eichhornia crassipes) for phytoremediation of ammoniacal nitrogen: a review. J Water Proc Eng 22:239–249

Toyama T, Nishimura Y, Ogata Y, Sei K, Mori K, Ike M (2015) Effects of planting *Phragmites australis* on nitrogen removal, microbial nitrogen cycling, and abundance of ammonia-oxidizing and denitrifying microorganisms in sediments. Environ Technol 37(4):478–485

Trias R, Ruiz-Rueda O, García-Lledó A, Vilar-Sanz A, López-Flores R, Quintana XD, Hallin S, Bañeras L (2012) Emergent macrophytes act selectively on ammonia-oxidizing bacteria and archaea. Appl Environ Microbiol 78(17):6352–6356

Umamaheswari R, Ramach S, Nobi EP (2009) Mapping the extend of seagrass meadows of Gulf of Mannar biosphere reserve, India using IRS ID satellite imagery. Int J Biodivers Conserv 1(5):187–193

Vaheeda KK, Thara KS (2014) Aquatic and semi-aquatic macrophytic flora of brackish waters of Kodungallur, Thrissur Dist, Kerala. Ann Plant Sci 3(06):748–751

Vladár P, Rusznyák A, Márialigeti K, Borsodi AK (2008) Diversity of sulfate-reducing bacteria inhabiting the rhizosphere of *Phragmites australis* in Lake Velence (Hungary) revealed by a combined cultivation-based and molecular approach. Microb Ecol 56(1):64–75

Zhang W, Wu X, Liu G, Chen T, Zhang G, Dong Z, Yang X, Hu P (2013) Pyrosequencing reveals bacterial diversity in the rhizosphere of three *Phragmites australis* ecotypes. Geomicrobiol J 30(7):593–599

Chapter 15
Spatial Identification of Vulnerable Coastal Ecosystems for Emerging Pollutants

Anuradha Kumari, Rahul Harshawardhan, Jyoti Kushawaha, and Ipsita Nandi

Abstract Coastal ecosystems play a crucial role in maintaining ecosystem services. These also harbor diverse groups of flora and fauna. Increased anthropogenic activities are degrading coastal ecosystem at a very fast pace. This in turn is adversely affecting species biodiversity as well as impacting human health and well-being. Among various pollutants affecting coastal ecosystem, certain contaminants known as emerging pollutant are causing great loss to its services and biodiversity. These contaminants are given undue concern in the past but are adversely affecting humans and marine biodiversity. These contaminants require different strategies for their detection, impact, as well as management. Hence it is required to have a complete insight into source, chemistry, and potential impact of these pollutants. In this chapter, a vulnerability map is created for the states along Indian coastline based on their potential sources and population of states. It was also observed that a wide knowledge gap exists among different coastal states regarding the occurrence of emerging pollutant. This study might act as an eye-opener for scientific community toward existing knowledge gap and further direct toward their investigation and management.

Keywords Coastal ecosystem · Anthropogenic activities · Population · Emerging pollutants · Knowledge gap

A. Kumari · J. Kushawaha
School of Environmental Science, Jawaharlal Nehru University, New Delhi, India

R. Harshawardhan
Office of Registrar General of India, Ministry of Home Affairs, New Delhi, India

I. Nandi (✉)
Institute of Environmental Science and Sustainable Development, Banaras Hindu University, Varanasi, India

S. Madhav et al. (eds.), *Coastal Ecosystems*, Coastal Research Library 38,
https://doi.org/10.1007/978-3-030-84255-0_15

1 Introduction

Oceans play vital role in providing valuable ecosystem services and planetary resilience to more than 70% of the Earth's surface (Heino et al. 2020). Coastal ecosystems play crucial role in global sustainability (Lu et al. 2018). Coastline areas gained huge importance due to its role in tourism, productivity, provisioning services, industrial development, transportation, waste disposal, petroleum production, and many other economic activities (Dudgeon 2014; Rani et al. 2015). These ongoing anthropogenic activities along the coast result in dynamic state of these ecosystems for equilibrium maintenance (Dudgeon 2014). Numerous developmental activities like urbanization and industrialization along the river basins and watersheds have severely threatened the coastal system in the form of sea level rise, ocean acidification, coastal erosion, etc. (Thompson 2014). Consequently, many of the world's coastal regions are under huge stress directly affecting their economic and ecological services (Kummu et al. 2016). United Nations Sustainable Development Goal 14, Life Below Water, and its 7 targets emphasize sustainable use and conservation of marine biodiversity while exploiting oceans, seas, and marine resources for sustainable development. It also emphasizes on actions for making oceans and seas more resilient and productive (Mugagga and Nabaasa 2016).

Indian coastal ecosystems are given due importance owing to high productivity, numerous ecological services, and densely populated coastal areas (Ayyam et al. 2019). However, changing scenarios such as global climate change, anthropogenic activities, global warming, mean sea level rise, natural calamities, and pollution are acting as multiple stressors increasing the threat and vulnerability of Indian coastal ecosystem (Dudgeon 2014).

Severe degradation of coastal ecosystems in recent times has adversely affected the global commercial market of marine and coastal fisheries (Dudgeon 2014). Though emerging organic pollutants have been detected in trace amount near coastal zones yet are harmful enough to adversely affect the aquatic species composition (Omar et al. 2018). Emerging pollutants are given due importance in recent times owing to their high rate of consumption estimated in tons per year (Petrie et al. 2015).

Emerging pollutants may be defined as "any synthetic or naturally occurring chemical or any microorganism that is not commonly monitored in the environment but has the potential to enter the environment and cause known or suspected adverse ecological and/or human health effects" (USGS 2017). Furthermore, a pollutant may be considered as "emerging" if it uses a new source or pathway for adversely affecting humans or employ a new detection method or a new treatment technology (DoD 2009).

Emerging pollutants have the property of altering the hormonal balance of the organism's endocrine system (Benotti et al. 2009; Nam et al. 2014). The alterations occur by blockage of the hormonal action through completion with the hormone receptor, mimicking or impersonating the endogenous hormones, or by decrease or increase in the level of hormonal activity (Bila and Dezotti 2003).

Emerging pollutants are usually found in various daily used products like resins, drugs, pesticides, plastics, cosmetics, detergents, fragrances, personal care products, and more (Pal et al. 2010; Zandaryaa and Frank-Kamenetsky 2015).

Additionally, emerging pollutants have drastically increased the threat against aquatic pollution as they are added into the environment through random nonpoint sources (Islam and Tanaka 2004). Because of their harmful impacts, this new generation pollutants on the ecosystem needs to be dealt with enhanced knowledge regarding their origin, transformation, and suitable mitigation strategies for their sustainable management (Gavrilescu et al. 2015). There is an immediate need to emphasize more on the emerging pollutants to mitigate coastal pollution. Therefore, mitigation and control of aquatic pollution and conservation of its species requires an in depth study regarding source, use, composition, and chemical nature of the emerging pollutants.

In purview of the above, this chapter aims to spatially identify the potential vulnerable areas for emerging pollutants affecting Indian coastal ecosystem. The chapter is dealt with following subheadings: (1) understanding significance of the coastal ecosystem; (2) insights to emerging pollutants, their sources, pathways, and impacts, (3) ecotoxicity of emerging pollutants; (4) global trend in coastal pollution; and (5) spatial identification of vulnerable areas of Indian coastal ecosystem.

2 Understanding the Coastal Ecosystem as a Prime Marine Resource

The coastal ecosystems are regions of very high productivity and accessibility (Friess et al. 2020). They are identified as encompassing broad range of habitat types and nurturing huge species and genetic diversity (Folke et al. 1998). Coastal ecosystem provides a wide array of goods and services with high economic and ecological value (Lewis et al. 2020). The goods and products from coastal and marine ecosystem include food supply for humans and aquatic animals like fish, krill, shellfish, etc.; minerals and oil resources; salt; construction materials like sand, coral, rock, wood, and lime; and biodiversity and genetic stock for various medicinal and biotechnological applications (World Resources Institute 2001). India is home to very wide range of coastal ecosystem, viz., mangroves, estuaries, lagoons, backwaters, rocky coasts, coral reefs, salt marshes, and sandy stretches (Rani et al. 2015). Elevated rate of loss of coastal and marine biodiversity over the past few decades has been identified as a matter of great global concern (Ravindran 2012). The ecological services provided by the coastal ecosystem can be broadly classified into five groups, namely, shoreline stabilization, biodiversity, water quality, food production, and recreation and tourism.

2.1 *Shoreline Stabilization*

Coastlines are constantly undergoing the process of erosion and accretion due to irregular and routine forces caused by waves, winds, tectonic processes, and storms (Mujabar and Chandrasekar 2013). In addition, the natural shoreline undergoes changes in response to the above mentioned forces and events like floods, tides, storms, fluctuations in sea levels, and human interventions in terms of developmental activities near coastal regions (Passeri et al. 2015). The coastal ecosystem enhances and facilitates shoreline stabilization and buffering services (Barbier et al. 2011). For instance, coral reefs, kelp beds, mangroves, and seagrasses reduce erosion due to the mitigating waves (Barbier et al. 2011). The rocky and sandy shores provide defense action against natural forces like strong winds and waves (Rahman and Rahman 2015). Moreover, the wetlands, mangroves, and seagrasses facilitate stabilization of soils by reducing sediment pollution (Barbier et al. 2011). The excessive exploitation of shorelines for economic activities like transportation, recreation, industries, and residential developments has led to profound impact on the coastal ecosystem and energy, material, and chemical cycles in the near-shoreline environment (Sundblad and Bergström 2014; Williams et al. 2018). Further developmental activities such as construction of dams altering river flow have disrupted natural sediment movement in adjacent shoreline areas, thereby accelerating shoreline erosion (Sundblad and Bergström 2014). Severe economic losses from disrupted shoreline have attracted the attention of policy makers leading to global mitigation effort for shoreline stabilization (Aminti et al. 1999; Crooks et al. 2011; Bilkovic et al. 2016; Guilfoyle et al. 2019).

2.2 *Biodiversity*

The marine biodiversity hosts broad range of species (Rishworth et al. 2020). Literature states that out of cataloged 1.7 million species nearly 250,000 are marine species (Heywood and Watson 1995). Furthermore among the 33 major animal phyla that include major kinds of organism, 32 are present in marine ecosystem, and of this 15 are exclusive to marine environment (Winston 1992; Norse 1993). Diverse range of marine organisms serves as source of broad range of medicines for bone growth and healing (Carson and Clarke 2018), dietary supplements (Barkia et al. 2019), antioxidants (Hamidi et al. 2020), anticancer drugs (Khalifa et al. 2019), as well as numerous biochemical products (Adnan et al. 2018; Reher et al. 2020).

2.3 *Water Quality*

Coastal ecosystems maintain water quality by absorbing nutrient inputs, filtering and degrading toxic contaminants, and regulating pathogens (Smith et al. 2013). Anthropogenic actions like deforestation of mangroves, conversions of wetlands, or destruction of seagrass beds severely degrade the capacity of coastal ecosystem to serve their ecological services (Pendleton et al. 2012). Additionally several contaminants entering coastal water system cause bioaccumulation and biomagnification of persistent chemical contaminants or emerging pollutants (Jitar et al. 2015;

Kim et al. 2020). This in turn disturbs the ecosystem balance due to high morbidity or mortality (Analuddin et al. 2017; Kim et al. 2020). Moreover, diseased or contaminated seafood, like fishes, crabs, shellfish, etc., causes various forms of lethal effect on human beings (Marques et al. 2010). Accelerated concentrations of pathogens in water further lead to deterioration of water quality causing several health implications on humans as well as economically important aquatic organisms (Marques et al. 2010).

2.4 *Food Production*

Seafood is one of the most consumed diets across the globe leading to high economic significance (Gephart et al. 2020). Fish and shellfish production owing to high degree of essential nutrient content plays a crucial role in ensuring food security (Youn et al. 2014). Studies indicate that more than 90% of the fish catch are derived from coastal ecosystems and relatively less percent of fishes came from the open ocean ecosystem (Hinrichsen 1999; Sherman 1993). Majority of the population living in developing nations are dependent on fish as their primary source of animal protein (Williams 1996). Out of 30 fish-dependent countries, 4 belong to developing nation clearly indicating the importance of seafood from coastal ecosystem (Laureti 1999). Furthermore, fish production overturns other major meat production in developing nation (Williams 1996). This is further substantiated by the overexploitation of Indian Ocean for fish production (Grainger and Garcia 1996; Lecomte et al. 2017).

2.5 *Recreation and Tourism*

Recreational and tourism industry is the fastest growing area of the global economy (Sofronov 2017). The coastal and marine ecosystem promotes growth of global tourism enhancing the national and international economy (Sutton-Grier et al. 2015). Coastal tourism has the highest share in GDP of many countries (Tan and Huang 2020; Pafi et al. 2020; Wei and Zhao 2020; Cherkasov et al. 2017). Coastal regions are the hotspots for tourism due to its high aesthetic values (Díaz-Asencio et al. 2011). The Mediterranean or the Caribbean coastal regions witness high economic gain from large inflow of tourist in the summer months (Gössling et al. 2018). Moreover, islands like Malta, Cyprus, the Balearic Islands, and Sicily have tourism as their main economic activity (Yunis 2001). Furthermore, excess use of coast for tourism has resulted into problems such as water pollution, loss of mangroves, land use and land cover changes, introduction of invasive species, overexploitation of resources, and industrial development (Díaz-Asencio et al. 2011). It is estimated that by the year 2020, the Mediterranean coast will amount to 346 million tourist arrivals which have environmental implications in the form of huge pressure on the coastal ecosystems (Yunis 2001). Tourism sector is known to provide quality jobs to majority of population (Achilov 2017). However, sustainable tourism is needed for the coastal zone to enhance their economic development and for their conservation (Le Tissier 2020).

3 Pollution in Coastal Ecosystem

The disposal of different types of waste into the oceans is a major source of pollution of coastal ecosystem (Borja et al. 2020). Dumping of the waste along the coast makes the coastal ecosystem vulnerable to various forms of health hazardous contaminants affecting both human and its biodiversity (Phelan et al. 2020). Human activities and mobilization of nutrients into different compartments of the hydrosphere have led to the high concentrations of phosphorus and nitrogen into oceans (Cloern 2001). Nutrient pollution leads to several problems such as eutrophication, algal bloom, loss of seagrass, dead zones, killing fishes, coral reef destruction, and death of seabirds and marine mammals (Howarth et al. 2000).

Change in pollution trend revealed major proportion of marine and coastal environment pollution is contributed by runoff pollutants from the land surface (Vikas and Dwarakish 2015). Nonpoint runoff pollutants are contributed from septic tanks, farms, forest areas, oils from the vehicle engines onto roads, and washed parking lots (Vikas and Dwarakish 2015). Microbial pollution of coastal ecosystem caused by warm-blooded animals and humans suggests that high population density near coastline areas created high environmental and human health risk (Mallin et al. 2001). In recent times, major constituent of coastal pollution is different types of emerging pollutants (González-Acevedo et al. 2019; Farré 2020). Consequently, most of the aquatic species are facing severe threat of pollution and in some cases extinction (Dias et al. 2019; Jabado et al. 2018).

4 Emerging Pollutants

The challenge of emerging pollutants has gained global consideration owing to its severe consequences on human health and environment. Emerging pollutants though are least monitored yet have potential to cause severe adverse ecological and human health implications (Geissen et al. 2015). Limited information with disharmonized sampling and analysis method makes managing these pollutants a tedious task (Geissen et al. 2015; Richardson and Kimura 2016). Furthermore, these compounds having nonpoint source of origin hence usually get unnoticed and undetected causing serious threats to coastal ecosystems (Sorensen et al. 2015). Usually emerging pollutants are introduced in the system by discharge of municipal, industrial, or pharmaceutical waste or surface runoff from agricultural areas (Verlicchi et al. 2012; Wolf et al. 2012; Duong et al. 2008; Sidhu et al. 2013).

4.1 *Types of Emerging Pollutants*

Amidst presence of large numbers of emerging pollutants, only few are toxicologically studied (Thomaidis et al. 2012; Noguera-Oviedo and Aga 2016; Richardson and Kimura 2019). Of these, six most abundantly found emerging pollutants of Indian coastal ecosystem are illustrated in Fig. 15.1 and are detailed in the following sections.

4.1.1 Dichlorodiphenyltrichloroethane (DDT)

Insecticides, pesticides, and their residuals are classified as one of the most devastating agents for aquatic ecosystem affecting different trophic levels (Duursma and Marchand 1976). DDT (1,1,1-trichloro-2,2-bis(4-chlorophenyl)ethane) is probably the best known and most useful organochlorine insecticide in the world (Mansouri et al. 2017). DDT is a widely used agrichemical as well as used for vector control for diseases like dengue, kala-azar, and malaria (Wilson et al. 2020; Mansouri et al. 2017; Van Den Berg et al. 2017). With long half-life period, DDT is one of the most persistent environmental pollutant (Mansouri et al. 2017). Biomagnification properties of DDT causes several adverse impacts such as thinning of eggshell in avian population, nervous system breakdown, liver fatigue, and carcinogenic effect on mammalian system (WHO 1979; Pavlikova et al. 2020).

4.1.2 Diclofenac

Pharmaceuticals are compounds extensively used in medicine, agriculture, drugs, antibiotics, and hormones (Lonappan et al. 2016). Pharmaceutically active compounds (PhACs) enter the environment by one route or another (Lonappan et al. 2016). Among PhACs, nonsteroidal anti-inflammatory drugs (NSAIDs) are widely

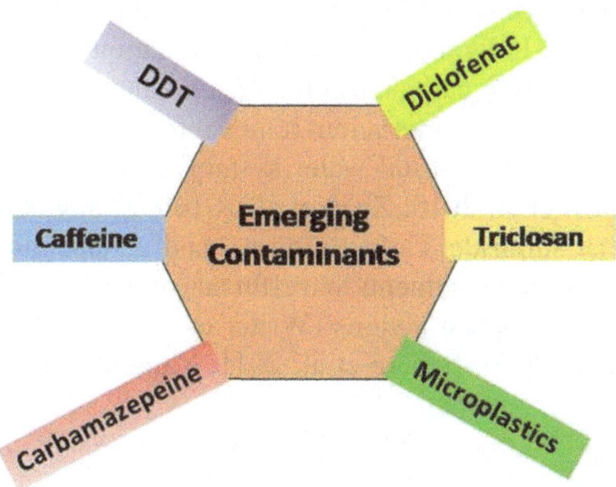

Fig. 15.1 Six most abundantly found emerging pollutants along the coast of India

used globally. These are detected at concentrations ranging from ng/L to low mg/L in different environmental compartments (Khetan and Collins 2007). Diclofenac is the most commonly used NSAID as pain killer in arthritis or acute injury. It also works as antiuricosuric and analgesic (McGettigan and Henry 2013). Diclofenac was discovered by Ciba-Geigy AG, a Swiss pharmaceutical company, in 1973, and its chemical name is 2-(2,6-dichloranilino) phenylacetic acid (Lonappan et al. 2016). Diclofenac is frequently detected in rivers, sediments, and sludges owing to inefficient treating systems finding its way to marine ecosystem (Kunkel and Radke 2012; Langford et al. 2011). Potential harmful effects to marine fauna at significant concentration include damaging renal and gastrointestinal tissue, induced lipid peroxidation (LPO), and tissue damage (Cleuvers 2004; Oaks et al. 2004; Schmidt et al. 2011).

However, the toxicity of diclofenac in the environment is poorly understood and needs further investigations (Lonappan et al. 2016).

4.1.3 Triclosan

Triclosan is an antimicrobial chemical and key ingredient in personal care product formulations like soaps, deodorants, shampoos, creams, moisturizers, lotions, face wash gels, cosmetics, etc. (Azeem et al. 2008). Several studies reported incomplete removal of triclosan from the wastewater treatment plant finding their way to marine environment (Ramaswamy et al. 2011; Zhao et al. 2010a). These antimicrobial and antifungal compounds are recognized as emerging pollutants of great concern because of their potential impact on changing species diversity (Tran et al. 2018).

4.1.4 Carbamazepine

Carbamazepine is a form of neuroactive drugs used as an antiepileptic and antipsychotic drug for treatment of depression in patients having epilepsy (Brodie et al. 2016). Around 11% of the US population consume antidepressant medications, and these drugs are third most prescribed for the age group of 18–44 years (Pratt et al. 2011). India is the second largest nation in Asian continent (approx 115.5 tons) in consuming carbamazepine (Zhang and Geissen 2010). Because of their high consumption and inefficient removal from wastewater treatment plants, these drugs find way to the coastal water systems as emerging pollutants (Ginebreda et al. 2010; Gros et al. 2010; Zhao et al. 2010b; Zhang et al. 2008). Some studies have shown that sulpiride, carbamazepine, and gabapentin are dominant and widely detected drugs in the influents and effluents of wastewater treatment plants of North American and European nations (Writer et al. 2013; Subedi et al. 2015; Kasprzyk-Hordern et al. 2009; Behera et al. 2011; Ying et al. 2017; Tran and Gin 2017; Gurke et al. 2015; Sun et al. 2016).

4.1.5 Microplastics

Microplastics are plastics made up of polystyrene (PS), polyethylene (PE), and polypropylene (PP) polymers having diameter of less than 5 mm (Law and Thompson 2014). Microplastics are major constituents in cleansers (Gregory 1996), scrubs (Fendall and Sewell 2009), toothpastes (Sharma and Chatterjee 2017), hand wash soaps (Napper et al. 2015), and biomedical products (Shi et al. 2009).

The plastic particles of varied size like nano, micro, and macro alter coastal ecosystem affecting human health (Hwang et al. 2020). Sources of microplastics are diverse, and it includes biomedical products, drinking water, food containers, and single-use plastic bottles, facial scrubs, and many more (Storck et al. 2015; Bruck and Ford 2018; Sussarellu et al. 2016; Schymanski et al. 2018). It is estimated that a typical daily use of exfoliating facial scrub of 5 ml quantity contains 4594 to 94,500 microplastic particles (Napper et al. 2015; Gregory 1996). These microplastic particles potentially pass into sewage system, and about 25% get filtered out of wastewater treatment plants (Napper et al. 2015; Carr et al. 2016). The impact of ingestion of microplastics has been studied on different aquatic species like fish and shellfish (Smith et al. 2018; Van Cauwenberghe and Janssen 2014). It was reported that over 20% of individual shellfish and fish have plastic debris and fibers into their gastrointestinal tract (Rochman et al. 2015).

Microplastics have emerged as pollutant severely affecting humans and aquatic environment as they cannot be digested; hence, their aggregation leads to gastrointestinal dysmotility or obstruction (Hwang et al. 2020).

4.1.6 Caffeine

Caffeine (1,3,7-trimethylxanthine) is the most consumed psychoactive substance in the world (de Paula and Farah 2019). Caffeine widely used as a diuretic, respiratory and cerebral stimulant as well as in food as beverages (Patay et al. 2017). Excess intake of caffeine results in "caffeinism," a syndrome characterized by a range of adverse reactions such as restlessness, nervousness, anxiety, irritability, agitation, muscle tremor, insomnia, headache, diuresis, tachycardia, arrhythmia, pulse irregularity and increased frequency, elevated respiration and gastrointestinal disturbances (e.g., nausea, vomiting, diarrhea), severe emesis, photophobia, palpitations, muscle twitching, convulsions, and unconsciousness (de Paula and Farah 2019). It is a potential chemical indicator for municipal wastewater pollution (Ogunseitan 1996; Piocos and De la Cruz 2000; Standley et al. 2000; Barber et al. 1996; Meade 1995; Seiler et al. 1999; Siegener and Chen 2002). Globally caffeine has been detected in surface water, wastewater, and groundwater (Mutiyar and Mittal 2012). Details of above discussed emerging pollutants with their potential sources and impacts are listed in Table 15.1.

Table 15.1 Details of emerging pollutants, their potential sources and adverse impacts on ecosystem and humans

Sl. no.	Pollutants	Sources	Impact on human	Impact on environment	References
1.	DDT	Agriculture (runoff)	Death, cancer, tumor	Affecting reproductive ability of top carnivore population through biomagnification	Duursma and Marchand (1976), Bradman et al. (1997), Rogan and Chen (2005), Mansouri et al. (2017)
2.	Diclofenac[a]	Nonsteroidal anti-inflammatory drugs (municipal wastewater)	Headache, vertigo, and diarrhea	Fish toxicity and loss of marine biodiversity	Shanmugam et al. (2014), Singh et al. (2014), Acuña et al. (2015), Gamarra et al. (2015)
3.	Triclosan[a]	Cosmetics Shampoos, soaps, medicated cosmetics (municipal wastewater)	Allergic reactions	Changes in phytoplanktonic composition and adverse impact on food chain	Goldstein (2014), Tran et al. (2018), Zhao et al. (2010a)
4.	Carbamazepine[a]	Antiepileptic drug (human urine and fecal municipal wastewater)	Nausea, dizziness, drowsiness, loss of balance and coordination	Retarded growth and development of lower invertebrates	Ferrari et al. (2003), Malarvizhi et al. (2012)
5.	Microplastics[a]	Plastic products, bags, polythene, food packaging, fiber, foam	Endocrine-disrupting substances and carcinogenic	Altering feeding habits, energy metabolism, and reproductive ability of marine fauna	Barboza et al. (2018), Sharma and Chatterjee (2017), Carbery et al. (2018), Smith et al. (2018), Van Cauwenberghe and Janssen (2014), Anbumani and Kakkar (2018), Naidu (2019)

(continued)

Table 15.1 (continued)

Sl. no.	Pollutants	Sources	Impact on human	Impact on environment	References
6.	Caffeine	Beverages (municipal wastewater)	Induce anxiety, tachycardia, restlessness, insomnia	Affecting the neural system of marine vertebrates showing shorter axons with abnormal branching with excessive synaptic vesicle; skeletal muscles lacking well-defined boundaries	Lieberman et al. (1987), Nawrot et al. (2003), Quadra et al. (2020)

[a]Though impact is not reported, these ECs when taken in excess are known to cause these side effects on humans

4.2 *Ecotoxicity of Emerging Pollutants*

Ecotoxicology serves as a tool to evaluate environmental quality and identify impacts caused by toxic pollutants present in the ecosystem (Posthuma et al. 2020). The organisms having direct exposure to the pollutant are considered as primary indicators of environmental health in ecotoxicity studies (Connon et al. 2012; Zuccato et al. 2006). Emerging pollutants like pharmaceuticals, personal care products, antibiotics, artificial sweeteners, hormones, and microplastics have emerged as new classes of contaminants, which have potential to cause severe adverse impact on aquatic ecosystems and human health (Tran et al. 2018). Moreover, some emerging pollutants are identified that may potentially cause hormonal imbalance in the endocrine system of organisms which is involved in sexual differentiation, brain organization, metabolism, organ coordination, and control of reproduction resulting in extinction of specific vulnerable species (Jondeau-Cabaton et al. 2013).

4.3 *Impact of Emerging Pollutants on Human*

Table 15.1 lists the details of impact of six emerging pollutants along the coastal ecosystem on human health. Out of the six pollutants under study, only two, viz., DDT and caffeine, have been reported to have direct health impacts on human health (Table 15.1). However, other contaminants though have not been reported to have health impact as pollutant, but they too are known to cause various forms of health implications with excess consumption (Table 15.1). These ECs also adversely affect marine biodiversity indirectly impacting human health and well-being (Table 15.1).

4.4 Pathway of Emerging Pollutant Affecting Environment and Human Health

The emerging pollutants under study belong to different groups based on their potential source of generation. The first groups of EPs are pesticides that are being released as agricultural runoff and wastes. They mainly follow bioaccumulation/bioconcentration pathway affecting human and other top carnivores of food chain (Rogan and Chen 2005; Mansouri et al. 2017). Second group of EPs belongs to those released from pharmaceuticals and personal care product's industrial waste. These EPs are mainly reported as pollutants affecting marine biodiversity, but they also have potential to cause several health implications in humans (Acuña et al. 2015; Tran et al. 2018; Malarvizhi et al. 2012). Third group of EPs is pollutant from beverage industries having direct impact on humans and marine biodiversity (Quadra et al. 2020) The fourth group is microplastics released from broad range of nonpoint sources having potential to cause adverse impact on ecosystem ultimately affecting humans (Smith et al. 2018). The entire pathway of EPs affecting marine ecosystem and ultimately human is illustrated in Fig. 15.2.

DDT an example from the pesticide group of emerging pollutant undergoes biomagnification in tissues and organs of fishes which upon consumption by top carnivores like birds leads to decreased reproductive ability and thinning of eggshells (Fig. 15.2). In human it is reported to cause several adverse impacts such as tumor, cancer, seizures, and death (Fig. 15.2). Diclofenac and carbamazepine from the pharmaceutical group have different pathway to the environment and humans. Diclofenac increases fish toxicity and leads to loss of marine biodiversity. The potential impact on humans upon excessive and unnecessary consumption results in headache, vertigo, and diarrhea (Fig. 15.2). Carbamazepine has adverse impact on growth and development of fish and crab population. Its potential effects on human with excess intake are nausea, dizziness, drowsiness, and loss of balance and coordination (Fig. 15.2). Triclosan under personal care products group affects the entire food chain of marine ecosystem by changing phytoplanktonic composition and reducing species richness (Fig. 15.2). The potential impacts on human under extreme exposure are allergic reactions (Fig. 15.2). Caffeine, contaminant from beverage industrial discharge group, has severe impact on marine vertebrates as it affects nervous and muscular system observed by occurrence of shorter axons with abnormal branching and excessive synaptic vesicles as well as skeletal muscles lacking well-defined boundaries (Fig. 15.2). The reported impacts on humans for caffeine pollution are anxiety, tachycardia, restlessness, and insomnia (Fig. 15.2). Microplastics are reported to cause alterations in feeding habits, energy metabolism, and reproductive ability of marine fauna, thereby having severe repercussion on marine food security. On humans, though not studied much as pollutant, they are reported to have potential impact as endocrine-disrupting and carcinogenic substance.

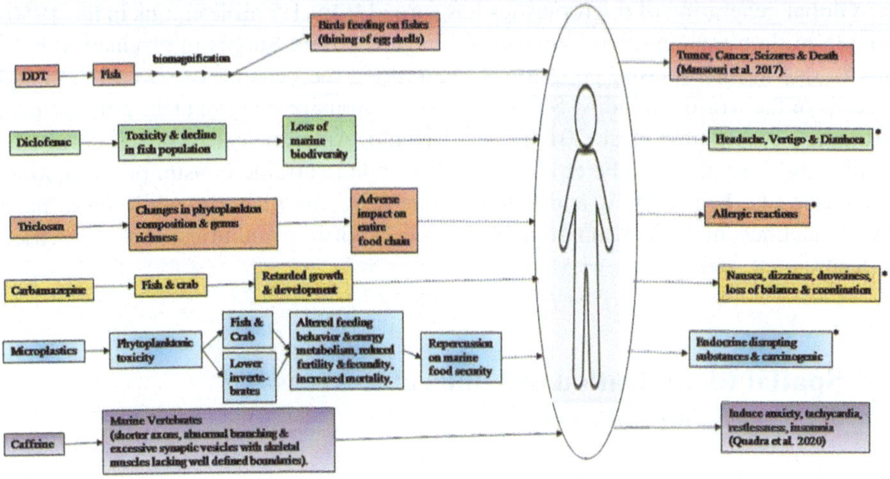

Fig. 15.2 Pathway for emerging pollutants affecting marine ecosystem. *Represent impact not reported as pollutant but are known as side effect on excess consumption

4.5 Global Pollution Trend

Diclofenac concentrations in the range of 2.5–13.48 μg/L are reported in groundwater and surface water sources in various countries like India, Luxembourg, France, Poland, Canada, Germany, Spain, Taiwan, and Serbia (Anumol et al. 2016; Singh et al. 2014; Carrara et al. 2008; Einsiedl et al. 2010; Sharma et al. 2019; Banzhaf et al. 2013; Kapelewska et al. 2018; Sathishkumar et al. 2020). Several studies have revealed that contamination of diclofenac in water bodies is caused by both non-point and point sources (López-Serna et al. 2013; Sathishkumar et al. 2020).

Triclosan as emerging pollutant has significantly contributed toward coastal ecosystem pollution. Concentration of triclosan present in rivers majorly depends on the efficiency and effectiveness of wastewater treatment plant (van Wijnen et al. 2018). Global TCS model suggests rise of import of triclosan by rivers in regions like Southeast Asia and decline in import in European countries. It is also estimated that in 2050 average annual concentrations of triclosan in rivers will increase twice the present concentration level (van Wijnen et al. 2018).

Carbamazepine is most frequently investigated emerging pollutant in North America, Europe, and Asia continents (Hughes et al. 2013). It is also estimated that in Netherlands, carbamazepine use has increased from 8400 kg in 2007 to 8990 kg by 2020 resulting its increase as an effluent in wastewater (Van der Aa et al. 2008; Moermond 2014).

Though global production and use of DDT have declined in agreement with the Stockholm Convention, it is still used in excess for mitigation and control of malaria and leishmaniasis (Van Den Berg et al. 2017).

Global generation of microplastics has surged from 1.5 million tons in the 1950s to 335 million tons in 2016 (Alimba and Faggio 2019). Surges in production have led to rise in microplastic pollution evident along the beaches of several seas and oceans of the world, including South Caribbean, Bonaire (Debrot et al. 2013), North Atlantic, USA (Ribic et al. 2010), Heard Island, Antarctica (Eriksen et al. 2013), Chile, and East Asia (Isobe et al. 2015). Per capita caffeine consumption in some countries like Brazil, Italy, and Ethiopia has risen significantly, while there have been declines in its consumption in coffee-exporting countries such as Africa (Quadra et al. 2020).

5 Spatial Identification of Vulnerable Areas for Emerging Pollutants

Developmental activities like industrialization and urbanization near the watersheds consequently led to excessive pollution load on the estuarine ecosystem serving as potential sink for the emerging pollutants (Freeman et al. 2019). Emerging pollutants gained attention owing to their ubiquitous presence and their potential to cause undesirable ecological effects (Ferrari et al. 2003; Cleuvers 2003; Al Aukidy et al. 2012; Verlicchi et al. 2012; Verlicchi and Zambello 2015). These pollutants are having numerous nonpoint sources of origin (Table 15.1) making their management a difficult task. Moreover, these pollutants are not given due consideration in the past; hence there is a paucity of insight regarding their effect on human and environment. Furthermore for better management of these pollutants, it is important to gain insight regarding vulnerability of states near the coastal ecosystem for these emerging pollutants.

5.1 Distribution of Emerging pollutant Among the States Along the Indian Coastline

The studied six emerging pollutants at the east and west coast of the Indian peninsular region as per reported in literature are detailed in Table 15.2. It was observed that among all the states present on the coastal boundaries, West Bengal and Tamil Nadu are the states reported to have maximum occurrence of emerging pollutants while Andhra Pradesh, Gujarat, Maharashtra, and Goa are reported with minimum number of emerging pollutants (Table 15.2 and Fig. 15.3).

Table 15.2 Details of occurrence of emerging pollutants in states along the east and west coast of India

Sl. no.	States	Emerging pollutants
1.	West Bengal	DDT, microplastics, caffeine, diclofenac, triclosan, carbamazepine
2.	Odisha	DDT, microplastic, diclofenac, triclosan, caffeine
3.	Andhra Pradesh	DDT, microplastics, diclofenac
4.	Tamil Nadu	DDT, microplastics, caffeine, diclofenac, triclosan, carbamazepine
5.	Gujarat	DDT, microplastics, diclofenac
6.	Maharashtra	DDT, microplastics, diclofenac
7.	Goa	Microplastics, DDT, caffeine
8.	Karnataka	Diclofenac, triclosan, carbamazepine, DDT, caffeine
9.	Kerala	Triclosan, DDT, caffeine, microplastics, carbamazepine

5.2 *Identification and Illustration of Vulnerable States for Emerging Pollutants*

Vulnerability of states toward emerging pollutant usually depends on their potential sources of discharge. In this study an attempt has been made to identify the vulnerable states. Based on literature, it was identified that there are vast variations among the occurrence of emerging pollutants and their potential sources of discharge. For instance, Maharashtra is reported to have occurrence of only three emerging pollutants, viz., DDT, microplastics, and diclofenac. However, it was observed to have potential sources for all six emerging pollutants as its coastal regions receive discharge from agriculture runoff, pharmaceutical industries, beverage industries, as well as direct exposure to discharge having microplastics and beauty care products. This shows an existing gap between studied emerging pollutant and ground level presence of emerging pollutants. Furthermore Maharashtra has the highest population compared to all other states under study, thereby making highest number of people vulnerable for emerging pollutants (Table 15.3). This raised the need to identify and classify states based on the gap between studied emerging pollutant and actual presence of pollutant based on potential sources. Therefore, in vulnerability classification of coastal states, two factors were taken into consideration. First is the gap between reported emerging pollutant and the actual presence based on potential sources. Second is the population of the states. The states with maximum reported emerging pollutant were considered least vulnerable, while those with least studied and reported emerging pollutant were considered highly vulnerable. Further population was considered as another factor for ranking vulnerability. Those states having same number of reported emerging pollutant are categorized based on their state population. For example, if two states have same number of reported emerging pollutant, then the one having maximum population is given higher rank compared to the one having lesser population.

It was observed that all the coastal states have potential sources for all six emerging pollutants (Table 15.3). But the reported emerging pollutant varied from state to state (Table 15.2). West Bengal and Tamil Nadu reported six emerging pollutants.

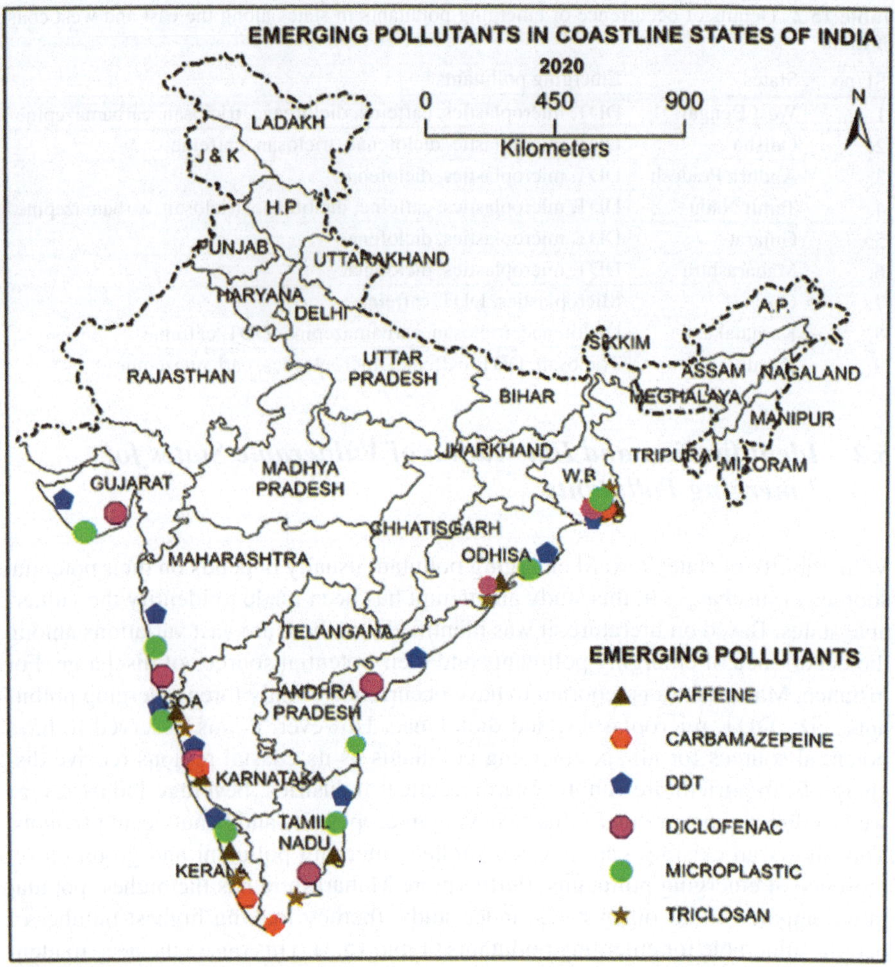

Fig. 15.3 Map of states showing occurrence of emerging pollutants along the east and west coast of India

Kerala, Karnataka, and Odisha have five reported emerging pollutants, while Maharashtra, Goa, Gujarat, and Andhra Pradesh have reported only three emerging pollutants (Table 15.3). Since Maharashtra, Goa, Gujarat, and Andhra Pradesh have least reported emerging pollutant, they were categorized as group of highly vulnerable states followed by Kerala, Karnataka, and Odisha with five reported emerging pollutants, and West Bengal and Tamil Nadu are categorized as group of least vulnerable as they have maximum reported emerging pollutant (Table 15.3). These groups are further categorized and ranked on the basis of population size. Among the highly vulnerable group, Maharashtra is given highest rank (9) as it has highest population load and with least reported emerging pollutant followed by Andhra Pradesh (8), then Gujarat (7), and then Goa (6). These groups were followed by

Table 15.3 States with different vulnerability rank against studied emerging pollutants

State	Population of state	Emerging pollutants		Rank in vulnerability
		Reported	Based on potential sources[a]	
Andhra Pradesh	84,580,777	3	6	8
Goa	1,458,545	3	6	6
Gujarat	60,439,692	3	6	7
Karnataka	61,095,297	5	6	5
Kerala	33,406,061	5	6	3
Maharashtra	112,374,333	3	6	9
Odisha	41,974,218	5	6	4
Tamil Nadu	72,147,030	6	6	1
West Bengal	91,276,115	6	6	2

Where 1 and 9 represent least and extreme vulnerability, respectively
[a]Population data source: https://censusindia.gov.in/ and https://www.census2011.co.in/

moderate vulnerable group with five reported emerging pollutants. In this Karnataka is given higher rank (5) followed by Odisha (4) and then Kerala (3) based on their state population. Among least vulnerable group, West Bengal is given rank 2 followed by Tamil Nadu (1) based on their population of state (Table 15.3).

Based on the above mentioned rank, the states were illustrated on a vulnerability map showing all the states with their vulnerability toward emerging pollutants (Fig. 15.4).

This vulnerability ranking is based purely on the gap in awareness based on literature and number of people exposed based on population. However, further studies are needed to accurately assess the risk of coastal state and its population toward these emerging pollutants.

6 Conclusion

Presently, due to several intensive anthropogenic activities, the marine pollution has emerged as a global challenge. It is likely to get intensified and exacerbate posing significant ecological risk and vulnerability near or around the coastal environment. Emerging pollutants have arisen as new generation pollutants and are not studied in detail but posed severe threat to the coastal ecosystem in various ways. Management of coastal ecosystem requires an insight to the category of pollutant affecting the coastal ecosystem and the degree of vulnerability of the coastal state toward them. In this study an attempt was made to demarcate vulnerable areas for commonly found emerging pollutants in the states located along east and west coast of India. Based on the distribution, abundance, and adverse impact of emerging pollutant, state population, as well as knowledge gap in scientific community, a vulnerability map was developed. This spatial analysis though tries to bring some insight into the

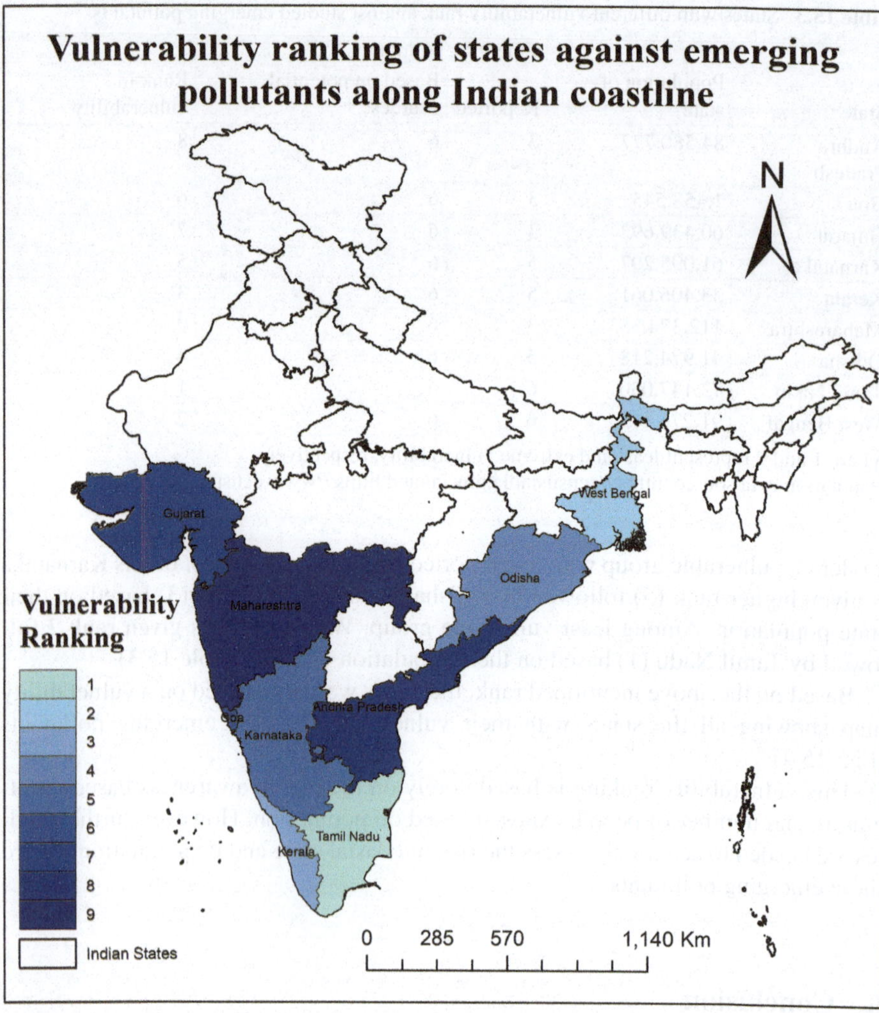

Fig. 15.4 Vulnerability map showing states along Indian coastline with different vulnerability toward emerging pollutants

vulnerable coastal states toward emerging pollution; however it needs to be substantiated with further studies. With more investigation in this regard, scientist can identify and categorize states based on their vulnerability toward emerging pollutants, thereby playing major role in its management and conservation. This may further help to take conservative steps for enrichment of coastal biodiversity. This sustainable approach may help design effective management strategies for protection and restoration of coastal and marine environment against the challenges like pollution, climate crisis, population explosion, and natural calamities at global level.

References

Achilov N (2017) Development of tourism industry: perspectives and advantages for growth as example in Kazakhstan. J Tourism Hosp 6(267):2167–0269

Acuña V, Ginebreda A, Mor JR, Petrovic M, Sabater S, Sumpter J, Barceló D (2015) Balancing the health benefits and environmental risks of pharmaceuticals: diclofenac as an example. Environ Int 85:327–333

Adnan M, Alshammari E, Patel M, Ashraf SA, Khan S, Hadi S (2018) Significance and potential of marine microbial natural bioactive compounds against biofilms/biofouling: necessity for green chemistry. PeerJ 6:e5049

Al Aukidy M, Verlicchi P, Jelic A, Petrovic M, Barcelò D (2012) Monitoring release of pharmaceutical compounds: occurrence and environmental risk assessment of two WWTP effluents and their receiving bodies in the Po Valley, Italy. Sci Total Environ 438:15–25

Alimba CG, Faggio C (2019) Microplastics in the marine environment: current trends in environmental pollution and mechanisms of toxicological profile. Environ Toxicol Pharmacol 68:61–74

Aminti P, Cipriani LE, Pranzini E (1999) An innovative, nonstructural, solution to beach erosion: costs less and delivers more benefit. Intercoast 33(7):34–35

Analuddin K, Sharma S, Septiana A, Sahidin I, Rianse U, Nadaoka K (2017) Heavy metal bioaccumulation in mangrove ecosystem at the coral triangle ecoregion, Southeast Sulawesi, Indonesia. Mar Pollut Bull 125(1–2):472–480

Anbumani S, Kakkar P (2018) Ecotoxicological effects of microplastics on biota: a review. Environ Sci Pollut Res 25(15):14373–14396

Anumol T, Vijayanandan A, Park M, Philip L, Snyder SA (2016) Occurrence and fate of emerging trace organic chemicals in wastewater plants in Chennai, India. Environ Int 92:33–42

Ayyam V, Palanivel S, Chandrakasan S (2019) Coastal regions of the tropics: an introduction. In: Coastal ecosystems of the tropics-adaptive management. Springer, Singapore, pp 3–20

Azeem A, Rizwan M, Ahmad FJ, Khan ZI, Khar RK, Aqil M, Talegaonkar S (2008) Emerging role of microemulsions in cosmetics. Recent Pat Drug Deliv Formul 2(3):275–289

Banzhaf S, Krein A, Scheytt T (2013) Using selected pharmaceutical compounds as indicators for surface water and groundwater interaction in the hyporheic zone of a low permeability riverbank. Hydrol Process 27(20):2892–2902

Barber LB, Leenheer JA, Pereira WE, Noyes TL, Brown GA, Tabor CF, Writer JH (1996) Organic contamination of the Mississippi River from municipal and industrial wastewater. US Geological Survey Circular USGS CIRC, pp 115–135

Barbier EB, Hacker SD, Kennedy C, Koch EW, Stier AC, Silliman BR (2011) The value of estuarine and coastal ecosystem services. Ecol Monogr 81(2):169–193

Barboza LGA, Vethaak AD, Lavorante BR, Lundebye AK, Guilhermino L (2018) Marine microplastic debris: an emerging issue for food security, food safety and human health. Mar Pollut Bull 133:336–348

Barkia I, Saari N, Manning SR (2019) Microalgae for high-value products towards human health and nutrition. Mar Drugs 17(5):304

Behera SK, Kim HW, Oh JE, Park HS (2011) Occurrence and removal of antibiotics, hormones and several other pharmaceuticals in wastewater treatment plants of the largest industrial city of Korea. Sci Total Environ 409(20):4351–4360

Benotti MJ, Trenholm RA, Vanderford BJ, Holady JC, Stanford BD, Snyder SA (2009) Pharmaceuticals and endocrine disrupting compounds in US drinking water. Environ Sci Technol 43(3):597–603

Bila DM, Dezotti M (2003) Pharmaceutical drugs in the environment. Química Nova 26(4):523–530

Bilkovic DM, Mitchell M, Mason P, Duhring K (2016) The role of living shorelines as estuarine habitat conservation strategies. Coast Manag 44(3):161–174

Borja A, White MP, Berdalet E, Bock N, Eatock C, Kristensen P, Leonard A, Lloret J, Pahl S, Parga M, Prieto JV (2020) Moving toward an agenda on ocean health and human health in Europe. Front Mar Sci 7:37

Bradman MA, Harnly ME, Draper W, Seidel S, Teran S, Wakeham D, Neutra R (1997) Pesticide exposures to children from California's Central Valley: results of a pilot study. J Expo Anal Environ Epidemiol 7(2):217–234

Brodie MJ, Besag F, Ettinger AB, Mula M, Gobbi G, Comai S, Aldenkamp AP, Steinhoff BJ (2016) Epilepsy, antiepileptic drugs, and aggression: an evidence-based review. Pharmacol Rev 68(3):563–602

Bruck S, Ford AT (2018) Chronic ingestion of polystyrene microparticles in low doses has no effect on food consumption and growth to the intertidal amphipod Echinogammarus marinus? Environ Pollut 233:1125–1130

Carbery M, O'Connor W, Palanisami T (2018) Trophic transfer of microplastics and mixed contaminants in the marine food web and implications for human health. Environ Int 115:400–409

Carr SA, Liu J, Tesoro AG (2016) Transport and fate of microplastic particles in wastewater treatment plants. Water Res 91:174–182

Carrara C, Ptacek CJ, Robertson WD, Blowes DW, Moncur MC, Sverko ED, Backus S (2008) Fate of pharmaceutical and trace organic compounds in three septic system plumes, Ontario, Canada. Environ Sci Technol 42(8):2805–2811

Carson MA, Clarke SA (2018) Bioactive compounds from marine organisms: potential for bone growth and healing. Mar Drugs 16(9):340

Cherkasov IL, Seredina MI, Mishurov OI, Adashova TA, Lebedeva OY (2017) The effect of international tourism on the development of global social-economic processes. J Environ Manag Tourism 8(6):1166–1170

Cleuvers M (2003) Aquatic ecotoxicity of pharmaceuticals including the assessment of combination effects. Toxicol Lett 142(3):185–194

Cleuvers M (2004) Mixture toxicity of the anti-inflammatory drugs diclofenac, ibuprofen, naproxen, and acetylsalicylic acid. Ecotoxicol Environ Saf 59(3):309–315

Cloern JE (2001) Our evolving conceptual model of the coastal eutrophication problem. Mar Ecol Prog Ser 210:223–253

Connon RE, Geist J, Werner I (2012) Effect-based tools for monitoring and predicting the ecotoxicological effects of chemicals in the aquatic environment. Sensors 12(9):12741–12771

Crooks S, Herr D, Tamelander J, Laffoley D, Vandever J (2011) Mitigating climate change through restoration and management of coastal wetlands and near-shore marine ecosystems: challenges and opportunities. World Bank, Washington, DC

de Paula J, Farah A (2019) Caffeine consumption through coffee: content in the beverage, metabolism, health benefits and risks. Beverages 5(2):37

Debrot AO, Bron PS, Leon R, Meesters HWG (2013) Marine debris in mangroves and on the seabed: largely-neglected litter problems. Mar Pollut Bull 72(1):1–1

Dias MP, Martin R, Pearmain EJ, Burfield IJ, Small C, Phillips RA, Yates O, Lascelles B, Borboroglu PG, Croxall JP (2019) Threats to seabirds: a global assessment. Biol Conserv 237:525–537

Díaz-Asencio M, Alvarado JC, Alonso-Hernández C, Quejido-Cabezas A, Ruiz-Fernández AC, Sanchez-Sanchez M, Gómez-Mancebo MB, Froidevaux P, Sanchez-Cabeza JA (2011) Reconstruction of metal pollution and recent sedimentation processes in Havana Bay (Cuba): a tool for coastal ecosystem management. J Hazard Mater 196:402–411

DoD (2009) Emerging contaminants (ECs). Department of Defense, Arlington

Dudgeon D (2014) Threats to freshwater biodiversity in a changing world. Global environmental change

Duong HA, Pham NH, Nguyen HT, Hoang TT, Pham HV, Pham VC, Berg M, Giger W, Alder AC (2008) Occurrence, fate and antibiotic resistance of fluoroquinolone antibacterials in hospital wastewaters in Hanoi, Vietnam. Chemosphere 72(6):968–973

Duursma EK, Marchand M (1976) Aspects of organic marine pollution (No. IAEA-TECDOC-187)

Einsiedl F, Radke M, Maloszewski P (2010) Occurrence and transport of pharmaceuticals in a karst groundwater system affected by domestic wastewater treatment plants. J Contam Hydrol 117(1–4):26–36

Eriksen M, Mason S, Wilson S, Box C, Zellers A, Edwards W, Farley H, Amato S (2013) Microplastic pollution in the surface waters of the Laurentian Great Lakes. Mar Pollut Bull 77(1–2):177–182

Farré M (2020) Remote and in-situ devices for the assessment of marine contaminants of emerging concern and plastic debris detection. Curr Opin Environ Sci Health 18:79–94

Fendall LS, Sewell MA (2009) Contributing to marine pollution by washing your face: microplastics in facial cleansers. Mar Pollut Bull 58(8):1225–1228

Ferrari B, Paxéus N, Giudice RL, Pollio A, Garric J (2003) Ecotoxicological impact of pharmaceuticals found in treated wastewaters: study of carbamazepine, clofibric acid, and diclofenac. Ecotoxicol Environ Saf 55(3):359–370

Folke C, Kautsky N, Berg H, Jansson Å, Troell M (1998) The ecological footprint concept for sustainable seafood production: a review. Ecol Appl 8(sp1):S63–S71

Freeman LA, Corbett DR, Fitzgerald AM, Lemley DA, Quigg A, Steppe CN (2019) Impacts of urbanization and development on estuarine ecosystems and water quality. Estuar Coasts 42(7):1821–1838

Friess DA, Yando ES, Alemu JB, Wong LW, Soto SD, Bhatia N (2020) Ecosystem services and disservices of mangrove forests and salt marshes. In: Oceanography and marine biology. CRC Press, Boca Raton, FL

Gamarra JS Jr, Godoi AFL, de Vasconcelos EC, de Souza KMT, de Oliveira CMR (2015) Environmental Risk Assessment (ERA) of diclofenac and ibuprofen: a public health perspective. Chemosphere 120:462–469

Gavrilescu M, Demnerová K, Aamand J, Agathos S, Fava F (2015) Emerging pollutants in the environment: present and future challenges in biomonitoring, ecological risks and bioremediation. New Biotechnol 32(1):147–156

Geissen V, Mol H, Klumpp E, Umlauf G, Nadal M, van der Ploeg M, van de Zee SE, Ritsema CJ (2015) Emerging pollutants in the environment: a challenge for water resource management. Int Soil Water Conserv Res 3(1):57–65

Gephart JA, Golden CD, Asche F, Belton B, Brugere C, Froehlich HE, Fry JP, Halpern BS, Hicks CC, Jones RC, Klinger DH (2020) Scenarios for global aquaculture and its role in human nutrition. Rev Fish Sci Aquacult 29:1–17

Ginebreda A, Muñoz I, de Alda ML, Brix R, López-Doval J, Barceló D (2010) Environmental risk assessment of pharmaceuticals in rivers: relationships between hazard indexes and aquatic macroinvertebrate diversity indexes in the Llobregat River (NE Spain). Environ Int 36(2):153–162

Goldstein WE (2014) Health impact of pharmaceuticals in the environment and in drinking water: analysis and research needs. In: Pharmaceutical accumulation in the environment: prevention, control, health effects, and economic impact. CRC Press, Boca Raton, FL, p 103

González-Acevedo ZI, García-Zarate MA, Flores-Lugo IP (2019) Emerging contaminants and nutrients in a saline aquifer of a complex environment. Environ Pollut 244:885–897

Gössling S, Hall CM, Scott D (2018) Coastal and ocean tourism. In: Handbook on marine environment protection. Springer, Cham, pp 773–790

Grainger RJ, Garcia SM (1996) Chronicles of marine fishery landings (1950–1994): trend analysis and fisheries potential. FAO, Rome, p 51

Gregory MR (1996) Plastic 'scrubbers' in hand cleansers: a further (and minor) source for marine pollution identified. Mar Pollut Bull 32(12):867–871

Gros M, Petrović M, Ginebreda A, Barceló D (2010) Removal of pharmaceuticals during wastewater treatment and environmental risk assessment using hazard indexes. Environ Int 36(1):15–26

Guilfoyle MP, Jung JF, Fischer RA Jr, Dickerson DD (2019) Developing best management practices for coastal engineering projects that benefit Atlantic Coast shoreline-dependent species

Gurke R, Rößler M, Marx C, Diamond S, Schubert S, Oertel R, Fauler J (2015) Occurrence and removal of frequently prescribed pharmaceuticals and corresponding metabolites in wastewater of a sewage treatment plant. Sci Total Environ 532:762–770

Hamidi M, Kozani PS, Kozani PS, Pierre G, Michaud P, Delattre C (2020) Marine bacteria versus microalgae: who is the best for biotechnological production of bioactive compounds with antioxidant properties and other biological applications? Mar Drugs 18(1):28

Heino J, Alahuhta J, Bini LM, Cai Y, Heiskanen AS, Hellsten S, Kortelainen P, Kotamäki N, Tolonen KT, Vihervaara P, Vilmi A (2020) Lakes in the era of global change: moving beyond single-lake thinking in maintaining biodiversity and ecosystem services. Biol Rev 96:89–106

Heywood VH, Watson RT (1995) Global biodiversity assessment, vol 1140. Cambridge university press, Cambridge

Hinrichsen D (1999) Coastal waters of the world: trends, threats, and strategies. Island Press, Washington, DC

Howarth RW, Anderson DB, Cloern JE, Elfring C, Hopkinson CS, Lapointe B, Malone T, Marcus N, McGlathery K, Sharpley AN, Walker D (2000) Nutrient pollution of coastal rivers, bays, and seas. Issues Ecol 7:1–16

Hughes SR, Kay P, Brown LE (2013) Global synthesis and critical evaluation of pharmaceutical data sets collected from river systems. Environ Sci Technol 47(2):661–677

Hwang J, Choi D, Han S, Jung SY, Choi J, Hong J (2020) Potential toxicity of polystyrene microplastic particles. Sci Rep 10(1):1–12

Islam MS, Tanaka M (2004) Impacts of pollution on coastal and marine ecosystems including coastal and marine fisheries and approach for management: a review and synthesis. Mar Pollut Bull 48(7–8):624–649

Isobe A, Uchida K, Tokai T, Iwasaki S (2015) East Asian seas: a hot spot of pelagic microplastics. Mar Pollut Bull 101(2):618–623

Jabado RW, Kyne PM, Pollom RA, Ebert DA, Simpfendorfer CA, Ralph GM, Al Dhaheri SS, Akhilesh KV, Ali K, Ali MH, Al Mamari TM (2018) Troubled waters: threats and extinction risk of the sharks, rays and chimaeras of the Arabian Sea and adjacent waters. Fish Fish 19(6):1043–1062

Jitar O, Teodosiu C, Oros A, Plavan G, Nicoara M (2015) Bioaccumulation of heavy metals in marine organisms from the Romanian sector of the Black Sea. New Biotechnol 32(3):369–378

Jondeau-Cabaton A, Soucasse A, Jamin EL, Creusot N, Grimaldi M, Jouanin I, Aït-Aïssa S, Balaguer P, Debrauwer L, Zalko D (2013) Characterization of endocrine disruptors from a complex matrix using estrogen receptor affinity columns and high performance liquid chromatography–high resolution mass spectrometry. Environ Sci Pollut Res 20(5):2705–2720

Kapelewska J, Kotowska U, Karpińska J, Kowalczuk D, Arciszewska A, Świrydo A (2018) Occurrence, removal, mass loading and environmental risk assessment of emerging organic contaminants in leachates, groundwaters and wastewaters. Microchem J 137:292–301

Kasprzyk-Hordern B, Dinsdale RM, Guwy AJ (2009) The removal of pharmaceuticals, personal care products, endocrine disruptors and illicit drugs during wastewater treatment and its impact on the quality of receiving waters. Water Res 43(2):363–380

Khalifa SA, Elias N, Farag MA, Chen L, Saeed A, Hegazy MEF, Moustafa MS, El-Wahed A, Al-Mousawi SM, Musharraf SG, Chang FR (2019) Marine natural products: a source of novel anticancer drugs. Mar Drugs 17(9):491

Khetan SK, Collins TJ (2007) Human pharmaceuticals in the aquatic environment: a challenge to green chemistry. Chem Rev 107(6):2319–2364

Kim JT, Choi YJ, Barghi M, Kim JH, Jung JW, Kim K, Kang JH, Lammel G, Chang YS (2020) Occurrence, distribution, and bioaccumulation of new and legacy persistent organic pollutants in an ecosystem on King George Island, maritime Antarctica. J Hazard Mater 405:124141

Kummu M, Guillaume JH, de Moel H, Eisner S, Flörke M, Porkka M, Siebert S, Veldkamp TI, Ward PJ (2016) The world's road to water scarcity: shortage and stress in the 20th century and pathways towards sustainability. Sci Rep 6:38495

Kunkel U, Radke M (2012) Fate of pharmaceuticals in rivers: deriving a benchmark dataset at favorable attenuation conditions. Water Res 46(17):5551–5565
Langford KH, Reid M, Thomas KV (2011) Multi-residue screening of prioritised human pharmaceuticals, illicit drugs and bactericides in sediments and sludge. J Environ Monit 13(8):2284–2291
Laureti E (1999) Fish and fishery products: world apparent consumption statistics based on food balance sheets (1961-1997) (No. F020. 061). FAO, Rome
Law KL, Thompson RC (2014) Microplastics in the seas. Science 345(6193):144–145
Le Tissier M (2020) Unravelling the relationship between ecosystem-based management, integrated coastal zone management and marine spatial planning. In: Ecosystem-based management, ecosystem services and aquatic biodiversity. Springer, Cham, pp 403–413
Lecomte M, Rochette J, Laurans Y, Lapeyre R (2017) Indian Ocean tuna fisheries: between development opportunities and sustainability issues
Lewis NS, Marois DE, Littles CJ, Fulford RS (2020) Projecting changes to coastal and estuarine ecosystem goods and services-models and tools. In: Ecosystem-based management, ecosystem services and aquatic biodiversity: theory, tools and applications. Springer, Cham, pp 235–254
Lieberman HR, Wurtman RJ, Emde GG, Roberts C, Coviella ILG (1987) The effects of low doses of caffeine on human performance and mood. Psychopharmacology 92(3):308–312
Lonappan L, Brar SK, Das RK, Verma M, Surampalli RY (2016) Diclofenac and its transformation products: environmental occurrence and toxicity—a review. Environ Int 96:127–138
López-Serna R, Jurado A, Vázquez-Suñé E, Carrera J, Petrović M, Barceló D (2013). Occurrence of 95 pharmaceuticals and transformation products in urban groundwaters underlying the metropolis of Barcelona, Spain. *Environ Pollut 174*:305-315.
Lu Y, Yuan J, Lu X, Su C, Zhang Y, Wang C, Cao X, Li Q, Su J, Ittekkot V, Garbutt RA (2018) Major threats of pollution and climate change to global coastal ecosystems and enhanced management for sustainability. Environ Pollut 239:670–680
Malarvizhi A, Kavitha C, Saravanan M, Ramesh M (2012) Carbamazepine (CBZ) induced enzymatic stress in gill, liver and muscle of a common carp, Cyprinus carpio. J King Saud Univ Sci 24(2):179–186
Mallin MA, Ensign SH, McIver MR, Shank GC, Fowler PK (2001) Demographic, landscape, and meteorological factors controlling the microbial pollution of coastal waters. In: The ecology and etiology of newly emerging marine diseases. Springer, Dordrecht, pp 185–193
Mansouri A, Cregut M, Abbes C, Durand MJ, Landoulsi A, Thouand G (2017) The environmental issues of DDT pollution and bioremediation: a multidisciplinary review. Appl Biochem Biotechnol 181(1):309–339
Marques A, Nunes ML, Moore SK, Strom MS (2010) Climate change and seafood safety: human health implications. Food Res Int 43(7):1766–1779
McGettigan P, Henry D (2013) Use of non-steroidal anti-inflammatory drugs that elevate cardiovascular risk: an examination of sales and essential medicines lists in low-, middle-, and high-income countries. PLoS Med 10(2):e1001388
Meade RH (ed) (1995) Contaminants in the Mississippi River, 1987-92, vol 1133. US Government Printing Office, Washington, DC
Moermond CTA (2014) Environmental risk limits for pharmaceuticals: derivation of WFD water quality standards for carbamazepine, metoprolol, metformin and amidotrizoic acid
Mugagga F, Nabaasa BB (2016) The centrality of water resources to the realization of Sustainable Development Goals (SDG). A review of potentials and constraints on the African continent. Int Soil Water Conserv Res 4(3):215–223
Mujabar PS, Chandrasekar N (2013) Shoreline change analysis along the coast between Kanyakumari and Tuticorin of India using remote sensing and GIS. Arab J Geosci 6(3):647–664
Mutiyar PK, Mittal AK (2012) Contaminants of emerging concern in Ganga river basin: pharmaceuticals personal care products (PPCPs). In: 3rd GEM IIT water seminar IFCPAR/CEFPIRA, New Delhi

Naidu SA (2019) Preliminary study and first evidence of presence of microplastics and colorants in green mussel, Perna viridis (Linnaeus, 1758), from southeast coast of India. Mar Pollut Bull 140:416–422
Nam SW, Jo BI, Yoon Y, Zoh KD (2014) Occurrence and removal of selected micropollutants in a water treatment plant. Chemosphere 95:156–165
Napper IE, Bakir A, Rowland SJ, Thompson RC (2015) Characterisation, quantity and sorptive properties of microplastics extracted from cosmetics. Mar Pollut Bull 99(1–2):178–185
Nawrot P, Jordan S, Eastwood J, Rotstein J, Hugenholtz A, Feeley M (2003) Effects of caffeine on human health. Food Addit Contam 20(1):1–30
Noguera-Oviedo K, Aga DS (2016) Lessons learned from more than two decades of research on emerging contaminants in the environment. J Hazard Mater 316:242–251
Norse EA (1993) Global marine biological diversity: a strategy for building conservation into decision making. Island Press, Washington, DC
Oaks JL, Gilbert M, Virani MZ, Watson RT, Meteyer CU, Rideout BA, Shivaprasad HL, Ahmed S, Chaudhry MJI, Arshad M, Mahmood S (2004) Diclofenac residues as the cause of vulture population decline in Pakistan. Nature 427(6975):630–633
Ogunseitan OA (1996) Removal of caffeine in sewage by Pseudomonas putida: implications for water pollution index. World J Microbiol Biotechnol 12(3):251–256
Omar TFT, Aris AZ, Yusoff FM, Mustafa S (2018) Occurrence, distribution, and sources of emerging organic contaminants in tropical coastal sediments of anthropogenically impacted Klang River estuary, Malaysia. Mar Pollut Bull 131:284–293
Pafi M, Flannery W, Murtagh B (2020) Coastal tourism, market segmentation and contested landscapes. Mar Policy 121:104189
Pal A, Gin KYH, Lin AYC, Reinhard M (2010) Impacts of emerging organic contaminants on freshwater resources: review of recent occurrences, sources, fate and effects. Sci Total Environ 408(24):6062–6069
Passeri DL, Hagen SC, Medeiros SC, Bilskie MV, Alizad K, Wang D (2015) The dynamic effects of sea level rise on low-gradient coastal landscapes: a review. Earth's Future 3(6):159–181
Patay EB, Fritea L, Antonescu A, Antonescu A, Dobjanschi L (2017) Coffea arabica: a plant with rich content in caffeine. In: The question of caffeine. InTech, Rijeka, p 27
Pavlikova N, Sramek J, Jelinek M, Halada P, Kovar J (2020) Markers of acute toxicity of DDT exposure in pancreatic beta-cells determined by a proteomic approach. PLoS One 15(10):e0229430
Pendleton L, Donato DC, Murray BC, Crooks S, Jenkins WA, Sifleet S, Craft C, Fourqurean JW, Kauffman JB, Marbà N, Megonigal P (2012) Estimating global "blue carbon" emissions from conversion and degradation of vegetated coastal ecosystems. PLoS One 7(9):e43542
Petrie B, Barden R, Kasprzyk-Hordern B (2015) A review on emerging contaminants in wastewaters and the environment: current knowledge, understudied areas and recommendations for future monitoring. Water Res 72:3–27
Phelan A, Ross H, Setianto NA, Fielding K, Pradipta L (2020) Ocean plastic crisis—mental models of plastic pollution from remote Indonesian coastal communities. PLoS One 15(7):e0236149
Piocos EA, De la Cruz AA (2000) Solid phase extraction and high performance liquid chromatography with photodiode array detection of chemical indicators of human fecal contamination in water. J Liq Chromatogr Relat Technol 23:1281–1292
Posthuma L, Zijp MC, De Zwart D, Van de Meent D, Globevnik L, Koprivsek M, Focks A, Van Gils J, Birk S (2020) Chemical pollution imposes limitations to the ecological status of European surface waters. Sci Rep 10(1):1–12
Pratt LA, Brody DJ, Gu Q (2011) Antidepressant use in persons aged 12 and over: United States, 2005-2008. US Department of Health and Human Services, Centers for Disease Control and Prevention, National Center for Health Statistics, Hyattsville, MD
Quadra GR, Paranaíba JR, Vilas-Boas J, Roland F, Amado AM, Barros N, Dias RJP, Cardoso SJ (2020) A global trend of caffeine consumption over time and related-environmental impacts. Environ Pollut 256:113343

Rahman MA, Rahman S (2015) Natural and traditional defense mechanisms to reduce climate risks in coastal zones of Bangladesh. Weather Clim Extremes 7:84–95
Ramaswamy BR, Shanmugam G, Velu G, Rengarajan B, Larsson DJ (2011) GC–MS analysis and ecotoxicological risk assessment of triclosan, carbamazepine and parabens in Indian rivers. J Hazard Mater 186(2–3):1586–1593
Rani NS, Satyanarayana ANV, Bhaskaran PK (2015) Coastal vulnerability assessment studies over India: a review. Nat Hazards 77(1):405–428
Ravindran S (2012) Environmental management for coastal hazard mitigation. In: Ecosystem approach to disaster risk reduction, p 65
Reher R, Kim HW, Zhang C, Mao HH, Wang M, Nothias LF, Caraballo-Rodriguez AM, Glukhov E, Teke B, Leao T, Alexander KL (2020) A convolutional neural network-based approach for the rapid annotation of molecularly diverse natural products. J Am Chem Soc 142(9):4114–4120
Ribic CA, Sheavly SB, Rugg DJ, Erdmann ES (2010) Trends and drivers of marine debris on the Atlantic coast of the United States 1997–2007. Mar Pollut Bull 60(8):1231–1242
Richardson SD, Kimura SY (2016) Water analysis: emerging contaminants and current issues. Anal Chem 88(1):546–582
Richardson SD, Kimura SY (2019) Water analysis: emerging contaminants and current issues. Anal Chem 92(1):473–505
Rishworth GM, Adams JB, Bird MS, Carrasco NK, Dänhardt A, Dannheim J, Lemley DA, Pistorius PA, Scheiffarth G, Hillebrand H (2020) Cross-continental analysis of coastal biodiversity change. Philos Trans R Soc B 375(1814):20190452
Rochman CM, Tahir A, Williams SL, Baxa DV, Lam R, Miller JT, Teh FC, Werorilangi S, Teh SJ (2015) Anthropogenic debris in seafood: plastic debris and fibers from textiles in fish and bivalves sold for human consumption. Sci Rep 5:14340
Rogan WJ, Chen A (2005) Health risks and benefits of bis(4-chlorophenyl)-1,1,1-trichloroethane (DDT). Lancet 366(9487):763–773
Sathishkumar P, Meena RAA, Palanisami T, Ashokkumar V, Palvannan T, Gu FL (2020) Occurrence, interactive effects and ecological risk of diclofenac in environmental compartments and biota-a review. Sci Total Environ 698:134057
Schmidt W, O'Rourke K, Hernan R, Quinn B (2011) Effects of the pharmaceuticals gemfibrozil and diclofenac on the marine mussel (Mytilus spp.) and their comparison with standardized toxicity tests. Mar Pollut Bull 62(7):1389–1395
Schymanski D, Goldbeck C, Humpf HU, Fürst P (2018) Analysis of microplastics in water by micro-Raman spectroscopy: release of plastic particles from different packaging into mineral water. Water Res 129:154–162
Seiler RL, Zaugg SD, Thomas JM, Howcroft DL (1999) Caffeine and pharmaceuticals as indicators of waste water contamination in wells. Groundwater 37(3):405–410
Shanmugam G, Sampath S, Selvaraj KK, Larsson DJ, Ramaswamy BR (2014) Non-steroidal anti-inflammatory drugs in Indian rivers. Environ Sci Pollut Res 21(2):921–931
Sharma S, Chatterjee S (2017) Microplastic pollution, a threat to marine ecosystem and human health: a short review. Environ Sci Pollut Res 24(27):21530–21547
Sharma BM, Bečanová J, Scheringer M, Sharma A, Bharat GK, Whitehead PG, Klánová J, Nizzetto L (2019) Health and ecological risk assessment of emerging contaminants (pharmaceuticals, personal care products, and artificial sweeteners) in surface and groundwater (drinking water) in the Ganges River Basin, India. Sci Total Environ 646:1459–1467
Sherman K (1993) Large marine ecosystems as global units for marine resources management—an ecological perspective. In: Large marine ecosystems: stress, mitigation, and sustainability. AAAS Press, Washington, DC, pp 3–14
Shi D, Cho HS, Chen Y, Xu H, Gu H, Lian J, Wang W, Liu G, Huth C, Wang L, Ewing RC (2009) Fluorescent polystyrene–Fe3O4 composite nanospheres for in vivo imaging and hyperthermia. Adv Mater 21(21):2170–2173

Sidhu JP, Ahmed W, Gernjak W, Aryal R, McCarthy D, Palmer A, Kolotelo P, Toze S (2013) Sewage pollution in urban stormwater runoff as evident from the widespread presence of multiple microbial and chemical source tracking markers. Sci Total Environ 463:488–496
Siegener R, Chen RF (2002) Caffeine in Boston Harbor seawater. Mar Pollut Bull 44(5):383–387. https://doi.org/10.1016/s0025-326x(00)00176-4
Singh KP, Rai P, Singh AK, Verma P, Gupta S (2014) Occurrence of pharmaceuticals in urban wastewater of north Indian cities and risk assessment. Environ Monit Assess 186(10):6663–6682
Smith P, Ashmore MR, Black HI, Burgess PJ, Evans CD, Quine TA, Thomson AM, Hicks K, Orr HG (2013) The role of ecosystems and their management in regulating climate, and soil, water and air quality. J Appl Ecol 50(4):812–829
Smith M, Love DC, Rochman CM, Neff RA (2018) Microplastics in seafood and the implications for human health. Curr Environ Health Rep 5(3):375–386
Sofronov B (2017) The economic impact on global tourism. Ann Spiru Haret Univ Econ Ser 17(2):127–139
Sorensen JPR, Lapworth DJ, Nkhuwa DCW, Stuart ME, Gooddy DC, Bell RA, Chirwa M, Kabika J, Liemisa M, Chibesa M, Pedley S (2015) Emerging contaminants in urban groundwater sources in Africa. Water Res 72:51–63
Standley LJ, Kaplan LA, Smith D (2000) Molecular tracers of organic matter sources to surface water resources. Environ Sci Technol 34(15):3124–3130
Storck FR, Kools SA, Rinck-Pfeiffer S (2015) Microplastics in fresh water resources. Global Water Research Coalition, Stirling, South Australia
Subedi B, Balakrishna K, Sinha RK, Yamashita N, Balasubramanian VG, Kannan K (2015) Mass loading and removal of pharmaceuticals and personal care products, including psychoactive and illicit drugs and artificial sweeteners, in five sewage treatment plants in India. J Environ Chem Eng 3(4):2882–2891
Sun Q, Li M, Ma C, Chen X, Xie X, Yu CP (2016) Seasonal and spatial variations of PPCP occurrence, removal and mass loading in three wastewater treatment plants located in different urbanization areas in Xiamen, China. Environ Pollut 208:371–381
Sundblad G, Bergström U (2014) Shoreline development and degradation of coastal fish reproduction habitats. Ambio 43(8):1020–1028
Sussarellu R, Suquet M, Thomas Y, Lambert C, Fabioux C, Pernet MEJ, Le Goïc N, Quillien V, Mingant C, Epelboin Y, Corporeau C (2016) Oyster reproduction is affected by exposure to polystyrene microplastics. Proc Natl Acad Sci 113(9):2430–2435
Sutton-Grier AE, Wowk K, Bamford H (2015) Future of our coasts: the potential for natural and hybrid infrastructure to enhance the resilience of our coastal communities, economies and ecosystems. Environ Sci Pol 51:137–148
Tan Y, Huang S (2020) Open Chinese tourism market for coastal destinations in Spain—taking Majorca Island as an example. In: Tourism product development in China, Asian and European countries. Springer, Singapore, pp 109–119
Thomaidis NS, Asimakopoulos AG, Bletsou AA (2012) Emerging contaminants: a tutorial minireview. Global NEST J 14(1):72–79
Thompson EL (2014) Indicators of anthropogenic change and biological risk in coastal aquatic environments
Tran NH, Gin KYH (2017) Occurrence and removal of pharmaceuticals, hormones, personal care products, and endocrine disrupters in a full-scale water reclamation plant. Sci Total Environ 599:1503–1516
Tran NH, Reinhard M, Gin KYH (2018) Occurrence and fate of emerging contaminants in municipal wastewater treatment plants from different geographical regions—a review. Water Res 133:182–207
USGS (2017) Contaminants of emerging concern in the environment. Environmental health—toxic substances hydrology program. U.S. Geological Survey
Van Cauwenberghe L, Janssen CR (2014) Microplastics in bivalves cultured for human consumption. Environ Pollut 193:65–70

Van Den Berg H, Manuweera G, Konradsen F (2017) Global trends in the production and use of DDT for control of malaria and other vector-borne diseases. Malaria J 16(1):401
Van der Aa NGFM, Kommer GJ, De Groot GM, Versteegh JFM (2008) Geneesmiddelen in bronnen voor drinkwater. Monitoring, toekomstig gebruik en beleidsmaatregelen. RIVM rapport 609715002
van Wijnen J, Ragas AM, Kroeze C (2018) River export of triclosan from land to sea: a global modelling approach. Sci Total Environ 621:1280–1288
Verlicchi P, Zambello E (2015) Pharmaceuticals and personal care products in untreated and treated sewage sludge: occurrence and environmental risk in the case of application on soil—a critical review. Sci Total Environ 538:750–767
Verlicchi P, Al Aukidy M, Zambello E (2012) Occurrence of pharmaceutical compounds in urban wastewater: removal, mass load and environmental risk after a secondary treatment—a review. Sci Total Environ 429:123–155
Vikas M, Dwarakish GS (2015) Coastal pollution: a review. Aquat Proc 4:381–388
Wei Q, Zhao Z (2020) Research on evaluation of coastal city water sports tourism resources based on set pair analysis. J Coast Res 115(SI):190–192
Williams M (1996) The transition in the contribution of living aquatic resources to food security, vol 13. Intl Food Policy Res Inst, Washington, DC
Williams AT, Rangel-Buitrago N, Pranzini E, Anfuso G (2018) The management of coastal erosion. Ocean Coast Manag 156:4–20
Wilson AL, Courtenay O, Kelly-Hope LA, Scott TW, Takken W, Torr SJ, Lindsay SW (2020) The importance of vector control for the control and elimination of vector-borne diseases. PLoS Negl Trop Dis 14(1):e0007831
Winston JE (1992) Systematics and marine conservation. In: Systematics, ecology, and the biodiversity crisis. Columbia University Press, New York, pp 144–168
Wolf L, Zwiener C, Zemann M (2012) Tracking artificial sweeteners and pharmaceuticals introduced into urban groundwater by leaking sewer networks. Sci Total Environ 430:8–19
World Health Organization (1979) DDT and its derivatives (No. 12). World Health Organization, Washington, DC
World Resources Institute (2001) Pilot analysis of global ecosystem: coastal ecosystem. World Resources Institute, Washington DC. ISBN: 1-56973-458-5. http://www.wri.org/wr2000. Accessed Sept 2020
Writer JH, Ferrer I, Barber LB, Thurman EM (2013) Widespread occurrence of neuro-active pharmaceuticals and metabolites in 24 Minnesota rivers and wastewaters. Sci Total Environ 461:519–527
Ying GG, Jiang YX, Yang YY, Yao L, Zhang Q-Q, Liu Y-S, Hu L-X (2017) Pharmaceuticals and personal care products (PPCPs) and artificial sweeteners (ASs) in surface and ground waters and their application as indication of wastewater contamination. Sci Total Environ 616–617:816–823
Youn SJ, Taylor WW, Lynch AJ, Cowx IG, Beard TD Jr, Bartley D, Wu F (2014) Inland capture fishery contributions to global food security and threats to their future. Glob Food Sec 3(3–4):142–148
Yunis E (2001) Sustainable development and management of tourism in coastal areas. World Tourism Organization, Madrid
Zandaryaa DS, Frank-Kamenetsky D (2015) Emerging pollutants in water and wastewater
Zhang Y, Geissen SU (2010) Prediction of carbamazepine in sewage treatment plant effluents and its implications for control strategies of pharmaceutical aquatic contamination. Chemosphere 80(11):1345–1352
Zhang Y, Geißen SU, Gal C (2008) Carbamazepine and diclofenac: removal in wastewater treatment plants and occurrence in water bodies. Chemosphere 73(8):1151–1161
Zhao JL, Ying GG, Liu YS, Chen F, Yang JF, Wang L (2010a) Occurrence and risks of triclosan and triclocarban in the Pearl River system, South China: from source to the receiving environment. J Hazard Mater 179(1–3):215–222

Zhao JL, Ying GG, Liu YS, Chen F, Yang JF, Wang L, Yang XB, Stauber JL, Warne MSJ (2010b) Occurrence and a screening-level risk assessment of human pharmaceuticals in the Pearl River system, South China. Environ Toxicol Chem 29(6):1377–1384

Zuccato E, Castiglioni S, Fanelli R, Reitano G, Bagnati R, Chiabrando C, Pomati F, Rossetti C, Calamari D (2006) Pharmaceuticals in the environment in Italy: causes, occurrence, effects and control. Environ Sci Pollut Res 13(1):15–21

Index

S. Madhav et al. (eds.), *Coastal Ecosystems*, Coastal Research Library 38,
https://doi.org/10.1007/978-3-030-84255-0

D

E

The manufacturer's authorised representative in the EU is Springer Nature Customer Service Centre GmbH, Europaplatz 3, 69115 Heidelberg, Germany. If you have any concerns regarding our products, please contact ProductSafety@springernature.com

Printed and bound by CPI Group (UK) Ltd, Croydon, CR0 4YY
30/09/2026
02225289-0010